T0310178

Advances in Fuzzy Clustering and its Applications

Advances in Fuzzy Clustering and its Applications

EDITED BY

J. Valente de Oliveira
University of Algarve, Portugal

W. Pedrycz
University of Alberta, Canada
Systems Research Institute of the Polish Academy of Sciences, Poland

John Wiley & Sons, Ltd

Telephone (+44) 1243 779777

Email (for orders and customer service enquiries): cs-books@wiley.co.uk
Visit our Home Page on www.wiley.com

Other Wiley Editorial Offices

John Wiley & Sons Inc., 111 River Street, Hoboken, NJ 07030, USA

Jossey-Bass, 989 Market Street, San Francisco, CA 94103-1741, USA

Wiley-VCH Verlag GmbH, Boschstr. 12, D-69469 Weinheim, Germany

John Wiley & Sons Australia Ltd, 42 McDougall Street, Milton, Queensland 4064, Australia

John Wiley & Sons (Asia) Pte Ltd, 2 Clementi Loop #02-01, Jin Xing Distripark, Singapore 129809

John Wiley & Sons Canada Ltd, 6045 Freemont Blvd, Mississauga, ONT, Canada L5R 4J3

Wiley also publishes its books in a variety of electronic formats. Some content that appears in print may not be available in electronic books.

Anniversary Logo Design: Richard J. Pacifico

British Library Cataloguing in Publication Data

A catalogue record for this book is available from the British Library

ISBN 978-0-470-02760-8 (HB)

Typeset in 9/11 pt Times Roman by Thomson Digital
Printed and bound in Great Britain by Antony Rowe Ltd, Chippenham, Wiltshire
This book is printed on acid-free paper responsibly manufactured from sustainable forestry in which at least two trees are planted for each one used for paper production.

Contents

List of Contributors

EDITORS

José Valente de Oliveira
The Ualg Informatics Lab
Faculty of Science and Technology
University of Algarve
Portugal

Witold Pedrycz
Department of Electrical and Computer Engineering University of Alberta, Canada
and **Systems Research Institute of the Polish Academy of Sciences Warsaw, Poland**

AUTHORS

János Abonyi
University of Veszprém
Department of Process Engineering
Hungary

Mark D. Alexiuk
Department of Electrical and
Computer Engineering
University of Manitoba
Canada

Palmen Angelov
Department of Communication
Systems
Lancaster University
UK

Jürgen Beringer
Fakultät für Informatik
Otto-von-Guericke-Universität Magdeburg
Germany

Michael R. Berthold
Department of Computer and
Information Science
University of Konstanz
Germany

Benjamin Bird
School of Chemistry
University of Nottingham
UK

Peter Bruza
School of Information Technology
Queensland University of Technology
Australia

Guihong Cao
Department d'Informatique
et Recherche operationnelle
Université de Montreal
Canada

Giovanna Castellano
Department of Computer Science
University of Bari
Italy

Mario G.C.A. Cimino
Dipartimento di Ingegneria dell'Informazione:
Elettronica, Informatica, Telecomunicazioni
University of Pisa
Italy

Christian Döring
School of Computer Science
Otto-von-Guericke-Universität Magdeburg
Germany

Pierpaolo D'Urso
Dipartimento di Scienze Economiche
Gestionali e Sociali
Università degli Studi del Molise
Italy

Anna M. Fanelli
Department of Computer Science
University of Bari
Italy

Balázs Feil
University of Veszprém
Department of Process Engineering
Hungary

Dimitar Filev
Ford Motor Company
Dearborn
USA

Hichem Frigui
Department of Computer Engineering
and Computer Science
University of Louisville
USA

Jonathan M. Garibaldi
School of Computer Science and Information
Technology
University of Nottingham
UK

Mike W. George
School of Chemistry
University of Nottingham
UK

Joydeep Ghosh
Department of Electrical and
Computer Engineering
University of Texas at Austin
Texas
USA

Fernando Gomide
State University of Campinas
Brazil

Patrick J.F. Groenen
Econometric Institute
Erasmus University Rotterdam
The Netherlands

Eyke Hüllermeier
Fakultät für Informatik
Otto-von-Guericke-Universität Magdeburg
Germany

Uzay Kaymak
Econometric Institute
Erasmus University Rotterdam
The Netherlands

Rudolf Kruse
School of Computer Science
Otto-von-Guericke-Universität Magdeburg
Germany

Raymond Lau
Department of Information Systems
City University of Hong Kong
Hong Kong SAR

Beatrice Lazzerini
Dipartimento di Ingegneria dell'Informazione:
Elettronica, Informatica, Telecomunicazioni
University of Pisa
Italy

Marie-Jeanne Lesot
School of Computer Science
Otto-von-Guericke-Universität Magdeburg
Germany

Francesco Marcelloni
Dipartimento di Ingegneria dell'Informazione:
Elettronica, Informatica, Telecomunicazioni
University of Pisa
Italy

Corrado Mencar
Department of Computer Science
University of Bari
Italy

Samia Nefti-Meziani
Department of Engineering and Technology
University of Manchester
UK

Mourad Oussalah
Electronics, Electrical and Computing Engineering
The University of Birmingham
UK

David E. Patterson
Department of Computer and
Information Science
University of Konstanz
Germany

Nick J. Pizzi
National Research Council
Institute for Biodiagnostics
Winnipeg
Canada

Kunal Punera
Department of Electrical and
Computer Engineering
University of Texas at Austin
USA

Thomas A. Runkler
Siemens AG
Corporate Technology Information and
Communications
München
Germany

Mika Sato-Ilic
Faculty of Systems and Information
Engineering
University of Tsukuba
Japan

Leila Roling Scariot da Silva
State University of Campinas
Brazil

Dawei Song
Knowledge Media Institute
The Open University
Milton Keynes
UK

George E. Tsekouras
Department of Cultural Technology and
Communication University of the Aegean
Mytilene
Greece

Joost van Rosmalen
Econometric Institute
Erasmus University Rotterdam
The Netherlands

Xiao-Ying Wang
School of Computer Science
and Information Technology
University of Nottingham
UK

Richard Weber
Department of Industrial Engineering
University of Chile
Santiago
Chile

Bernd Wiswedel
Department of Computer and
Information Science
University of Konstanz
Germany

Ronald Yager
Iona College
New Rochelle
USA

Foreword

Well, here I am writing a foreword for this book. Here is the (free dictionary, Farlex) definition:

'foreword - a short introductory essay preceding the text of a book.'

An essay about fuzzy clustering? For inspiration, I looked at the forewords in my first two books. When I wrote my first book about fuzzy clustering (Bezdek, 1981), I asked Lotfi Zadeh to write a foreword for it. By then, Lotfi and I were friends, so he did it, and I was happy. But why? Was it to prove to you that I could get him to do it? Was it because he would say things that had never been said about fuzzy models? Was it a promotional gimmick that the publisher thought would get more buyers interested? Was it . . . hmmm, I still didn't know, so I read more carefully.

Lotfi speculated on a variety of possibilities for fuzzy clustering in that foreword. The most interesting sentence (Bezdek, 1981, p. 5) was perhaps:

"Although the results of experimental studies reported in this book indicate that fuzzy clustering techniques often have significant advantages over more conventional methods, universal acceptance of the theory of fuzzy sets as a natural basis for pattern recognition and cluster analysis is not likely to materialize in the very near future."

In short, his foreword was careful, and it was cautionary – Lotfi speculated that fuzzy clustering might not assume a central place in clustering, but this seems overshadowed by his more general worry about the role of fuzzy models in computation.

My second book (Bezdek and Pal, 1992) was much more similar to this volume than my first, because the 1981 effort was a one-author text, while the 1992 book was a collection of 51 papers (the "chapters") that Pal and I put together (we were editors, just like de Oliveira and Pedrycz) that seemed to provide a state-of-the-art "survey" of what was happening with fuzzy models in various pattern recognition domains in 1992. Perhaps the principal difference between these two books is that fuzzy clustering was only one of the five topics of our 1992 book, whereas the current volume is *only* about fuzzy clustering. The other noticeable difference was that the papers we collected had already been published elsewhere, whereas the chapters in this book have not.

I am looking at the foreword to our 1992 book right now, again written by Lotfi. Well, a lot of positive things happened for fuzzy sets in the 11 years that separated these two forewords (read, Japan builds fuzzy controllers), and Lotfi's 1992 foreword was both more historical and more confident than the 1981 offering. Here is the first sentence of that 1992 forward:

"To view the contents of this volume in a proper perspective it is of historical interest to note that the initial development of the theory of fuzzy sets was motivated in large measure by problems in pattern recognition and cluster analysis."

Did you notice that Lotfi used *exactly* the same term "pattern recognition and cluster analysis" in *both* forewords? In contradistinction, I believe that most people today view clustering as one of many topics encompassed by the much broader field of pattern recognition (classifier design, feature selection, image processing, and so on). My guess is that Lotfi probably used the term pattern recognition almost as a synonym for classification. This is a small point, but in the context of this volume, an interesting one, because to this day, Lotfi contends that the word *cluster* is ill defined, and hence cluster analysis is not really a topic at all. Nonetheless, you have in your hands a new book about fuzzy cluster analysis.

What should I point out to you in 2006 about this topic? Well, the main point is that fuzzy clustering is now a pretty mature field. I just "googled" the index term "fuzzy cluster analysis," and the search returned this statistic at 1 p.m. on September 6, 2006:

> "Results 1–10 of about 1 640 000 for fuzzy cluster analysis (0.34 seconds)."

Never mind duplication, mixed indexing, and all the other false positives represented by this statistic. The fact is fuzzy clustering is a pretty big field now. There are still some diehard statisticians out there who deny its existence, much less its value to real applications, but by and large this is no longer a controversial undertaking, nor is its real value to practitioners questionable. Moreover, I can pick any chapter in this book and get returns from Google that amaze me. Example: Chapter 4 has the somewhat exotic title "Fuzzy Clustering with Minkowski Distance Functions." What would you guess for this topic – 12 papers? Here is the return:

> "Results 1–10 of about 20 000 for Fuzzy Clustering with Minkowski distance functions (0.37 seconds)."

There aren't 20 000 papers out there about this topic, but there are probably a few hundred, and this is what makes the current book useful. Most of these chapters offer an encapsulated survey of (some of) the most important work on their advertised contents. This is valuable, because I don't want to sift through 20 000 entries to find the good stuff about Minkowski-based fuzzy clustering – I want the experts to guide me to 20 or 30 papers that have it.

Summary. We no longer need worry whether the *topics* in this fuzzy clustering book *are* good stuff – they are. What we need that these chapters provide is a quick *index to the good stuff.* And for this, you should be grateful (and buy the book, for which de Oliveira and Pedrycz will be grateful!), because if you rely on "google," you can spend the rest of your life sifting through the chaff to find the grain.

Jim Bezdek
Pensacola, USA

Preface

Clustering has become a widely accepted synonym of a broad array of activities of exploratory data analysis and model development in science, engineering, life sciences, business and economics, defense, and biological and medical disciplines. Areas such as data mining, image analysis, pattern recognition, modeling, and bio-informatics are just tangible examples of numerous pursuits that vigorously exploit the concepts and algorithms of clustering treated as essential tools for problem formulation and development of specific solutions or a vehicle facilitating interpretation mechanisms. The progress in the area happens at a high pace and these developments concern the fundamentals, algorithmic enhancements, computing schemes, and validation practices. The role of fuzzy clustering becomes quite prominent within the general framework of clustering. This is not surprising given the fact that clustering helps gain an interesting insight into data structure, facilitate efficient communication with users and data analysts, and form essential building blocks for further modeling pursuits. The conceptual underpinnings of fuzzy sets are particularly appealing, considering their abilities to quantify a level of membership of elements to detected clusters that are essential when dealing with the inherent phenomenon of partial belongingness to the group. This feature is of particular interest when dealing with various interpretation activities.

Even a very quick scan of the ongoing research reveals how dynamic the area of fuzzy clustering really is. For instance, a simple query on Science Direct "fuzzy clustering" returns slightly under 400 hits (those are the papers published since 2000). A similar search on ISI Web of Knowledge returns more than 500 hits. In IEEE Xplore one can find around 800 hits. More than half of these entries have been published after 2000. These figures offer us an impression about the rapid progress in the area and highlight a genuine wealth of the applications of the technology of fuzzy clustering.

This volume aims at providing a comprehensive, coherent, and in depth state-of-the-art account on fuzzy clustering. It offers an authoritative treatment of the subject matters presented by leading researchers in the area. While the volume is self-contained by covering some fundamentals and offering an exposure to some preliminary material on algorithms and practice of fuzzy clustering, it offers a balanced and broad coverage of the subject including theoretical fundamentals, methodological insights, algorithms, and case studies.

The content of the volume reflects the main objectives we intend to accomplish. The organization of the overall material helps the reader to proceed with some introductory material, move forward with more advanced topics, become familiar with recent algorithms, and finally gain a detailed knowledge of various application-driven facets.

The contributions have been organized into five general categories: *Fundamentals, Visualization, Algorithms and Computational Aspects, Real-time and Dynamic Clustering,* and *Applications and Case Studies*. They are fairly reflective of the key pursuits in the area.

Within the section dealing with the fundamentals, we are concerned with the principles of clustering as those are seen from the perspective of fuzzy sets. We elaborate on the role of fuzzy sets in data analysis, discuss the principles of data organization, and present fundamental algorithms and their augmentations. Different paradigms of unsupervised learning along with so-called knowledge-based clustering and data organization are also addressed in detail. This part is particularly aimed at the readers who would intend to gather some background material and have a quick yet carefully organized look at the essential of the methodology of fuzzy clustering.

In fuzzy clustering, visualization is an emerging subject. Due to its huge potential to address interpretation and validation issues visualization deserves to be treated as a separate topic.

The part entitled *Algorithms and Computational Aspects* focuses on the major lines of pursuits on the algorithmic and computational augmentations of fuzzy clustering. Here the major focus is on the demonstration of effectiveness of the paradigm of fuzzy clustering in high-dimensional problems, distributed problem solving, and uncertainty management.

The chapters arranged in the group entitled *Real-time and Dynamic Clustering* describe the state-of-the-art algorithms for dynamical developments of clusters, i.e., for clustering built for data gathered over time. Since new observations are available at each time instant, a dynamic update of clusters is required.

The *Applications and Case Studies* part is devoted to a series of applications in which fuzzy clustering plays a pivotal role. The primary intent is to discuss its role in the overall design process in various tasks of prediction, classification, control, and modeling. Here it becomes highly instructive to highlight at which phase of the design clustering is of relevance, what role it plays, and how the results – information granules – facilitate further detailed development of models or enhance interpretation aspects.

PART I FUNDAMENTALS

The part on *Fundamentals* consists of four chapters covering the essentials of fuzzy clustering and presenting a rationality and a motivation, basic algorithms and their various realizations, and cluster validity assessment.

Chapter 1 starts with an introduction to basic clustering algorithms including hard, probabilistic, and possibilistic ones. Then more advanced methods are presented, including the Gustafson–Kessel algorithm and kernel-based fuzzy clustering. Variants on a number of algorithm components as well as on problem formulations are also considered.

Chapter 2 surveys the most relevant methods of relational fuzzy clustering, i.e., fuzzy clustering for relational data. A distinction between object and relational data is presented and the consequences of this distinction on clustering algorithms are thoroughly analyzed. A most useful taxonomy for relational clustering algorithms together with some guidelines for selecting clustering schemes for a given application can also be found in this chapter.

In Chapter 3 the authors offer a contribution that deals with another fundamental issue in clustering: distance functions. The focus is on fuzzy clustering problems and algorithms using the Minkowski distance – definitely an interesting and useful idea.

In Chapter 4 the authors discuss the combination of multiple partitioning obtained from independent clustering runs into a consensus partition – a topic that is gaining interest and importance. A relevant review of commonly used approaches, new consensus strategies (including one based on information-theoretic K-means), as well as a thorough experimental evaluation of these strategies are presented.

PART II VISUALIZATION

Visualization is an important tool in data analysis and interpretation. Visualization offers the user the possibility of quickly inspecting a huge volume of data, and quickly selecting data space regions of interest for further analysis. Generally speaking, this is accomplished by producing a low-dimensional graphical representation of the clusters. The part of the book on *Visualization* consists of two major contributions.

Chapter 5 reviews relevant approaches to validity and visualization of clustering results. It also presents novel tools that allow the visualization of multi-dimensional data points in terms of bi-dimensional plots which facilitates the assessment of clusters' goodness. The chapter ends with an appendix with a comprehensive description of cluster validity indexes.

Chapter 6 aims at helping the user to visually explore clusters. The approach consists of the construction of local, one-dimensional neighborhood models, the so-called neighborgrams. An algorithm is included that generates a subset of neighborgrams from which the user can manage potential cluster candidates during the clustering process. This can be viewed as a form of integrating user domain knowledge into the clustering process.

PART III ALGORITHMS AND COMPUTATIONAL ASPECTS

This part provides the major lines of work on algorithmic and computational augmentations of fuzzy clustering with the intention of demonstrating its effectiveness in high-dimensional problems, distributed problem solving and uncertainty handling. Different paradigms of unsupervised learning along with so-called knowledge-based clustering and data organization are also addressed.

Chapter describes and evaluates a clustering algorithm based on the Yager's participatory learning rule. This learning rule pays special attention to current knowledge as it dominates the way in which new data are used for learning. In participatory clustering the number of clusters is not given a priori as it depends on the cluster structure that is dynamically built by the algorithm.

Chapter 8 offers a comprehensive and in-depth study on fuzzy clustering of fuzzy data.

The authors of Chapter 9 also address the problem of clustering fuzzy data. In this case, clustering is based on the amount of mutual inclusion between fuzzy sets, especially between data and cluster prototypes.

Extraction of semantically valid rules from data is an active interdisciplinary research topic with foundations in computer and cognitive sciences, psychology, and philosophy. Chapter 10 addresses this topic from the clustering perspective. The chapter describes a clustering framework for extracting interpretable rules for medical diagnostics.

Chapter 11 focuses on the combination of regression models with fuzzy clustering. The chapter describes and evaluates several regression models for updating the partition matrix in clustering algorithms. The evaluation includes an analysis of residuals and reveals the interesting characteristics of this class of algorithm.

Hierarchical fuzzy clustering is discussed in Chapter 12. The chapter presents a clustering-based systematic approach to fuzzy modeling that takes into account the following three issues: (1) the number of clusters required a priori in fuzzy clustering; (2) initialization of fuzzy clustering methods, and (3) the trade off between accuracy and interpretability.

Chapter 13 deals with the process of inferring dissimilarity relations from data. For this, two methods are analyzed with respect to factors such as generalization and computational complexity. The approach is particularly interesting for applications where the nature of dissimilarity is conceptual rather than metric.

Chapter 14 describes how clustering and feature selection can be unified to improve the discovery of more relevant data structures. An extension of the proposed algorithm for dealing with an unknown number of clusters is also presented. Interesting applications on image segmentation and text categorization are included.

PART IV REAL-TIME AND DYNAMIC CLUSTERING

Real-time and dynamic clustering deals with clustering with time-varying or noisy data and finds its applications in areas as distinct as video or stock market analysis. Three chapters focus on this timely topic.

Chapter 15 provides a review of dynamic clustering emphasizing its relationship with the area of data mining. Data mining is a matter of paramount relevance today and this chapter shows how dynamic clustering can be brought into the picture. The chapter also describes two novel approaches to dynamic clustering.

Chapter 16 describes the development of an efficient online version of the fuzzy C-means clustering for data streams, i.e., data of potentially unbound size whose continuous evolution is not under the control of the analyzer.

Chapter 17 presents two approaches to real-time clustering and generation of rules from data. The first approach concerns a density-driven approach with its origin stemming from the techniques of mountain and subtractive clustering while the second one looks at the distance based with foundations in the *k*-nearest neighbors and self-organizing maps.

PART V APPLICATIONS AND CASE STUDIES

The last part of the book includes three chapters describing various applications and interesting case studies in which fuzzy clustering plays an instrumental role. The function of fuzzy clustering is discussed in the overall design process in a variety of tasks such as prediction, classification, and modeling.

Chapter 18 presents a novel clustering algorithm that incorporates spatial information by defining multiple feature partitions and shows its application to the analysis of magnetic resonance images.

Chapter 19 exploits both the *K*-means and the fuzzy C-means clustering algorithms as the means to identify correlations between words in texts, using the hyperspace analogue to language (HAL) model.

Another bio-medical application is provided in Chapter 20 where fuzzy clustering techniques are used in the identification of cancerous cells.

FINAL REMARKS

All in all, fuzzy clustering forms a highly enabling technology of data analysis. The area is relatively mature and exhibits a rapid expansion in many different directions including a variety of new concepts, methodologies, algorithms, and innovative and highly advanced applications.

We do hope that the contributions compiled in this volume will bring the reader a fully updated and highly comprehensive view of the recent developments in the fundamentals, algorithms, and applications of fuzzy clustering.

Our gratitude goes to all authors for sharing their expertise and recent research outcomes and reviewers whose constructive criticism was of immense help in producing a high quality volume. Finally, our sincere thanks go to the dedicated and knowledgeable staff at John Wiley & Sons, Ltd, who were highly instrumental in all phases of the project.

Part I
Fundamentals

1 Fundamentals of Fuzzy Clustering

Rudolf Kruse, Christian Döring, and Marie-Jeanne Lesot

Department of Knowledge Processing and Language Engineering,
University of Magdeburg, Germany

1.1 INTRODUCTION

Clustering is an unsupervised learning task that aims at decomposing a given set of *objects* into subgroups or *clusters* based on similarity. The goal is to divide the data-set in such a way that objects (or example cases) belonging to the same cluster are as similar as possible, whereas objects belonging to different clusters are as dissimilar as possible. The motivation for finding and building classes in this way can be manifold (Bock, 1974). Cluster analysis is primarily a tool for discovering previously hidden structure in a set of unordered objects. In this case one assumes that a 'true' or natural grouping exists in the data. However, the assignment of objects to the classes and the description of these classes are unknown. By arranging similar objects into clusters one tries to reconstruct the unknown structure in the hope that every cluster found represents an actual type or category of objects. Clustering methods can also be used for data reduction purposes. Then it is merely aiming at a simplified representation of the set of objects which allows for dealing with a manageable number of homogeneous groups instead of with a vast number of single objects. Only some mathematical criteria can decide on the composition of clusters when classifying data-sets automatically. Therefore clustering methods are endowed with distance functions that measure the dissimilarity of presented example cases, which is equivalent to measuring their similarity. As a result one yields a partition of the data-set into clusters regarding the chosen dissimilarity relation.

All clustering methods that we consider in this chapter are partitioning algorithms. Given a positive integer c, they aim at finding the best partition of the data into c groups based on the given dissimilarity measure and they regard the space of possible partitions into c subsets only. Therein partitioning clustering methods are different from hierarchical techniques. The latter organize data in a nested sequence of groups, which can be visualized in the form of a dendrogram or tree. Based on a dendrogram one can decide on the number of clusters at which the data are best represented for a given purpose. Usually the number of (true) clusters in the given data is unknown in advance. However, using the partitioning methods one is usually required to specify the number of clusters c as an input parameter. Estimating the actual number of clusters is thus an important issue that we do not leave untouched in this chapter.

Advances in Fuzzy Clustering and its Applications Edited by J. Valente de Oliveira and W. Pedrycz

A common concept of all described clustering approaches is that they are prototype-based, i.e., the clusters are represented by *cluster prototypes* $C_i, i = 1, \ldots, c$. Prototypes are used to capture the structure (distribution) of the data in each cluster. With this representation of the clusters we formally denote the set of prototypes $C = \{C_1, \ldots, C_c\}$. Each prototype C_i is an n-tuple of parameters that consists of a *cluster center* $\mathbf{c}_i$ (location parameter) and maybe some additional parameters about the size and the shape of the cluster. The cluster center $\mathbf{c}_i$ is an instantiation of the attributes used to describe the domain, just as the data points in the data-set to divide. The size and shape parameters of a prototype determine the extension of the cluster in different directions of the underlying domain. The prototypes are constructed by the clustering algorithms and serve as prototypical representations of the data points in each cluster.

The chapter is organized as follows: Section 1.2 introduces the basic approaches to hard, fuzzy, and possibilistic clustering. The objective function they minimize is presented as well as the minimization method, the alternating optimization (AO) scheme. The respective partition types are discussed and special emphasis is put on a thorough comparison between them. Further, an intuitive understanding of the general properties that distinguish their results is presented. Then a systematic overview of more sophisticated fuzzy clustering methods is presented. In Section 1.3, the variants that modify the used distance functions for detecting specific cluster shapes or geometrical contours are discussed. In Section 1.4 variants that modify the optimized objective functions for improving the results regarding specific requirements, e.g., dealing with noise, are reviewed. Lastly, in Section 1.5, the alternating cluster estimation framework is considered. It is a generalization of the AO scheme for cluster model optimization, which offers more modeling flexibility without deriving parameter update equations from optimization constraints. Section 1.6 concludes the chapter pointing at related issues and selected developements in the field.

1.2 BASIC CLUSTERING ALGORITHMS

In this section, we present the fuzzy C-means and possibilistic C-means, deriving them from the hard c-means clustering algorithm. The latter one is better known as k-means, but here we call it (hard) C-means to unify the notation and to emphasize that it served as a starting point for the fuzzy extensions. We further restrict ourselves to the simplest form of cluster prototypes at first. That is, each prototype only consists of the center vectors, $C_i = (\mathbf{c}_i)$, such that the data points assigned to a cluster are represented by a prototypical point in the data space. We consider as a distance measure d an inner product norm induced distance as for instance the Euclidean distance. The description of the more complex prototypes and other dissimilarity measures is postponed to Section 1.3, since they are extensions of the basic algorithms discussed here.

All algorithms described in this section are based on *objective functions* J, which are mathematical criteria that quantify the goodness of cluster models that comprise prototypes and data partition. Objective functions serve as cost functions that have to be minimized to obtain optimal cluster solutions. Thus, for each of the following cluster models the respective objective function expresses desired properties of what should be regarded as "best" results of the cluster algorithm. Having defined such a criterion of optimality, the clustering task can be formulated as a function optimization problem. That is, the algorithms determine the best decomposition of a data-set into a predefined number of clusters by minimizing their objective function. The steps of the algorithms follow from the optimization scheme that they apply to approach the optimum of J. Thus, in our presentation of the hard, fuzzy, and possibilistic c-means we discuss their respective objective functions first. Then we shed light on their specific minimization scheme.

The idea of defining an objective function and have its minimization drive the clustering process is quite universal. Aside from the basic algorithms many extensions and modifications have been proposed that aim at improvements of the clustering results with respect to particular problems (e.g., noise, outliers). Consequently, other objective functions have been tailored for these specific applications. We address the most important of the proposed objective function variants in Section 1.4. However, regardless of the specific objective function that an algorithm is based on, the objective function is a goodness measure.

Thus it can be used to compare several clustering models of a data-set that have been obtained by the same algorithm (holding the number of clusters, i.e., the value of c, fixed).

In their basic forms the hard, fuzzy, and possibilistic C-means algorithms look for a predefined number of c clusters in a given data-set, where each of the clusters is represented by its center vector. However, hard, fuzzy, and possibilistic C-means differ in the way they assign data to clusters, i.e., what type of data partitions they form. In classical (hard) cluster analysis each datum is assigned to exactly one cluster. Consequently, the hard C-means yield exhaustive partitions of the example set into non-empty and pairwise disjoint subsets. Such hard (crisp) assignment of data to clusters can be inadequate in the presence of data points that are almost equally distant from two or more clusters. Such special data points can represent hybrid-type or mixture objects, which are (more or less) equally similar to two or more types. A crisp partition arbitrarily forces the full assignment of such data points to one of the clusters, although they should (almost) equally belong to all of them. For this purpose the fuzzy clustering approaches presented in Sections 1.2.2 and 1.2.3 relax the requirement that data points have to be assigned to one (and only one) cluster. Data points can belong to more than one cluster and even with different degrees of membership to the different clusters. These gradual cluster assignments can reflect present cluster structure in a more natural way, especially when clusters overlap. Then the memberships of data points at the overlapping boundaries can express the ambiguity of the cluster assignment.

The shift from hard to gradual assignment of data to clusters for the purpose of more expressive data partitions founded the field of fuzzy cluster analysis. We start our presentation with the hard C-means and later on we point out the relatedness to the fuzzy approaches that is evident in many respects.

1.2.1 Hard c-means

In the classical C-means model each data point $\mathbf{x}_j$ in the given data-set $X = \{\mathbf{x}_1, \ldots, \mathbf{x}_n\}$, $X \subseteq \mathbb{R}^p$ is assigned to exactly one cluster. Each cluster Γ_i is thus a subset of the given data-set, $\Gamma_i \subset X$. The set of clusters $\Gamma = \{\Gamma_1, \ldots, \Gamma_c\}$ is required to be an exhaustive partition of the data-set X into c non-empty and pairwise disjoint subsets Γ_i, $1 < c < n$. In the C-means such a data partition is said to be optimal when the sum of the squared distances between the cluster centers and the data points assigned to them is minimal (Krishnapuram and Keller, 1996). This definition follows directly from the requirement that clusters should be as homogeneous as possible. Hence the objective function of the hard C-means can be written as follows:

$$J_h(X, U_h, C) = \sum_{i=1}^{c} \sum_{j=1}^{n} u_{ij} d_{ij}^2, \tag{1.1}$$

where $C = \{C_1, \ldots, C_c\}$ is the set of cluster prototypes, d_{ij} is the distance between $\mathbf{x}_j$ and cluster center $\mathbf{c}_i$, U is a $c \times n$ binary matrix called partition matrix. The individual elements

$$u_{ij} \in \{0, 1\} \tag{1.2}$$

indicate the assignment of data to clusters: $u_{ij} = 1$ if the data point $\mathbf{x}_j$ is assigned to prototype C_i, i.e., $\mathbf{x}_j \in \Gamma_i$; and $u_{ij} = 0$ otherwise. To ensure that each data point is assigned exactly to one cluster, it is required that:

$$\sum_{i=1}^{c} u_{ij} = 1, \quad \forall j \in \{1, \ldots, n\}. \tag{1.3}$$

This constraint enforces exhaustive partitions and also serves the purpose to avoid the trivial solution when minimizing J_h, which is that no data is assigned to any cluster: $u_{ij} = 0, \forall i, j$. Together with $u_{ij} \in \{0, 1\}$ it is possible that data are assigned to one or more clusters while there are some remaining clusters left empty. Since such a situation is undesirable, one usually requires that:

$$\sum_{j=1}^{n} u_{ij} > 0, \quad \forall i \in \{1, \ldots, c\}. \tag{1.4}$$

J_h depends on the two (disjoint) parameter sets, which are the cluster centers c and the assignment of data points to clusters U. The problem of finding parameters that minimize the C-means objective function is NP-hard (Drineas *et al.*, 2004). Therefore, the hard C-means clustering algorithm, also known as *ISODATA algorithm* (Ball and Hall, 1966; Krishnapuram and Keller, 1996), minimizes J_h using an alternating optimization (AO) scheme.

Generally speaking, AO can be applied when a criterion function cannot be optimized directly, or when it is impractical. The parameters to optimize are split into two (or even more) groups. Then one group of parameters (e.g., the partition matrix) is optimized holding the other group(s) (e.g., the current cluster centers) fixed (and vice versa). This iterative updating scheme is then repeated. The main advantage of this method is that in each of the steps the optimum can be computed directly. By iterating the two (or more) steps the joint optimum is approached, although it cannot be guaranteed that the global optimum will be reached. The algorithm may get stuck in a local minimum of the applied objective function J. However, alternating optimization is the commonly used parameter optimization method in clustering algorithms. Thus for each of the algorithms in this chapter we present the corresponding parameter update equations of their alternating optimization scheme.

In the case of the hard C-means the iterative optimization scheme works as follows: at first initial cluster centers are chosen. This can be done randomly, i.e., by picking c random vectors that lie within the smallest (hyper-)box that encloses all data; or by initializing cluster centers with randomly chosen data points of the given data-set. Alternatively, more sophisticated initialization methods can be used as well, e.g., Latin hypercube sampling (McKay, Beckman and Conover, 1979). Then the parameters C are held fixed and cluster assignments U are determined that minimize the quantity of J_h. In this step each data point is assigned to its closest cluster center:

$$u_{ij} = \begin{cases} 1, & \text{if } i = \operatorname{argmin}_{l=1}^{c} d_{lj} \\ 0, & \text{otherwise} \end{cases} . \tag{1.5}$$

Any other assignment of a data point than to its closest cluster would not minimize J_h for fixed clusters. Then the data partition U is held fixed and new cluster centers are computed as the mean of all data vectors assigned to them, since the mean minimizes the sum of the square distances in J_h. The calculation of the mean for each cluster (for which the algorithm got its name) is stated more formally:

$$\mathbf{c}_i = \frac{\sum_{j=1}^{n} u_{ij}\mathbf{x}_j}{\sum_{j=1}^{n} u_{ij}}. \tag{1.6}$$

The two steps (1.5) and (1.6) are iterated until no change in C or U can be observed. Then the hard C-means terminates, yielding final cluster centers and data partition that are possibly locally optimal only.

Concluding the presentation of the hard C-means we want to mention its expressed tendency to become stuck in local minima, which makes it necessary to conduct several runs of the algorithm with different initializations (Duda and Hart, 1973). Then the best result out of many clusterings can be chosen based on the values of J_h.

We now turn to the fuzzy approaches, that relax the requirement $u_{ij} \in \{0, 1\}$ that is placed on the cluster assignments in classical clustering approaches. The extensions are based on the concepts of fuzzy sets such that we arrive at gradual memberships. We will discuss two major types of gradual cluster assignments and fuzzy data partitions altogether with their differentiated interpretations and standard algorithms, which are the (probabilistic) fuzzy C-means (FCM) in the next section and the possibilistic fuzzy C-means (PCM) in Section 1.2.3.

1.2.2 Fuzzy c-means

Fuzzy cluster analysis allows gradual memberships of data points to clusters measured as degrees in [0,1]. This gives the flexibility to express that data points can belong to more than one cluster. Furthermore, these membership degrees offer a much finer degree of detail of the data model. Aside from assigning a data point to clusters in shares, membership degrees can also express how ambiguously or definitely a data point should belong to a cluster. The concept of these membership degrees is substantiated by the definition and interpretation of fuzzy sets (Zadeh, 1965). Thus, fuzzy clustering allows fine grained solution spaces in the form of fuzzy partitions of the set of given examples $X = \{\mathbf{x}_1, \ldots, \mathbf{x}_n\}$. Whereas the clusters Γ_i of data partitions have been classical subsets so far, they are represented by the fuzzy sets μ_{Γ_i} of the data-set X in the following. Complying with fuzzy set theory, the cluster assignment u_{ij} is now the membership degree of a datum $\mathbf{x}_j$ to cluster Γ_i, such that: $u_{ij} = \mu_{\Gamma_i}(\mathbf{x}_j) \in [0, 1]$. Since memberships to clusters are fuzzy, there is not a single label that is indicating to which cluster a data point belongs. Instead, fuzzy clustering methods associate a fuzzy label vector to each data point $\mathbf{x}_j$ that states its memberships to the c clusters:

$$\mathbf{u}_j = (u_{1j}, \ldots, u_{cj})^T. \tag{1.7}$$

The $c \times n$ matrix $U = (u_{ij}) = (\mathbf{u}_1, \ldots, \mathbf{u}_n)$ is then called a fuzzy partition matrix. Based on the fuzzy set notion we are now better suited to handle ambiguity of cluster assignments when clusters are badly delineated or overlapping.

So far, the general definition of fuzzy partition matrices leaves open how assignments of data to more than one cluster should be expressed in form of membership values. Furthermore, it is still unclear what degrees of belonging to clusters are allowed, i.e., the solution space (set of allowed fuzzy partitions) for fuzzy clustering algorithms is not yet specified. In the field of fuzzy clustering two types of fuzzy cluster partitions have evolved. They differ in the constraints they place on the membership degrees and how the membership values should be interpreted. In our discussion we begin with the most widely used type, the probabilistic partitions, since they have been proposed first. Notice, that in literature they are sometimes just called fuzzy partitions (dropping the word 'probabilistic'). We use the subscript f for the probabilistic approaches and, in the next section, p for the possibilistic models. The latter constitute the second type of fuzzy partitions.

Let $X = \{\mathbf{x}_1, \ldots, \mathbf{x}_n\}$ be the set of given examples and let c be the number of clusters $(1 < c < n)$ represented by the fuzzy sets μ_{Γ_i}, $(i = 1, \ldots, c)$. Then we call $U_f = (u_{ij}) = (\mu_{\Gamma_i}(\mathbf{x}_j))$ a *probabilistic cluster partition* of X if

$$\sum_{j=1}^{n} u_{ij} > 0, \quad \forall i \in \{1, \ldots, c\}, \qquad \text{and} \tag{1.8}$$

$$\sum_{i=1}^{c} u_{ij} = 1, \quad \forall j \in \{1, \ldots, n\} \tag{1.9}$$

hold. The $u_{ij} \in [0, 1]$ are interpreted as the membership degree of datum $\mathbf{x}_j$ to cluster Γ_i relative to all other clusters.

Constraint (1.8) guarantees that no cluster is empty. This corresponds to the requirement in classical cluster analysis that no cluster, represented as (classical) subset of X, is empty (see Equation (1.4)). Condition (1.9) ensures that the sum of the membership degrees for each datum equals 1. This means that each datum receives the same weight in comparison to all other data and, therefore, that all data are (equally) included into the cluster partition. This is related to the requirement in classical clustering that partitions are formed exhaustively (see Equation (1.3)). As a consequence of both constraints no cluster can contain the full membership of all data points. Furthermore, condition (1.9) corresponds to a normalization of the memberships per datum. Thus the membership degrees for a given datum *formally resemble* the probabilities of its being a member of the corresponding cluster.

Example: Figure 1.1 shows a (probabilistic) fuzzy classification of a two-dimensional symmetric data-set with two clusters. The grey scale indicates the strength of belonging to the clusters. The darker shading in the image indicates a high degree of membership for data points close to the cluster centers, while membership decreases for data points that lie further away from the clusters. The membership values of the data points are shown in Table 1.1. They form a probabilistic cluster partition according to the definition above. The following advantages over a conventional clustering representation can be noted: points in the center of a cluster can have a degree equal to 1, while points close to boundaries can be

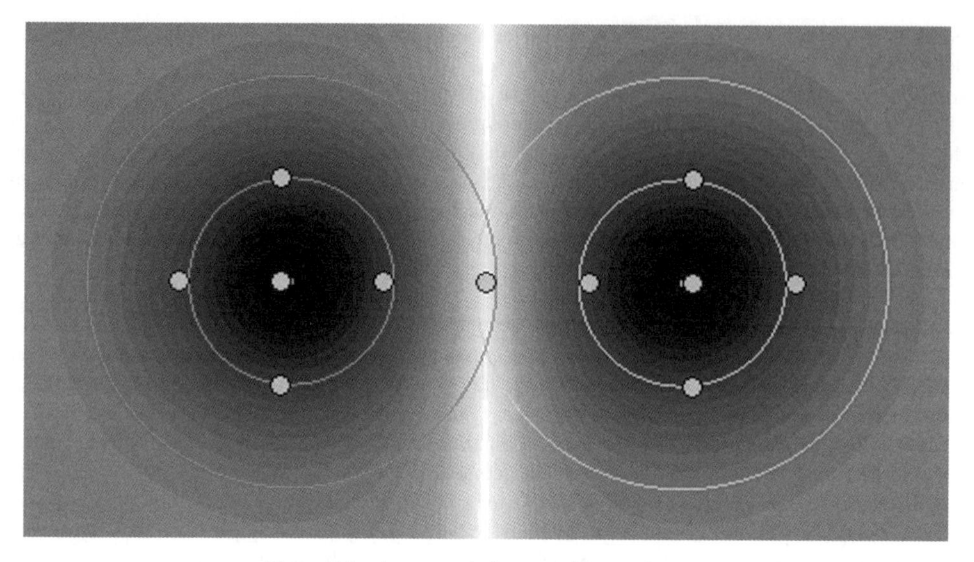

Figure 1.1 A symmetric data-set with two clusters.

identified as such, since their membership degree to the cluster they are closer to is considerably smaller than 1. Points on class boundaries may be classified as undetermined with a degree of indeterminacy proportional to their similarity to core points. The equidistant data point $\mathbf{x}_5$ in the middle of the figure would have to be arbitrarily assigned with full weight to one of the clusters if only classical ('crisp') partitions were allowed. In this fuzzy partition, however, it can be associated with the equimembership vector $(0.5, 0.5)^T$ to express the ambiguity of the assignment. Furthermore, crisp data partitions cannot express the difference between data points in the center and those that are rather at the boundary of a cluster. Both kinds of points would be fully assigned to the cluster they are most similar to. In a fuzzy cluster partition they are assigned degrees of belonging depending on their closeness to the centers.

After defining probabilistic partitions we can turn to developing an objective function for the fuzzy clustering task. Certainly, the closer a data point lies to the center of a cluster, the higher its degree of membership should be to this cluster. Following this rationale, one can say that the distances between the cluster centers and the data points (strongly) assigned to it should be minimal. Hence the problem to divide a given data-set into c clusters can (again) be stated as the task to minimize the squared distances of the data points to their cluster centers, since, of course, we want to maximize the degrees of membership. The probabilistic fuzzy objective function J_f is thus based on the least sum of squared distances just as J_h

Table 1.1 A fuzzy partition of the symmetric data-set.

j	x	y	u_{0j}	u_{1j}
0	−3	0	0.93	0.07
1	−2	0	0.99	0.01
2	−1	0	0.94	0.06
3	−2	1	0.69	0.31
4	−2	−1	0.69	0.31
5	0	0	**0.50**	**0.50**
6	1	0	0.06	0.94
7	2	0	0.01	0.99
8	3	0	0.07	0.93
9	2	1	0.31	0.69
10	2	−1	0.31	0.69

of the hard C-means. More formally, a fuzzy cluster model of a given data-set X into c clusters is defined to be optimal when it minimizes the objective function:

$$J_f(\mathrm{X}, \mathrm{U}_f, \mathrm{C}) = \sum_{i=1}^{c} \sum_{j=1}^{n} u_{ij}^m d_{ij}^2, \tag{1.10}$$

under the constraints (1.8) and (1.9) that have to be satisfied for probabilistic membership degrees in U_f. The condition (1.8) avoids the trivial solution of minimization problem, i.e., $u_{ij} = 0, \forall i, j$. The normalization constraint (1.9) leads to a 'distribution' of the weight of each data point over the different clusters. Since all data points have the same fixed amount of membership to share between clusters, the normalization condition implements the known partitioning property of any probabilistic fuzzy clustering algorithm. The parameter $m, m > 1$, is called the *fuzzifier* or *weighting exponent*. The exponentiation of the memberships with m in J_f can be seen as a function g of the membership degrees, $g(u_{ij}) = u_{ij}^m$, that leads to a generalization of the well-known least squared error functional as it was applied in the hard c-means (see Equation (1.1)). The actual value of m then determines the 'fuzziness' of the classification. It has been shown for the case $m = 1$ (when J_h and J_f become identical), that cluster assignments remain hard when minimizing the target function, even though they are allowed to be fuzzy, i.e., they are not constrained in $\{0, 1\}$ (Dunn, 1974b). For achieving the desired fuzzification of the resulting probabilistic data partition the function $g(u_{ij}) = u_{ij}^2$ has been proposed first (Dunn, 1974b). The generalization for exponents $m > 1$ that lead to fuzzy memberships has been proposed in (Bezdek, 1973). With higher values for m the boundaries between clusters become softer, with lower values they get harder. Usually $m = 2$ is chosen. Aside from the standard weighting of the memberships with u_{ij}^m other functions g that can serve as fuzzifiers have been explored. Their influence on the memberships will be discussed in Section 1.4.2.

The objective function J_f is alternately optimized, i.e., first the membership degrees are optimized for fixed cluster parameters, then the cluster prototypes are optimized for fixed membership degrees:

$$\mathrm{U}_\tau = j_U(\mathrm{C}_{\tau-1}), \quad \tau > 0 \quad \text{and} \tag{1.11}$$

$$\mathrm{C}_\tau = j_C(\mathrm{U}_\tau). \tag{1.12}$$

In each of the two steps the optimum can be computed directly using the parameter update equations j_U and j_C for the membership degrees and the cluster centers, respectively. The update formulae are derived by simply setting the derivative of the objective function J_f w.r.t. the parameters to optimize equal to zero (taking into account the constraint (1.9)). The resulting equations for the two iterative steps form the fuzzy C-means algorithm.

The membership degrees have to be chosen according to the following update formula that is independent of the chosen distance measure (Bezdek, 1981; Pedrycz, 2005):

$$u_{ij} = \frac{1}{\sum_{l=1}^{c} \left(\frac{d_{ij}^2}{d_{lj}^2} \right)^{\frac{1}{m-1}}} = \frac{d_{ij}^{-\frac{2}{m-1}}}{\sum_{l=1}^{c} d_{lj}^{-\frac{2}{m-1}}}. \tag{1.13}$$

In this case there exists a cluster i with zero distance to a datum $\mathbf{x}_j$, $u_{ij} = 1$ and $u_{lj} = 0$ for all other clusters $l \neq i$. The above equation clearly shows the relative character of the probabilistic membership degree. It depends not only on the distance of the datum $\mathbf{x}_j$ to cluster i, but also on the distances between this data point and other clusters.

The update formulae j_C for the cluster parameters depend, of course, on the parameters used to describe a cluster (location, shape, size) and on the chosen distance measure. Therefore a general update formula cannot be given. In the case of the basic fuzzy C-means model the cluster center vectors serve as prototypes, while an inner product norm induced metric is applied as distance measure. Consequently the derivations of J_f w.r.t. the centers yield (Bezdek, 1981):

$$\mathbf{c}_i = \frac{\sum_{j=1}^{n} u_{ij}^m \mathbf{x}_j}{\sum_{j=1}^{n} u_{ij}^m}. \tag{1.14}$$

The choice of the optimal cluster center points for fixed memberships of the data to the clusters has the form of a generalized mean value computation for which the fuzzy C-means algorithm has its name.

The general form of the AO scheme of coupled equations (1.11) and (1.12) starts with an update of the membership matrix in the first iteration of the algorithm ($\tau = 1$). The first calculation of memberships is based on an initial set of prototypes C_0. Even though the optimization of an objective function could mathematically also start with an initial but valid membership matrix (i.e., fulfilling constraints (1.8) and (1.9)), a C_0 initialization is easier and therefore common practice in all fuzzy clustering methods. Basically the fuzzy C-means can be initialized with cluster centers that have been randomly placed in the input space. The repetitive updating in the AO scheme can be stopped if the number of iterations τ exceeds some predefined number of maximal iterations τ_{max}, or when the changes in the prototypes are smaller than some termination accuracy. The (probabilistic) fuzzy C-means algorithm is known as a stable and robust classification method. Compared with the hard C-means it is quite insensitive to its initialization and it is not likely to get stuck in an undesired local minimum of its objective function in practice (Klawonn, 2006). Due to its simplicity and low computational demands, the probabilistic fuzzy C-means is a widely used initializer for other more sophisticated clustering methods. On the theoretical side it has been proven that either the iteration sequence itself or any convergent subsequence of the probabilistic FCM converges in a saddle point or a minimum – but not in a maximum – of the objective function (Bezdek, 1981).

1.2.3 Possibilistic c-means

Although often desirable, the 'relative' character of the probabilistic membership degrees can be misleading (Timm, Borgett, Döring and Kruse, 2004). Fairly high values for the membership of datum in more than one cluster can lead to the impression that the data point is typical for the clusters, but this is not always the case. Consider, for example, the simple case of two clusters shown in Figure 1.2. Datum $\mathbf{x}_1$ has the same distance to both clusters and thus it is assigned a membership degree of about 0.5. This is plausible. However, the same degrees of membership are assigned to datum $\mathbf{x}_2$ even though this datum is further away from both clusters and should be considered less typical. Because of the normalization, however, the sum of the memberships has to be 1. Consequently $\mathbf{x}_2$ receives fairly high membership degrees to both clusters. For a correct interpretation of these memberships one has to keep in mind that they are rather degrees of sharing than of typicality, since the constant weight of 1 given to a datum must be distributed over the clusters. A better reading of the memberships, avoiding misinterpretations, would be (Höppner, Klawonn, Kruse and Runkler 1999): 'If the datum $\mathbf{x}_j$ has to be assigned to a cluster, then with the probability u_{ij} to the cluster i'.

The normalization of memberships can further lead to undesired effects in the presence of noise and outliers. The fixed data point weight may result in high membership of these points to clusters, even though they are a large distance from the bulk of data. Their membership values consequently affect the clustering results, since data point weight attracts cluster prototypes. By dropping the normalization constraint (1.9) in the following definition one tries to achieve a more intuitive assignment of degrees of membership and to avoid undesirable normalization effects.

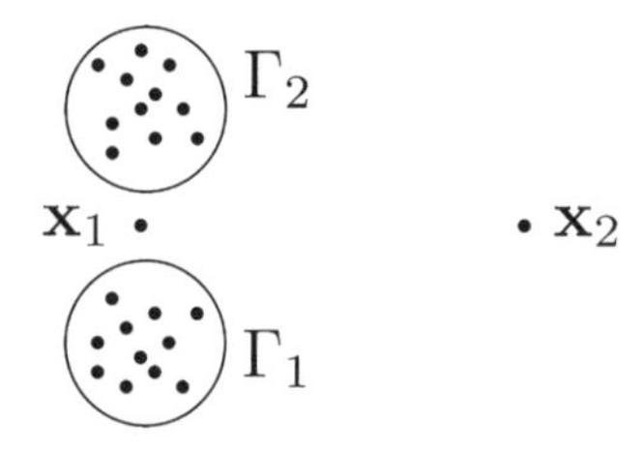

Figure 1.2 A situation in which the probabilistic assignment of membership degrees is counterintuitive for datum $\mathbf{x}_2$.

Let $X = \{\mathbf{x}_1, \ldots, \mathbf{x}_n\}$ be the set of given examples and let c be the number of clusters $(1 < c < n)$ represented by the fuzzy sets μ_{Γ_i}, $(i = 1, \ldots, c)$. Then we call $U_p = (u_{ij}) = (\mu_{\Gamma_i}(\mathbf{x}_j))$ a possibilistic cluster partition of X if

$$\sum_{j=1}^{n} u_{ij} > 0, \quad \forall i \in \{1, \ldots, c\} \tag{1.15}$$

holds. The $u_{ij} \in [0, 1]$ are interpreted as the degree of representativity or typicality of the datum $\mathbf{x}_j$ to cluster Γ_i.

The membership degrees for one datum now *resemble* the possibility (in the sense of possibility theory (Dubois and Prade, 1988) of its being a member of the corresponding cluster (Davé and Krishnapuram, 1997; Krishnapuram and Keller, 1993).

The objective function J_f that just minimizes the squared distances between clusters and assigned data points would not be appropriate for possibilistic fuzzy clustering. Dropping the normalization constraint leads to the mathematical problem that the objective function would reach its minimum for $u_{ij} = 0$ for all $i \in \{1, \ldots, c\}$ and $j \in \{1, \ldots, n\}$, i.e., data points are not assigned to any cluster and all clusters are empty. In order to avoid this trivial solution (that is also forbidden by constraint (1.15)), a penalty term is introduced, which forces the membership degrees away from zero. That is, the objective function J_f is modified to

$$J_p(X, U_p, C) = \sum_{i=1}^{c} \sum_{j=1}^{n} u_{ij}^m d_{ij}^2 + \sum_{i=1}^{c} \eta_i \sum_{j=1}^{n} (1 - u_{ij})^m, \tag{1.16}$$

where $\eta_i > 0\,(i = 1, \ldots, c)$ (Krishnapuram and Keller, 1993). The first term leads to a minimization of the weighted distances. The second term suppresses the trivial solution since this sum rewards high memberships (close to 1) that make the expression $(1 - u_{ij})^m$ become approximately 0. Thus the desire for (strong) assignments of data to clusters is expressed in the objective function J_p. In tandem with the first term the high membership can be expected especially for data that are close to their clusters, since with a high degree of belonging the weighted distance to a closer cluster is smaller than to clusters further away. The cluster specific constants η_i are used balance the contrary objectives expressed in the two terms of J_p. It is a reference value stating at what distance to a cluster a data point should receive higher membership to it. These considerations mark the difference to probabilistic clustering approaches. While in probabilistic clustering each data point has a constant weight of 1, possibilistic clustering methods have to learn the weights of data points.

The formula for updating the membership degrees that is derived from J_p by setting its derivative to zero is (Krishnapuram and Keller, 1993):

$$u_{ij} = \frac{1}{1 + \left(\frac{d_{ij}^2}{\eta_i}\right)^{\frac{1}{m-1}}}. \tag{1.17}$$

First of all, this update equation clearly shows that the membership of a datum $\mathbf{x}_j$ to cluster i depends only on its distance d_{ij} to this cluster. Small distance corresponds to high degree of membership whereas larger distances (i.e., strong dissimilarity) results in low membership degrees. Thus the u_{ij} have typicality interpretation.

Equation (1.17) further helps to explain the parameters η_i of the clusters. Considering the case $m = 2$ and substituting η_i for d_{ij}^2 yields $u_{ij} = 0.5$. It becomes obvious that η_i is a parameter that determines the distance to the cluster i at which the membership degree should be 0.5. Since that value of membership can be seen as definite assignment to a cluster, the permitted extension of the cluster can be controlled with this parameter. Depending on the cluster's shape the η_i have different geometrical interpretation. If hyperspherical clusters as in the possibilistic C-means are considered, $\sqrt{\eta_i}$ is their mean diameter. In shell clustering $\sqrt{\eta_i}$ corresponds to the mean thickness of the contours described by the cluster prototype information (Hööppner, Klawonn, Kruse and Runkler 1999) (see Section 1.3.2). If such properties of the clusters to search for are known prior to the analysis of the given data, η_i can be set to the desired value. If all clusters have the same properties, the same value can be chosen for all clusters. However, the information on the actual shape property described by η_i is often not known in advance. In that case these parameters must be estimated. Good estimates can be found using a probabilistic clustering model of the given data-set. The η_i are then estimated by the fuzzy intra-cluster distance using the fuzzy memberships matrix U_f as it has been determined by the probabilistic counterpart of the chosen possibilistic algorithm (Krishnapuram and Keller, 1993). That is, for all clusters ($i = 1, \ldots, n$):

$$\eta_i = \frac{\sum_{j=1}^{n} u_{ij}^m d_{ij}^2}{\sum_{j=1}^{n} u_{ij}^m}. \tag{1.18}$$

Update equations j_C for the prototypes are as well derived by simply setting the derivative of the objective function J_p w.r.t. the prototype parameters to optimize equal to zero (holding the membership degrees U_p fixed). Looking at both objective functions J_f and J_p it can be inferred that the update equations for the cluster prototypes in the possibilistic algorithms must be identical to their probabilistic counterparts. This is due to the fact that the second, additional term in J_p vanishes in the derivative for fixed (constant) memberships u_{ij}. Thus the cluster centers in the possibilistic C-means algorithm are re-estimated as in Equation (1.14).

1.2.4 Comparison and Respective Properties of Probabilistic and Possibilistic Fuzzy c-means

Aside from the different interpretation of memberships, there are some general properties that distinguish the behaviour and the results of the possibilistic and probabilistic fuzzy clustering approaches.

Example: Figures 1.3 and 1.4 illustrate a probabilistic and a possibilistic fuzzy C-means classification of the Iris data-set into three clusters (Blake and Merz, 1998; Fisher, 1936). The displayed partitions of the data-set are the result of alternatingly optimizing J_f and J_p, respectively (Timm, Borgelt, Döring and Kruse, 2004). The grey scale indicates the membership to the closest cluster. While probabilistic memberships rather divide the data space, possibilistic membership degrees only depend on the typicality to the respective closest clusters. On the left, the data-set is divided into three clusters. On the right, the possibilistic fuzzy C-means algorithm detects only two clusters, since two of the three clusters in the upper right of Figure 1.4 are identical. Note that this behaviour is specific to the possibilistic approach. In the probabilistic counterpart the cluster centers are driven apart, because a cluster, in a way, 'seizes' part of the weight of a datum and thus leaves less that may attract other cluster centers. Hence sharing a datum between clusters is disadvantageous. In the possibilistic approach there is nothing that corresponds to this effect.

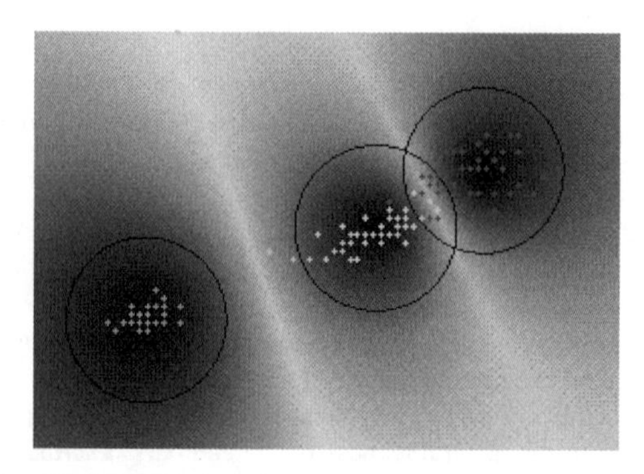

Figure 1.3 Iris data-set classified with probalistic fuzzy C-means algorithm. Attributes petal length and petal width.

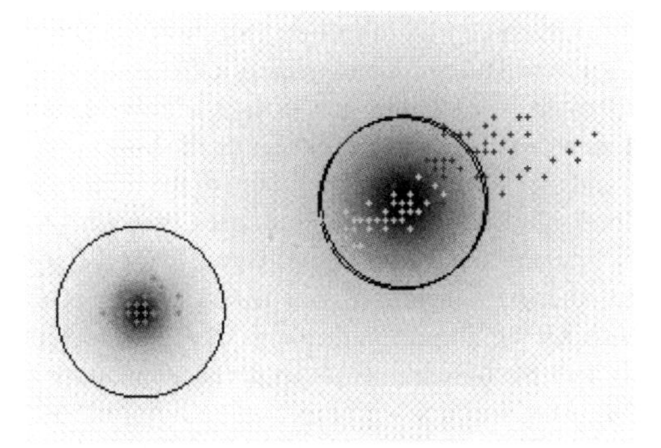

Figure 1.4 Iris data-set classifed with possibilistic fuzzy C-means algorithm. Attribtes petal length and petal width.

1.2.4.1 Cluster Coincidence

One of the major characteristics in which the approaches differ lies in the fact that probabilistic algorithms are forced to partition the data exhaustively while the corresponding possibilistic approaches are not compelled to do so. The former distribute the total membership of the data points (sums up to one) whereas the latter are rather required to determine the data point weights by themselves. Probabilistic algorithms attempt to cover all data points with clusters, since sharing data point weight is disadvantageous. In the possibilistic case, there is no interaction between clusters. Thus the found clusters in possibilistic models can be located much closer to each other than in a probabilistic clustering. Clusters can even coincide, which has been widely observed (Barni, Cappellini and Mecocci, 1996; Krishnapuram and Keller, 1996). This leads to solutions in which one cluster being actually present in a data-set can be represented by two clusters in the possibilistic model. In worse cases there is data left in other regions of the input space that has cluster structure, but which is not covered by clusters in the model. Then possibilistic algorithms show the tendency to interpret data points in such left over regions as outliers by assigning low memberships for these data to all clusters (close to 0) instead of further adjusting the possibly non-optimal cluster set (Höppner, Klawonn, Jruse and Runkler, 1999).

This described behaviour is exhibited, since J_p treats each cluster independently. Every cluster contributes to some extent to the value of the objective function J_p regardless of other clusters. The resulting behaviour has been regarded by stating that possibilistic clustering is a rather mode-seeking technique, aimed at finding meaningful clusters (Krishnapuram and Keller, 1996). The number c of known or desired clusters has been interpreted as an upper bound, since cluster coincidence in effect leads to a smaller number of clusters in the model (Höppner, Klawonn, Kruse and Runkler, 1999). For reducing the tendency of coinciding clusters and for a better coverage of the entire data space usually a probabilistic analysis is carried out before (exploiting its partitional property). The result is used for the prototype initialization of the first run of the possibilistic algorithm as well as for getting the initial guesses of the η_i (and c). After the first possibilistic analysis has been carried out, the values of the η_i are re-estimated once more using the first possibilistic fuzzy partition. The improved estimates are used for running the possibilistic algorithm a second time yielding the final cluster solution (Höppner, Klawonn, Jruse and Runkler, 1999).

1.2.4.2 Cluster Repulsion

Dealing with the characteristics of the possibilistic clustering techniques as above is a quite good measure. However, there are theoretical results, which put forth other developments. We discovered that the objective function J_p is, in general, truly minimized only if all cluster centers are identical (Timm, Borgelt, Döring and Kruse, 2004). The possibilistic objective function can be decomposed into c independent terms, one for each cluster. This is the amount by which each cluster contributes to the value of J_p. If there is a single optimal point for a cluster center (as will usually be the case, since multiple optimal points would require a high symmetry in the data), all cluster centers moved to that point results in the lowest value of J_p for a given data-set. Consequently other results than all cluster centers being identical are achieved only because the algorithm gets stuck in a local minimum of the objective function. In the example of the PCM model in Figure 1.4 the cluster on the lower left in the figure has been found, because it is well separated and thus forms a local minimum of the objective function. This, of course, is not a desirable situation. Good solutions w.r.t the minimization of J_p unexpectedly do not correspond to what we regard as a good solution of the clustering problem. Hence the possibilistic algorithms can be improved by modifying the objective function in such a way that the problematic property examined above is removed (see Section 1.4.4). These modifications of J_p lead to better detection of the shape of very close or overlapping clusters. Such closely located point accumulations have been problematic, since possibilistic clusters 'wander' in the direction where most of the data can be found in their η_i environment, which easily leads to cluster coincidence. Nevertheless, the modified possibilistic techniques should also be initialized with the corresponding probabilistic algorithms as described in the last paragraph. It is a good measure for improving the chances that all data clouds will be regarded in the resulting possibilistic model leaving no present cluster structure unclassified. Recent developments that try to alleviate the problematic properties of the possibilistic clustering algorithms propose using a combination of both fuzzy and possibilistic memberships (see Section 1.4.4).

1.2.4.3 Recognition of Positions and Shapes

The possibilistic models do not only carry problematic properties. Memberships that depend only on the distance to a cluster while being totally independent from other clusters lead to prototypes that better reflect human intuition. Calculated based on weights that reflect typicality, the centers of possibilistic clusters as well as their shape and size better fit the data clouds compared to their probabilistic relatives. The latter ones are known to be unable to recognize cluster shapes as perfectly as their possibilistic counterparts. This is due to the following reasons: if clusters are located very close or are even overlapping, then they are separated well because sharing membership is disadvantageous (see upper right in Figure 1.3). Higher memberships to data points will be assigned in directions pointing away from the overlap. Thus the centers are repelling each other. If complex prototypes are used, detected cluster shapes are likely to be slightly distorted compared to human intuition. Noise and outliers are another reason for little prototype distortions. They have weight in probabilistic partitions and therefore attract clusters which can result in small prototype deformations and less intuitive centers. Possibilistic techniques are less sensitive to outliers and noise. Low memberships will be assigned due to greater distance. Due to this property and the more intuitive determination of positions and shapes, possibilistic techniques are attractive tools in image processing applications. In probabilistic fuzzy clustering, noise clustering techniques are widely appreciated (see Section 1.4.1). In one of the noise handling approaches, the objective function J_f is modified such that a virtual noise cluster "seizes" parts of the data point weight of noise points and outliers. This leads to better detection of actual cluster structure in probabilistic models.

1.3 DISTANCE FUNCTION VARIANTS

In the previous section, we considered the case where the distance between cluster centers and data points is computed using the Euclidean distance, leading to the standard versions of fuzzy C-means and possibilistic C-means. This distance only makes it possible to identify spherical clusters. Several variants have been proposed to relax this constraint, considering other distances between cluster centers and data points. In this section, we review some of them, mentioning the fuzzy Gustafson–Kessel algorithm, fuzzy shell clustering algorithms and kernel-based variants. All of them can be applied both in the fuzzy probabilistic and possibilistic framework.

Please note that a more general algorithm is provided by the fuzzy relational clustering algorithm (Hathaway and Bezdek, 1994) that takes as input a distance matrix. In this chapter, we consider the variants that handle object data and do not present the relational approach.

1.3.1 Gustafson-Kessel Algorithm

The Gustafson–Kessel algorithm (Gustafson and Kessel, 1979) replaces the Euclidean distance by a cluster-specific Mahalanobis distance, so as to adapt to various sizes and forms of the clusters. For a cluster i, its associated Mahalanobis distance is defined as

$$d^2(\mathbf{x}_j, C_i) = (\mathbf{x}_j - \mathbf{c}_i)^T \Sigma_i^{-1} (\mathbf{x}_j - \mathbf{c}_i), \tag{1.19}$$

where Σ_i is the covariance matrix of the cluster. Using the Euclidean distance as in the algorithms presented in the previous section is equivalent to assuming that $\forall i, \Sigma_i = I$, i.e., all clusters have the same covariance that equals the identity matrix. Thus it only makes it possible to detect spherical clusters, but it cannot identify clusters having different forms or sizes.

The Gustafson–Kessel algorithm models each cluster Γ_i by both its center $\mathbf{c}_i$ and its covariance matrix $\Sigma_i, i = 1, \ldots, c$. Thus cluster prototypes are tuples $C_i = (\mathbf{c}_i, \Sigma_i)$ and both $\mathbf{c}_i$ and Σ_i are to be learned. The eigenstructure of the positive definite $p \times p$ matrix Σ_i represents the shape of cluster i. Specific constraints can be taken into account, for instance restricting to axis-parallel cluster shapes, by considering only diagonal matrices. This case is usually preferred when clustering is applied for the generation of fuzzy rule systems (Höppner, Klawonn, Kruse, and Runkler, 1999). The sizes of the clusters, if known in advance, can be controlled using the constants $\varrho_i > 0$ demanding that $\det(\Sigma_i) = \varrho_i$. Usually the clusters are assumed to be of equal size setting $\det(\Sigma_i) = 1$.

The objective function is then identical to the fuzzy C-means (see Equation (1.10)) or the possibilistic one (see Equation (1.16)), using as distance the one represented above in Equation (1.19). The update equations for the cluster centers $\mathbf{c}_i$ are not modified and are identical to those indicated in Equation (1.14). The update equations for the membership degrees are identical to those indicated in Equation (1.13) and Equation (1.17) for the FCM and PCM variants respectively, replacing the Euclidean distance by the cluster specific distance given above in Equation (1.19). The update equations for the covariance matrices are

$$\Sigma_i = \frac{\Sigma_i^*}{\sqrt[p]{\det(\Sigma_i^*)}}, \quad \text{where} \quad \Sigma_i^* = \frac{\sum_{j=1}^n u_{ij} (\mathbf{x}_j - \mathbf{c}_i)(\mathbf{x}_j - \mathbf{c}_i)^T}{\sum_{j=1}^n u_{ij}}. \tag{1.20}$$

They are defined as the covariance of the data assigned to cluster i, modified to incorporate the fuzzy assignment information.

The Gustafson–Kessel algorithm tries to extract much more information from the data than the algorithms based on the Euclidean distance. It is more sensitive to initialization, therefore it is recommended to initialize it using a few iterations of FCM or PCM depending on the considered partition type. Compared with FCM or PCM, the Gustafson–Kessel algorithm exhibits higher computational demands due to the matrix inversions. A restriction to axis-parallel cluster shapes reduces computational costs.

1.3.2 Fuzzy Shell Clustering

The clustering approaches mentioned up to now search for convex "cloud-like" clusters. The corresponding algorithms are called *solid* clustering algorithms. They are "specially useful" in data analysis applications. Another area of application of fuzzy clustering algorithms is image recognition and analysis. Variants of FCM and PCM have been proposed to detect lines, circles or ellipses on the data-set, corresponding to more complex data substructures; the so-called *shell* clustering algorithms (Klawonn, Kruse, and Timm, 1997) extract prototypes that have a different nature than the data points. They need to modify the definition of the distance between a data point and the prototype and replace the Euclidean by other distances. For instance the fuzzy c-varieties (FCV) algorithm was developed for the recognition of lines, planes, or hyperplanes; each cluster is an affine subspace characterized by a point and a set of orthogonal unit vectors, $C_i = (\mathbf{c}_i, \mathbf{e}_{i1}, \ldots, \mathbf{e}_{iq})$ where q is the dimension of the affine subspace. The distance between a data point $\mathbf{x}_j$ and cluster i is then defined as

$$d^2(\mathbf{x}_j, C_i) = ||\mathbf{x}_j - \mathbf{c}_i||^2 - \sum_{l=1}^{q} (\mathbf{x}_j - \mathbf{c}_i)^T \mathbf{e}_{il}.$$

The fuzzy c-varieties (FCV) algorithm is able to recognize lines, planes or hyperplanes (see Figure 1.5). These algorithms can also be used for the construction of locally linear models of data with underlying functional interrelations.

Other similar FCM and PCM variants include the adaptive fuzzy c-elliptotypes algorithm (AFCE) that assigns disjoint line segments to different clusters (see Figure 1.6). Circle contours can be detected by the fuzzy c-shells and the fuzzy c-spherical shells algorithm. Since objects with circle-shaped boundaries in are projected into the picture plane the recognition of ellipses can be necessary. The fuzzy c-ellipsoidal shells algorithm is able to solve this problem. The fuzzy c-quadric shells algorithm (FCQS) is furthermore able to recognize hyperbolas, parabolas, or linear clusters. Its flexibility can be observed in Figures 1.7 and 1.8. The shell clustering techniques have also been extended to non-smooth structures such as rectangles and other polygons. Figures 1.9 and 1.10 illustrate results obtained with the fuzzy c-rectangular (FCRS) and fuzzy c-2-rectangular shells (FC2RS) algorithm. The interested reader may be referred to Höppner, Klawonn, Kruse, and Runkler (1999) and Bezdek, Keller, Krishnapuram, and Pal (1999) for a complete discussion of this branch of methods.

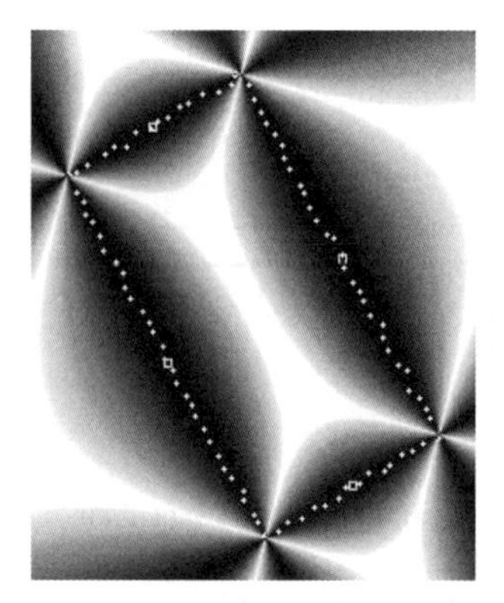

Figure 1.5 FCV analysis.

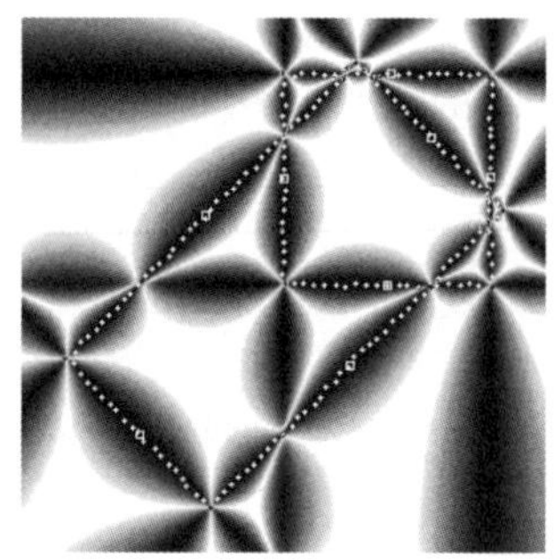

Figure 1.6 AFCE analysis.

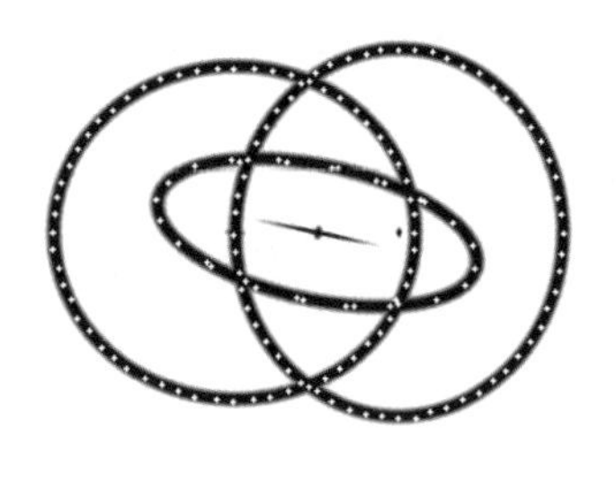

Figure 1.7 FCQS analysis.

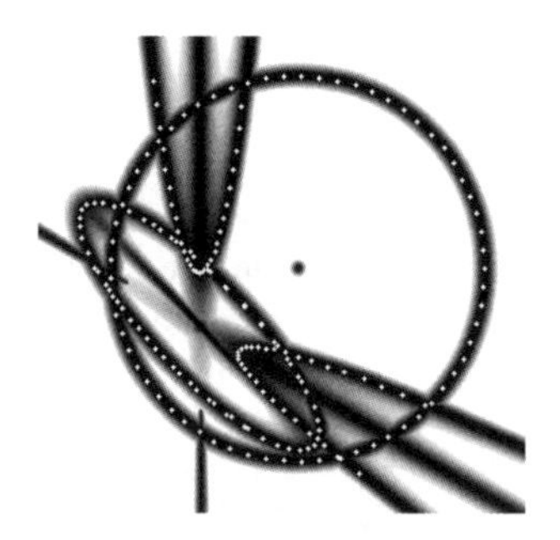

Figure 1.8 FCQS analysis.

Figure 1.9 FCRS analysis.

Figure 1.10 FC2RS analysis.

1.3.3 Kernel-based Fuzzy Clustering

The kernel variants of fuzzy clustering algorithms further modify the distance function to handle non-vectorial data, such as sequences, trees, or graphs, without needing to modify completely the algorithms themselves. Generally speaking, kernel learning methods (see e.g., Schölkopf and Smola (2002); Vapnik (1995)) constitute a set of machine learning algorithms that make it possible to extend, in a formal framework, classic linear algorithms. This extension addresses a double aim: on the one hand, it makes it possible to address tasks that require a richer framework than the linear one, while still preserving this generally simple formalism. On the other hand, it makes it possible to apply algorithms to data that are not described in a vectorial form, but as more complex objects, such as sequences, trees or graphs. More generally, kernel methods can be applied independently of the data nature, without needing to adapt the algorithm. In this section, data points can be vectorial or not, therefore we denote them x_j instead of $\mathbf{x}_j$.

1.3.3.1 Principle

Kernel methods are based on an *implicit* data representation transformation $\phi : \mathcal{X} \to \mathcal{F}$ where $\mathcal{X}$ denotes the input space and $\mathcal{F}$ is called the *feature space*. $\mathcal{F}$ is usually of high or even infinite dimension and is only constrained to be a Hilbert space, i.e., to dispose of a scalar product. The second principle of kernel methods is that data are not handled directly in the feature space, which could lead to expensive costs given its dimension; they are only handled through their scalar products that are computed using the initial representation. To that aim, the so-called *kernel function* is used: it is a function $k : \mathcal{X} \times \mathcal{X} \to \mathbb{R}$, such that $\forall x, y \in \mathcal{X}, \langle \phi(x), \phi(y) \rangle = k(x, y)$ Thus the function ϕ is not needed to be known explicitly, scalar products in the feature space only depend on the initial representation.

In order to apply this *kernel trick*, kernel methods are algorithms written only in terms of scalar products between the data. The data representation enrichment then comes from using a scalar product based on an implicit transformation of the data, instead of being only the Euclidean one. The possibility to apply the algorithm to non-vectorial data only depends on the availability of a function $k : \mathcal{X} \times \mathcal{X} \to \mathbb{R}$ having the properties of a scalar product (Schölkopf and Smola, 2002).

1.3.3.2 Kernel Fuzzy Clustering

The kernel framework has been applied to fuzzy clustering and makes it possible to consider other distances than the Euclidean one. It is to be underlined that fuzzy shell clustering, discussed in Section 1.3.2, also takes into account other metrics, but it has an intrinsic difference: it aims at extracting prototypes that have a different nature than the data points, and thus it modifies the distance between points and cluster prototypes. In the kernel approach, the similarity is computed between pairs of data points and does not involve cluster centers; the kernel function influences more directly that points are to be grouped in the same clusters, and does not express a comparison with a cluster representative. Usually, cluster representatives have no explicit representation as they belong to the feature space. Thus the kernel approach can be applied independently of the data nature whereas fuzzy shell algorithms must be specified for each desired prototype nature. On the other hand, kernel methods do not have an explicit representative of the cluster and cannot be seen as prototype-based clustering methods.

The kernel variant of fuzzy clustering (Wu, Xie, and Yu, 2003) consists of transposing the objective function to the feature space, i.e., applying it to the transformed data $\phi(x)$. The cluster centers then belong to the feature space, we therefore denote them $c_i^\phi, i = 1, \ldots, c$ ($c_i^\phi \in \mathcal{F}$). They are looked for in the form of linear combinations of the transformed data, as

$$c_i^\phi = \sum_{r=1}^{n} a_{ir}\phi(x_r). \tag{1.21}$$

This formulation is coherent with the solution obtained with standard FCM. Optimization must then provide the a_{ir} values, together with the membership degrees. Due to the previous form of the centers, the Euclidean distance between points and centers in the feature space can be computed as

$$d\phi_{ir}^2 = \|\phi(x_r) - c_i^\phi\|^2 = k_{rr} - 2\sum_{s=1}^{n} a_{is}k_{rs} + \sum_{s,t=1}^{n} a_{is}a_{it}k_{st}, \tag{1.22}$$

where we denote $k_{rs} = k(x_r, x_s) = \langle\phi(x_r), \phi(x_s)\rangle$. Thus, the objective function becomes

$$J^\phi = \sum_{i=1}^{c}\sum_{r=1}^{n} u_{ir}^m \left(k_{rr} - 2\sum_{s=1}^{n} a_{is}k_{rs} + \sum_{s,t=1}^{n} a_{is}a_{it}k_{st} \right). \tag{1.23}$$

The minimization conditions then lead to the following update equations

$$u_{ir} = \frac{1}{\sum_{l=1}^{c}\left(\frac{d\phi_{ir}^2}{d\phi_{lr}^2}\right)^{\frac{1}{m-1}}}, \qquad a_{ir} = \frac{u_{ir}^m}{\sum_{s=1}^{n} u_{is}^m}, \qquad \text{i.e.,} \quad c_i^\phi = \frac{\sum_{r=1}^{n} u_{ir}^m \phi(x_r)}{\sum_{s=1}^{n} u_{is}^m}. \tag{1.24}$$

Thus the update equations, as well as the objective function, can be expressed solely in terms of the kernel function, i.e., in terms of scalar products. Equation (1.24) shows that membership degrees have the same form as in the standard FCM (see Equation (1.13)), replacing the Euclidean distance by the distance in the feature space, as defined in Equation (1.22). The expression of the cluster centers is comparable to the standard case (see Equation (1.14)), as the weighted mean of the data. The difference is that cluster centers belong to the feature space and have no explicit representation, only the weighting coefficients are known.

There exist other variants for the kernelization of the fuzzy C-means, as for instance the one proposed by Zhang and Chen (2003a,b). The latter is specific insofar as it only considers the Gaussian kernel $k(x, y) = \exp(-d(x, y)^2/\sigma^2)$ and exploits its properties to simplify the algorithm. More precisely it makes the hypothesis that cluster centers can be looked for explicitly in the input space ($c_i \in \mathcal{X}$), and considers its transformation to the feature space $\phi(c_i)$. This differs from the general case, as presented above, where cluster centers are only defined in the feature space. The objective function then becomes

$$J = \sum_{i=1}^{c}\sum_{r=1}^{n} u_{ir}^m \|\phi(c_i) - \phi(x_j)\|^2 = 2\sum_{i=1}^{c}\sum_{r=1}^{n} u_i^m (1 - \mathrm{e}^{-d(c_i, x_j)^2/\sigma^2}), \tag{1.25}$$

exploiting the fact that the Gaussian kernel leads to $d\phi^2(x, y) = k(x, x) + k(y, y) - 2k(x, y) = 2(1 - k(x, y))$. Thus this method constitutes a special case of the FCM kernelization and cannot be applied to any type of data independently of their nature. It is to be noted that this objective function (Equation (1.25)) is identical to the one proposed by Wu and Yang (2002) in the framework of robust variants of FCM, as described in the next section.

It should be noticed that the application of a kernel method needs to select the kernel and its parameters, which may be difficult. This task can be seen as similar to the problem of feature selection and data representation choice in the case of non-kernel methods.

1.4 OBJECTIVE FUNCTION VARIANTS

The previous variants of fuzzy C-means are obtained when considering different distance functions that lead to a rewrite of the objective functions and in some cases modify the update equations. In this section, we consider other variants that are based on deeper modifications of the objective functions. The modifications aim at improving the clustering results in specific cases, for instance when dealing with noisy data. It is to be noticed that there exists a very high number of variants for fuzzy clustering algorithms, we only mention some of them.

We organized them in the following categories: some variants are explicitly aimed at handling noisy data. Others study at a theoretical level the role of the fuzzifier m in the objective function (see notations in Equation (1.10)) and propose some modifications. Other variants introduce new terms in the objective function so as to optimize the cluster number instead of having it fixed at the beginning of the process. Lastly, we mention some variants that are aimed at improving the possibilistic C-means, in particular with respect to the coinciding cluster problem (see Section 1.2.4).

It is to be noted that the limits between these categories are not clear-cut and that for instance the modification of the fuzzifier can influence the noise handling properties. We categorize the methods according to their major characteristics and underline their other properties.

When giving update equations for cluster prototypes, we consider only the case where the Euclidean distance is used and when prototypes are reduced to cluster centers. Most methods can be generalized to other representations, in particular those including size and form parameters. The interested reader is referred to the original papers.

1.4.1 Noise Handling Variants

The first variants of fuzzy C-means we consider aim at handling noisy data. It is to be noticed that PCM is a solution to this problem, but it has difficulty of its own as mentioned in Section 1.2.4 (cluster coincidence problem, sensitivity to initialization). Therefore other approaches take FCM as the starting point and modify it so as to enable it to handle noisy data. When giving the considered objective functions, we do not recall the constraints indicated in Equations (1.8) and (1.9) that apply in all cases.

The aim of these variants is then to define robust fuzzy clustering algorithms, i.e., algorithms whose results do not depend on the presence or absence of noisy data points or outliers[1] in the data-set. Three approaches are mentioned here: the first one is based on the introduction of a specific cluster, the so-called noise cluster that is used to represent noisy data points. The second method is based on the use of robust estimators, and the third one reduces the influence of noisy data points by defining weights denoting the point representativeness.

1.4.1.1 Noise Clustering

The noise clustering (NC) algorithm was initially proposed by Davé (1991) and was later extended (Davé and Sen, 1997, 1998). It consists in adding, beside the c clusters to be found in a data-set, the so-called noise cluster; the latter aims at grouping points that are badly represented by normal clusters, such as noisy data points or outliers. It is not explicitly associated to a prototype, but directly to the distance between an implicit prototype and the data points: the center of the noise cluster is considered to be at a constant distance, δ, from all data points. This means that all points have a priori the same 'probability' of belonging to the noise cluster. During the optimization process, this 'probability' is then adapted as a function of the probability according to which points belong to normal clusters. The noise cluster is then introduced in the objective function, as any other cluster, leading to

$$J = \sum_{i=1}^{c} \sum_{j=1}^{n} u_{ij}^m d_{ij}^2 + \sum_{k=1}^{n} \delta^2 \left(1 - \sum_{i=1}^{c} u_{ik} \right)^m . \tag{1.26}$$

The added term is similar to the terms in the first sum: the distance to the cluster prototype is replaced by δ and the membership degree to this cluster is defined as the complement to 1 of the sum of all membership degrees to the standard clusters. This in particular implies that outliers can have low membership degrees to the standard clusters and high degree to the noise cluster, which makes it possible to reduce their influence

[1]Outliers correspond to atypical data points, that are very different from all other data, for instance located at a high distance from the major part of the data. More formally, according to Hawkins (1980), an outlier is 'an observation that deviates so much from other observations as to arouse suspicion that it was generated by a different mechanism'.

on the standard cluster: as PCM, the noise clustering approach relaxes the FCM normalization constraint expressed in Equation (1.9) according to which membership degrees to good clusters must sum to 1.

Further comparison between NC and PCM (see Equations (1.26) and (1.16) shows that the algorithms are identical in the case of a single cluster, with δ^2 corresponding to η (Davé and Sen, 1997, 1998). In the case $c > 1$, the difference is that PCM considers one η_i per cluster, whereas a single parameter is defined in the NC case. This means that PCM has the advantage of having one noise class per good cluster, whereas NC has only one (the NC generalization described hereafter overcomes this drawback). As a consequence, the membership degrees to the noise cluster differ for the two methods: in the PCM case, they are, for each noise cluster, the complement to 1 to the membership to the associated good cluster. In noise clustering, as there is a single noise cluster, the membership degree to it is the complement to the sum of all other memberships.

Another difference between PCM and NC comes from the fact that the PCM cost function can be decomposed into c independent terms (one per cluster), whereas in the noise clustering approach such a decomposition is not possible. This decomposition is one of the reasons why PCM leads to coinciding clusters. Thus Davé and Krishnapuram (1997) interpret NC as a robustified FCM, whereas PCM behaves like c independent NC algorithms.

The objective function (1.26) requires the setting of parameter δ. In the initial NC algorithm, it was set to

$$\delta^2 = \lambda \frac{1}{c \cdot n} \left(\sum_{i=1}^{c} \sum_{j=1}^{n} d_{ij}^2 \right), \tag{1.27}$$

i.e., its squared value is a proportion of the mean of the squared distances between points and other cluster prototypes, with λ a user-defined parameter determining the proportion: the smaller the λ, the higher the proportion of points that are considered as outliers.

Noise clustering has been generalized to allow the definition of several δ, and to define a noise scale per cluster. To that aim, each point is associated to a noise distance $\delta_j, j = 1, \ldots, n$, the latter being defined as the size of the cluster the point maximally belongs to, as in PCM: $\delta_j = \eta_{i*}$ for $i* = \arg\max_l u_{lj}$ (Davé and Sen, 1997, 1998). In this case, the difference between PCM and NC about distance scale vanishes, the only remaining difference is the independence of clusters in the PCM objective function that does not appear in the noise clustering case.

1.4.1.2 Robust Estimators

Another approach to handle noisy data-sets is based on the exploitation of robust estimators: as indicated in Section 1.2.2, the fuzzy C-means approach is based on a least square objective function. It is well known that the least square approach is highly sensitive to aberrant points, which is why FCM gives unsatisfactory results when applied to data-sets contaminated with noise and outliers. Therefore, it has been proposed to introduce a robust estimator in the FCM classic objective function (see Equation (1.10)), leading to consider

$$J = \sum_{i=1}^{c} \sum_{j=1}^{n} u_{ij}^m \rho_i(d_{ij}), \tag{1.28}$$

where ρ_i are robust symmetric positive definite functions having their minimum in 0 (Frigui and Krishnapuram, 1996). According to the robust M-estimator framework, ρ should be chosen such that $\rho(z) = \log(J(z)^{-1})$ represents the contribution of error z to the objective function and J the distribution of these errors. Choosing $\rho(z) = z^2$ as it is usually the case is equivalent to assuming a normal distribution of the errors z and leads to constant weighting functions. That is, big errors have the same weight as small errors and play too important a role on the correction applied to the parameters, making the objective function sensitive to outliers. Therefore it is proposed to use another ρ, whose weighting functions tend to 0 for big values of z. Frigui and Krishnapuram (1996) design their own robust estimator to adapt to the desired behaviour, defining the robust c-prototypes (RCP) algorithm.

In the case where clusters are represented only by centers and a probabilistic partition is looked for (i.e., with constraint (1.9)), the update equations for the membership degrees and cluster prototypes derived from Equation (1.28) then become (Frigui and Krishnapuram, 1996)

$$\mathbf{c}_i = \frac{\sum_{j=1}^{n} u_{ij}^m f_{ij} \mathbf{x}_j}{\sum_{j=1}^{n} u_{ij}^m f_{ij}}, \qquad u_{ij} = \frac{1}{\sum_{k=1}^{c} \left[\frac{\rho(d_{ij}^2)}{\rho(d_{kj}^2)} \right]^{\frac{1}{m-1}}}, \tag{1.29}$$

where $f_{ij} = f(d_{ij})$ and $f = \frac{\mathrm{d}\rho(z)}{\mathrm{d}z}$. It is to be noted that outliers still have membership degrees $u_{ij} = 1/c$ for all clusters. The difference and advantage as compared with FCM comes from their influence on the center, which is reduced through the f_{ij} coefficient (see Frigui and Krishnapuram (1996) for the f_{ij} expression).

Other robust clustering algorithms include the method proposed by Wu and Yang (2002) that consider the modified objective function

$$J = \sum_{i=1}^{c} \sum_{j=1}^{n} u_{ij}^m (1 - \mathrm{e}^{-\beta d_{ij}^2}), \tag{1.30}$$

where β is a user-defined parameter that the authors propose to set to the inverse of the sample covariance matrix. This function is first proposed as a replacement of the Euclidean distance by the more robust exponential metric; yet, as pointed out by Zhang and Chen (2004), the mapping $(x, y) \mapsto \exp(-\beta d(x, y))$ is not a metric. Still, the analysis of the above objective function in the robust estimator framework holds and shows that this function leads to a robust fuzzy clustering algorithm that can handle noisy data-sets Wu and Yang (2002).

Davé and Krishnapuram (1996, 1997) show that PCM can be interpreted in this robust clustering framework based on the M-estimator. They consider a slightly different formalization, where the objective function for each cluster is written

$$J = \sum_{j=1}^{n} \rho(\mathbf{x}_j - \mathbf{c}), \quad \text{leading to} \quad \mathbf{c} = \frac{\sum_{j=1}^{n} w(d_{ij}) \mathbf{x}_j}{\sum_{j=1}^{n} w(d_{ij})}, \quad \text{where} \quad w(z) = \frac{1}{z} \frac{\mathrm{d}\rho}{\mathrm{d}z}. \tag{1.31}$$

Comparing with the update equations of PCM, this makes it possible to identify a weight function w and by integration to deduce the associated estimator ρ. Davé and Krishnapuram (1996, 1997) show the obtained ρ is indeed a robust function. This justifies at a formal level the qualities of PCM as regards noise handling.

1.4.1.3 Weight Modeling

A third approach to handle outliers is exemplified by Keller (2000). It consists of associating each data point a weight to control the influence it can have on the cluster parameters. The considered objective function is

$$J = \sum_{i=1}^{c} \sum_{j=1}^{n} u_{ij}^m \frac{1}{\omega_j^q} d_{ij}^2, \quad \text{under constraint} \quad \sum_{j=1}^{n} \omega_j = \omega, \tag{1.32}$$

where the factor ω_j represents the weight for data point j, q a parameter to control the influence of the weighting factor and ω a normalizing coefficient. The minimization conditions of this objective function lead to the following update equations:

$$u_{ij} = \frac{1}{\sum_{l=1}^{c} \left(\frac{d_{ij}^2}{d_{lj}^2} \right)^{\frac{1}{m-1}}}, \qquad \mathbf{c}_i = \frac{\sum_{j=1}^{n} \frac{u_{ij}^m}{\omega_j^q} \mathbf{x}_j}{\sum_{j=1}^{n} \frac{u_{ij}^m}{\omega_j^q}}, \qquad \omega_j = \frac{\left(\sum_{i=1}^{c} u_{ij}^m d_{ij}^2 \right)^{\frac{1}{q+1}}}{\sum_{l=1}^{n} \left(\sum_{i=1}^{c} u_{il}^m d_{il}^2 \right)^{\frac{1}{q+1}}} \omega.$$

Thus, the membership degrees are left unchanged, whereas the cluster centers take into account the weights; points with high representativeness play a more important role than outliers. Representativeness depends on the weighted average distance to cluster centers.

1.4.2 Fuzzifier Variants

Another class of FCM variants is based on the study of the fuzzifier, i.e., the exponent m in Equation (1.10): as indicated in Section 1.2.2, FCM can be derived from the hard C-means algorithm by relaxing the partition constraints, so that membership degrees belong to [0,1] and not $\{0,1\}$. To prevent membership degrees from being restricted to the two values 0 and 1, the objective function must be modified and the m fuzzifier is introduced.

Now as can be observed and proved (Klawonn and Höppner, 2003b; Rousseeuw, Trauwaert, and Kautman, 1995), actually membership degrees do not exactly cover the range [0,1]: they never equal 0 or 1 (except in the special case where a data point coincides with a cluster center), i.e., in fact they belong to]0,1[. In other words, membership functions have a core reduced to a single point (the cluster center) and unbounded support. This is a drawback in the case of noisy data-sets, as in the case of clusters with different densities (Klawonn and Höppner, 2003b; Rousseeuw, Trauwaert and Kautman, 1995): high density clusters tend to influence or completely attract other prototypes (note that this problem can be handled by using other distances than the Euclidean one).

To overcome this problem, Rousseeuw, Trauwaert and Kaufman, (1995) proposed replacing the objective function by

$$J = \sum_{i=1}^{c} \sum_{j=1}^{n} [\alpha u_{ij} + (1 - \alpha) u_{ij}^2] d_{ij}^2, \tag{1.33}$$

where α is a user-defined weight determining the influence of each component. When $\alpha = 1$, the objective function reduces to the hard C-means function (see Equation (1.1)), leading to a maximal contrast partition (membership degrees take only values 0 or 1). On the contrary, $\alpha = 0$ leads to the fuzzy C-means with $m = 2$ and a low contrast partition (outliers for instance have the same membership degree as all clusters). α makes it possible to obtain a compromise situation, where membership degrees in]0,1[are reserved for points whose assignment is indeed unclear, whereas the others, and in particular outliers, have degrees 0 or 1.

Klawonn and Höppner, (2003a,b) also take as their starting point the observation that membership degrees actually never take the values 0 and 1. They perform the analysis in a more formal framework that allows more general solutions: they proposed considering as objective function

$$J = \sum_{i=1}^{c} \sum_{j=1}^{n} g(u_{ij}) d_{ij}^2. \tag{1.34}$$

Note that robust approaches proposed applying a transformation to the distances, whereas here a transformation is applied to the membership degrees. Taking into account the constraints on u_{ij} normalization (see Equation (1.9)), and setting the derivative to 0, the partial derivative of the associated Lagrangian leads to

$$g'(u_{ij}) d_{ij}^2 - \lambda_j = 0, \tag{1.35}$$

where λ_j is the Lagrange multiplier associated with the normalization constraint concerning x_j. As it is independent of i, this equation implies $g'(u_{ij}) d_{ij}^2 = g'(u_{kj}) d_{kj}^2$ for all i, k. This explains why zero membership degrees can never occur: the standard function $g(u) = u^m$ yields $g'(0) = 0$. Thus, in order to balance the two terms, no matter how large d_{ij}^2 and how small d_{kj}^2 are, u_{ij} cannot be 0.

Therefore, they proposed replaceing the standard g function with other ones. The conditions g must satisfy are $g(0) = 0$ and $g(1) = 1$, increasing and differentiable. Further, the derivative g' must be increasing and must satisfy $g'(0) \neq 0$. Klawonn and Höppner, (2003b) consider the same function as Rousseeuw, Trauwaert, and Kautman (1995), i.e., $g(u) = \alpha u^2 + (1 - \alpha)u$. Gaussian functions $g(u) = (e^{\alpha u} - 1)/(e^{\alpha} - 1)$ were also suggested, since the parameter α has a similar effect to the fuzzifier m in the standard fuzzy clustering: the smaller the α, the crisper the partition tends to be (Klawonn and Höppner, 2003a). Klawonn (2004) proposesd dropping the differentiability condition and considering a piecewise linear transformation to obtain more flexibility than with a single parameter α. For instance, non-increasing functions that are flatter around 0.5 make it possible to avoid ambiguous membership degrees forcing them to tend to 0 or 1.

1.4.3 Cluster Number Determination Variants

Partitioning clustering algorithms consist of searching for the optimal fuzzy partition of the data-set into c clusters, where c is given as input to the algorithm. In most real data mining cases, this parameter is not known in advance and must be determined. Due to the cluster merging phenomenon, the definition of an appropriate c value for PCM is not so important as for FCM. Yet, as mentioned earlier, at a theoretical level, PCM relies on an ill-posed optimization problem and other approaches should be considered. They usually consist of testing several c values and comparing the quality of the obtained partition using so-called validity criteria (see for instance Halkidi, Batistakis, and Vazirgiannis (2002); this solution is computationally expensive. Other approaches, presented in this section, consist of considering the c value as a parameter to be optimized.

Now with this respect the FCM objective function is minimal when $c = n$, i.e., each cluster contains a single point as in this case $d_{ij} = 0$. Thus a regularization term is added, that is minimal when all points belong to the same cluster, so as to penalize high c values. Then the combination of terms in the objective function makes it possible to find the optimal partition in the smallest possible number of clusters.

Following this principle, Frigui and Krishnapuram (1997) proposed the competitive agglomeration (CA) algorithm based on the objective function

$$J = \sum_{i=1}^{c} \sum_{j=1}^{n} u_{ij}^m d_{ij}^2 - \alpha \sum_{i=1}^{c} \left(\sum_{j=1}^{n} u_{ij} \right)^2. \tag{1.36}$$

The additional term is the sum of squares of cardinalities of the clusters, which is indeed minimal when all points are assigned to a single cluster and all others are empty. The optimization process for this function does not exactly follow the AO scheme and involves competition between clusters, based on their sizes and distances to the points. Small clusters are progressively eliminated. A robust extension to CA has been proposed in Frigui and Krishnapuram (1999): the first term in Equation (1.36) is then replaced by the term provided in Equation (1.28) to exploit the robust estimator properties.

Sahbi amd Boujemaa (2005) proposed using as regularizer an entropy term, leading to

$$J = \sum_{i=1}^{c} \sum_{j=1}^{n} u_{ij}^m d_{ij}^2 - \alpha \frac{1}{n} \sum_{j=1}^{n} - \sum_{i=1}^{c} u_{ij} \log_2(u_{ij}).$$

To verify the constraints on the memberships $u_{ij} \in [0, 1]$, they proposed considering Gaussian membership functions in the form $u_{ij} = \exp(-\mu_{ij})$ and estimating the μ_{ij} parameters. α then intervenes in the parameter of the exponential and is to be interpreted as a scaling factor: when it is underestimated, each point is a cluster; when it is overestimated, the membership functions are approximately constant, and one gets a single big cluster. The number of clusters is then indirectly determined.

1.4.4 Possibilistic c-means Variants

As indicated in Section 1.24, the possibilistic C-means may lead to unsatisfactory results, insofar as the obtained clusters may be coincident. This is due to the optimized objective function, whose global minimum is obtained when all clusters are identical (see Section 1.2.4). Hence the possibilistic C-means can be improved by modifying its objective function. We mention here two PCM variants, based on the adjunction of a penalization term in the objective function and the combination of PCM with FCM.

1.4.4.1 Cluster Repulsion

In order to hinder cluster merging, Timm and Kruse (2002) and Timm, Borgelt, Döreing, and Kruse (2004) proposed including in the objective function a term expressing repulsion between clusters, so as to force them to be distinct: the considered objective function is written

$$J = \sum_{i=1}^{c}\sum_{j=1}^{n} u_{ij}^m d_{ij}^2 + \sum_{i=1}^{c} \eta_i \sum_{i=1}^{n} (1 - u_{ij})^m + \sum_{i=1}^{c} \gamma_i \sum_{k=1, k \neq i}^{c} \frac{1}{\xi d(\mathbf{c}_i, \mathbf{c}_j)^2}. \tag{1.37}$$

The first two terms constitute the PCM objective function (see Equation (1.16)), the last one expresses the repulsion between clusters: it is all the bigger as the distance between clusters is small. γ_i is a parameter that controls the strength of the cluster repulsion: it balances the two clustering objectives, namely the fact that clusters should be both compact and distinct. This coefficient depends on clusters so that repulsion can get stronger when the number of points associated with cluster i increases (Timm, Borgelt, Döring, and Kruse, 2004). Parameter ξ makes repulsion independent of the normalization of data attributes. The minimization conditions lead to the update equation

$$\mathbf{c}_i = \frac{\sum_{j=1}^{n} u_{ij}^m \mathbf{x}_j - \gamma_i \sum_{k=1, k \neq i}^{c} \frac{1}{d(\mathbf{c}_i, \mathbf{c}_k)^4} \mathbf{c}_k}{\sum_{j=1}^{n} u_{ij}^m - \gamma_i \sum_{k=1, k \neq i}^{c} \frac{1}{d(\mathbf{c}_i, \mathbf{c}_k)^4}} \tag{1.38}$$

(the update equation for the membership degrees is not modified and is identical to Equation (1.17)). Equation (1.38) shows the effect of repulsion between clusters: a cluster is attracted by the data assigned to it and it is simultaneously repelled by the other clusters.

1.4.4.2 PCM Variants Based on Combination with FCM

Pal, Pal, and Bezdek (1997) and Pal, Pal, Keller, and Bezdek (2004) proposed another approach to overcome the problems encountered with the possibilistic C-means: they argued that both possibilistic degrees and membership degrees are necessary to perform clustering. Indeed, possibilistic degrees make it possible to reduce the influence of outliers whereas membership degrees are necessary to assign points. Likewise, Davé and Sen (1998) underlined that a good clustering result requires both the partitioning property of FCM and the modeseeking robust property of PCM.

In Pal, Pal, and Bezdek (1997) the combination of FCM and PCM is performed through the optimization of the following objective function:

$$J = \sum_{i=1}^{c}\sum_{j=1}^{n} (u_{ij}^m + t_{ij}^\eta) d_{ij}^2, \quad \text{under the constraints} \quad \begin{cases} \forall j & \sum_{i=1}^{c} u_{ij} = 1 \\ \forall i & \sum_{j=1}^{n} t_{ij} = 1 \end{cases}. \tag{1.39}$$

This means that u_{ij} is a membership degree, whereas t_{ij} corresponds to a possibilistic coefficient. Indeed, it is not submitted to the normalization constraint on the sum across the clusters. The normalization constraint it must hold aims at preventing the trivial result where $t_{ij} = 0$ for all i, j. As pointed out in several papers (Davé and Sen, 1998; Pal, Pal, Keller, and Bezdek, 2004) the problem is that the relative scales of probabilistic and possibilistic coefficients are then different and the membership degrees dominate the equations. Moreover, the possibilistic coefficients take very small values in the case of big data-sets.

Therefore Pal, Pal, Keller and Bezdek (2004) proposed another combination method, in the form

$$J = \sum_{i=1}^{c}\sum_{j=1}^{n}(au_{ij}^{m} + bt_{ij}^{\eta})d_{ij}^{2} + \sum_{i=1}^{c}\eta_i\sum_{j=1}^{n}(1 - t_{ij})^{\eta}, \tag{1.40}$$

which uses the same constraint for t_{ij} as in the standard PCM (second term in J), and combines possibilistic and membership degrees. a and b are user-defined parameters that rule the importance the two terms must play. In the case where the Euclidean distance is used, the update equations are then

$$u_{ij} = \frac{1}{\sum_{l=1}^{c}\left(\frac{d_{ij}^2}{d_{lj}^2}\right)^{\frac{1}{m-1}}}, \qquad t_{ij} = \frac{1}{1 + \left(\frac{b}{\eta_i}d_{ij}^2\right)^{\frac{1}{\eta-1}}}, \qquad \mathbf{c}_i = \frac{\sum_{j=1}^{n}(au_{ij}^{m} + bt_{ij}^{\eta})\mathbf{x}_j}{\sum_{j=1}^{n}(au_{ij}^{m} + bt_{ij}^{\eta})}.$$

Thus u_{ij} are similar to the membership degrees of FCM (see Equation (1.13)), and t_{ij} to the possibilistic coefficients of PCM when replacing η_i with η_i/b (see Equation (1.17)). Cluster centers then depend on both coefficients, parameters a, b, m, and η rule their relative influence. This shows that if b is higher than a the centers will be more influenced by the possibilistic coefficients than the membership degrees. Thus, to reduce the influence of outliers, a bigger value for b than a should be used. Still, it is to be noticed that these four parameters are to be defined by the user and that their influence is correlated, making it somewhat difficult to determine their optimal value. Furthermore the problem of this method is that it loses the interpretation of the obtained coefficients; in particular, due to their interaction, t_{ij} cannot be interpreted as typicality anymore.

1.5 UPDATE EQUATION VARIANTS: ALTERNATING CLUSTER ESTIMATION

In this section, we study the fuzzy clustering variants that generalize the alternating optimization scheme used by the methods presented up to now. The notion alternating cluster estimation (ACE) stands for a distinguished methodology to approach clustering tasks with the aim of having the flexibility to tailor new clustering algorithms that better satisfy application-specific needs. Instead of reformulating the clustering task as a minimization problem by defining objective functions, the data analyst chooses cluster prototypes that satisfy some desirable properties as well as cluster membership functions that have better suited shapes for particular applications. This is possible, since the ACE framework generalizes the iterative updating scheme for cluster models that stems from the alternating optimization approaches (Equation (1.11 and 1.12)). However, the purpose of minimizing objective functions with expressions for j_U and j_C is abandoned. Instead, the user chooses heuristic equations to re-estimate partitions and cluster parameters by which the resulting algorithm iteratively refines the cluster model. Thus the classification task is directly described by the chosen update equations, which do not necessarily reflect the minimization of some criterion anymore.

Alternating cluster estimation is justified by the observation that convergence is seldom a problem in practical examples (local minima or saddle points can be avoided). The ACE framework is particularly useful when cluster models become too complex to minimize them analytically or when the objective function lacks differentiability (Höppner, Klawonn, Kruse, and Runkler 1999). However, it is to be noted that the ACE framework also encompasses all those algorithms that follow from the minimization of objective functions as long as their respective update equations are chosen (which follow from the necessary conditions for a minimum).

When clustering is applied to the construction of fuzzy rule-based systems, the flexibility of ACE framework in choosing among different update equations is of particular interest. In such applications the fuzzy sets carry semantic meaning, e.g., they are assigned linguistic labels like "low", "approximately zero" or "high". Consequently the fuzzy sets, in fuzzy controllers for instance, are required to be convex, or even monotonous (Zadeh, 1965). Furthermore, they have to have limited support, i.e., membership degrees different from zero are allowed only within a small interval of their universe. ACE provides the flexibility to define fuzzy clustering algorithms that produce clusters Γ_i whose corresponding fuzzy sets μ_{Γ_i} fulfil these requirements. The clusters and membership degrees $\mu_{\Gamma_i}(\mathbf{x}_j) = u_{ij}$ obtained with the objective function-based clustering techniques contrarily do not carry the desired properties. The u_{ij} obtained by AO as in the previous section can be interpreted as discrete samples of continuous membership functions $\mu_i : \mathbb{R}^p \rightarrow [0, 1]$ for each cluster. The actual shape that is taken on by these membership functions results from the respective update equations for the membership degrees. For the probabilistic fuzzy AO algorithms the continuous membership function follows from Equation (1.13), with d_{ij} being the Euclidian distance $|| \cdot ||$:

$$\mu_i(\mathbf{x}) = \frac{||\mathbf{x} - \mathbf{c}_i||^{-\frac{2}{m-1}}}{\sum_{l=1}^{c} ||\mathbf{x} - \mathbf{c}_l||^{-\frac{2}{m-1}}}. \tag{1.41}$$

Figure 1.11 shows the membership functions that would result from the probabilistic FCM algorithm for two clusters. Obviously, the membership functions μ_i are not convex ($i = \{1, 2\}$). The membership for data points at first decreases the closer they are located to the other cluster center, but beyond the other center membership to the first cluster increases again due to normalization constraint. Possibilistic membership functions that result from a continuous extension according to Equation (1.17) are convex, but they are not restricted to local environments around their centers (i.e., the memberships will never reach zero for larger distances). Thus, if fuzzy sets with limited support as in fuzzy controllers are desired, possibilistic membership functions are inadequate as well. The transformation of the membership functions of the objective function-based techniques into the desired forms for the particular application is possible, but often leads to approximation errors and less accurate models.

Therefore ACE allows you to choose other membership functions aside from those that stem from an objective function-based AO scheme. Desired membership function properties can easily be incorporated in ACE. The user can choose from parameterized Gaussian, trapezoidal, Cauchy, and triangular functions (Höppner, Klawonn, Kruse, and Runkler, 1999). We present the triangular shaped fuzzy set as an example in Figure 1.12, since it has all the desired properties considered above:

$$\mu_i(\mathbf{x}) = \begin{cases} 1 - \left(\frac{||\mathbf{x}-\mathbf{c}_i||}{r_i}\right)^{\alpha} & \text{if} \quad ||\mathbf{x} - \mathbf{c}_i|| \leq r_i \\ 0 & \text{otherwise,} \end{cases} \tag{1.42}$$

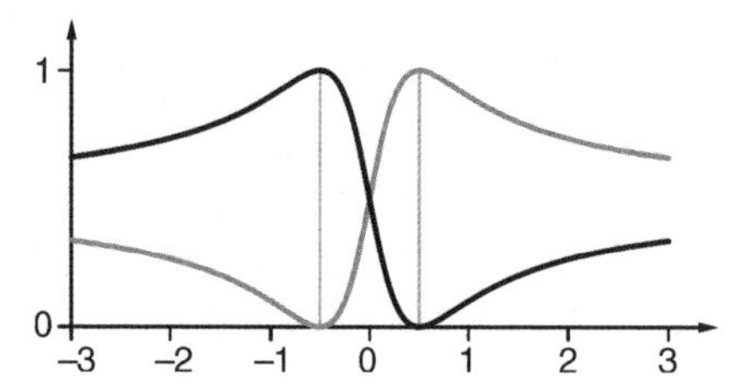

Figure 1.11 The membership functions obtained by probabilistic AO for two clusters at −0.5 and 0.5.

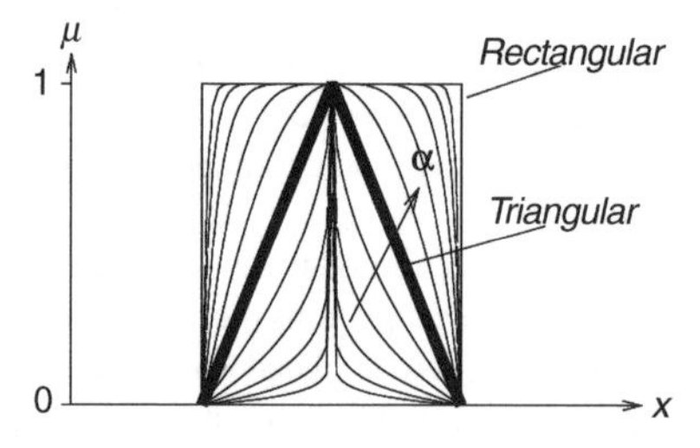

Figure 1.12 The parameterized triangular fuzzy set.

where r_i are the radii of the clusters, $\alpha \in \mathbb{R}_{>0}$. In an ACE algorithm using hypercone shaped clusters ($\alpha = 1$) the memberships of data to fixed clusters are estimated using the above equation, such that $u_{ij} = \mu_i(\mathbf{x}_j)$.

Deviating from alternating optimization of objective functions the user can also choose between alternative update equations for the cluster prototypes. In ACE, a large variety of parameterized equations stemming from defuzzification methods are offered for the re-estimation of cluster centers for fixed memberships. The reference to defuzzification techniques arises, since a "crisp" center is computed from fuzzily weighted data points. Also higher-order prototypes like lines, line segments, and elliptotypes have been proposed for the ACE scheme (Höppner, Klawonn, Kruse, and Runkler, 1999). In the simplest case, however, when clusters are represented by their centers only, new centers vectors could be calculated as the weighted mean of data points assigned to them (like in the FCM; see Equation (1.14)).

After the user has chosen the update equations for U and C, memberships and cluster parameters are alternatingly estimated (or updated, but not necessarily optimized w.r.t. some criterion function) as defined. This leads to a sequence $\{(\mathrm{U}_1, C_1), (\mathrm{U}_2, C_2), \ldots\}$ that is terminated after a predefined number of iterations $t_{\max}$ or when the C_t have stabilized. Some instances of the ACE might be sensitive to the initialization of the cluster centers. Thus determining C_0 with some iterations of the probabilistic FCM might be recommended. Notice that all conventional objective function-based algorithms can be represented as instances of the more general ACE framework by selecting their membership functions as well as their prototype update equations. An experimental comparison between 'real' ACE algorithms that do not reflect the minimization of an objective function and classical AO algorithms as presented above can be found in (Höppner, Klawonn, Kruse, and Runkler, 1999).

1.6 CONCLUDING REMARKS

In this chapter we attempted to give a systematic overview of the fundamentals of fuzzy clustering, starting from the basic algorithms and underlining the difference between the probabilistic and possibilistic paradigms. We then described variants of the basic algorithms, adapted to specific constraints or expectations. We further pointed out major research directions associated with fuzzy clustering. The field is so broad that it is not possible to mention all of them. In this conclusion we briefly point out further research directions that we could not address in the main part of the chapter due to length constraints.

1.6.1 Clustering Evaluation

An important topic related to clustering is that of cluster evaluation, i.e., the assessment of the obtained clusters quality: clustering is an unsupervised learning task, which means data points are not associated with labels or targets that indicate the desired output. Thus no reference is provided to which the obtained results can be compared. Major cluster validity approaches include the evaluation of the trade off between cluster compactness and cluster separability (Dunn 1974a; Rezaee, Lehieveldt and Reiber, 1998; Xie and Beni, 1991) and stability based approaches (see e.g., Ben-Hur, Elisseeff, and Guyon (2002)).

Some criteria are specifically dedicated to fuzzy clustering: the partition entropy criterion for instance computes the entropy of the obtained membership degrees,

$$PE = -\sum_{i,j} u_{ij} \log u_{ij},$$

and must be minimized (Bezdek, 1975). Indeed, it takes into account that the fuzzy membership degrees are degrees of freedom that simplify the optimization of the objective function, but that the desired clustering output is still a crisp partition. A data partition that is too fuzzy rather indicates a bad adequacy between the cluster number and the considered data-set and it should be penalized. Other fuzzy clustering dedicated criteria can be found in Bezder (1974) or Windham (1981).

Such criteria can be used to evaluate quantitatively the clustering quality and to compare algorithms one with another. They can also be applied to compare the results obtained with a single algorithm, when the parameter values are changed. In particular they can be used in order to select the optimal number of clusters: applying the algorithm for several c values, the value c^* leading to the optimal decomposition according to the considered criterion is selected.

1.6.2 Shape and Size Regularization

As presented in Section 1.3.1, some fuzzy clustering algorithms make it possible to identify clusters of ellipsoidal shapes and with various sizes. This flexibility implies that numerous cluster parameters are to be adjusted by the algorithms. The more parameters are involved the more sensitive the methods get to their initialization. Furthermore, the additional degrees of freedom lead to a lack of robustness.

Lately, a new approach has been proposed (Borgelt and Kruse, 2005) that relies on regularization to introduce shape and size constraints to handle the higher degrees of freedom effectively. With a time-dependent shape regularization parameter, this method makes it possible to perform a soft transition from the fuzzy C-means (spherical clusters) to the Gustafson–Kessel algorithm (general ellipsoidal clusters).

1.6.3 Co-clustering

Co-clustering, also called bi-clustering, two mode clustering, two way clustering or subspace clustering, has the specific aim of simultaneously identifying relevant subgroups in the data and relevant attributes for each subgroup: it aims at performing both clustering and local attribute selection. It is in particular applied in the bio-informatics domain, so as to detect groups of similar genes and simultaneously groups of experimental conditions that justify the gene grouping. Other applications include text mining, e.g., for the identification of both document clusters and their characteristic keywords (Kummamuru, Dhawale, and Krishnapuram, 2003). Many dedicated clustering algorithms have been proposed, including fuzzy clustering methods as for instance Frigui and Nasraoui (2000).

1.6.4 Relational Clustering

The methods described in this chapter apply to object data, i.e., consider the case where a description is provided for each data point individually. In other cases, this information is not available, the algorithm input takes the form of a pairwise dissimilarity matrix. The latter has size $n \times n$, each of its elements indicates the dissimilarity between point couples. Relational clustering aims at identifying clusters exploiting this input. There exists a large variety of fuzzy clustering techniques for such settings (Bezdek, Keller, Krishnapuram, and Pal, 1999; Hathaway and Bezdek, 1994) that are also based on objective function optimization or the ACE scheme (runkler and Bezdek, 2003). The interested reader is also referred to the respective chapter in Bezdek, Keller, Krishnapuram, and Pal (1999).

1.6.5 Semisupervised Clustering

Clustering is an unsupervised learning task. Yet it may be the case that the user has some a priori knowledge about couples of points that should belong to the same cluster. Semisupervised clustering is concerned with this learning framework, where some partial information is available : the clustering results must then verify additional constraints, implied by these pieces of information. Specific clustering algorithms have been proposed to handle these cases; the interested reader is referred to chapter 7 in this book.

ACKNOWLEDGEMENTS

Marie-Jeanne Lesot was supported by a Lavoisier grant from the French Ministère des Affaires Etrangères.

REFERENCES

Ball, G. and Hall, D. (1966) 'Isodata an iterative method of multivariate data analysis and pattern classification'. *IEEE Int. Comm. Conf. (Philadelphia, PA)*, vol. 2715 (2003) of *Lecture Notes in Artifical Intelligence*. IEEE Press, Piscataway, NJ, USA,

Barni, M. Cappellini, V. and Mecocci, A. (1996) 'Comments on a possibilistic approach to clustering'. *IEEE Transactions on Fuzzy Systems* **4**, 393–396.

Ben-Hur, A., Elisseeff, A. and Guyon, I. (2002) 'A stability based method for discovering structure in clustered data' In *Pacific Symposium on Biocomputing* (ed. Scientific W), vol. 7, pp. 6–17.

Bezdek, J. (1973) *Fuzzy Mathematics in Pattern Classification* PhD thesis Applied Math. Center, Cornell University, Ithaca, USA.

Bezdek, J. (1974) 'Cluster validity with fuzzy sets'. *Journal of Cybernetics* **3**(3), 58–73.

Bezdek, J. (1975) 'Mathematical models for systematics and taxonomy' *Proc. of the 8th Int. Conf. on Numerical Taxonomy*, pp. 143–166. Freeman.

Bezdek, J. (1981) *Pattern Recognition With Fuzzy Objective Function Algorithms*. Plenum Press, New York.

Bezdek, J. C., Keller, J., Krishnapuram, R. and Pal, NR. (1999) *Fuzzy Models and Algorithms for Pattern Recognition and Image Processing* Kluwer Boston, London chapter 3. Cluster Analysis for Relational Data, pp. 137–182.

Blake, CL. and Merz, C. J. (1998) UCI repository of machine learning databases.

Bock, H. H. 1974 *Automatische Klassifikation*. Vadenhoeck & Ruprecht, Göttingen, Zürich.

Borgelt, C. and Kruse, R. (2005) 'Fuzzy and probabilistic clustering with shape and size constraints' *Proc. 11th Int. Fuzzy Systems Association World Congress (IFSA'05, Beijing, China)*, pp. 945–950. Tsinghua University Press and Springer-Verlag, Beijing, China, and Heidelberg, Germany.

Davé, R. (1991) 'Characterization and detection of noise in clustering'. *Pattern Recognition Letters* **12**, 657–664.

Davé, R. and Krishnapuram, R. (1996) 'M-estimators and robust fuzzy clustering' In *Proc. of the Int. Conf. of the North Americam Fuzzy Information Processing Society, NAFIPS'96* (ed. Smith M, Lee M, Keller J and Yen J), pp. 400–404. IEEE.

Davé, R. and Krishnapuram, R. (1997) 'Robust clustering methods: a unified view'. *IEEE Transactions on Fuzzy Systems* **5**, 270–293.

Davé, R. and Sen, S. (1997) 'On generalising the noise clustering algorithms' *Proc. of the 7th IFSA World Congress, IFSA'97*, pp. 205–210.

Davé, R. and Sen, S. (1998) 'Generalized noise clustering as a robust fuzzy c-m-estimators model' *Proc. of the 17th Int. Conference of the North American Fuzzy Information Processing Society: NAFIPS'98*, pp. 256–260.

Drineas, P., *et al.* (2004) 'Clustering large graphs via the singular value decomposition'. *Machine Learning* **56**, 933.

Dubois, D. and Prade, H. (1988) *Possibility Theory*. Plenum Press, New York, NY, USA.

Duda, R. and Hart, P. (1973) *Pattern Classification and Scene Analysis*. J. Wiley & Sons, Inc., New York, NY, USA.

Dunn, J. (1974a) 'Well separated clusters and optimal fuzzy partitions'. *Journal of Cybernetics* **4**, 95–104.

Dunn, J. C. (1974b) 'A fuzzy relative of the isodata process and its use in detecting compact, well separated clusters'. *Journal of Cybernetics* **3**, 95–104.

Fisher, R. A. (1936) 'The use of multiple measurements in taxonomic problems'. *Annals of Eugenics* **7**(2), 179–188.

Frigui, H. and Krishnapuram, R. (1996) 'A robust algorithm for automatic extraction of an unknown number of clusters from noisy data'. *Pattern Recognition Letters* **17**, 1223–1232.

Frigui, H. and Krishnapuram, R. (1997) 'Clustering by competitive agglomeration'. *Pattern Recognition* **30**(7), 1109–1119.

Frigui, H. and Krishnapuram, R. (1999) 'A robust competitive clustering algorithm with applications in computer vision'. *IEEE Transactions on Pattern Analysis and Machine Intelligence* **21**(5), 450–465.

Frigui, H. and Nasraoui, O. (2000) 'Simultaneous clustering and attribute discrimination' *Proc. of the 9th IEEE Int. Conf. on Fuzzy Systems, Fuzz-IEEE'00*, pp. 158–163.

Gustafson, E. E. and Kessel , W. C. (1979) 'Fuzzy clustering with a fuzzy covariance matrix' *Proc. of the IEEE Conference on Decision and Control, San Diego, Californien*, pp. 761–766. IEEE Press, Piscataway, NJ.

Halkidi, M., Batistakis, Y. and Vazirgiannis, M. (2002) 'Cluster validity methods: Part I and part II'. *SIGMOD Record* **31**(2), 19–27 and 40–45.

Hathaway, R. and Bezdek, J. (1994) 'Nerf c-means: Non-euclidean relational fuzzy clustering'. *Pattern Recognition* **27**(3), 429–437.

Hawkins, D. (1980) *Identification of Outliers*. Chapman and Hall, London.

Höppner, F., Klawonn, F., Kruse, R, and Runkler, T. (1999) *Fuzzy Cluster Analysis*. J. Wiley & Sons, Ltd, Chichester, United Kingdom.

Keller, A. (2000) 'Fuzzy clustering with outliers' In *Proc. of the 19th Int. Conf. of the North American Fuzzy Information Processing Society, NAFIPS'00* (ed. Whalen T), pp. 143–147.

Klawonn, F. (2004) 'Fuzzy clustering: Insights and a new approach'. *Mathware and soft computing* **11**, 125–142.

Klawonn, F. (2006) 'Understanding the membership degrees in fuzzy clustering'. In *Proc. of the 29th Annual Conference of the German Classification Society, GfKl 2005* (ed. Spiliopoulou M, Kruse R, Borgelt C, Nrnberger A and Gaul W), pp. 446–454 Studies in Classification, Data Analysis, and Knowledge Organization. Springer.

Klawonn, F. and Höppner, F. (2003a) 'An alternative approach to the fuzzifier in fuzzy clustering to obtain better clustering results' *Proceedings 3rd Eusflat*, pp. 730–734.

Klawonn, F. and Höppner, F. (2003b) 'What is fuzzy about fuzzy clustering? – understanding and improving the concept of the fuzzifier' In *Advances in Intelligent Data Analysis V* (ed. Berthold M, Lenz HJ, Bradley E, Kruse R and Borgelt C), pp. 254–264. Springer.

Klawonn, F., Kruse, R. and Timm, H. (1997) 'Fuzzy shell cluster analysis' In *Learning, networks and statistics* (ed. della Riccia, G., Lenz, H. and Kruse, R.) Springer pp. 105–120.

Krishnapuram, R. and Keller, J. (1993) 'A possibilistic approach to clustering'. *IEEE Transactions on Fuzzy Systems* **1**, 98–110.

Krishnapuram, R. and Keller, J. (1996) 'The possibilistic c-means algorithm: insights and recommendations'. *IEEE Trans. Fuzzy Systems* **4**, 385–393.

Kummamuru, K., Dhawale, A. K. and Krishnapuram, R. (2003) 'Fuzzy co-clustering of documents and keywords' *Proc. of the IEEE Int. Conf. on Fuzzy Systems, Fuzz-IEEE'03.*

McKay, M. D , Beckman, R. J. and Conover, W. J. (1979) 'A comparison of three methods for selecting values of input variables in the analysis of output from a computer code'. *Technometrics* **21**(2), 239–245.

Pal, N., Pal, K. and Bezdek, J. (1997) 'A mixed c-means clustering model' *Proc. of FUZZ'IEEE'97*, pp. 11–21.

Pal, N., Pal, K., Keller, J. and Bezdek, J. (2004) 'A new hybrid c-means clustering model' *Proc. of FUZZ IEEE'04*, pp. 179–184.

Pedrycz, W. (2005) *Knowledge-Based Clustering: From Data to Information Granules*. J. Wiley & Son Inc., Holboken, USA.

Rezaee, M., Lelieveldt, B. and Reiber, J. (1998) 'A new cluster validity index for the fuzzy C-means'. *Pattern Recognition Letters* **19**, 237–246.

Rousseeuw, P., Trauwaert, E. and Kaufman, L. (1995) 'Fuzzy clustering with high contrast'. *Journal of Computational and Applied Mathematics* **64**, 81–90.

Runkler, T. A. and Bezdek, J. C. (2003) 'Web mining with relational clustering'. *Int. Jo. Approx. Reasoning* **32**(2–3), 217–236.

Sahbi, H. and Boujemaa, N. (2005) 'Validity of fuzzy clustering using entropy regularization' *Proc. of the IEEE Int. Conf. on Fuzzy Systems.*

Schölkopf, B. and Smola, A. (2002) *Learning with Kernels.* MIT Press.

Timm, H. and Kruse, R. (2002) 'A modification to improve possibilistic fuzzy cluster analysis' *Proc. of FUZZ-IEEE'02.*

Timm, H., Borgelt, C., Döring, C. and Kruse, R. (2004) 'An extension to possibilistic fuzzy cluster analysis'. *Fuzzy Sets and Systems* **147**, 3–16.

Vapnik, V. (1995) *The Nature of Statistical Learning Theory.* Springer, New York, USA.

Windham, M. P. (1981) 'Cluster validity for fuzzy clustering algorithm'. *Fuzzy Sets and Systems* **5**, 177–185.

Wu, K. and Yang, M. (2002) 'Alternating c-means clustering algorithms'. *Pattern Recognition* **35**, 2267–2278.

Wu, Z., Xie, W. and Yu, J. (2003) 'Fuzzy c-means clustering algorithm based on kernel method' *Proc. of ICCIMA'03*, pp. 1–6.

Xie, X. and Beni, G. (1991) 'A validity measure for fuzzy clustering'. *IEEE Transactions on pattern analysis and machine intelligence* **13**(4), 841–846.

Zadeh, L. A. (1965) 'Fuzzy sets'. *Information Control* **8**, 338–353.

Zhang, D. and Chen, S. (2003a) 'Clustering incomplete data using kernel-based fuzzy c-means algorithm'. *Neural Processing Letters* **18**(3), 155–162.

Zhang, D. and Chen, S. (2003b) 'Kernel-based fuzzy and possibilistic c-means' *Proc. of ICANN'03*, pp. 122–125.

Zhang, D. and Chen, S. (2004) 'A comment on 'alternative c-means clustering algorithms''. *Pattern Recognition* **37**(2), 179–174.

2

Relational Fuzzy Clustering

Thomas A. Runkler

Siemens AG, München, Germany

2.1 INTRODUCTION

Clustering is a method used to partition a set of objects into subsets, the so-called clusters. We consider the case that the set of objects is specified by *data*. We distinguish between *object* data and *relational* data. Object data contain a numerical vector of features for each object, for example, the length, width, height, and weight of an object. Relational data quantify the relation between each pair of objects, for example, the similarity of the two objects. For some sets of objects the object data representation is more appropriate, for other sets of objects the relational data representation is more appropriate. In general, object data can be transformed into relational data by applying a (pairwise) relational operator. For example, the Euclidean distances between pairs of feature vectors can be interpreted as relations between the corresponding objects. In the same way, each relational data-set can at least approximately be represented by object data. For example, objects can be placed on a two-dimensional diagram, so that more similar pairs of objects are placed closer together than less similar pairs of objects. Most chapters of this book exclusively deal with object data. In this chapter we will explicitly focus on relational data and how to find clusters in relational data. This chapter is organized as follows. In Section 2.2 we introduce object and relational data in a more formalized way. In Section 2.3 we briefly review object clustering models. In Section 2.4 we introduce the relational duals of these object clustering models and their extensions. In Section 2.5 we consider relational clustering with non-spherical, higher-order prototypes. In Section 2.6 we show that relational data can be clustered by just interpreting them as object data. In Section 2.7 we present some illustrative application examples. Finally, in Section 2.8 we give some conclusions.

2.2 OBJECT AND RELATIONAL DATA

A set of objects $O = \{o_1, \ldots, o_n\}, n \in \mathbb{N}^+$, can be numerically specified by an *object* data-set $X = \{x_1, \ldots, x_n\} \subset \mathbb{R}^p, p \in \mathbb{N}^+$. In this object data-set each of the n objects is numerically described by p real-valued features. Notice that features on any subset of $\mathbb{R}$ are included here, in particular (discrete) features on (subsets of) $\mathbb{N}$. Each object is represented as a point in the p-dimensional feature space, so an

Advances in Fuzzy Clustering and its Applications Edited by J. Valente de Oliveira and W. Pedrycz

object data-set X can be visualized by plotting the feature vectors. If $p \in \{1,2,3\}$ then the vectors in X can be directly plotted. If $p > 3$, then linear or nonlinear projection methods can be applied to produce a visualization of X.

A set of objects $O = \{o_1, \ldots, o_n\}$, $n \in \mathbb{N}^+$, can also be numerically specified by a *relational* dataset $R \subset \mathbb{R}^{n \times n}$. In this relational data-set each pair of objects is numerically described by a real-valued relation. Again, relations on any subset of $\mathbb{R}$ are included here, in particular (discrete) relations on (subsets of) $\mathbb{N}$. For convenience we do not distinguish between data-sets and matrices. For example, we denote the ith component of the element x_k from the *set* X as the element x_{ki} of the matrix X. Each entry $r_{jk}, j, k \in \{1, \ldots, n\}$ in the $n \times n$ relation *matrix* R then quantifies the relation between the pair of objects (o_j, o_k). Since a relational data-set R is equivalent to a square matrix, it can be easily visualized by a three-dimensional plot on a rectangular $n \times n$ grid, where the third dimension is given by the elements of the matrix R.

Often we consider *positive* relations where $r_{jk} \geq 0$ for all $j, k = 1, \ldots, n$, and *symmetric* relations where $r_{jk} = r_{kj}$ for all $j, k = 1, \ldots, n$. We distinguish between *similarity* and *dissimilarity* relations, and require *similarity* relations to be *reflexive*, $r_{jj} = 1$ for all $j = 1, \ldots, n$, and *dissimilarity* relations to be *irreflexive*, $r_{jj} = 0$ for all $j = 1, \ldots, n$.

A popular example for a relational dataset was given by Johnson and Wichern (1992). Consider the words for the numbers 1 ('one') to 10 ('ten') in the 11 languages $O = \{$ English, Norwegian, Danish, Dutch, German, French, Spanish, Italian, Polish, Hungarian, Finnish$\}$. The words for the same number in two different languages are called *concordant*, if they have the same first letter. For example, the English word 'seven' and the German 'sieben' are concordant, while the English 'eight' and the German 'acht' are not. English and German have four concordant words for numbers between 1 and 10: 5, 6, 7, and 9 – so the concordance between English and German is four. Since the concordances between two languages are integers between 0 and 10, the distance between two languages is defined as 10 minus the concordance. For O we obtain the distance matrix

$$D = \begin{pmatrix} 0 & 2 & 2 & 7 & 6 & 6 & 6 & 6 & 7 & 9 & 9 \\ 2 & 0 & 1 & 5 & 4 & 6 & 6 & 6 & 7 & 8 & 9 \\ 2 & 1 & 0 & 6 & 5 & 6 & 5 & 5 & 6 & 8 & 9 \\ 7 & 5 & 6 & 0 & 5 & 9 & 9 & 9 & 10 & 8 & 9 \\ 6 & 4 & 5 & 5 & 0 & 7 & 7 & 7 & 8 & 9 & 9 \\ 6 & 6 & 6 & 9 & 7 & 0 & 2 & 1 & 5 & 10 & 9 \\ 6 & 6 & 5 & 9 & 7 & 2 & 0 & 1 & 3 & 10 & 9 \\ 6 & 6 & 5 & 9 & 7 & 1 & 1 & 0 & 4 & 10 & 9 \\ 7 & 7 & 6 & 10 & 8 & 5 & 3 & 4 & 0 & 10 & 9 \\ 9 & 8 & 8 & 8 & 9 & 10 & 10 & 10 & 10 & 0 & 8 \\ 9 & 9 & 9 & 9 & 9 & 9 & 9 & 9 & 9 & 8 & 0 \end{pmatrix}. \tag{2.1}$$

Notice that D is positive, symmetric, and irreflexive. In Johnson and Wichern (1992) the following crisp clusters were reported:

c	cluster structure
3	$\{\{E, N, Da, Fr, I, Sp, P\}, \{Du, G\}, \{H, Fi\}\}$
4	$\{\{E, N, Da, Fr, I, Sp, P\}, \{Du, G\}, \{H\}, \{Fi\}\}$
5	$\{\{E, N, Da\}, \{Fr, I, Sp, P\}, \{Du, G\}, \{H\}, \{Fi\}\}$
7	$\{\{E, N, Da\}, \{Fr, I, Sp\}, \{P\}, \{Du\}, \{G\}, \{H\}, \{Fi\}\}$
9	$\{\{E\}, \{N, Da\}, \{Fr, I\}, \{Sp\}, \{P\}, \{Du\}, \{G\}, \{H\}, \{Fi\}\}$

Notice that these cluster structures are intuitively reasonable, even if they are only based on the very simple concordance measure.

In general, relational data-sets can be manually or automatically generated. If the relational data-set is manually generated, then a human expert has to rate (quantify) the relation between each pair of objects based on some more or less subjective criteria. This is only feasible for a small number of objects. For a larger number of objects this process is too expensive, because for n objects we have to quantify $n \times n$ relation values. If the relational dataset is automatically generated, some (numerical) features have to be identified to provide the necessary information about the relation between each pair of objects. These features may be relational, but they may also be *object features*. To compute relational data from object feature data, any norm $||.|| : \mathbb{R}^p \rightarrow \mathbb{R}$ can be used, for example,

$$r_{jk} = ||x_j - x_k||. \tag{2.2}$$

In this case the relational data are *dissimilarities*. If we use a similarity measure (like cosine) instead, then the relational data are *similarities*. The concordance measure presented above is an example of a dissimilarity measure based on the character representation of the objects. In the following we will focus on dissimilarities on numerical data, but dissimilarities on other data or similarities can be handled in a similar way.

If we add constant vectors to each element of X, and/or rotate the vectors around an arbitrary center in $\mathbb{R}^p$, then the resulting relational dataset R produced by (2.2) will stay the same. Therefore, even if a relational data-set R is generated from an object data-set X, then the original data-set X can in general not be reconstructed from R. Moreover, in the general case, for a given data-set R there might not even exist an object data-set X and a norm $||.||$ that produces R using (2.2). It is, however, possible to at least *approximately* produce a corresponding object data-set X from a given relational data-set R and a norm $||.||$ by *Sammon mapping* (Sammon, 1969). Sammon mapping produces an object data-set $Y = \{y_1, \ldots, y_n\} \subset \mathbb{R}^q$ so that the distances

$$d_{jk} = ||y_j - y_k||, \tag{2.3}$$

$j, k = 1, \ldots, n$, are as close as possible to the corresponding relational data r_{jk}, i.e., $d_{jk} \approx r_{jk}$ for all $j, k = 1, \ldots, n$. If R is computed from an object data-set X, then this implies that $X \approx Y$. Sammon mapping is done by minimizing the error functional

$$E_{\text{Sammon}} = \frac{1}{\sum_{j=1}^{n}\sum_{k=j+1}^{n} r_{jk}} \sum_{j=1}^{n}\sum_{k=j+1}^{n} \frac{(d_{jk} - r_{jk})^2}{r_{jk}}. \tag{2.4}$$

Minimization of E_{Sammon} can be done by gradient descent or Newton's algorithm. The derivatives of E_{Sammon} that are needed for these numerical optimization algorithms are

$$\frac{\partial E_{\text{Sammon}}}{\partial x_j} = \frac{2}{\sum_{j=1}^{n}\sum_{k=j+1}^{n} r_{jk}} \sum_{j \neq k} \frac{d_{jk} - r_{jk}}{r_{jk}} \frac{y_k - y_j}{d_{jk}} \tag{2.5}$$

and

$$\frac{\partial^2 E_{\text{Sammon}}}{\partial x_j^2} = \frac{2}{\sum_{j=1}^{n}\sum_{k=j+1}^{n} r_{jk}} \sum_{j \neq k} \left(\frac{1}{(r_{jk})^2} - \frac{1}{r_{jk} d_{jk}} + \frac{(y_k - y_j)^2}{r_{jk}(d_{jk})^3} \right). \tag{2.6}$$

Notice that with each update of each object data vector $x_j, j = 1, \ldots, n$, all the distances $d_{jk} = ||x_j - x_k||, k = 1, \ldots, n$, also have to be updated. In principle, Sammon mapping allows any object data processing algorithm to be extended to relational data. In particular, we can apply Sammon mapping to R and then apply a clustering algorithm to the resulting X. This approach was presented in (Pal, Eluri, and Mandal, 2002) for the case of fuzzy clustering. This indirect approach will not be pursued further here; instead we will focus on methods that *explicitly* process relational data in Section 2.4. Before that, however, we will give a brief review of object clustering methods.

2.3 OBJECT DATA CLUSTERING MODELS

Clustering partitions a set of objects $O = \{o_1, \ldots, o_n\}$ into $c \in \{2, \ldots, n-1\}$ non-empty and pairwise disjoint subsets $C_1, \ldots, C_c \subset O$, so $C_i \neq \emptyset$ for all $i \in \{1, \ldots, c\}$, $C_i \cup C_j = \emptyset$ for all $i, j \in \{1, \ldots, c\}$, and $C_1 \cup \ldots \cup C_c = O$. The sets $C_1, \ldots, C_c$ can equivalently be described by a (hard) partition matrix $U \in M_{\text{hcn}}$, where

$$M_{\text{hcn}} = \left\{ U \in \{0,1\}^{c \times n} | \sum_{i=1}^{c} u_{ik} = 1, k = 1, \ldots, n, \sum_{k=1}^{n} u_{ik} > 0, i = 1, \ldots, c \right\}. \tag{2.7}$$

2.3.1 Sequential Agglomerative Hierarchical Clustering

For object datasets $X = \{x_1, \ldots, x_n\} \subset \mathbb{R}^p$ finding good cluster partitions is guided by the similarities between feature vectors. Objects with similar feature vectors should belong to the same cluster, and objects with different feature vectors should belong to different clusters. A simple clustering algorithm is *sequential agglomerative hierarchical nonoverlapping* (SAHN) clustering (Sneath and Sokal, 1973). The SAHN algorithm starts with n clusters, each with one data point, so $C_1 = \{x_1\}, \ldots, C_n = \{x_n\}$. In each step SAHN merges the pair of clusters with the lowest distance from each other, until the desired number of clusters is achieved, or until finally all points are agglomerated in one single cluster. Depending on the measure to compute the distances between pairs of clusters we distinguish three variants: *single linkage* uses the minimum distance between all pairs of points from different clusters, *complete linkage* uses the maximum distance, and *average linkage* uses the average distance.

2.3.2 (Hard) c-means

The drawback of all SAHN variants is their complexity that grows quadratically with the number of objects. Hence, SAHN is not efficient for large data-sets. The complexity of SAHN can be significantly reduced, if we do not compute all distances between pairs of points but if we represent each cluster by a *cluster center* and only compute the distances between pairs of these points and centers. In this case, the cluster structure is additionally specified by a set $V = \{v_1, \ldots, v_c\} \subset \mathbb{R}^p$, where v_i is the center of cluster $i, i \in \{1, \ldots, c\}$. For a good cluster partition each data point should be as close as possible to the center of the cluster it belongs to. This is the idea of the *(hard) c-means (HCM)* clustering model (Ball and Hall, 1965) that minimizes the objective function

$$J_{\text{HCM}}(U, V; X) = \sum_{i=1}^{c} \sum_{k=1}^{n} u_{ik} \, ||x_k - v_i||^2. \tag{2.8}$$

Optimization of the HCM clustering model can be done by *alternating optimization* (AO) through the necessary conditions for extrema of $J_{\text{HCM}}(U, V; X)$.

$$u_{ik} = \begin{cases} 1 & \text{if} ||x_k - v_i|| = \min\{||x_k - v_1||, \ldots, ||x_k - v_c||\} \\ 0 & \text{otherwise,} \end{cases} \tag{2.9}$$

$$v_i = \frac{\sum_{k=1}^{n} u_{ik} x_k}{\sum_{k=1}^{n} u_{ik}}, \tag{2.10}$$

$i = 1, \ldots, c, k = 1, \ldots, n$. There exist two versions of AO. The first version randomly initializes V and then repeatedly updates U and V until some termination criterion on V holds, and the second version randomly initializes U and then repeatedly updates V and U until some termination criterion on U holds. If $p < n$ then the first version is more efficient and if $n < p$ then the second version is more efficient. Some alternative ways to optimize clustering models are genetic algorithms (Bezdek and Hataway, 1994), artificial life techniques (Runkler and Bezdek, 1997), ant colony optimization (ACO) (Runkler, 2005a), and particle swarm optimization (PSO) (Runkler and Katz, 2006).

2.3.3 Fuzzy c-means

The main disadvantage of HCM is that it has to assign each point to exactly one cluster, also points that partially belong to several overlapping clusters. This disadvantage is overcome by *fuzzy clustering*. In analogy to (2.7) the set of fuzzy partitions is defined as

$$M_{\text{fcn}} = \left\{ U \in [0,1]^{c \times n} \middle| \sum_{i=1}^{c} u_{ik} = 1, k = 1, \ldots, n, \sum_{k=1}^{n} u_{ik} > 0, i = 1, \ldots, c \right\}. \tag{2.11}$$

In analogy to (2.8) the *fuzzy c-means (FCM)* clustering model (Bezdek, 1981) is defined by the objective function

$$J_{\text{FCM}}(U, V; X) = \sum_{i=1}^{c} \sum_{k=1}^{n} u_{ik}^{m} \, ||x_k - v_i||^2 \tag{2.12}$$

with a fuzziness parameter $m \in (1, \infty)$ with a typical value of $m = 2$. Optimization of the FCM clustering model can be done by AO through the necessary conditions for extrema of $J_{\text{FCM}}(U, V; X)$.

$$u_{ik} = 1 \Bigg/ \sum_{j=1}^{c} \left(\frac{||x_k - v_i||}{||x_k - v_j||} \right)^{\frac{2}{m-1}}, \tag{2.13}$$

$$v_i = \frac{\sum\limits_{k=1}^{n} u_{ik}^{m} x_k}{\sum\limits_{k=1}^{n} u_{ik}^{m}}. \tag{2.14}$$

Notice the similarity between Equations (2.10) and (2.14), and notice the difference between Equations (2.9) and (2.13)!

2.3.4 Possibilistic c-means and Noise Clustering

A drawback of fuzzy partitions is that the some of memberships in all clusters has to be one for each data point, also for noise points or remote outliers. This drawback is overcome by possibilistic partitions

$$M_{\text{pcn}} = \left\{ U \in [0,1]^{c \times n} \middle| \sum_{k=1}^{n} u_{ik} > 0, i = 1, \ldots, c \right\}. \tag{2.15}$$

Two clustering models that produce possibilistic partitions are *possibilistic c-means* (PCM) (Krishnapuram and Keller, 1993) and *noise clustering* (NC) (Davé, 1991). The PCM objective function is

$$J_{\text{PCM}}(U, V; X) = \sum_{i=1}^{c} \sum_{k=1}^{n} u_{ik}^{m} \, ||x_k - v_i||^2 - \sum_{i=1}^{c} \eta_i \sum_{k=1}^{n} (1 - u_{ik})^m, \tag{2.16}$$

with the cluster radii $\eta_1, \ldots, \eta_c \in \mathbb{R}^+$. The necessary conditions for optima of PCM are Equation (2.14) and (the Cauchy function)

$$u_{ik} = 1 \Bigg/ \left(1 + \left(\frac{||x_k - v_i||^2}{\eta_i}\right)^{\frac{1}{m-1}}\right). \tag{2.17}$$

The NC objective function is

$$J_{\text{NC}}(U, V; X) = \sum_{i=1}^{c} \sum_{k=1}^{n} u_{ik}^m \, ||x_k - v_i||^2 + \sum_{k=1}^{n} \delta^2 \left(1 - \sum_{j=1}^{c} u_{jk}\right)^m, \tag{2.18}$$

with the parameter $\delta \in \mathbb{R}^+$ that represents the distance between each point and the center of a "*noise*" cluster. The necessary conditions for optima of NC are Equation (2.14) and

$$u_{ik} = 1 \Bigg/ \left(\sum_{j=1}^{c} \left(\frac{||x_k - v_i||}{||x_k - v_j||}\right)^{\frac{2}{m-1}} + \left(\frac{||x_k - v_i||}{\delta}\right)^{\frac{2}{m-1}}\right). \tag{2.19}$$

2.3.5 Alternating Cluster Estimation

The AO algorithms of all four (HCM, FCM, PCM, and NC) and many other clustering models are special cases of the *alternating cluster extimation* (ACE) (Runkler and Bezdek, 1999a). There exist two versions of ACE, corresponding to the two versions of AO introduced before. The first version initializes V and then alternatingly updates U and V. The second version initializes U and then alternatingly updates V and U. Both ACE versions are defined by the update equations for U and V. The update equations for U may be Equations (2.9), (2.13), (2.17), and (2.19) or, for example, exponential functions

$$u_{ik} = \mathrm{e}^{-\left(\frac{||x_k - v_i||}{\sigma_i}\right)^{\alpha}}, \tag{2.20}$$

$\sigma_1, \ldots, \sigma_c, \alpha > 0$, or hyperconic functions *(dancing cones)*

$$u_{ik} = \begin{cases} 1 - \left(\frac{||x_k - v_i||}{r_i}\right)^{\alpha} & \text{for} ||x_k - v_i|| \leq r_i, \\ 0 & \text{otherwise,} \end{cases} \tag{2.21}$$

$r_1, \ldots, r_c, \alpha > 0$ (Runkler and Bezdek, 1999b). For $X \subset [-2, 2], V = \{-1, 1\}, m \in \{1.1, 1.5, 2, 3\}$, $\alpha \in \{0.5, 1, 2, 4\}, \eta_1 = \eta_2 = \sigma_1 = \sigma_2 = r_1 = r_2 = 1$, the four classes of membership functions (2.13), (2.17), (2.20), and (2.21) are displayed in Figure 2.1. Notice that these membership function families share many similarities. For example, for $m \to 1$ and for $\alpha \to 0$, they all become equal to the HCM partitions $u_{1k} = 1$ and $u_{2k} = 0$ for $x < 0$ and $u_{1k} = 0$ and $u_{2k} = 1$ for $x > 0$, and for $m \to \infty$ and for $\alpha \to \infty$, they all have single peaks at v_1 and v_2. The update equations for V may be Equations (2.10), (2.14) (which corresponds to a *basic defuzzification distribution* (BADD) (Filev and Yager, 1991)), or other defuzzification operators like *semi-linear defuzzification* (SLIDE) (Yager and Filev, 1993) or *extended center of area* (XCOA) (Runkler and Glesner, 1993). For a more extensive overview on defuzzification see (Runkler, 1997). Notice that ACE is a family of *generalized* clustering algorithms that may or may not optimize any clustering models. Depending on the choice of the update equations for U and V, ACE may obtain different characteristics. For example, hyperconic membership functions lead to a low sensitivity to noise and outliers.

2.3.6 Non-spherical Prototypes

All of the clustering models presented up to here are based on distances $||x_k - v_i||$ between data points and cluster centers. If $||.||$ is the Euclidean distance, then *hyperspherical* clusters are found. To extend a

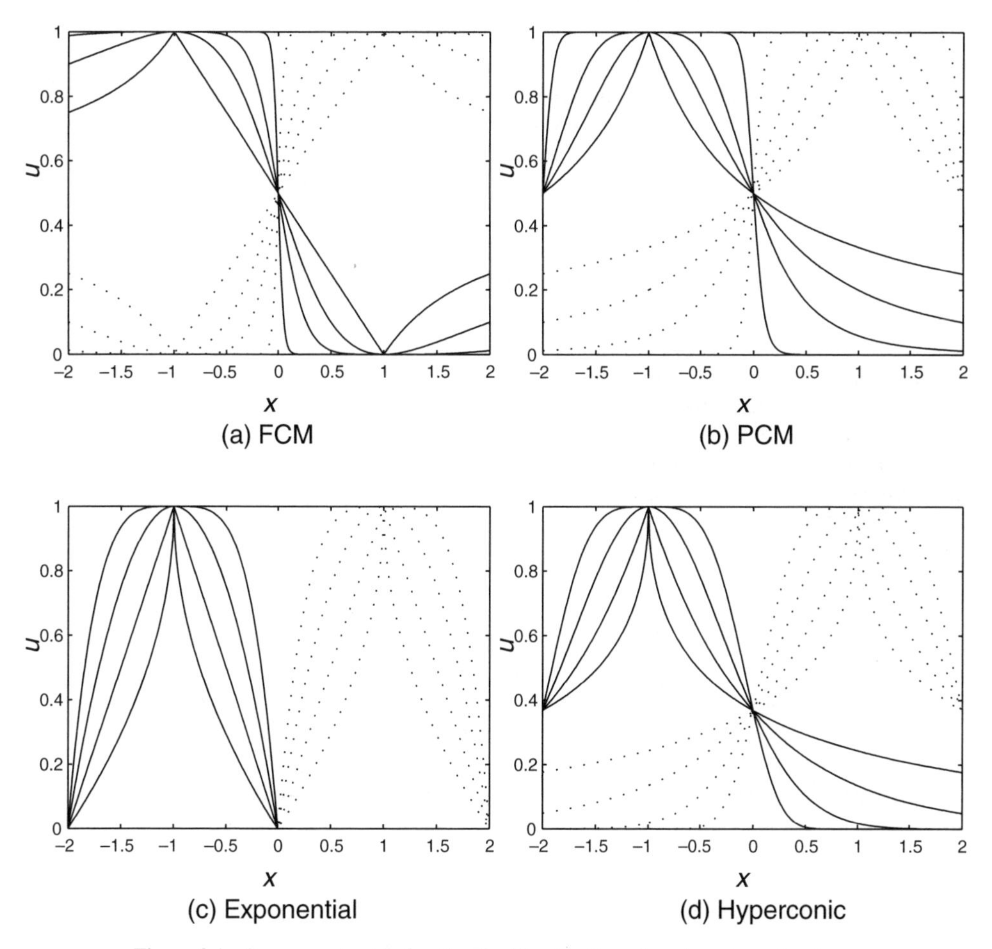

Figure 2.1 Some membership function families for alternating clustering estimation.

(hyperspherical) clustering model to other cluster shapes, the Euclidean distance between data points and cluster centers has to be replaced by a different distance measure. For example, instead of the Euclidean distance we can use the local Mahalanobis distance. This means that for each cluster $i = 1, \ldots, c$ we compute the local (fuzzy) covariance matrix

$$S_i = \sum_{k=1}^{n} u_{ik}^m \cdot (v_i - x_k)^T \cdot (v_i - x_k), \tag{2.22}$$

compute the corresponding norm inducing matrix

$$A_i = (\rho_i \det S_i)^{1/p} \cdot (S_i^T \cdot S_i) \cdot S_i^T \tag{2.23}$$

with the cluster volumes $\rho_1, \ldots, \rho_c > 0$, and set $||x_k - v_i||$ as the matrix norm $(v_i - x_k) \cdot A_i \cdot (v_i - x_k)^T$. This distance measure leads to clustering models that find *hyperellipsoidal* clusters. If we use this distance measure in the FCM model, then we call the resulting model the *Gustafson–Kessel* (GK) model (Gustafson and Kessel, 1979). However, the same distance measure can be used in all of the clustering models presented so far.

In the GK model, each cluster $i = 1, \ldots, c$ is represented by a cluster center v_i and a norm inducing matrix A_i. More generally, we can allow arbitrary geometrical parameters and the corresponding distance measures. This leads to more complicated cluster prototypes. As examples for these more complicated prototypes we briefly present linear varieties, so-called elliptotypes and circles. A linear variety can be represented by an anchor point and one or more direction vectors. So, the prototypes of a *c-varieties* model are $(v_i, d_{i,1}, \ldots, d_{i,q}) \in \mathbb{R}^{p\times(q+1)}, i = 1, \ldots, c, q \in \{1, \ldots, p-1\}$. The distance between data point x_k and the ith variety is

$$\sqrt{||x_k - v_i||^2 - \sum_{j=1}^{q}((x_k - v_i)^T d_{ij})^2}, \tag{2.24}$$

where the direction vectors d_{ij} are the largest eigenvectors of the local covariance matrices S_i (2.22). If these prototypes are used in the FCM model, then we obtain the *fuzzy c-varieties* (FCV) model (Bezdek, Coray, Gunderson, and Watson, 1981a).

An *elliptotype* is a fuzzy set whose α cuts are (hyper-)ellipsoidals. An elliptotype is specified by the same parameters as a variety, $(v_i, d_{i,1}, \ldots, d_{i,q}), i = 1, \ldots, c, q \in \{1, \ldots, p-1\}$, but the distance between x_k and the ith elliptotype is

$$\sqrt{||x_k - v_i||^2 - \alpha \cdot \sum_{j=1}^{q}((x_k - v_i)^T d_{ij})^2}, \tag{2.25}$$

$\alpha \in [0, 1]$, which is a linear combination of the Euclidean distance between x_k and v_i and the distance between x_k and the variety according to (2.24). If these prototypes are used in the FCM model, then we obtain the *fuzzy c-elliptotypes* (FCE) model (Bezdek, Coray, Gunderson, and Watson, 1981b).

As the last example for more complicated (object data) prototypes we present circles. A circle is represented by a center and a radius, which yields to circle prototypes $(v_i, r_i), i = 1, \ldots, c$, $v_i \in \mathbb{R}^p, r_i \in \mathbb{R}$. The distance between x_k and the ith circle is

$$|||x_k - v_i|| - r_i| \tag{2.26}$$

and the radii can be updated by

$$r_i = \frac{\sum_{k=1}^{n} u_{ik}^m ||x_k - v_i||}{\sum_{k=1}^{n} u_{ik}^m}, \tag{2.27}$$

$i = 1, \ldots, c$. If these prototypes are used in the FCM model, then we obtain the *fuzzy c-shells* (FCS) model (Davé, 1990).

In accordance with the fuzzy clustering models for these three cluster prototypes (FCV, FCE, and FCS), we can define possibilistic clustering models, noise clustering models, and alternating cluster estimation: PCV, PCE, PCS, NC–V, NC–E, NC–S, ACE–V, ACE–E, and ACE–S. The three prototype families (varieties, elliptotypes, and shells) are only examples for possible prototypes. Clustering algorithms for many other geometric forms can be obtained by just specifying the geometric cluster prototype, the distance between data points and prototypes, and an algorithm to compute the free cluster prototypes.

2.4 RELATIONAL CLUSTERING

The clustering models reviewed in the previous section are tailored to find clusters in object data. For all of them, similar clustering models for relational data can be developed. In this section we review some of these clustering models for relational data.

Let us first restrict ourselves to relational data clustering models specified by objective functions. The relational data-set is denoted as $R \subset \mathbb{R}^{n\times n}$, and the goal is to compute a partition matrix U that minimizes the objective function $J(U;R)$. Notice that neither object data X nor cluster centers V are available in relational clustering. Hence, the distances $||x_k - v_i||$ between data points and cluster prototypes that appear in each of the clustering models from the previous section can in general not explicitly be computed. There are (at least) three ways to overcome this problem: the restriction of the solution space, the implicit computation of object data, and the explicit computation of object data. In the following we present (at least) one representative for each of these three ways.

2.4.1 Relational Fuzzy c-medoids

The relational matrix R can be interpreted to contain distances between pairs of data points from X. In the clustering models from the previous section we need to compute distances between points from X and cluster centers from V. Hence, if we require $V \subset X$, i.e., each cluster center has to be on one of the data points, then we can immediately use the information in R for the required distances. The cluster centers $V \subset X$ are called *medoids*. Therefore, the corresponding clustering model is called the *(relational) fuzzy c-medoids* ((R)FCMdd) model (Krishnapuram, Joshi, Nasraoui, and Yi, 2001). (R)FCMdd has the same objective function as FCM (2.12), but the additional restriction $V \subset X$. (R)FCMdd can be optimized by alternating optimization. For simplicity we present the object data version FCMdd here, which can be easily converted into the relational data version RFCMdd. The partition matrix U is computed according to FCM using Equation (2.13), but the choice of $V \subset X$ is a discrete optimization problem that has to be solved using exhaustive search. The contribution of the ith cluster to the objective function $J_{\text{FCMdd}} = J_{\text{FCM}}$ is

$$J_i^* = \sum_{k=1}^{n} u_{ik}^m ||x_k - v_i||^2, \tag{2.28}$$

so $J = \sum_{i=1}^{c} J_i^*$. If say $v_i = x_j$, then we have $||v_i - x_k|| = r_{jk}$, and so

$$J_i^* = J_{ij} = \sum_{k=1}^{n} u_{ik}^m r_{jk}^2. \tag{2.29}$$

So the best choice for each cluster center is $v_i = x_{w_i}, i = 1, \ldots, n$, with

$$w_i = \text{argmin}\{J_{i1}, \ldots, J_{in}\}. \tag{2.30}$$

The exhaustive search is computationally expensive, so (R)FCMdd has a relatively high computational complexity.

2.4.2 Relational Fuzzy c-means

The second approach to transform an objective function for object clustering into an objective function for relational data clustering is to compute implicitly the cluster prototypes. This can be done by inserting the equation to compute the cluster prototypes in AO into the objective function (for object data). This process, called *reformulation* (Hathaway and Bezdek, 1995), yields an objective function for relational data. For example, a reformulation of the FCM model is obtained by inserting Equation (2.14) into (2.12), which leads to the *relational fuzzy c-means* (RFCM) (Bezdek and Hathaway, 1987) model with the objective function

$$J_{\text{RFCM}}(U;R) = \sum_{i=1}^{c} \frac{\sum_{j=1}^{n} \sum_{k=1}^{n} u_{ij}^m u_{ik}^m r_{jk}^2}{\sum_{j=1}^{n} u_{ij}^m}. \tag{2.31}$$

Optimization of J_{RFCM} can be done by randomly initializing and then iteratively updating U using the necessary conditions for extrema of RFCM:

$$u_{ik} = 1 \Bigg/ \sum_{j=1}^{n} \frac{\displaystyle\sum_{s=1}^{n} \frac{u_{is}^m r_{sk}}{\sum_{r=1}^{n} u_{ir}^m} - \sum_{s=1}^{n}\sum_{t=1}^{n} \frac{u_{is}^m u_{it}^m r_{st}}{2\left(\sum_{r=1}^{n} u_{ir}^m\right)^2}}{\displaystyle\sum_{s=1}^{n} \frac{u_{js}^m r_{sk}}{\sum_{r=1}^{n} u_{jr}^m} - \sum_{s=1}^{n}\sum_{t=1}^{n} \frac{u_{js}^m u_{jt}^m r_{st}}{2\left(\sum_{r=1}^{n} u_{jr}^m\right)^2}}, \tag{2.32}$$

$i = 1, \ldots, c, k = 1, \ldots, n$, until some termination criterion holds. Notice that the optimization of the RFCM is no *alternating* optimization (AO), but simply *optimization*, since the family V of optimization variables has disappeared, so there remains only one family of optimization variables: U. Relational duals for other object clustering models can be found in Hathaway, Davenport, and Bezdek (1989).

2.4.3 Non-Euclidean Relational Fuzzy c-means

If the relational data-set R is explicitly computed from object data X using the same norm $||.||$ that is used in the clustering model, then minimizing $J_{\text{FCMdd}}(U, V; X)$ will produce the same partition matrix U as minimizing $J_{\text{RFCMdd}}(U; R)$, and minimizing $J_{\text{FCM}}(U, V; X)$ will produce the same partition matrix U as minimizing $J_{\text{RFCM}}(U; R)$. However, different norms might have been used in the computation of R and in clustering. Or the relational data-set may be obtained without any underlying object data-set, for example, by manual rating. In these cases, we cannot match the object data versions with the relational data versions of these algorithms any more. The RFCM model might even yield partition matrices $U \notin M_{\text{fcn}}$, in particular we can have some $u_{ik} < 0$ or $u_{ik} > 1$, for non-Euclidean relational data-sets. To fix this problem, Hathaway and Bezdek (1994) transformed the non-Euclidean distance matrix D into a Euclidean distance matrix D_β by applying a so-called β-spread transform

$$D_\beta = D + \beta \cdot B, \tag{2.33}$$

with a suitable $\beta \in \mathbb{R}^+$, where $B \in [0, 1]^{n \times n}$ is the off-diagonal matrix with $b_{ij} = 1$ for all $i, j = 1, \ldots, n, i \neq j$, and $b_{ii} = 0$ for all $i = 1, \ldots, n$. In the *non-Euclidean relational fuzzy c-means* (NERFCM) algorithm (Hathaway and Bezdek, 1994) the value of β is sucessively increased, i.e., higher values of β are added to the off-diagonal elements of R, until the Euclidean case is achieved, and we finally have $U \in M_{\text{fcn}}$.

2.4.4 Relational Alternating Cluster Estimation

In the same way as we extended AO of object clustering models to ACE, we can also extend the optimization of relational data clustering models to *relational alternating cluster estimation* (RACE) (Runkler and Bezdek, 1998). With the relational clustering methods presented above, the partitions were computed from the relational data using the necessary conditions for extrema of an objective function (for clustering relational data). For example, in RFCM we minimize $J_{\text{RFCM}}(U; R)$ (2.31) by subsequently computing $U(R)$ by (2.32), where (2.32) is derived from (2.31). In RACE, we abandon the objective function and define a (relational) clustering algorithm by simply *specifying* a function $U(R)$ to compute the partition. The partition function used in (Runkler and Bezdek, 1998) was the Cauchy (or RPCM) membership function

$$u_{ik} = 1 \Bigg/ \left(1 + \left(\frac{r_{jk}^2}{\eta_i}\right)^{\frac{1}{m-1}}\right). \tag{2.34}$$

This model uses a medoid approach again. To find out which distance r_{jk} corresponds to the membership u_{ik}, i.e., which point x_k is equal to which cluster center v_j, an exhaustive search is applied again. Here, we choose $x_k = v_j$, so that the sum of the memberships of o_k in all the other clusters is as low as possible.

$$k = \operatorname{argmin}\left\{ \sum_{i=1,\, i\neq j}^{n} u_{jk} \right\}, \tag{2.35}$$

$k = 1, \ldots, n$. Notice the similarity between Equations (2.34) and (2.17) and the similarity between Equations (2.35) and (2.30) with (2.29). The choice of this partition function is just one out of infinitely many possible RACE instances. Just as ACE, RACE may or may not optimize any clustering model, depending on the choice of $U(R)$.

2.5 RELATIONAL CLUSTERING WITH NON-SPHERICAL PROTOTYPES

The relational clustering models presented in the previous section are derived from object clustering models that find spherical clusters. Therefore, they will find clusters that correspond to spherical clusters in the associated object data. Notice that we assume that we can (at least approximately) associate any relational data-set to an object data-set. If the (object) data-set contains clusters with more complicated shapes, then the relational clustering models from the previous section will probably fail. In order to find clusters in relational data that correspond to non-spherical prototypes in the associated object data we present three approaches in this section: kernelization, projection, and local estimation of object data.

2.5.1 Kernelized Relational Clustering

Kernelization has gained a lot of interest in pattern recognition as an efficient way to transform a data-set $X = \{x_1, \ldots, x_n\} \in \mathbb{R}^p$ to a higher dimensional data-set $Y = \{y_1, \ldots, y_n\} \in \mathbb{R}^q, q > p$, so that the data structure in Y is simpler than in X. For example, in *support vector machines* (SVM) (Müller *et al.*, 2001), the transformation theoretically maps not linearly separable data X to linearly separable data Y. The idea in kernelized clustering is to map X to Y, so that the clusters can be better found in Y than in X. More particularly, if X contains non-spherical clusters, then X can be mapped to Y, so that Y (theoretically) contains spherical clusters. In this way, *non-spherical* clusters in X can be found by applying a *spherical* clustering algorithm to Y.

According to Mercer's theorem (Mercer, 1909; Schölkopf and Smola, 2002) there is a mapping $\varphi : \mathbb{R}^p \rightarrow \mathbb{R}^q$ such that

$$k(x_j, x_k) = \langle \varphi(x_j), \varphi(x_k) \rangle \tag{2.36}$$

with the generalized dot product

$$\langle y_j, y_k \rangle = \sum_{i=1}^{q} y_{ji} y_{ki}. \tag{2.37}$$

This means that a dot product in Y can be simply computed by evaluating a kernel function in X. Hence, replacing dot products with kernels is generally called the *kernel trick*. Some common kernel functions are Gaussian kernels

$$k(x_j, x_k) = e^{-\frac{||x_j - x_k||^2}{\sigma^2}}, \tag{2.38}$$

$\sigma \in \mathbb{R}^+$, hyperbolic tangent function kernels

$$k(x_j, x_k) = 1 - \tanh\left(\frac{||x_j - x_k||^2}{\sigma^2} \right), \tag{2.39}$$

polynomial kernels

$$k(x_j, x_k) = \langle x_j, x_k \rangle^\alpha, \tag{2.40}$$

$\alpha \in \mathbb{N}^+$, and *radial basis function* (RBF) (Powell, 1985) kernels

$$k(x_j, x_k) = f(||x_j - x_k||). \tag{2.41}$$

Notice that Gaussian and hyperbolic tangent function kernels are special cases of RBF kernels. A comparison between Equations (2.38) and (2.20) shows that kernel functions can be used as prototype functions in ACE and vice versa.

Kernelization of a clustering model can be done by replacing dot products by kernels. This has been done for HCM (Girolami, Smola, and Müller, 2002; Schölkopf, 1998; Zhang and Rudnicky, 2002), FCM (Wu, Xie, and Yu, 2003; Zhang and Chen, 2002, 2003a), PCM (Zhang and Chen, 2003b), and NERFCM (Hathaway, Huband, and Bezdek, 2005). Each entry in a relational data-set is a distance between a pair of points. Following Mercer's theorem this distance can be computed in Y by a kernel in X:

$$r_{jk}^2 = ||\varphi(x_j), \varphi(x_k)||^2 \tag{2.42}$$
$$= (\varphi(x_j), \varphi(x_k))^T (\varphi(x_j), \varphi(x_k)) \tag{2.43}$$
$$= \varphi(x_j)^T \varphi(x_j) - 2\,\varphi(x_j)^T \varphi(x_k) + \varphi(x_k)^T \varphi(x_k) \tag{2.44}$$
$$= k(x_j, x_j) - 2 \cdot k(x_j, x_k) + k(x_k, x_k) \tag{2.45}$$
$$= 2 - 2 \cdot k(x_j, x_k), \tag{2.46}$$

where we assume that $k(x, x) = 0$ for all $x \in \mathbb{R}^p$. This means that kernelization of a relational clustering algorithm is equivalent to transforming the relational data-set R into a relational data-set R' by Equation (2.46) and then simply applying the (unchanged) relational clustering algorithm to R'. In particular, for Gaussian kernels (2.38) we obtain

$$r'_{jk} = \sqrt{2 - 2 \cdot e^{-\frac{r_{jk}^2}{\sigma^2}}}, \tag{2.47}$$

and for hyperbolic tangent function kernels (2.39) we obtain

$$r'_{jk} = \sqrt{2 \cdot \tanh\left(\frac{r_{jk}^2}{\sigma^2}\right)}. \tag{2.48}$$

These two functions are visualized in Figure 2.2 for the parameters $\sigma \in \{0.5, 1, 2, 4\}$. Notice the close similarity between the Gaussian and the hyperbolic tangent function kernels. Apparently, the effect of kernelization in relational clustering is that large distances are clipped at the threshold of $\sqrt{2}$, and that smaller distances are scaled by an almost constant factor. We assume that the relational clustering algorithm is invariant against scaling of the relational data by a constant factor $c \in \mathbb{R}^+$. This means that we obtain the same results whether we cluster the original data-set R or a scaled data-set R^* with $r_{jk}^* = c \cdot r_{jk}$ for all $j, k = 1, \ldots, n$. In this case we can scale each of the transformation curves in Figures 2.2(a) and (b) by a constant $c(\sigma)$, so that the slope in the origin becomes equal to one, i.e., small distances remain (almost) unchanged. This scaling yields the curves shown in Figures 2.2(c) and (d), which indicate that the effect of the kernelization is essentially a clipping of the large distances at the threshold σ.

2.5.2 Fuzzy Nonlinear Projection

At the end of Section 2.2 we presented an indirect relational clustering approach that first explicitly produces a complete (approximate) object data-set by Sammon mapping and then applies object clustering. We will not pursue this approach further here, but consider relational clustering methods than either implicitly or locally construct approximate object data.

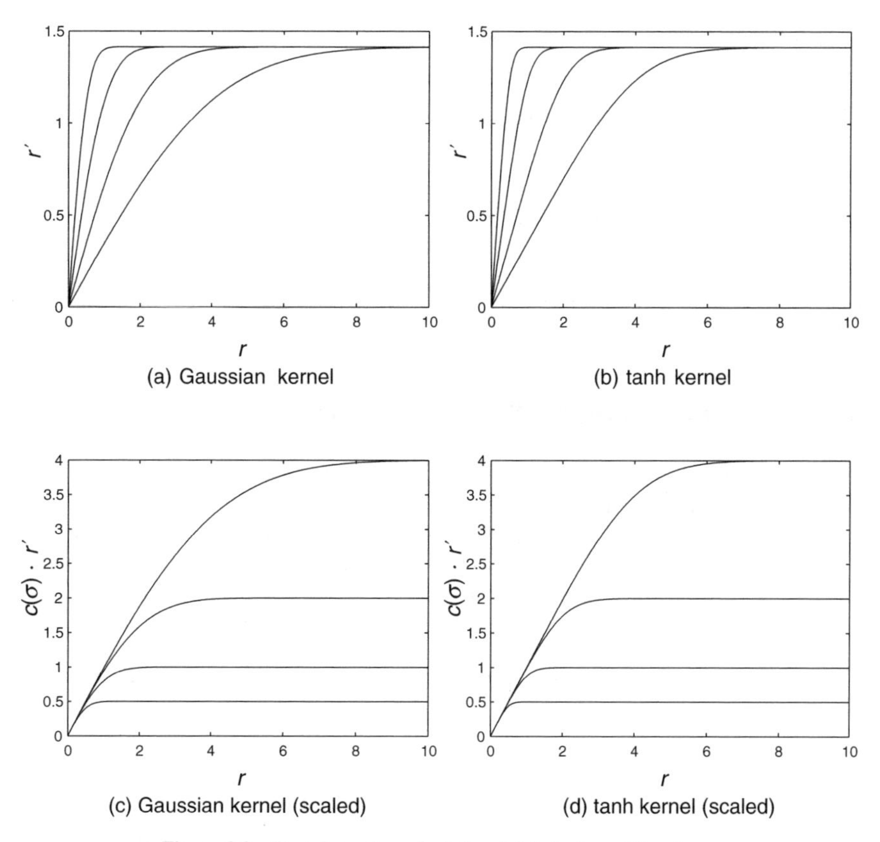

Figure 2.2 Transformation of relational data by kernel functions.

An approach to implicitly construct approximate object data in clustering is *fuzzy nonlinear projection* (FNP) (Runkler, 2003). The idea in FNP is to combine the objective functions of Sammon mapping (2.4) and FCM (2.12). This yields the FNP objective function

$$J_{\text{FNP}}(U, Y; R) = \frac{1}{\sum_{j=1}^{n} \sum_{k=j+1}^{n} r_{jk}} \sum_{i=1}^{c} \sum_{j=1}^{n} \sum_{k=j+1}^{n} u_{ij}^m u_{ik}^m \frac{(d_{jk} - r_{jk})^2}{r_{jk}}, \tag{2.49}$$

where $U \in M_{fcn}$ (2.11) and d_{jk} as in (2.3). FNP maps the relational data-set R to an object data-set Y so that objects belonging to the same cluster approximately have the same distance in X as in Y. Minimization of $J_{\text{FNP}}(U, Y; R)$ can be done by alternatingly performing one step of Newton optimization with respect to Y

$$y_k = y_k - \left(\frac{\partial J_{\text{FNP}}}{\partial y_k} \right) \Bigg/ \left(\frac{\partial^2 J_{\text{FNP}}}{\partial y_k^2} \right), \tag{2.50}$$

where

$$\frac{\partial J_{\text{FNP}}}{\partial y_k} = \frac{2}{\sum_{j=1}^{n} \sum_{k=j+1}^{n} r_{jk}} \sum_{i=1}^{c} \sum_{j \neq k} u_{ij}^m u_{ik}^m \frac{d_{jk} - r_{jk}}{r_{jk}} \frac{y_k - y_j}{d_{jk}}, \tag{2.51}$$

$$\frac{\partial^2 J_{\text{FNP}}}{\partial y_k^2} = \frac{2}{\sum_{j=1}^{n} \sum_{k=j+1}^{n} r_{jk}} \sum_{i=1}^{c} \sum_{j \neq k} u_{ij}^m u_{ik}^m \left(\frac{1}{(r_{jk})^2} - \frac{1}{r_{jk} d_{jk}} + \frac{(y_k - y_j)^2}{r_{jk} (d_{jk})^3} \right), \tag{2.52}$$

and finding the corresponding optimal U by solving

$$\partial J_{\text{FNP}} / \partial u_{ik} = 0, \tag{2.53}$$

where $U \in M_{\text{fcn}}$, which leads to the update equation

$$u_{ik} = 1 \Bigg/ \sum_{l=1}^{c} \left(\frac{\sum_{j \neq k} u_{ij}^m \frac{(d_{jk} - r_{jk})^2}{r_{jk}}}{\sum_{j \neq k} u_{lj}^m \frac{(d_{jk} - r_{jk})^2}{r_{jk}}} \right)^{\frac{1}{m-1}}. \tag{2.54}$$

Notice that in FNP, both Y and U have to be initialized, even if the order of the computation of Y and U is reversed.

2.5.3 Relational Gustafson–Kessel Clustering

FNP performs an implicit construction of approximate object data. In this section we present an approach to *explicitly* construct approximate object data, but only in a local environment: *relational Gustafson–Kessel clustering with medoids* (RGKMdd) (Runkler, 2005b). For simplicity we present this algorithm only for the two-dimensional case $q = 2$. Just as FCMdd, RGKMdd requires that only data points can be cluster centers, $V \subset X$. Therefore, the Euclidean distances between points and cluster centers

$$||x_k - v_i|| = \sqrt{(x_{k1} - v_{i1}) \cdot (x_{k1} - v_{i1}) + (x_{k2} - v_{i2}) \cdot (x_{k2} - v_{i2})} \tag{2.55}$$

can be simply taken from R. If we want to use local Mahalanobis distances as in the GK model, however, then the computation of the local covariance matrices S_i (2.22) needs to compute

$$\begin{pmatrix} (x_{k1} - v_{i1}) \cdot (x_{k1} - v_{i1}) & (x_{k1} - v_{i1}) \cdot (x_{k2} - v_{i2}) \\ (x_{k2} - v_{i2}) \cdot (x_{k1} - v_{i1}) & (x_{k2} - v_{i2}) \cdot (x_{k2} - v_{i2}) \end{pmatrix}, \tag{2.56}$$

and the entries of this matrix cannot be directly taken from R. They can, however, be locally computed by a triangulation approach (Lee, Slagle, and Blum, 1977). In addition to x_k and v_i we then consider another data point $x_j \in X$, where $x_j \neq v_i$ and $x_j \neq x_k$, and form a triangle (Figure 2.3), so we have

$$|x_{k1} - v_{i1}| = r_{ik} \cdot \cos \alpha_{ik}, \tag{2.57}$$

$$|x_{k2} - v_{i2}| = r_{ik} \cdot \sin \alpha_{ik}, \tag{2.58}$$

where, following the cosine theorem,

$$\cos \alpha = \frac{d_{ik}^2 + d_{ij}^2 - r_{jk}^2}{2 \cdot d_{ik} \cdot d_{ij}}. \tag{2.59}$$

To determine the signs of the distance vectors another data point $x_l \in X$ is chosen, where $x_l \neq v_i, x_l \neq x_k$, and $x_l \neq x_j$. Now the sign of $(x_{k1} - v_{i1}) \cdot (x_{k2} - v_{i2})$ is positive, if and only if

$$|\alpha + \beta - \pi| \geq \gamma, \tag{2.60}$$

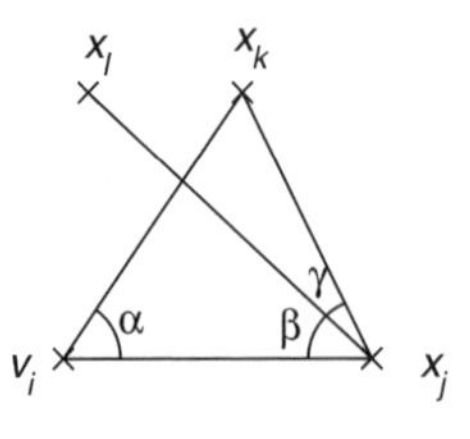

Figure 2.3 Triangulation for the (R)GKMdd model.

where the angles β and γ can be computed using the cosine theorem again,

$$\cos\beta = \frac{r_{jl}^2 + d_{ij}^2 - d_{il}^2}{2 \cdot r_{jl} \cdot d_{ij}}, \tag{2.61}$$

$$\cos\gamma = \frac{r_{jl}^2 + r_{jk}^2 - r_{kl}^2}{2 \cdot r_{jl} \cdot r_{jk}}. \tag{2.62}$$

2.6 RELATIONAL DATA INTERPRETED AS OBJECT DATA

In the previous sections we have presented clustering algorithms for object and for relational data. We assumed that object data are clustered by object clustering, and relational data are clustered by relational clustering. In this section, however, we follow the idea introduced in Katz, Runkler, and Heesche (2005) and apply object clustering models to relational data. This approach is illustrated for a simple example here. The suitability of this approach to real world data is proven in the orginal paper (Katz, Runkler, and Heesche, 2005). Consider a data-set with five objects, where objects o_1, o_2, and o_4 have distance zero to each other, objects o_3 and o_5 also have distance zero, and the distances between all the other pairs of objects are equal to one. This yields the following (crisp) relational data-set:

$$R = \begin{pmatrix} 0 & 0 & 1 & 0 & 1 \\ 0 & 0 & 1 & 0 & 1 \\ 1 & 1 & 0 & 1 & 0 \\ 0 & 0 & 1 & 0 & 1 \\ 1 & 1 & 0 & 1 & 0 \end{pmatrix}. \tag{2.63}$$

Let us now interpret R as an object dataset

$$\begin{aligned} X = \{&(0,0,1,0,1)^T, \\ &(0,0,1,0,1)^T, \\ &(1,1,0,1,0)^T, \\ &(0,0,1,0,1)^T, \\ &(1,1,0,1,0)^T\}. \end{aligned} \tag{2.64}$$

Obviously, for $c = 2$ we expect any object clustering algorithm to find the cluster centers

$$\begin{aligned} V = \{&(0,0,1,0,1)^T, \\ &(1,1,0,1,0)^T\}. \end{aligned} \tag{2.65}$$

Notice that $V = X$! This result corresponds to the partition matrix

$$U_s = \begin{pmatrix} 1 & 1 & 0 & 1 & 0 \\ 0 & 0 & 1 & 0 & 1 \end{pmatrix}, \tag{2.66}$$

which corresponds to the natural clusters $\{o_1, o_2, o_4\}$ and $\{o_3, o_4\}$. Notice the correspondence between the rows of U and the cluster centers that always occurs when binary relational data-sets are clustered! This example shows that clusters in relational data cannot only be found by relational clustering but also by object clustering. If object clustering is applied to relational data, then each row (or column) of the relational dataset is interpreted as a "*relation pattern*" that indicates the correspondence to all other objects. This leads to the correlation between the rows of the partition matrix and the cluster centers. Notice that there is no straightforward approach to apply relational clustering to object data!

□ non–kernelized	□ kernelized (k)
□ Euclidean	□ non–Euclidean (NE)
□ object	□ relational (R)
□ F □ P □ N	□ ACE
□ CM □ CMdd □ GK	□ GKMdd □ CV □ CE □ CS □ other
□ non–projection	□ projection (P)

Figure 2.4 A clustering form: check one box in each row to specify a clustering algorithm!

2.7 SUMMARY

In this chapter we have provided a survey of existing approaches for relational clustering. The various relational clustering models can be distinguished using the following six criteria:

- The *partition function* used might be derived from a given clustering model and the corresponding set of admissible partitions, or it might be specified by the user. The fuzzy, possibilistic, and noise clustering models lead to the fuzzy (RF), possibilistic (RP), and noise (RN) partition functions. In general, the alternating cluster estimation (RACE) scheme allows arbitrary functions such as Gaussian or triangular functions to be used as partition functions.
- *Non-Euclidean* relational data might lead to problems in relational clustering. These problems can be avoided by applying a β-spread transformation to the relational data until they become Euclidean.
- *Kernelization* corresponds to another transformation of the relational data that can be interpreted as a transformation of the corresponding object data to a (much) higher-imensional space.
- *Nonlinear projection* combines clustering with projection. Objects that belong to the same cluster have the same distance in the original as in the projected space. Nonlinear projection not only provides a cluster partition but also object data that correspond to the original relational data.
- Various cluster *prototypes* can be used in relational clustering, for example, points or C-means (CM), c-medoids (CMdd), points with local covariance matrices or Gustafson–Kessel (GK) prototypes, Gustafson–Kessel medoids (GKMdd), c-varieties (CV), c-elliptotypes (CE), or c-shells (CS).
- *Relational* clustering models were designed for relational data, but even object clustering algorithms may be used for clustering relational data.

Using this taxonomy, a clustering algorithm can be simply specified using the check box form shown in Figure 2.4. For example, if you tick the second box in rows 1 through 3, and the first box in row 4 through 6, then you obtain kNERFCM. This check box form allows you to specify $2 \cdot 2 \cdot 2 \cdot 4 \cdot 7 \cdot 2 = 448$ different relational clustering algorithms, and most of them may be useful. Notice that using this taxonomy the RFNP algorithm should actually be called RFCMP, but for convenience we keep the original name RFNP here.

2.8 EXPERIMENTS

In order to illustrate the different algorithms for relational fuzzy clustering described in this chapter we select 10 out of the 448 algorithms whose taxonomy was presented in the previous section and apply these 10 algorithms to the language concordance data introduced in Section 2.2. In particular, we consider the RACE, NERFCM, RFNP, RGKMdd, and FCM algorithms and their kernelized variants kRACE, kNERFCM, kRFNP, kRGKMdd, and kFCM. Kernelization was done using Gaussian kernels (2.47) using $\sigma = 10$. This maps the relational data from $[0, 10]$ to $[0, \sqrt{2 - 2 \cdot e^{-1}}] \approx [0, 1.124]$, so the effect is similar to a *normalization*. In each experiment the algorithm searched for $c = 3$ clusters. Remember that in the original paper by Johnson and Wichern (1992) the optimal *crisp* solution was given as $\{\{E, N, Da, Fr, I, Sp, P\}, \{Du, G\}, \{H, Fi\}\}$. Each algorithm was started with random initialization and then run for $t = 100$ steps. A quick analysis of the objective functions showed that for the language

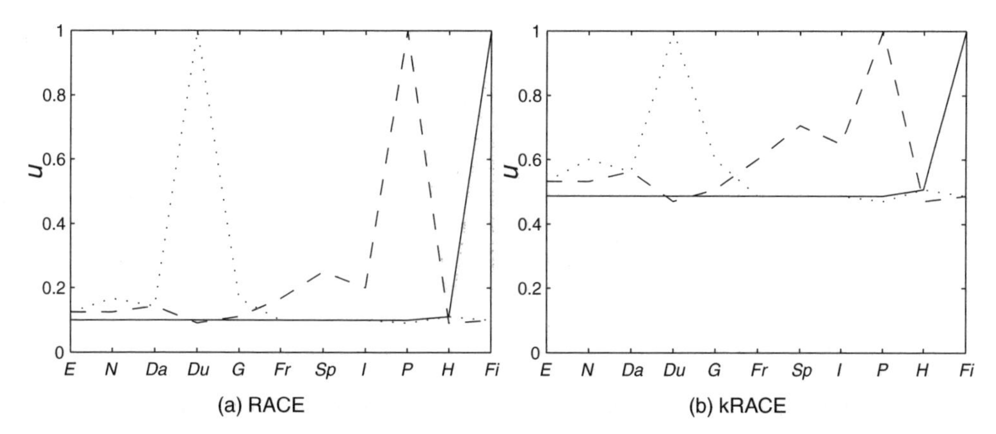

Figure 2.5 RACE and kRACE membership functions obtained for the language concordance data.

concordance data-set all of the considered algorithms had come close to convergence after 100 steps. Notice that this analysis could not be performed for (k)RACE, since the ACE algorithms usually do not possess an objective function to be minimized. Fuzziness was always set to $m = 2$, and the projection methods (k)FNP and (k)RGKMdd used two-dimensional projections.

Figure 2.5 shows the results for RACE and kRACE. The three membership functions are shown as solid, dashed, and dotted curves over the 11 objects (languages). RACE produces one cluster that is dominated by *Du* and has the second largest membership for *G*, one cluster that is dominated by *P* and has relatively large memberships for *Fr*, *Sp*, and *I*, and one cluster that is dominated by *Fi* and has the second largest membership for *H*. Notice that this correlates quite well with the originally proposed crisp partion above. Also notice that this is obviously not a *fuzzy* partition in the sense that the sum of memberships is equal to one for each datum, $U \notin M_{\text{fcn}}$ with M_{fcn} as in (2.11). The kRACE memberships in Figure 2.5(b) are very similar to the RACE memberships in Figure 2.5(a), but the memberships are almost linearly transformed from $[0, 1]$ to $[0.5, 1]$. This effect is caused by the normalization of the memberships due to kernelization.

Figure 2.6 shows the results for NERFCM and kNERFCM. NERFCM produces one cluster that mainly includes *E*, *N*, and *Da* and one cluster that mainly includes *Fr*, *Sp*, *I*, and *P*, so these two clusters correspond to the big crisp cluster reported above. The third cluster basically contains the remaining objects (language) and thus corresponds to the two small crisp clusters. The kNERFCM membership

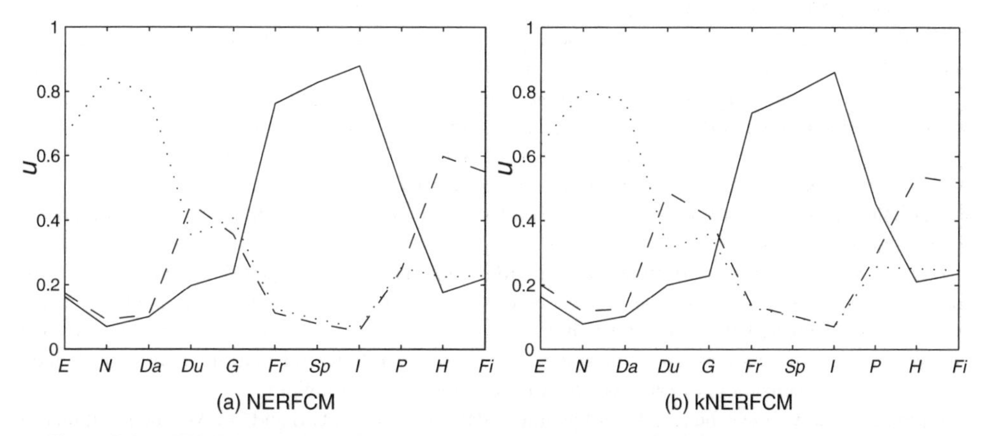

Figure 2.6 NERFCM and kNERFCM membership functions obtained for the language concordance data.

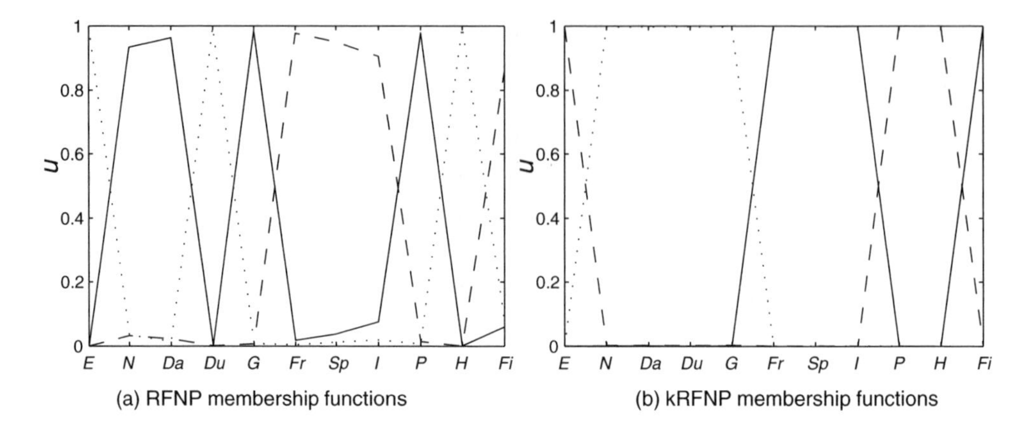

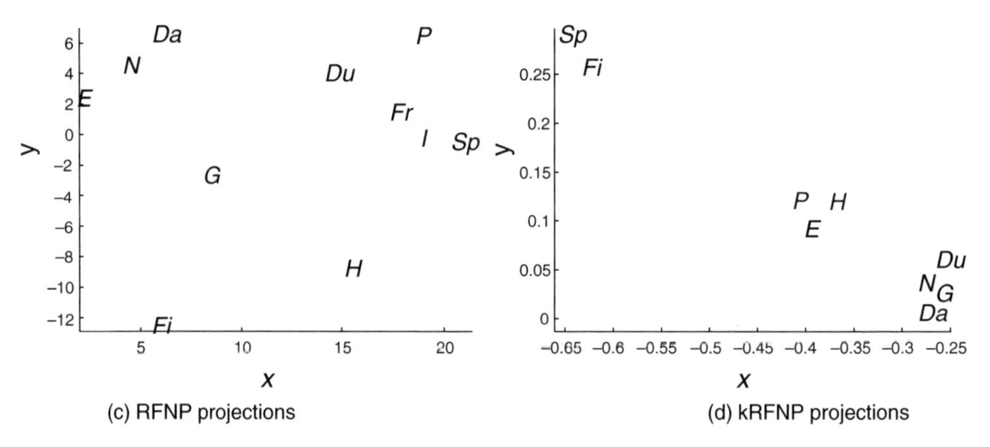

Figure 2.7 RFNP and kRFNP results obtained for the language concordance data.

functions are almost the same as the NERFCM membership functions, so for $\sigma = 10$, kernelization does not change the NERFCM results much.

Figure 2.7 shows the results for RFNP and kRFNP. The RFNP and kRFNP partitions in Figure 2.7(a) and (b), respectively, are almost crisp and do not match the crisp clusters presented in the original work. However, if we look at the projections produced by RFNP in Figure 2.7(c), then we can observe a meaningful arrangement of the objects *E*, *N*, and *Da* at the top left corner, *Fr*, *I*, *Sp*, and *P* at the top-right corner, and the remaining objects distributed around these, with *Fi* and *H* furthest away. This correlates well with the (k)NERFCM results. In the projections produced by kRFNP in Figure 2.7(d) the objects are moved much closer together, so apparently the effect of kernelization is a higher concentration of the RFNP object projections.

Figure 2.8 shows the results for RGKMdd and kRGKMdd. The RGKMdd partition is crisp and does not match our expectations. The reason is that one of the local covariance matrices becomes degenerate because it is not covered by sufficient data points. This is a general problem with all members of the GK family when applied to very small data-sets. However, the kRGKMdd partition looks much more reasonable, so kernelization seems to be able to overcome this GK problem.

Figure 2.9 finally shows the results for the object–data algorithms FCM and kFCM. The partitions are very similar to the partitions that were obtained by RFCM and kRFCM which corroborates the claim in

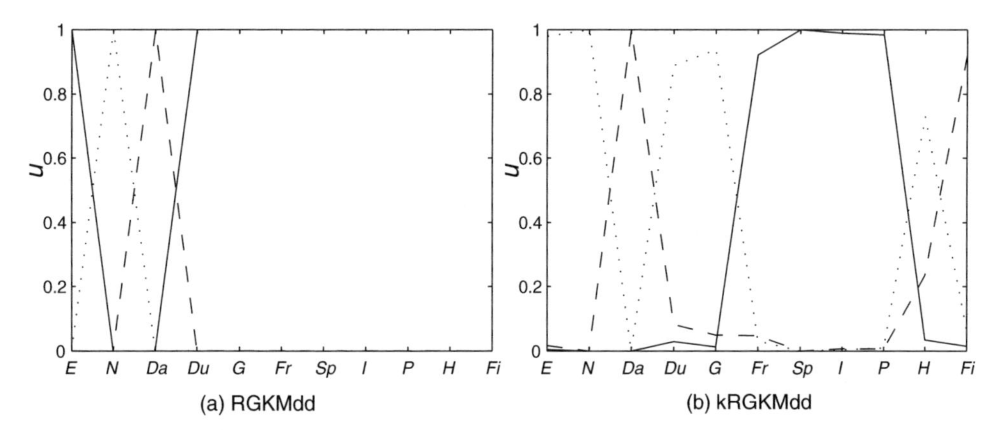

Figure 2.8 RGKMdd and kRGKMdd membership functions obtained for the language concordance data.

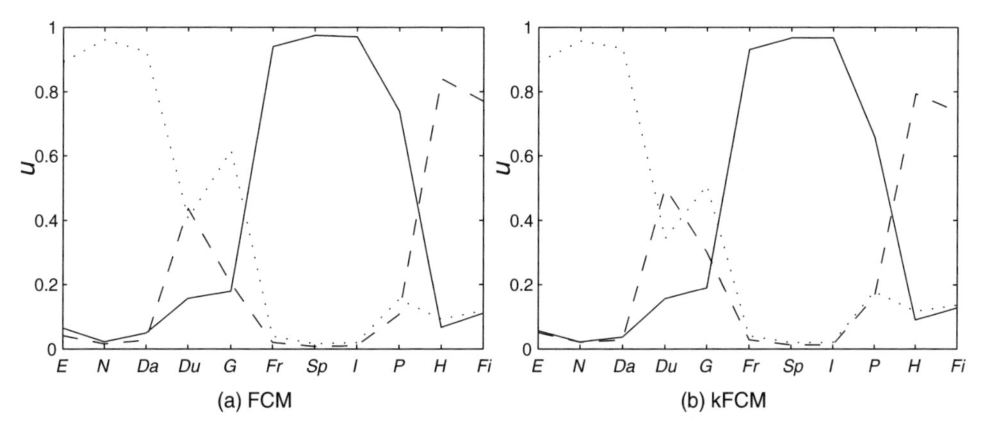

Figure 2.9 FCM and kFCM membership functions obtained for the language concordance data.

Katz, Runkler, and Heesche (2005) that object data clustering can produce reasonable results even in relational clustering.

2.9 CONCLUSIONS

In this chapter we have presented a multitude of different algorithms for clustering relational data and discussed the specific characteristics of some of these. Which of these algorithms is most appropriate for a given application is usually a highly application-specific question. Here are some general guidelines for selecting a clustering model:

- In general it is useful to start with a simple clustering model and then successively try more complicated clustering models. This means that in the beginning the checks in Figure 2.4 should be as far left as possible, and only after some experiments should be moved to the right.
- If the data are very noisy, then possibilistic (P), noise (N), or other ACE models should be preferred.
- If the data contain outliers, possibilistic (P), noise (N), or other ACE models might be more suitable that fuzzy (F) models. Moreover, kernelization is recommendable, since it clips high distances.

- If the data come from a non-Euclidean process, then the β transform should be used, i.e., a non-Euclidean (NE) version.
- If the clusters in the data are assumed to be non-hyperspherical, then there are several ways to exploit this information: if the geometrical cluster shape is explicitly known, then the corresponding cluster prototypes should be used, such as c-varieties (CV), c-elliptotypes (CE), or c-shells (CS). If the clusters are roughly (but not exactly) hyperspherical, then local covariance matrices as in the Gustafson–Kessel (GK) model are recommendable. If no information about the cluster shape is available, then kernelization or projection might be a good choice.

REFERENCES

Ball, G.B. and Hall, D.J. (1965) Isodata, an iterative method of multivariate analysis and pattern classification *IFIPS Congress*.

Bezdek, J.C. (1981) *Pattern Recognition with Fuzzy Objective Function Algorithms*. Plenum Press, New York.

Bezdek, J.C. and Hathaway, R.J. (1987) 'Clustering with relational C-means partitions from pairwise distance data'. *International Journal of Mathematical Modelling* **8**, 435–439.

Bezdek, J.C. and Hathaway, R.J. (1994) 'Optimization of fuzzy clustering criteria using genetic algorithms' *IEEE Conference on Evolutionary Computation, Orlando*, vol. 2, pp. 589–594.

Bezdek, J.C., Coray, C., Gunderson, R. and Watson, J. (1981a) 'Detection and characterization of cluster substructure', I. Linear structure: Fuzzy c-lines. *SIAM Journal on Applied Mathematics* **40**(2), 339–357.

Bezdek, J.C., Coray, C., Gunderson, R. and Watson, J. (1981b) 'Detection and characterization of cluster substructure', II. Fuzzy c-varieties and convex combinations thereof. *SIAM Journal on Applied Mathematics* **40**(2), 358–372.

Davé, R.N. (1990) 'Fuzzy shell clustering and application to circle detection in digital images'. *International Journal on General Systems* **16**, 343–355.

Davé, R.N. (1991) 'Characterization and detection of noise in clustering'. *Pattern Recognition Letters* **12**, 657–664.

Filev, D.P. and Yager, R.R. (1991) 'A generalized defuzzification method via BAD distributions'. *International Journal of Intelligent Systems* **6**, 687–697.

Girolami, M. (2002) 'Mercer kernel-based clustering in feature space'. *IEEE Transactions on Neural Networks* **13**, 780–784.

Gustafson, E.E. and Kessel, W.C. (1979) 'Fuzzy clustering with a covariance matrix' *IEEE International Conference on Decision and Control, San Diego*, pp. 761–766.

Hathaway, R.J. and Bezdek, J.C. (1994) 'NERF C-means: Non-Euclidean relational fuzzy clustering'. *Pattern Recognition* **27**, 429–437.

Hathaway, R.J. and Bezdek, J.C. (1995) 'Optimization of clustering criteria by reformulation'. *IEEE Transactions on Fuzzy Systems* **3**(2), 241–245.

Hathaway, R.J., Davenport, J.W. and Bezdek, J.C. (1989) 'Relational duals of the C-means algorithms'. *Pattern Recognition* **22**, 205–212.

Hathaway, R.J., Huband, J.M. and Bezdek, J.C. (2005) 'Kernelized non-Euclidean relational fuzzy C-means algorithm' *IEEE International Conference on Fuzzy Systems*, pp. 414–419, Reno.

Johnson, R.A. and Wichern, D.W. (1992) *Applied Multivariate Statistical Analysis* Prentice Hall, Englewood Cliffs, NJ, USA, pp. 582–589.

Katz C, Runkler, T.A. and Heesche, K. (2005) 'Fuzzy clustering using similarity measures: Clustering relational data by object data methods' *GMA/GI Workshop Fuzzy Systems and Computational Intelligence, Dortmund, Germany*, pp. 46–58.

Krishnapuram, R. and Keller, J.M. (1993) 'A possibilistic approach to clustering'. *IEEE Transactions on Fuzzy Systems* **1**(2), 98–110.

Krishnapuram R, Joshi A, Nasraoui, O. and Yi, L. (2001) 'Low-complexity fuzzy relational clustering algorithms for web mining'. *IEEE Transactions on Fuzzy Systems* **9**(4), 595–607.

Lee, R.C.T., Slagle, J.R. and Blum, H. (1977) 'A triangulation method for the sequential mapping of points from n-space to two-space'. *IEEE Transactions on Computers* **26**(3), 288–292.

Mercer, J. (1909) 'Functions of positive and negative type and their connection with the theory of integral equations'. *Philosophical Transactions of the Royal Society A* **209**, 415–446.

Müller, K.R. *et al.* (2001) 'An introduction to kernel-based learning algorithms'. *IEEE Transactions on Neural Networks* **12**, 181–201.

Pal, N.R., Eluri, V.K. and Mandal, G.K. (2002) 'Fuzzy logic approaches to structure preserving dimensionality reduction'. *IEEE Transactions on Fuzzy Systems* **10**(3), 277–286.

Powell, M.J.D. (1985) 'Radial basis functions for multi–variable interpolation: a review' *IMA Conference on Algorithms for Approximation of Functions and Data*, pp. 143–167, Shrivenham, UK.

Runkler, T.A. (1997) 'Selection of appropriate defuzzification methods using application specific properties'. *IEEE Transactions on Fuzzy Systems* **5**(1), 72–79.

Runkler, T.A. (2003) 'Fuzzy nonlinear projection' *IEEE International Conference on Fuzzy Systems*, St. Louis, USA.

Runkler, T.A. (2005a) 'Ant colony optimization of clustering models'. *International Journal of Intelligent Systems* **20**(12), 1233–1261.

Runkler, T.A. (2005b) 'Relational Gustafson Kessel clustering using medoids and triangulation' *IEEE International Conference on Fuzzy Systems*, pp. 73–78, Reno.

Runkler, T.A. and Bezdek, J.C. (1997) 'Living clusters: An application of the El Farol algorithm to the fuzzy C-means model' *European Congress on Intelligent Techniques and Soft Computing*, pp. 1678–1682, Aachen, Germany.

Runkler, T.A. and Bezdek, J.C. (1998) 'RACE: Relational alternating cluster estimation and the wedding table problem' In *Fuzzy-Neuro-Systems '98, München* (ed. Brauer W), vol. 7 of *Proceedings in Artificial Intelligence*, pp. 330–337.

Runkler, T.A. and Bezdek, J.C. (1999a) 'Alternating cluster estimation: A new tool for clustering and function approximation'. *IEEE Transactions on Fuzzy Systems* **7**(4), 377–393.

Runkler, T.A. and Bezdek, J.C. (1999b) 'Function approximation with polynomial membership functions and alternating cluster estimation'. *Fuzzy Sets and Systems* **101**(2), 207–218.

Runkler, T.A. and Glesner, M. (1993) 'Defuzzification with improved static and dynamic behavior: Extended center of area' *European Congress on Intelligent Techniques and Soft Computing*, pp. 845–851, Aachen, Germany.

Runkler, T.A. and Katz, C. (2006) 'Fuzzy clustering by particle swarm optimization' *IEEE International Conference on Fuzzy Systems*, Vancouver, Canada.

Sammon, J.W. (1969) 'A nonlinear mapping for data structure analysis'. *IEEE Transactions on Computers* **C-18**(5), 401–409.

Schölkopf, B. and Smola, A. (2002) *Learning with Kernels*. MIT Press, Cambridge, USA.

Schölkopf, B., Smola, A. and Müller, K.R. (1998) 'Nonlinear component analysis as a kernel eigenvalue problem'. *Neural Computation* **10**, 1299–1319.

Sneath, P. and Sokal, R. (1973) *Numerical Taxonomy*. Freeman, San Francisco, USA.

Wu, Z.D., Xie, W.X. and Yu, J.P. (2003) 'Fuzzy C-means clustering algorithm based on kernel method' *International Conference on Computational Intelligence* and Multimedia Applications, pp. 49–54, Xi'an, China.

Yager, R.R. and Filev, D.P. (1993) SLIDE: 'A simple adaptive defuzzification method'. *IEEE Transactions on Fuzzy Systems* **1**(1), 69–78.

Zhang, D.Q. and Chen, S.C. (2002) 'Fuzzy clustering using kernel method' *International Conference on Control and Automation*, pp. 123–127.

Zhang, D.Q. and Chen, S.C. (2003a) 'Clustering incomplete data using kernel-based fuzzy C-means algorithm'. *Neural Processing Letters* **18**, 155–162.

Zhang, D.Q. and Chen, S.C. (2003b) 'Kernel-based fuzzy and possibilistic C-means clustering' *International Conference on Artificial Neural Networks*, pp. 122–125, Istanbul, Turkey.

Zhang, R. and Rudnicky, A. (2002) 'A large scale clustering scheme for kernel k-means' *International Conference on Pattern Recognition*, pp. 289–292, Quebec, Canada.

3
Fuzzy Clustering with Minkowski Distance Functions

Patrick J.F. Groenen, Uzay Kaymak, and Joost van Rosmalen

Econometric Institute, Erasmus University Rotterdam, Rotterdam, The Netherlands

3.1 INTRODUCTION

Since Ruspini (1969) first proposed the idea of fuzzy partitions, fuzzy clustering has grown to be an important tool for data analysis and modeling. Especially after the introduction of the fuzzy c-means algorithm (Bezdek, 1973; Dunn, 1973), objective function-based fuzzy clustering has received much attention from the scientific community as well as the practitioners of fuzzy set theory (Baraldi and Blonda, 1999a,b; Bezdek and Pal, 1992; Höppner, Klawonn, Kruse and Runkler, 1999; Yang, 1993). Consequently, fuzzy clustering has been applied extensively for diverse tasks such as pattern recognition (Santoro, Prevete, Cavallo, and Catanzariti, 2006), data analysis (D'Urso, 2005), data mining (Crespo and Weber, 2005), image processing (Yang, Zheng and Lin, 2005), and engineering systems design (Sheu, 2005). Objective function-based fuzzy clustering has also become one of the key techniques in fuzzy modeling, where it is used for partitioning the feature space from which the rules of a fuzzy system can be derived (Babuška, 1998).

In general, objective function-based fuzzy clustering algorithms partition a data-set into overlapping groups by minimizing an objective function derived from the distance between the cluster prototypes and the data points (or objects). The clustering results are largely influenced by how this distance is computed, since it determines the shape of the clusters. The success of fuzzy clustering in various applications may depend very much on the shape of the clusters. As a result, there is a significant amount of literature on fuzzy clustering, which is aimed at investigating the use of different distance functions in fuzzy clustering, leading to different cluster shapes.

One way of influencing the shape of the clusters is to consider prototypes with a geometric structure. The fuzzy c-varieties (FCV) algorithm uses linear subspaces of the clustering space as prototypes (Bezdek, Coray, Gunderson, and Watson, 1981a), which is useful for detecting lines and other linear structures in the data. The fuzzy c-elliptotypes (FCE) algorithm takes convex combinations of fuzzy

Advances in Fuzzy Clustering and its Applications Edited by J. Valente de Oliveira and W. Pedrycz

c-varieties prototypes with fuzzy c-means prototypes to obtain localized clusters (Bezdek, Coray, Gunderson, and Watson, 1981b). Kaymak amd Setnes (2002) proposed using volumes in the clustering space as the cluster prototypes. Liang, Chou, and Han (2005) introduced a fuzzy clustering algorithm that can also deal with fuzzy data.

Another way for influencing the shape of the clusters is modifying the distance measure that is used in the objective function. Distances in the well known fuzzy c-means algorithm of Bezdek (1973) are measured by the squared Euclidean distance. Gustafson and Kessel (1979) use the quadratic Mahanalobis norm to measure the distance. Jajuga (1991) proposed using the L_1-distance and Bobrowski and Bezdek (1991) also used the L_∞-distance. Bargiela and Pedrycz (2005) applied the L_∞-distance to model granular data. Further, Hathaway, Bezdek, and Hu (2000) studied the Minkowski semi-norm as the dissimilarity function.

In this chapter, we consider fuzzy clustering with the more general case of the Minkowski distance and the case of using a root of the squared Minkowski distance. The Minkowski norm provides a concise, parametric distance function that generalizes many of the distance functions used in the literature. The advantage is that mathematical results can be shown for a whole class of distance functions, and the user can adapt the distance function to suit the needs of the application by modifying the Minkowski parameter. By considering the additional case of the roots of the squared Minkowski distance, we introduce an extra parameter that can be used to control the behavior of the clustering algorithm with respect to outliers. This root provides an additional way of dealing with outliers, which is different from the "noise cluster" approach proposed in Dave (1991).

Our analysis follows the approach that Groenen and Jajuga (2001) introduced previously. Minimization of the objective function is partly done by iterative majorization. One of the advantages of iterative majorization is that it is a guaranteed descent algorithm, so that every iteration reduces the objective function until convergence is reached. The algorithm in Groenen and Jajuga (2001) was limited to the case of a Minkowski parameter between 1 and 2, that is, between the L_1-distance and the Euclidean distance. Here, we extend their majorization algorithm to any Minkowski distance with Minkowski parameter greater than (or equal to) 1. This extension also includes the case of the L_∞-distance. We also explore the behaviour of our algorithm with an illustrative example using real-world data.

The outline of the chapter is as follows. We expose the formalization of the clustering problem in Section 3.2. The majorizing algorithm for fuzzy c-means with Minkowski distances is given in Section 3.3, while the influence of a robustness parameter is considered in Section 3.4. We discuss in Section 3.5 the behavior of our algorithm by using an illustrative example based on empirical data concerning attitudes about the Internet. Finally, conclusions are given in Section 3.6.

3.2 FORMALIZATION

In this chapter, we focus on the fuzzy clustering problem that uses a root of the squared Minkowski distance. This problem can be formalized by minimizing the objective (or loss) function

$$L(\mathbf{F}, \mathbf{V}) = \sum_{i=1}^{n} \sum_{k=1}^{K} f_{ik}^{s} d_{ik}^{2\lambda}(\mathbf{V}) \tag{3.1}$$

under the constraints

$$\begin{aligned} &0 \leq f_{ik} \leq 1, \quad i = 1, \ldots, n \quad k = 1, \ldots, K \\ &\textstyle\sum_{k=1}^{K} f_{ik} = 1, \quad i = 1, \ldots, n \end{aligned} \tag{3.2}$$

where n is the number of objects, K is the number of fuzzy clusters, f_{ik} is the membership grade of object i in fuzzy cluster k, s is the weighting exponent larger than 1. The distance between object i given by the ith row of the $n \times m$ data matrix $\mathbf{X}$ and fuzzy cluster k of the $K \times m$ cluster coordinate matrix $\mathbf{V}$ is given by

$$d_{ik}^{2\lambda}(\mathbf{V}) = \left(\sum_{j=1}^{m} |x_{ij} - v_{kj}|^{p} \right)^{2\lambda/p}, \quad 1 \leq p \leq \infty, \quad 0 \leq \lambda \leq 1, \tag{3.3}$$

where λ is the root of the squared Minkowski distance $d_{ik}^{2\lambda}(\mathbf{V})$ with $1 \leq p \leq \infty$.

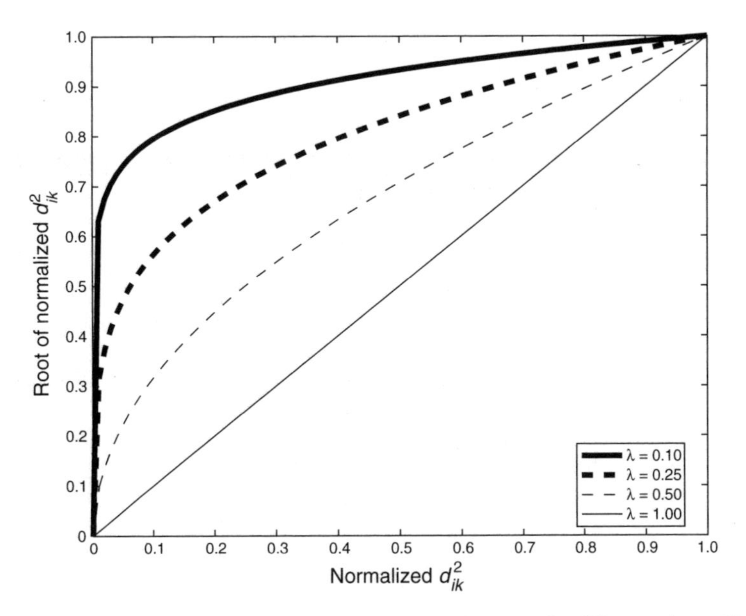

Figure 3.1 Root of the normalized squared Minkowski distance for different values of λ.

The introduction of the root λ allows the control of the loss function against outliers. Figure 3.1 shows how the root of the (normalized) squared Minkowski distance varies for different values of λ. For large λ, e.g., $\lambda = 1$, the difference between the large distance values and the small distance values is emphasized. Hence, outliers may dominate the loss function, whereas the loss function will be more robust if λ is small, because the relative difference between the large distance values and small distance values is reduced.

The use of Minkowski distances allows you to vary the assumptions of the shape of the clusters by varying p. The most often used value is $p = 2$, which assumes a circular cluster shape. Using $p = 1$ assumes that the clusters are in the shape of a (rotated) square in two dimensions or a diamond like shape in three or more dimensions. For $p = \infty$, the clusters are assumed to be in the form of a box with sides parallel to the axes. Both $p = 1$ and $p = \infty$ can be used in cases where the data structures have "boxy" shapes, that is, shapes with sharp "edges" (Bobrowski and Bezdek, 1991). A summary of combinations of λ and p and some properties of the distances are presented in Table 3.1 (taken from Groenen and Jajuga, 2001).

Groenen and Jajuga (2001) note that (3.1) has several known fuzzy clustering models as a special case. For example, for $p = 2$ and $\lambda = 1$, the important member of fuzzy ISO DATA, a well-known family of fuzzy clustering, is obtained that corresponds to squared Euclidean distances (while assuming the identity metric). A fuzzy clustering objective function that is robust against outliers can be obtained by choosing $\lambda = 1/2$ and $p = 1$ so that the L_1-norm is used. Note that this choice implicitly assumes a "boxy" shape of the clusters. A robust version of fuzzy clustering with a circular shape can be specified by $\lambda = 1/2$ and $p = 2$, which implies the unsquared Euclidean distance. Thus, λ takes care of robustness

Table 3.1 Special distances obtained by specific choice of λ and p and some of their properties.

p	λ	Distance	Assumed cluster shape	Robust
1	1.0	Squared L_1	Box/diamond	No
1	0.5	L_1	Box/diamond	Yes
2	1.0	Squared Euclidean	Circular	No
2	0.5	Unsquared Euclidean	Circular	Yes
∞	1.0	Squared dominance	Box	No
∞	0.5	Dominance	Box	Yes

issues and p of the shape of the clusters. Dodge and Rousson (1998) named the cluster centroids for $\lambda = 1/2$ and $p = 1$ "L_1-medians," for $\lambda = 1$ and $p = 1$ "L_1-means," for $\lambda = 1/2$ and $p = 2$ "L_2-medians," and for $\lambda = 1$ and $p = 2$ the well known "L_2-means."

3.3 THE MAJORIZING ALGORITHM FOR FUZZY C-MEANS WITH MINKOWSKI DISTANCES

Depending on the particular function, the minimization method of iterative majorization has some nice properties. The most important one is that in each iteration of the iterative majorization the loss function is decreased until this value converges. Such guaranteed descent methods are useful because no step in the wrong direction can be taken. Note that this property does not imply that a global minimum is found unless the function exhibits a special property such as convexity. Some general papers on iterative majorization are De Leeuw (1994), Heiser (1995), Lange, Hunter and Yang (2000), Kiers (2002), and Hunter and Lange (2004). An introduction can be found in Borg and Groenen (2005).

The majorization algorithm of Groenen and Jajuga (2001) worked for all $1 \leq p \leq 2$. Below we expand their majorization algorithm to the situation of all $p \geq 1$. Each iteration of their algorithm consists of two steps: (1) update the cluster memberships $\mathbf{F}$ for fixed centers $\mathbf{V}$ and (2) update $\mathbf{V}$ for fixed $\mathbf{F}$. For Step (2) we use majorization. Below, we start by explaining some basic ideas of iterative majorization. Then, the update of the cluster memberships is given. This is followed by some derivations for the update of the cluster centers $\mathbf{V}$ in the case of $1 \leq p \leq 2$. Then, the update is derived for $2 < p < \infty$ and a special update for the case of $p = \infty$.

3.3.1 Iterative Majorization

Iterative majorization can be seen as a gradient method with a fixed step size. However, iterative majorization can also be applied to functions that are at some points nondifferentiable. Central to iterative majorization is the use of an auxiliary function similar to the first-order Taylor expansion used as an auxiliary function in a gradient method and second-order expansion for Newton's method. The unique feature of the auxiliary function in iterative majorization – the so-called majorizing function – is that it touches the original function or is located above it. In contrast, the auxiliary functions of the gradient method or Newton's method can be partially below and above the original function.

Let the original function be presented by $\varphi(\mathbf{X})$, the majorizing function by $\hat{\varphi}(\mathbf{X}, \mathbf{Y})$, where $\mathbf{Y}$ is the current known estimate. Then, a majorizing function has to fulfil the following three requirements: (1) $\hat{\varphi}(\mathbf{X}, \mathbf{Y})$ is a more simple function in $\mathbf{X}$ than $\varphi(\mathbf{X})$, (2) it touches $\varphi(\mathbf{X})$ at the known supporting point $\mathbf{Y}$ so that $\varphi(\mathbf{Y}) = \hat{\varphi}(\mathbf{Y}, \mathbf{Y})$, and (3) $\hat{\varphi}(\mathbf{X}, \mathbf{Y})$ is never smaller than $\varphi(\mathbf{X})$, that is, $\varphi(\mathbf{X}) \leq \hat{\varphi}(\mathbf{X}, \mathbf{Y})$ for all $\mathbf{X}$. Often, the majorizing function is either linear or quadratic.

To see how a single iteration reduces $\varphi(\mathbf{X})$, consider the following. Let $\mathbf{Y}$ be some known point and let the minimum of the majorizing function $\hat{\varphi}(\mathbf{X}, \mathbf{Y})$ be given by $\mathbf{X}^+$. Note that for a majorizing algorithm to be sufficiently fast, it should be easy to compute $\mathbf{X}^+$. Because the $\hat{\varphi}(\mathbf{X}, \mathbf{Y})$ is always larger than or equal to the $\varphi(\mathbf{X})$, we must have $\varphi(\mathbf{X}^+) \leq \hat{\varphi}(\mathbf{X}^+, \mathbf{Y})$. This property is essential for the so-called sandwich inequality, that is, the chain

$$\varphi(\mathbf{X}^+) \leq \hat{\varphi}(\mathbf{X}^+, \mathbf{Y}) \leq \hat{\varphi}(\mathbf{Y}, \mathbf{Y}) = \varphi(\mathbf{Y}), \tag{3.4}$$

which proves that the update $\mathbf{X}^+$ never increases the original function. For the next iteration, we simply set $\mathbf{Y}$ equal to $\mathbf{X}^+$ and compute a new majorizing function. For functions that are bounded from below or are sufficiently constrained, the majorization algorithm always gives a convergent sequence of non-increasing function values, see, for example, Borg and Groenen (2005).

One property that we use here is that if a function consists of a sum of functions and each of these functions can be majorized, then the sum of the majorizing functions also majorizes the original functions. For example, suppose that $\varphi(\mathbf{X}) = \sum_i \varphi_i(\mathbf{X})$ and $\varphi_i(\mathbf{X}) \leq \hat{\varphi}_i(\mathbf{X}, \mathbf{Y})$ then $\varphi(\mathbf{X}) \leq \sum_i \hat{\varphi}_i(\mathbf{X}, \mathbf{Y})$.

3.3.2 Updating the Cluster Membership

For fixed cluster centers **V**, Groenen and Jajuga (2001) derive the update of the cluster memberships **F** as

$$f_{ik} = \frac{\left(d_{ik}^{2\lambda}(\mathbf{V})\right)^{-1/(s-1)}}{\sum_{l=1}^{K}\left(d_{il}^{2\lambda}(\mathbf{V})\right)^{-1/(s-1)}} \tag{3.5}$$

for fixed **V** and $s > 1$, see also Bezdek (1973). These memberships are derived by taking the Lagrangian function, setting the derivatives equal to zero, and solving the equations.

There are two special cases. The first one occurs if s is large. The larger s, the closer $-1/(s-1)$ approaches zero. As a consequence $[d_{ik}^{2\lambda}(\mathbf{V})]^{-1/(s-1)} \approx 1$ for all ik so that update (3.5) will yield $f_{ik} \approx 1/K$. Numerical accuracy can produce equal cluster memberships, even for not too large s, such as $s = 10$. If this happens for all f_{ik}, then all cluster centers collapse into the same point and the algorithm gets stuck. Therefore, in practical applications s should be chosen quite small, say $s \leq 2$. The second special case occurs if s approaches 1 from above. In that case, update (3.5) approaches the update for hard clustering, that is, setting

$$f_{ik} = \begin{cases} 1 & \text{if } d_{ik} = \min_l d_{il} \\ 0 & \text{if } d_{ik} \neq \min_l d_{il}, \end{cases} \tag{3.6}$$

where it is assumed that $\min_l d_{il}$ is unique.

3.3.3 Updating the Cluster Coordinates

We follow the majorization approach of Groenen and Jajuga (2001) for finding an update of the cluster coordinates **V** for fixed **F**. Our loss function $L(\mathbf{F}, \mathbf{V})$ may be seen as a weighted sum of the λth root of squared Minkowski distances. Because the weights f_{ik}^{s} are nonnegative, it is enough for now to consider $d_{ik}^{2\lambda}(\mathbf{V})$, the root of squared Minkowski distances. Let us focus on the root for a moment. Groenen and Heiser (1996) proved that for root λ of a, with $0 \leq \lambda \leq 1$, $a \geq 0$ and $b > 0$, the following majorization inequality holds:

$$a^{\lambda} \leq (1-\lambda)b^{\lambda} + \lambda b^{\lambda-1}a, \tag{3.7}$$

with equality if $a = b$. Using (3.7), we can obtain the majorizing inequality

$$d_{ik}^{2\lambda}(\mathbf{V}) \leq (1-\lambda)d_{ik}^{2\lambda}(\mathbf{W}) + \lambda d_{ik}^{2(\lambda-1)}(\mathbf{W})d_{ik}^{2}(\mathbf{V}), \tag{3.8}$$

where **W** is the estimate of **V** from the previous iteration and we assume for the moment that $d_{ik}(\mathbf{W}) > 0$. Thus, the root λ of a squared Minkowski distance can be majorized by a constant plus a positive weight times the squared Minkowski distance.

The next step is to majorize the squared Minkowski distance. To do so, we distinguish three cases: (a) $1 \leq p \leq 2$, (b) $2 < p < \infty$, and (c) $p = \infty$.

For the case of $1 \leq p \leq 2$, Groenen and Jajuga (2001) use Hölder's inequality to prove that

$$\begin{aligned} d_{ik}^{2}(\mathbf{V}) &\leq \frac{\sum_{j=1}^{m}(x_{ij}-v_{kj})^2|x_{ij}-w_{kj}|^{p-2}}{d_{ik}^{p-2}(\mathbf{V})} \\ &= \sum_{j=1}^{m} a_{ijk}^{(1\leq p\leq 2)}(x_{ij}-v_{kj})^2, \\ &= \sum_{j=1}^{m} a_{ijk}^{(1\leq p\leq 2)}v_{kj}^2 - 2\sum_{j=1}^{m} b_{ijk}^{(1\leq p\leq 2)}v_{kj} + c_{ik}^{(1\leq p\leq 2)}, \end{aligned} \tag{3.9}$$

where

$$a_{ijk}^{(1\leq p\leq 2)} = \frac{|x_{ij} - w_{kj}|^{p-2}}{d_{ik}^{p-2}(\mathbf{V})},$$
$$b_{ijk}^{(1\leq p\leq 2)} = a_{ijk}^{(1\leq p\leq 2)} x_{ij},$$
$$c_{ik}^{(1\leq p\leq 2)} = \sum_{j=1}^{m} a_{ijk}^{(1\leq p\leq 2)} x_{ij}^2.$$

For $p \geq 2$, (3.9) is reversed, so that it cannot be used for majorization. However, Groenen, Heiser, and Meulman (1999) have developed majorizing inequalities for squared Minkowski distances with $2 < p < \infty$ and $p = \infty$. We first look at $2 < p < \infty$. They proved that the Hessian of the squared Minkowski distance always has the largest eigenvalue smaller than $2(p-1)$. By numerical experimentation they even found a smaller maximum eigenvalue of $(p-1)2^{1/p}$ but they were unable to prove this. Knowing an upper bound of the largest eigenvalue of the Hessian is enough to derive a majorizing inequality if it is combined with the requirement of touching at the supporting point (that is, at this point the gradients of the squared Minkowski distance and the majorizing function must be equal and the same must hold for their function values).

This majorizing inequality can be derived as follows. For notational simplicity, we express the squared Minkowski distance as $d^2(\mathbf{t}) = (\sum_j |t_j|^p)^{2/p}$. The first derivative $\partial d^2(\mathbf{t})/\partial t_j$ can be expressed as $2t_j|t_j|^{p-2}/d^{p-2}(\mathbf{t})$. Knowing that the largest eigenvalue of the Hessian of $d^2(\mathbf{t})$ is bounded by $2(p-1)$, a quadratic majorizing function can be found (Groenen, Heiser, and Meulman, 1999) of the form

$$d^2(\mathbf{t}) \leq 4(p-1)\sum_{j=1}^{m} t_j^2 - 2\sum_{j=1}^{m} t_j b_j + c,$$

with

$$b_j = 4(p-1)u_j - \frac{1}{2}\frac{\partial d^2(\mathbf{u})}{\partial u_j} = u_j\left[4(p-1) - \frac{|u_j|^{p-2}}{d^{p-2}(\mathbf{u})}\right],$$
$$c = d^2(\mathbf{u}) + 4(p-1)\sum_{j=1}^{m} u_j^2 - \sum_{j=1}^{m} u_j \frac{\partial d^2(\mathbf{u})}{\partial u_j} = 4(p-1)\sum_{j=1}^{m} u_j^2 - d^2(\mathbf{u}),$$

and $\mathbf{u}$ the known current estimate of $\mathbf{t}$. Substituting $t_j = x_{ij} - v_{kj}$ and $u_j = x_{ij} - w_{kj}$ gives the majorizing inequality

$$\begin{aligned} d_{ik}^2(\mathbf{V}) \leq\ & 4(p-1)\sum_{j=1}^{m}(x_{ij} - v_{kj})^2 \\ & - 2\sum_{j=1}^{m}(x_{ij} - v_{kj})(x_{ij} - w_{kj})[4(p-1) - |x_{ij} - w_{kj}|^{p-2}/d_{ik}^{p-2}(\mathbf{W})] \\ & + 4(p-1)\sum_{j=1}^{m}(x_{ij} - w_{kj})^2 - d_{ik}(\mathbf{W}). \end{aligned}$$

Some rewriting yields

$$d_{ik}^2(\mathbf{V}) \leq a^{(2<p<\infty)}\sum_{j=1}^{m} v_{kj}^2 - 2\sum_{j=1}^{m} b_{ijk}^{(2<p<\infty)} v_{kj} + \sum_{j=1}^{m} c_{ijk}^{(2<p<\infty)}, \tag{3.10}$$

where

$$
\begin{aligned}
a^{(2<p<\infty)} &= 4(p-1), \\
b_{ijk}^{(2<p<\infty)} &= a^{(2<p<\infty)} w_{kj} - (x_{ij} - w_{kj})|x_{ij} - w_{kj}|^{p-2} / d_{ik}^{p-2}(\mathbf{W}), \\
c_{ik}^{(2<p<\infty)} &= a^{(2<p<\infty)} \sum_{j=1}^{m} w_{kj}^2 - d_{ik}^2(\mathbf{W}) + 2\sum_{j=1}^{m} x_{ij}(x_{ij} - w_{kj})|x_{ij} - w_{kj}|^{p-2} / d_{ik}^{p-2}(\mathbf{W}).
\end{aligned}
$$

If p gets larger, $a^{(2<p<\infty)}$ also becomes larger, thereby making the majorizing function steeper. As a result, the steps taken per iteration will be smaller. For the special case of $p = \infty$, Groenen, Heiser, and Meulman (1999) also provided a majorizing inequality. This one can be (much) faster than using (3.10) with a large p. It depends on the difference between the two largest values of $|x_{ij} - w_{kj}|$ over the different j.

Let us for the moment focus on $d^2(\mathbf{t})$ again. And let φ_j be an index that orders the values $|t_j|$ decreasingly, so that $|t_{\varphi_1}| \leq |t_{\varphi_2}| \leq \ldots \leq |t_{\varphi_m}|$. The majorizing function for $p = \infty$ becomes

$$
d^2(\mathbf{t}) \leq a \sum_{j=1}^{m} t_j^2 - 2\sum_{j=1}^{m} t_j b_j + c,
$$

with

$$
\begin{aligned}
a &= \begin{cases} \dfrac{|u_{\varphi_1}|}{|u_{\varphi_1}| - |u_{\varphi_2}|} & \text{if } |u_{\varphi_1}| - |u_{\varphi_2}| > \varepsilon, \\ \dfrac{\epsilon + |u_{\varphi_1}|}{\varepsilon} & \text{if} |u_{\varphi_1}| - |u_{\varphi_2}| \leq \varepsilon, \end{cases} \\
b_j &= \begin{cases} a \dfrac{|u_{\varphi_2}| u_j}{|u_j|} & \text{if } j = \varphi_1, \\ a u_j & \text{if } j \neq \varphi_1, \end{cases} \\
c &= d^2(\mathbf{u}) + 2\sum_j u_j b_j - a \sum_j u_j^2.
\end{aligned}
$$

Note that the definition of a for $|u_{\varphi_1}| - |u_{\varphi_2}| \leq \varepsilon$ takes care of ill conditioning, that is, values of a getting too large. Strictly speaking, majorization is not valid anymore, but for small enough ϵ the monotone convergence is retained.

Backsubstitution of $t_j = x_{ij} - v_{kj}$ and $u_j = x_{ij} - w_{kj}$ gives the majorizing inequality

$$
\begin{aligned}
d_{ik}^2(\mathbf{V}) &\leq a_{ik}^{(p=\infty)} \sum_j (x_{ij} - v_{kj})^2 - 2\sum_j (x_{ij} - v_{kj}) b_{ijk}^{(p=\infty)} + c_{ik}^{(p=\infty)} \\
&= a_{ik}^{(p=\infty)} \sum_j v_{kj}^2 - 2\sum_j v_{kj} b_{ijk}^{(p=\infty)} + c_{ik}^{(p=\infty)},
\end{aligned}
\tag{3.11}
$$

where

$$
\begin{aligned}
a_{ik}^{(p=\infty)} &= \begin{cases} \dfrac{|x_{i\varphi_1} - w_{k\varphi_1}|}{|x_{i\varphi_1} - w_{k\varphi_1}| - |x_{i\varphi_2} - w_{k\varphi_2}|} & \text{if } |x_{i\varphi_1} - w_{k\varphi_1}| - |x_{i\varphi_2} - w_{k\varphi_2}| > \varepsilon, \\ \dfrac{\varepsilon + |x_{i\varphi_1} - w_{k\varphi_1}|}{\varepsilon} & \text{if } |x_{i\varphi_1} - w_{k\varphi_1}| - |x_{i\varphi_2} - w_{k\varphi_2}| \leq \varepsilon, \end{cases} \\
b_{ijk}^{(p=\infty)} &= \begin{cases} a_{ik}^{(p=\infty)} \left[x_{ij} - \dfrac{|x_{i\varphi_2} - w_{k\varphi_2}|(x_{i\varphi_1} - w_{k\varphi_1})}{|x_{i\varphi_1} - w_{k\varphi_1}|} \right] & \text{if } j = \varphi_1, \\ a_{ik}^{(p=\infty)} w_{kj} & \text{if } j \neq \varphi_1, \end{cases} \\
c_{ik}^{(p=\infty)} &= d_{ik}^2(\mathbf{W}) - 2\sum_j b_{ijk}^{(p=\infty)} (x_{ij} - w_{kj}) - \sum_j a_{ik}^{(p=\infty)} w_{kj}^2 + 2\sum_j a_{ik}^{(p=\infty)} x_{ij}^2.
\end{aligned}
$$

Recapitulating, the loss function is a weighted sum of the root of the squared Minkowski distance. The root can be majorized by (3.8) that yields a function of squared Minkowski distances. For the case $1 \leq p \leq 2$, (3.9) shows how the squared Minkowski distance can be majorized by a quadratic function in $\mathbf{V}$ (see Figure (3.2), (3.10) shows how this can be done for $2 < p < \infty$ and (3.11) for $p = \infty$. These results can be combined to obtain the following majorizing function for $L(\mathbf{F}, \mathbf{V})$, that is,

$$L(\mathbf{F}, \mathbf{V}) \leq \lambda \sum_{j=1}^{m} \sum_{k=1}^{K} v_{kj}^2 \sum_{i=1}^{n} a_{ijk} - 2\lambda \sum_{j=1}^{m} \sum_{k=1}^{K} v_{kj} \sum_{i=1}^{n} b_{ijk} + c + \sum_{i=1}^{n} \sum_{k=1}^{K} c_{ik}, \tag{3.12}$$

where

$$a_{ijk} = \begin{cases} f_{ik}^s d_{ik}^{2(\lambda-1)}(\mathbf{W}) a_{ijk}^{(1 \leq p \leq 2)} & \text{if } 1 \leq p \leq 2, \\ f_{ik}^s d_{ik}^{2(\lambda-1)}(\mathbf{W}) a^{(2<p<\infty)} & \text{if } 2 < p < \infty, \\ f_{ik}^s d_{ik}^{2(\lambda-1)}(\mathbf{W}) a_{ik}^{(p=\infty)} & \text{if } p = \infty, \end{cases}$$

$$b_{ijk} = \begin{cases} f_{ik}^s d_{ik}^{2(\lambda-1)}(\mathbf{W}) b_{ijk}^{(1 \leq p \leq 2)} & \text{if } 1 \leq p \leq 2, \\ f_{ik}^s d_{ik}^{2(\lambda-1)}(\mathbf{W}) b_{ijk}^{(2<p<\infty)} & \text{if } 2 < p < \infty, \\ f_{ik}^s d_{ik}^{2(\lambda-1)}(\mathbf{W}) b_{ijk}^{(p=\infty)} & \text{if } p = \infty, \end{cases}$$

$$c_{ik} = \begin{cases} f_{ik}^s d_{ik}^{2(\lambda-1)}(\mathbf{W}) c_{ik}^{(1 \leq p \leq 2)} & \text{if } 1 \leq p \leq 2, \\ f_{ik}^s d_{ik}^{2(\lambda-1)}(\mathbf{W}) c_{ik}^{(2<p<\infty)} & \text{if } 2 < p < \infty, \\ f_{ik}^s d_{ik}^{2(\lambda-1)}(\mathbf{W}) c_{ik}^{(p=\infty)} & \text{if } p = \infty, \end{cases}$$

$$c = \sum_{i=1}^{n} \sum_{k=1}^{K} f_{ik}^s (1 - \lambda) d_{ik}^{2\lambda}(\mathbf{W}).$$

It is easily recognized that (3.12) is a quadratic function in the cluster coordinate matrix $\mathbf{V}$ that reaches its minimum for

$$v_{kj}^{+} = \frac{\sum_{i=1}^{n} b_{ijk}}{\sum_{i=1}^{n} a_{ijk}}. \tag{3.13}$$

3.3.4 The Majorization Algorithm

The majorization algorithm can be summarized as follows.

1. Given a data-set $\mathbf{X}$. Set $0 \leq \lambda \leq 1$, $1 \leq p \leq \infty$, and $s \geq 1$. Choose ε, a small positive constant.
2. Set the membership grades $\mathbf{F} = \mathbf{F}^0$ with $0 \leq f_{ik}^0 \leq 1$ and $\sum_{k=1}^{K} f_{ik}^0 = 1$ and the cluster coordinate matrix $\mathbf{V} = \mathbf{V}_0$. Compute $L_{\text{prev}} = L(\mathbf{F}, \mathbf{V})$.
3. Update $\mathbf{F}$ by (3.5) if $s > 1$ or by (3.6) if $s = 1$.
4. Set $\mathbf{W} = \mathbf{V}$. Update $\mathbf{V}$ by (3.13).
5. Stop if $(L_{\text{prev}} - L(\mathbf{F}, \mathbf{V}))/L(\mathbf{F}, \mathbf{V})) < \varepsilon$.
6. Set $L_{\text{prev}} = L(\mathbf{F}, \mathbf{V})$ and go to Step 3.

3.4 THE EFFECTS OF THE ROBUSTNESS PARAMETER λ

The parameter λ in the fuzzy clustering algorithm determines how robust the algorithm is with respect to outliers. To show the effects of λ on the solution, we generated a two-dimensional artificial data-set of 21

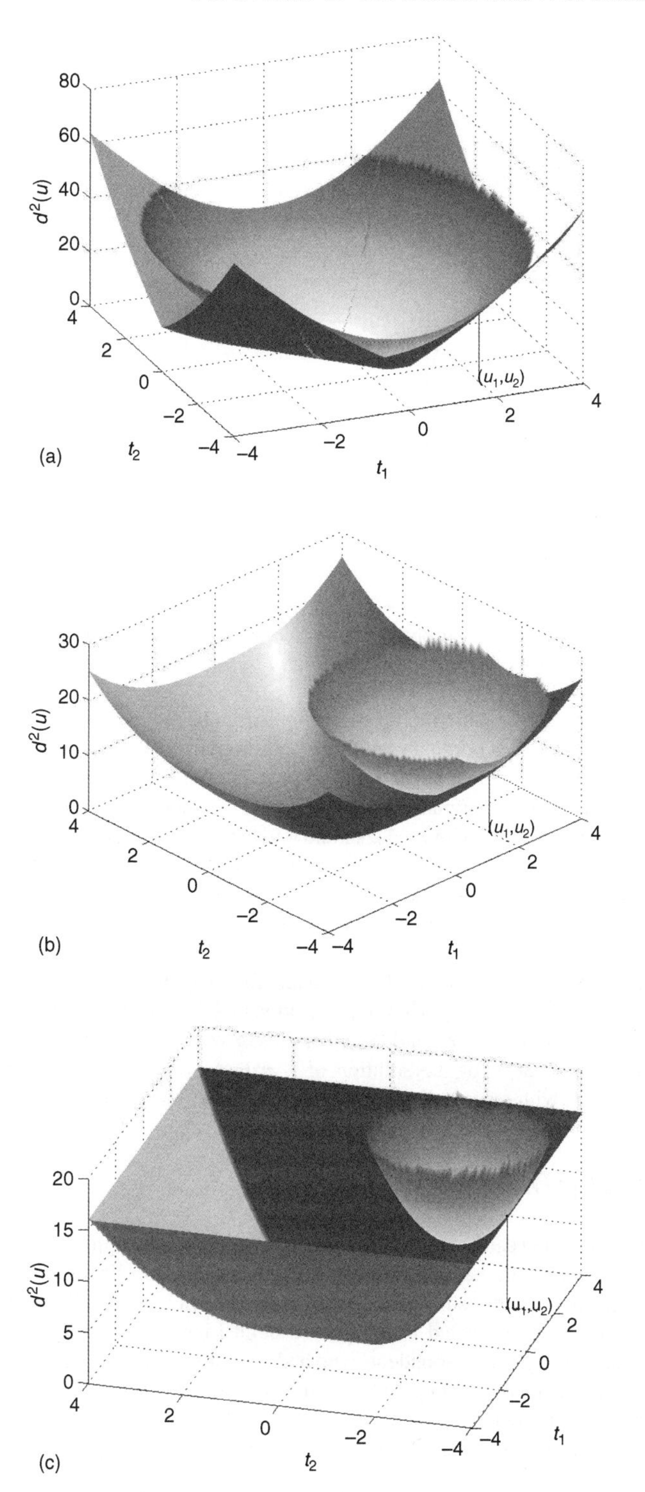

Figure 3.2 The original function $d^2(\mathbf{t})$ and the majorizing functions for $p = 1$, $p = 3$, and $p = \infty$ using supporting point $\mathbf{u} = [2, -3]'$.

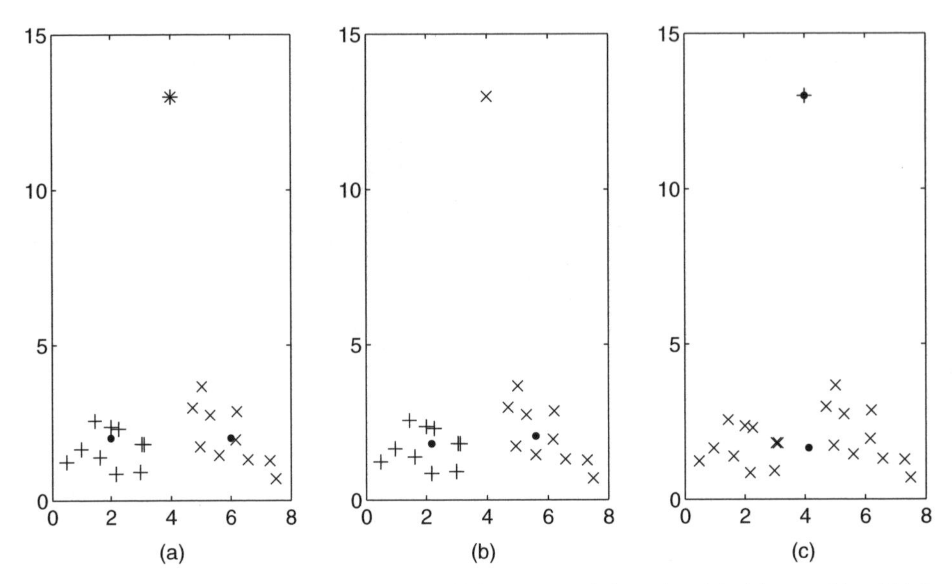

Figure 3.3 Results of fuzzy clustering algorithm in artificial data-set. (a) shows the original data-set. (b) and (c) show the results for $\lambda = 0.8$ and $\lambda = 1.0$, respectively. The cluster centers are marked using dots and the data points using $+$ and $\times$ signs, depending on which cluster they belong to.

observations as follows. Two clusters of 10 observations each have their cluster centers at coordinates (2,2) and (6,2), respectively. The coordinates of the observations in these clusters are normally distributed with a variance of 1 and mean equal to their cluster centers. The final observation is an outlier and is located at (4,13). The resulting configuration of observations is shown in Figure 3.3(a).

Any robust clustering algorithm with $K = 2$ should find the two clusters that were used to generate this data set. However, if the clustering algorithm is not robust, one cluster may have its center at the outlying observation and the other cluster may contain the remaining data points. To study for what values of λ our algorithm is robust, we ran our fuzzy clustering algorithm for $\lambda = 0.01, 0.02, \ldots, 0.99, 1.00$ using $p = 2$ and $s = 1.2$.

For values of $\lambda < 0.87$, we find that the cluster centers found by the algorithm are close to the centers used to construct the data, see Figure 3.3(b) for the solution of $\lambda = .8$. For $\lambda \geq 0.87$, we found one cluster center at the coordinates of the outlier, and the other cluster was located at the center of the remaining observations, see Figure 3.3(c) for the solution of $\lambda = 1$. As expected, we find that the clustering algorithm is more robust with respect to outliers if the parameter λ is set to a low value.

3.5 INTERNET ATTITUDES

To show how fuzzy clustering can be used in practice, we apply it to an empirical data-set. Our data-set is based on a questionnaire on attitudes toward the Internet.[1] It consists of evaluations of 22 statements about the Internet by respondents gathered around 2002 before the wide availability of broadband Internet access. The statements were evaluated using a seven-point Likert scale, ranging from 1 (completely disagree) to 7 (completely agree). Respondents who had a missing value on at least one of the statements were removed from the analysis yielding 193 respondents in our analysis.

The respondents are clustered using the fuzzy clustering algorithm to study their attitudes toward the Internet. We use $K = 3$. The convergence criterion ε of the majorization algorithm was set to 10^{-8}. The

---

[1]We would like to thank Peter Verhoef for making these data available. The data can be found at http://people.few.eur.nl/groenen/Data.

Table 3.2 Results of fuzzy clustering for Internet data-set using $K = 3$.

λ	s	p	$L(\mathbf{F}, \mathbf{V})$	Cluster volumes		
				Cluster 1	Cluster 2	Cluster 3
0.5	1.2	1	4087	1.315	1.411	1.267
0.5	1.2	2	1075	1.227	1.326	1.268
0.5	1.2	∞	421	1.355	1.408	1.341
0.5	1.5	1	2983	0.965	0.965	0.979
0.5	1.5	2	773	0.920	0.920	0.920
0.5	1.5	∞	310	0.973	0.973	0.973
1	1.2	1	103 115	1.281	1.363	1.236
1	1.2	2	7358	1.177	1.328	1.173
1	1.2	∞	1123	1.257	1.588	1.284
1	1.5	1	83 101	0.965	0.997	0.979
1	1.5	2	5587	0.920	0.920	0.920
1	1.5	∞	878	0.977	0.977	0.977

monotone convergence of the majorization algorithm generally leads to a local minimum. However, depending on the data and the different settings of p, s, and λ, several local minima may exist. Therefore, in every analysis, we applied 10 random starts and report the best one. We tried three different values of p $(1, 2, \infty)$ to examine the cluster shape, two values of s (1.2, 1.5) to study the sensitivity for the fuzziness parameter s, and two values for λ (0.5, 1.0) to check the sensitivity for outliers.

Table 3.2 shows some results for this data-set using different values for λ, s, and p. The final value of the loss function and the volumes of the three clusters are calculated in every instance. As there is no natural standardization for $L(\mathbf{F}, \mathbf{V})$, the values can only be used to check for local minima within a particular choice of λ, s, and p.

The labeling problem of clusters refers to possible permutations of the clusters among different runs. To avoid this problem, we took the $\mathbf{V}$ obtained by $\lambda = 1, p = 1$, and $s = 1.2$ as a target solution $\mathbf{V}^*$ and tried all permutation matrices $\mathbf{P}$ of the rows of $\mathbf{V}$ (with $\mathbf{V}^{(\text{Perm})} = \mathbf{PV}$) for other combinations of λ, p, and s and chose the one that minimizes the sum of the squared residuals

$$\sum_k \sum_j (v^*_{kj} - v^{(\text{Perm})}_{kj})^2 = ||\mathbf{V}^* - \mathbf{PV}||^2. \tag{3.14}$$

The permutation $\mathbf{P}$ that minimizes (3.14) is also applied to the cluster memberships, so that $\mathbf{F}^{(\text{Perm})} = \mathbf{FP}'$. By using this strategy, we assume that the clusters are the same among the different analyses.

To compare the size of the clusters in solutions for different settings of λ, s, and p, we do not want to use $L(\mathbf{F}, \mathbf{V})$ as it depends on these settings. Therefore, we define a volume measure for the size of a cluster. To do so, we need the cluster covariance matrix with elements

$$\mathbf{G}_k = \frac{\sum_{i=1}^n f_{ik}^s (\mathbf{x}_i - \mathbf{v}_k)'(\mathbf{x}_i - \mathbf{v}_k)}{\sum_{i=1}^n f_{ik}^s},$$

where $\mathbf{x}_i$ is the $1 \times j$ row vector of row i of $\mathbf{X}$ and $\mathbf{v}_k$ row i of $\mathbf{V}$. Then, as a measure of the volume of cluster k one can use $\det(\mathbf{G}_k)$. However, we take $\det(\mathbf{G}_k)^{1/m}$, which can be interpreted as the geometric mean of the eigenvalues of $\mathbf{G}_k$ and has the advantage that it is not sensitive to m. Note that $\mathbf{G}_k$ still depends on s so that it is only fair to compare cluster volumes for fixed s. If outliers are a problem in this data-set, we expect that the cluster volumes will be larger for the nonrobust case of $\lambda = 1$ than for the robust case of $\lambda = 0$.

Table 3.2 shows that for $s = 1.5$ the cluster volumes are nearly all the same with a slight difference among the clusters of $p = 1$. For $s = 1.2$, Cluster 2 is generally the largest and the other two have about the same size. The more robust setting of $\lambda = 0.5$ generally shows slightly larger clusters, but the effect does not seem large. Therefore, outliers do not seem to be a problem in this data-set.

To interpret the clusters, we have to look at $\mathbf{V}$. As it is impossible to show the clusters in a 22-dimensional space, they are represented by parallel coordinates (Inselberg, 1981, 1997). Every cluster

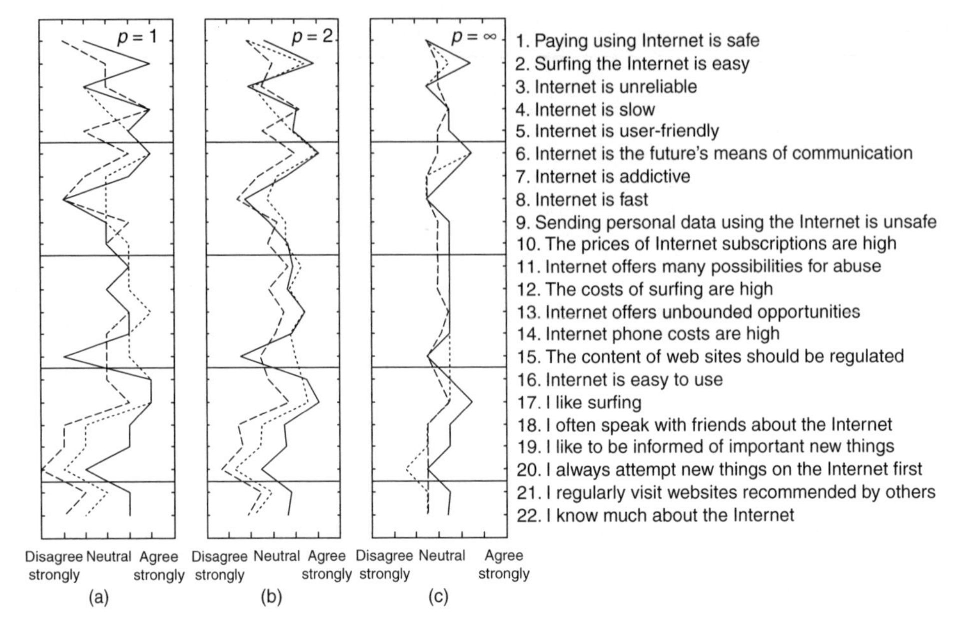

Figure 3.4 Parallel coordinates representation of clusters with $\lambda = 1$, $p = 1$, and $s = 1.2$. The lines correspond to Cluster 1 (solid line), Cluster 2 (dashed line), and Cluster 3 (dotted line).

k defines a line through the cluster centers v_{kj}, see Figure 3.4 for $s = 1.2$ and $\lambda = 1$. Note that the order of the variables is unimportant. This figure can be interpreted by considering the variables that have different scores for the clusters. The patterns for $p = 1, 2$ and ∞ are similar and $p = 1$ shows them the clearest.

For $p = 1$ and $\lambda = 1$, each cluster center is a (weighted) median of a cluster. Because all elements of the Internet data-set are integers, the cluster centers necessarily have integer values. Figure 3.4 shows the parallel coordinates for $p = 1$. The solid line represents Cluster 1 and is characterized by respondents saying that the Internet is easy, safe, addictive and who seem to form an active user community (positive answers to variables 16 to 22). However, the strongest difference of Cluster 1 to the others is given by their total rejection of regulation of content on the Internet. We call this cluster the experts. Cluster 2 (the dashed line) refers to respondents that are not active users (negative answers to variables 18 to 22), find the Internet not user-friendly, unsafe to pay, not addictive, and they are neutral on the issue of regulation of, the content of Web sites. This cluster is called the novices. Cluster 3 looks in some respects like Cluster 1 (surfing is easy, paying is not so safe) but those respondents do not find the Internet addictive, are neutral on the issue of the speed of the Internet connection, and seem to be not such active users as those of Cluster 1. They are mostly characterized by finding the costs of Internet high and allowing for some content regulation. This cluster represents the cost-aware Internet user.

As we are dealing with three clusters and the cluster memberships sum to one, they can be plotted in a triangular two-dimensional scatterplot – called a triplot – as in Figure 3.5. To reconstruct the fuzzy memberships from this plot, the following should be done. For Cluster 1, one has to project a point along a line parallel to the axis of Cluster 3 onto the axis of Cluster 1. We have done this with dotted lines for respondent 112 for the case of $p = 1$, $s = 1.2$, and $\lambda = 1$. We can read from the plot that this respondent has fuzzy memberships f_{i1} of about 0.20. Similarly, for Cluster 2, we have to draw a line horizontally (parallel to the axis of Cluster 1) and project it onto the axis of Cluster 2 showing f_{i2} of about 0.65. Finally, f_{i3} is obtained by projecting onto the axis of Cluster 3 along a line parallel to Cluster 2, yielding f_{i3} of about 0.15. In four decimals, these values are 0.2079, 0.6457, and 0.1464. Thus, a point located close to a corner implies that this respondent has been almost solely assigned to this cluster. Also, a point exactly in the middle of the triangle implies an equal memberships of 1/3 to all three clusters. Finally, points that are on a straight line from a corner orthogonal to a cluster axis have equal cluster memberships of two clusters. For

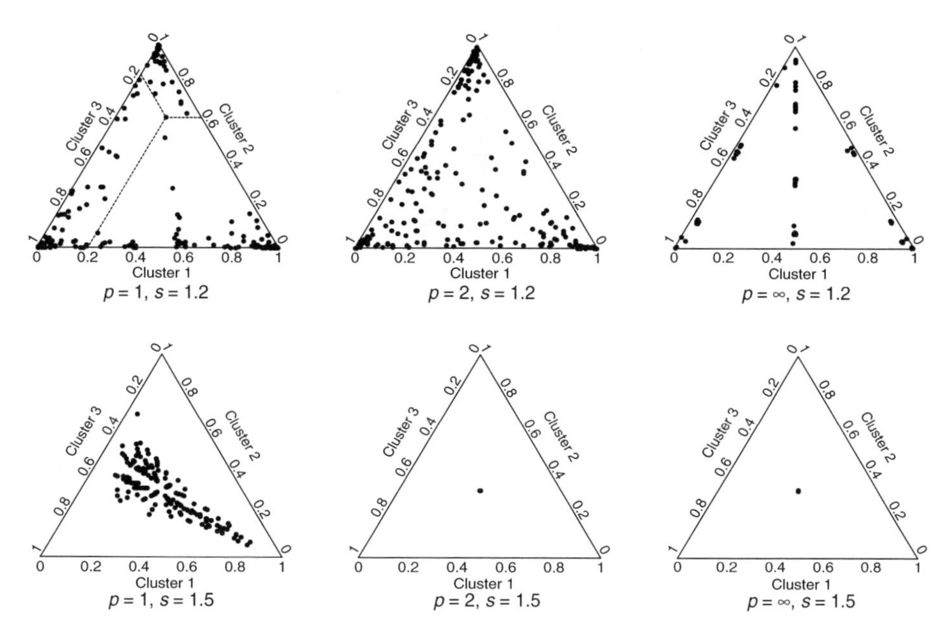

Figure 3.5 Triplot showing the cluster membership in **F** for each respondent for $s = 1.2, 1.5, \lambda = 1$, and $p = 1, 2, \infty$.

the case of $p = \infty$, Figure 3.5 shows a vertical line (starting in Cluster 2 and orthogonal to the Cluster 1 axis), implying that the memberships for Clusters 1 and 3 are the same for those respondents.

For the choice $s = 1.5$ and $p = 2$ or ∞, all clusters centers are in close proximity to each other in the center. In other words, all fuzzy memberships are about 1/3 and consequently the three cluster centers are the same. Therefore, $s = 1.5$ is too large for $p = 2$ or ∞. This finding is an indication of overlapping clusters. For a value of $s = 1.2$, the triplot for $p = 1$ shows more pronounced clusters because most of the respondents are in the corners. For $p = 2$ and $s = 1.2$, the memberships are more evenly distributed over the triangle although many respondents are still located in the corners. For $p = \infty$ and $s = 1.2$, some respondents are on the vertical line (combining equal memberships to Clusters 1 and 3 for varying membership of Cluster 2). The points that are located close to the Cluster 1 axis at 0.5 have a membership of 0.5 for Clusters 1 and 3, those close to 0.5 at the Cluster 2 axis have 0.5 for Clusters 1 and 2, those close to the Cluster 3 axis at 0.5 have 0.5 for Clusters 2 and 3.

For the robust case of $\lambda = 1/2$, the triplots of the fuzzy memberships are given in Figure 3.6. One of the effects of setting $\lambda = 1/2$ seems to be that the f_{ik} are more attracted to the center and, hence, respondents are less attracted to a single cluster than in the case of $\lambda = 1$. Again, for $s = 1.5$ and $p = 2$ and ∞, all clusters merge into one cluster and the parallel coordinates plots of the clusters would show a single line. For $s = 1.2$, the parallel coordinates plots of the clusters resemble Figure 3.5 reasonably well. For $s = 1.2$ and $p = 2$, the lines in the parallel coordinates plot are closer together than for $\lambda = 1$.

For this data-set, the clusters cannot be well separated because for a relatively small s of 1.5, the clusters coincide (except for $p = 1$). The cluster centers seem to be better separated when p is small, especially for $p = 1$. The fuzziness parameter s needs to be chosen small in this data-set to avoid clusters collapsing into a single cluster. The effect of varying λ seems to be that the cluster memberships are less extreme for $\lambda = 1/2$ than for $\lambda = 1$.

3.6 CONCLUSIONS

We have considered objective function based fuzzy clustering algorithms using a generalized distance function. In particular, we have studied the extension of the fuzzy c-means algorithm to the case of the

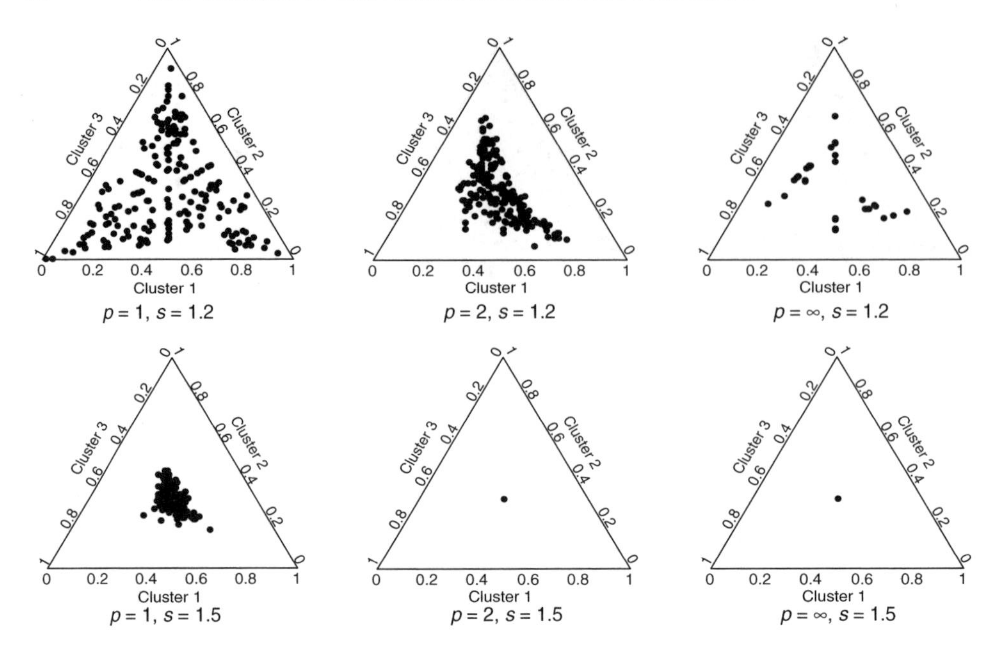

Figure 3.6 Triplot showing the cluster membership in **F** for each respondent for $s = 1.2, 1.5$, $\lambda = 0.5$, and $p = 1, 2, \infty$.

parametric Minkowski distance function and to the case of the root of the squared Minkowski distance function. We have derived the optimality conditions for the membership values from the Lagrangian function. For cluster centers, however, we have used iterative majorization to derive the optimality conditions. One of the advantages of iterative majorization is that it is a guaranteed descent algorithm, so that every iteration reduces the objective function until convergence is reached. We have derived suitable majorization functions for the distance function that we studied. Extending results from Groenen and Jajuga (2001), we have given a majorization algorithm for *any* Minkowski distance with Minkowski parameter greater than (or equal to) 1. This extension also included the case of the L_∞-distance and the roots of the squared Minkowski distance.

By adapting the Minkowski parameter p, the user influences the distance function to take specific cluster shapes into account. We have also introduced an additional parameter λ for computing the roots of the squared Minkowski distance. This parameter can be used to protect the clustering algorithm against outliers. Hence, more robust clustering results can be obtained.

We have illustrated some key aspects of the behavior of our algorithm using empirical data regarding attitudes about the Internet. With this particular data-set, we have observed extremely overlapping clusters, already with a fuzziness parameter of $s = 1.5$. This finding deviates from the general practice in fuzzy clustering, where this parameter is often selected equal to 2. Apparently, the choice of s and p has to be done with some care for a given data-set.

REFERENCES

Babuška, R. (1998) *Fuzzy Modeling for Control*. Kluwer Academic Publishers, Boston, MA.

Baraldi, A. and Blonda, P. (1999a) 'A survey of fuzzy clustering algorithms for pattern recognition' — part I. *IEEE Transactions on Systems, Man and Cybernetics, Part B* **29**(6), 778–785.

Baraldi, A. and Blonda, P. (1999b) 'A survey of fuzzy clustering algorithms for pattern recognition' – part II. *IEEE Transactions on Systems, Man and Cybernetics, Part B* **29**(6), 786–801.

Bargiela, A. and Pedrycz, W. (2005) 'A model of granular data: a design problem with the Tchebyschev FCM'. *Soft Computing* **9**(3), 155–163.

Bezdek, J.C. (1973) *Fuzzy mathematics in pattern classification* PhD thesis Cornell University Ithaca.

Bezdek, J.C. and Pal, S.K. (1992) *Fuzzy Models for Pattern Recognition.* IEEE Press, New York.

Bezdek, J.C., Coray, C., Gunderson, R. and Watson, J. (1981a) 'Detection and characterization of cluster substructure, I. linear structure: fuzzy *c*-lines'. *SIAM Journal of Applied Mathematics* **40**(2), 339–357.

Bezdek, J.C., Coray, C., Gunderson, R. and Watson, J. (1981b) 'Detection and characterization of cluster substructure, II. fuzzy *c*-varieties and convex combinations thereof'. *SIAM Journal of Applied Mathematics* **40**(2), 358–372.

Bobrowski, L. and Bezdek, J.C. (1991) '*c*-Means clustering with the l_1 and l_∞ norms'. *IEEE Transactions on Systems, Man and Cybernetics* **21**, 545–554.

Borg, I. and Groenen, P.J.F. (2005) *Modern Multidimensional Scaling: Theory and Applications* 2nd edn. Springer, New York.

Crespo, F. and Weber, R. (2005) 'A methodology for dynamic data mining based on fuzzy clustering'. *Fuzzy Sets and Systems* **150**(2), 267–284.

Dave, R.N. (1991) 'Characterization and detection of noise in clustering'. *Pattern Recognition Letters* **12**(11), 657–664.

De Leeuw, J. (1994) 'Block relaxation algorithms in statistics' In *Information Systems and Data Analysis* (ed. H.-H. Bock, Lenski, W. and Richter MM) pp. 308–324. Springer Berlin.

Dodge, Y. and Rousson, V. (1998) 'Multivariate L_1 mean' In *Advances in Data Science and Classification* (ed. Rizzi A, Vichi, M. and Bock H), pp. 539–546. Springer, Berlin.

Dunn, J. (1973) 'A fuzzy relative of the ISODATA process and its use in detecting compact, well-separated clusters'. *Journal of Cybernetics* **3**(3), 32–57.

D'Urso, P. (2005) 'Fuzzy clustering for data time arrays with inlier and outlier time trajectories'. *IEEE Transactions on Fuzzy Systems* **13**(5), 583–604.

Groenen, P.J.F. and Heiser, W.J. (1996) 'The tunneling method for global optimization in multidimensional scaling'. *Psychometrika* **61**, 529–550.

Groenen, P.J.F. and Jajuga, K. (2001) 'Fuzzy clustering with squared Minkowski distances'. *Fuzzy Sets and Systems* **120**, 227–237.

Groenen, P.J.F., Heiser, W.J. and Meulman, J.J. (1999) 'Global optimization in least-squares multidimensional scaling by distance smoothing'. *Journal of Classification* **16**, 225–254.

Gustafson, D.E. and Kessel, W.C. (1979) 'Fuzzy clustering with a fuzzy covariance matrix' *Proc. IEEE CDC*, pp. 761–766, San Diego, USA.

Hathaway, R.J., Bezdek, J.C. and Hu, Y. (2000) 'Generalized fuzzy *c*-means clustering strategies using l_p norm distances'. *IEEE Transactions on Fuzzy Systems* **8**(5), 576–582.

Heiser, W.J. (1995) *Convergent Computation by Iterative Majorization: Theory and Applications in Multidimensional Data Analysis* pp. 157–189. Oxford University Press, Oxford, UK.

Höppner, F., Klawonn, F., Kruse, R. and Runkler, T. (1999) *Fuzzy Cluster Analysis: Methods for Classification, Data Analysis and Image Recognition.* John Wiley & Sons, Inc., New York, USA.

Hunter, D.R. and Lange, K. (2004) 'A tutorial on MM algorithms'. *The American Statistician* **39**, 30–37.

Inselberg, A. (1981) '*N*-dimensional graphics, part I: Lines and hyperplanes. Technical Report G320-2711', IBM Los Angeles Scientific Center, Los Angeles (CA).

Inselberg, A. (1997) 'Multidimensional detective' *Proc. IEEE Symp. Information Visualization*, pp. 100–107.

Jajuga, K. (1991) 'L_1-norm based fuzzy clustering'. *Fuzzy Sets and Systems* **39**, 43–50.

Kaymak, U. and Setnes, M. (2002) 'Fuzzy clustering with volume prototypes and adaptive cluster merging'. *IEEE Transactions on Fuzzy Systems* **10**(6), 705–712.

Kiers, H.A.L. (2002) 'Setting up alternating least squares and iterative majorization algorithms for solving various matrix optimization problems'. *Computational Statistics and Data Analysis* **41**, 157–170.

Lange, K., Hunter, D.R. and Yang, I. (2000) 'Optimization transfer using surrogate objective functions'. *Journal of Computational and Graphical Statistics* **9**, 1–20.

Liang, G., Chou, T. and Han, T. (2005) 'Cluster analysis based on fuzzy equivalence relation'. *European Journal of Operational Research* **166**(1), 160–171.

Ruspini, E. (1969) 'A new approach to clustering'. *Information and Control* **15**, 22–32.

Santoro, M., Prevete, R., Cavallo, L. and Catanzariti, E. (2006) 'Mass detection in mammograms using Gabor filters and fuzzy clustering'. *Lecture Notes in Artificial Intelligence* **3849**, 334–343.

Sheu, H.B. (2005) 'A fuzzy clustering approach to real-time demand-responsive bus dispatching control'. *Fuzzy Sets and Systems* **150**(3), 437–455.

Yang, M.S. (1993) 'A survey of fuzzy clustering'. *Mathematical and Computer Modelling* **18**(11), 1–16.

Yang, Y, Zheng, C.X. and Lin, P. (2005) 'Fuzzy clustering with spatial constraints for image thresholding'. *Optica Applicata* **35**(4), 943–954.x

4 Soft Cluster Ensembles

Kunal Punera and Joydeep Ghosh

Department of Electrical and Computer Engineering, University of Texas at Austin, Texas, USA

4.1 INTRODUCTION

Cluster ensembles is a 'knowledge reuse' framework for combining multiple clusterings of a set of objects without accessing the original features of the objects. This problem was first proposed in Strehl and Ghosh (2002) where the authors applied it for improving the quality and robustness of clustering, and in distributed clustering. A related problem of consensus clustering also exists in the marketing literature (Kreiger and Green, 1999) where often a set of individuals is segmented in multiple ways based on different criteria (needs-based, demographics, etc.) and one is interested in obtaining a single, unified segmentation.

The idea of combining multiple models is well established in the classification and regression scenarios where it has led to impressive improvements in a wide variety of domains (Breiman, 1999; Freund and Schapire, 1996; Ghosh, 2002). Combining clusterings is, however, a more difficult problem than combining the results of multiple classifiers, since clusterings are invariant to cluster label permutations. In other words, all partitions of a set of objects that differ only in the cluster labeling are identical. As a result, before combining the clusterings one has to identify which clusters from different clusterings correspond to each other. This sub-problem of identifying cluster correspondences is further complicated by the fact that the number of clusters in the individual solutions might vary significantly. These differences, along with wide variations in the clustering algorithms and features of data used for underlying clustering algorithms, make solving cluster ensembles a very challenging problem. Even so, the ability to combine clusterings in an ensemble is very useful.

Cluster ensembles have been shown to be useful in many application scenarios. Some of the principal ones are

- *Knowledge reuse*. An important application of cluster ensembles is combining knowledge encoded in multiple clusterings. An example of this is exploiting the knowledge in legacy clusterings while re-clustering the data. We might not have access to the features that were originally used while creating the legacy clusterings; they might even have been created manually by a domain expert. Also, in many cases the number of clusters in the original data might have changed or new features might now be

Advances in Fuzzy Clustering and its Applications Edited by J. Valente de Oliveira and W. Pedrycz

available. In these cases, re-clustering all the data with the new features may not be possible. Cluster ensembles can be employed to combine multiple clusterings in these feature/object distributed scenarios (Ghosh, Strehl, and Merugu, 2002; Strehl and Ghosh, 2002).

- *Multi-view clustering.* A set of objects can be clustered multiple times using different attributes/criteria. For example, in marketing applications, customers can be segmented based on their needs, psychographic or demographic profiles, brand choices, etc.. Consensus clustering can be used to combine all such partitions into one, which is often easier to act on (Kreiger and Green, 1999).
- *Distributed computing.* In many applications, the data to be clustered is distributed over many sites, and data sharing is prohibited. In the case of distributed computing, communication costs make sharing all the data with a central site prohibitively expensive, but communicating clustering results is cheaper. In other cases, while sharing actual features of data might be prohibited because of privacy reasons, the sharing of clustering results might be permissible, as in Merugu and Ghosh (2003). Both these scenarios can be handled by locally clustering data present at each site, and then transferring only the clustering solutions to a central site. Cluster ensemble algorithms can then be used to combine these clusterings into a composite clustering at the central site.
- *Improved quality of results.* Each clustering algorithm has its own search biases and uses different types of features. Combining the results of multiple different clusterings algorithms could give improvements over their individual solutions, as the combined solution would take into account all their biases. It has been seen that using cluster ensembles to combine diverse clustering solutions leads to more accurate results on average (Hadjitodorov, Kuncheva, and Todorova, 2006; Kuncheva and Hadjitodorov, 2004).
- *Robust solutions.* Many clustering algorithms suffer from initialization problems, often finding local minima of their objective functions. The cluster ensembles framework can be used to alleviate these problems of unstable clustering results. Multiple runs of a clustering algorithm, obtained with different initializations or with different sets of features, can be combined in order to obtain a robust final solution (Fern and Brodley, 2003; Fred and Jain, 2002).

There have been several attempts to solve cluster ensembles in the recent past. Strehl and Ghosh (2002) proposed three graph-theoretic approaches for finding the consensus clustering. A bipartite graph partitioning based approach has been proposed by Fern and Brodley (2004). Topchy, Jain, and Punch (2004) proposed the use of a mixture of multinomial distributions to model the ensemble of labels along the lines of classical latent class analysis in marketing literature. Some of these approaches will be described in detail in Section 4.2. While these techniques are very varied in the algorithms they employ, there is a common thread that they only work with hard constituent clusterings. It is the goal of this chapter to investigate *soft* cluster ensembles.

4.1.1 Ensembles of Soft Clusterings

There are several clustering algorithms, such as EM (Dempster, Laird, and Rubin, 1977) and fuzzy c-means (Bezdek and Pal, 1992; Dunn, 1973), that naturally output soft partitions of data. A soft partition assigns a value for the degree of association of each instance to each output cluster. So instead of a label vector for all the instances, we have a matrix of values in which each instance is associated with every cluster with some membership value; often these values are the posterior probabilities and add up to one. In order to solve an ensemble formed of soft clusterings using one of the existing algorithms mentioned above, we would have to 'harden' the clusterings. This process involves completely assigning each instance to the cluster to which it is most associated. This results in the loss of the information contained in the uncertainties of the cluster assignments. This is especially true for application settings where underlying clustering algorithms access partial views of the data, such as in distributed data mining. A landmark work on 'collaborative' fuzzy clustering was done by Pedrycz (2002). The author considered a vertical partitioning scenario, and captured the collaboration between multiple partitionings via pairwise interaction coefficients. This resulted in an extended cost function to accommodate the collaboration effect in the optimization process. This approach is restricted in scope in many ways: each partition needs to have the same number of clusters; the difficult cluster correspondence problem is assumed to be already solved; and the distances between each point and its representative in each of the solutions need to be known. Despite these constraints, it was illustrated that, at least for simple two and three cluster problems, collaboration had a positive effect on cluster quality. This further motivates the present study, where we propose flexible frameworks for combining multiple soft clusterings directly without 'hardening' the individual solutions first. We introduce a new consensus function (ITK) based on the information-theoretic k-means algorithm (Dhillon, Mallela, and Kumar, 2003b) that is more efficient and effective than existing approaches. For evaluation purposes, we create a large number of ensembles of varying degrees of difficulty, and report clustering results achieved by the various existing and new algorithms on them. In order to objectively evaluate ITK we extend existing algorithms to operate on soft cluster ensembles as well.

4.1.2 Organization of this Chapter

In Section 4.2 we first define the *hard* cluster ensemble problem formally, and then go on to describe the various consensus functions that have been proposed in literature. The *soft* cluster ensembles are then formally introduced in Section 4.3 followed by several new consensus functions that operate on them. The experimental setup for our extensive evaluation of these algorithms and the empirical results then follow in Section 4.4 and Section 4.5, respectively. Finally, in Section 4.6 we conclude the chapter and briefly mention possible directions for future research.

4.2 CLUSTER ENSEMBLES

In this section, we will first define the *hard* cluster ensemble problem formally, and then present graph-theoretic solutions proposed by Strehl and Ghosh (2002) and Fern and Brodley (2004). We will also present some related work on robust clustering by Fred and Jain (2002), and on generative models for ensembles by Topchy, Jain, and Punch (2004). Other methods such as Voting-Merging (Dimitriadou, Weingessel, and Hornik, 2001) and GA-Search (Gablentz, Koppen, and Dimitriadou, 2000) are not presented as they are either not competitive or too restrictive in their scope. We will end the section with a brief discussion on past work on the role of diversity in the cluster ensembles problem.

4.2.1 The Hard Cluster Ensemble Problem

Let $X = \{x_1, x_2, \ldots, x_n\}$ denote a set of instances/objects. Each partitioning of the data (called a clustering) is represented as a vector of labels over the data. Let $\lambda^{(q)} \in \{1, 2, \ldots k^{(q)}\}^n$ denote the label vector of the qth constituent clustering of X; i.e., $\lambda_i^{(q)}$ is the label of x_i in the qth partitioning. A set of r such clusterings $\lambda^{(1,2,\ldots,r)}$ is called a cluster ensemble (for an example, see Table 4.1). The goal is to find a consensus function Γ that would combine the r clusterings $\lambda^{(1,2,\ldots,r)}$ into a single clustering/labeling λ.

It is instructive, for presentation later in this section, to consider that every hard clustering can be mapped to a hypergraph. A hypergraph consists of vertices and hyperedges. While an edge connects two vertices of a graph, a hyperedge can connect any number of vertices. For each clustering vector $\lambda^{(q)}$ a binary indicator matrix $H^{(q)}$ can be defined with n rows and $k^{(q)}$ columns. $H_{i,j}^{(q)}$ is 1 if x_i was placed in cluster j in clustering $\lambda^{(q)}$. The entire ensemble of clusterings can hence be represented by a concatenation of individual indicator matrices as $H = (H^{(1)}, \ldots, H^{(r)})$. The matrix H, now, defines a hypergraph with n vertices and $\sum_{q=1}^{r} k^{(q)}$ hyperedges. Each hyperedge connects all the vertices that have a value 1 in the corresponding column. This transformation of $\lambda^{(1,2,\ldots,r)}$ to H is shown in Tables 4.1 and 4.2.

Table 4.1 A set of three clusterings.

	$\lambda^{(1)}$	$\lambda^{(2)}$	$\lambda^{(3)}$
x_1	1	2	1
x_2	1	2	1
x_3	1	3	2
x_4	2	3	2
x_5	2	3	3
x_6	3	1	3
x_7	3	1	3

4.2.2 Graph-theoretic Approaches

Upon formulating the cluster ensemble problem, Strehl and Ghosh (2002) proposed three graph-theoretic approaches (CSPA, HGPA, and MCLA) for finding the consensus clustering. Later Fern and Brodley (2004) proposed the HBGF algorithm that is based on bipartite graph partitioning. All these approaches use the efficient graph partitioning algorithm METIS by Karypis and Kumar (1998) to partition graphs induced by the cluster ensemble and find the consensus clustering. Note that there is implicitly an additional constraint in these solutions, namely that the consensus clusters obtained should be of comparable size. We describe these and other algorithms in the following subsections.

4.2.2.1 Cluster-based Similarity Partitioning Algorithm (CSPA)

In CSPA the similarity between two data points is defined to be directly proportional to the number of constituent clusterings of the ensemble in which they are clustered together. The intuition is that the more similar two data points are the higher is the chance that constituent clusterings will place them in the same cluster. Hence, in this approach an $n \times n$ similarity matrix is computed as $W = \frac{1}{r}HH^T$. This similarity matrix (graph) can be clustered using any reasonable pairwise similarity based clustering algorithm to obtain the final clustering. In CSPA the authors chose METIS to partition the similarity graph to obtain the desired number of clusters. Because CSPA constructs a fully connected graph its computational and storage complexity are $\mathcal{O}(n^2)$. Hence it is more expensive in terms of resources than algorithms that will be introduced below.

4.2.2.2 Hypergraph Partitioning Algorithm (HGPA)

The HGPA algorithm seeks directly to partition the hypergraph defined by the matrix H in Table 4.2. Hypergraph partitioning seeks to cluster the data by eliminating the minimal number of hyperedges. This

Table 4.2 Hypergraph representation of clusterings.

	$H^{(1)}$			$H^{(2)}$			$H^{(3)}$		
	h_1	h_2	h_3	h_4	h_5	h_6	h_7	h_8	h_9
v_1	1	0	0	0	1	0	1	0	0
v_2	1	0	0	0	1	0	1	0	0
v_3	1	0	0	0	0	1	0	1	0
v_4	0	1	0	0	0	1	0	1	0
v_5	0	1	0	0	0	1	0	0	1
v_6	0	0	1	1	0	0	0	0	1
v_7	0	0	1	1	0	0	0	0	1

partitioning is performed by the package HMETIS by Karypis, Aggarwal, Kumar, and Shekhar (1997). In the HGPA algorithm all the vertices and hyperedges are weighted equally. In our experiments, HGPA displayed a lack of robustness and routinely performed worse than the CSPA and MCLA algorithms. Hence, we will not discuss this algorithm or report any results for it in the remainder of this chapter.

4.2.2.3 Meta-clustering Algorithm (MCLA)

The MCLA algorithm takes a slightly different approach to finding the consensus clustering than the previous two methods. First, it tries to solve the cluster correspondence problem and then uses voting to place data points into the final consensus clusters. The cluster correspondence problem is solved by grouping the clusters identified in the individual clusterings of the ensemble.

As we have seen earlier, the matrix H represents each cluster as n-length binary vectors. In MCLA, the similarity of cluster c_i and c_j is computed based on the number of data points that are clustered into both of them, using the Jaccard measure $W_{i,j} = \frac{|c_i \cap c_j|}{|c_i \cup c_j|}$. This similarity matrix (graph), with clusters as nodes, is partitioned into *meta-clusters* using METIS.

The final clustering of instances is produced in the following fashion. All the clusters in each meta-cluster are collapsed to yield an association vector for the meta-cluster. This association vector for a meta-cluster is computed by averaging the association of instances to each of the constituent clusters of that meta-cluster. The instance is then clustered into the meta-cluster that it is most associated to.

Computing the cluster similarity matrix exhibits a quadratic time complexity on the number of clusters in the ensemble. This is often significantly less than n^2. Furthermore, the averaging and voting operations are linear in n. This makes MCLA computationally very efficient.

4.2.2.4 Hybrid Bipartite Graph Formulation (HBGF)

This method was introduced by Fern and Brodley (2004) with an aim to model the instances and clusters simultaneously in a graph. The CSPA algorithm models the ensemble as a graph with the vertices representing instances in the data, while the MCLA algorithm models the ensemble as a graph of clusters. The HBGF technique combines these two ideas and represents the ensemble by a bipartite graph in which the individual data points and the clusters of the constituent clusterings are both vertices. The graph is bipartite because there are no edges between vertices that are both either instances or clusters. The complete set of rules to assign the weights on the edges is as follows:

- $W(i,j) = 0$ if i,j are both clusters or both instances;
- $W(i,j) = 0$ if instance i doesn't belong to cluster j;
- $W(i,j) = 1$ if instance i belongs to cluster j.

This bipartite graph is partitioned into k parts yielding the consensus clustering. The clustering is performed using METIS and spectral clustering (Ng, Jordan, and Weiss, 2001). The clusters in the consensus clustering contain both instances and the original clusters. Hence, the method yields a co-clustering solution. This method has also been previously used to simultaneously cluster words and documents by Dhillon (2001).

The computational complexity of HBGF is $\mathcal{O}(n \times t)$, where t is the total number of clusters in the ensemble. While this is significantly less than quadratic in the number of instances (as in CSPA), in practice we observe the algorithm to be fairly resource hungry both in terms of CPU time and storage.

4.2.2.5 Evidence Accumulation Framework

Evidence accumulation (Fred and Jain, 2001, 2002) is a simple framework, very similar to the cluster ensemble framework, for combining the results of multiple weak clusterings in order to increase robustness of the final solution. The framework uses a k-means type algorithm to produce several clusterings each with a random initialization. The number of clusters specified in each k-means clustering is typically much larger than the true number of clusters desired. The data instances are then mapped into the similarity space where the similarity between two instances i and j is the fraction of clusterings in which they ended up in the same cluster. A minimum spanning-tree based clustering algorithm is then used to obtain the final clustering. In practice any appropriate clustering technique could be employed. This framework and the consensus function that it uses are very similar to the cluster ensemble framework and the CSPA algorithm (Strehl and Ghosh, 2002).

A similar framework for obtaining robust clustering solutions has been proposed by Frossyniotis, Pertselakis, and Stafylopatis (2002). The actual consensus function used in this algorithm only works on heavily restricted type of ensembles; each constituent clustering has the same number of clusters. Also, Fern and Brodley (2003) extended this approach to accept soft clusterings as input. The details of this approach are presented in Section 4.3.4.

4.2.3 Ensemble as a Mixture of Multinomials

Topchy, Jain, and Punch (2004) modeled the ensemble, $\lambda^{(1,2,\ldots,r)}$, using a generative model and used EM to estimate the parameters of the model. The EM procedure along with the parameters provides us with a soft final clustering.

In this approach, it is assumed that the ensemble has been generated from a mixture of multidimensional multinomial distributions. Each data point is generated by first picking a multinomial distribution according to the priors. After picking a component of the mixture, the cluster label in each clustering is picked from a multinomial distribution over the cluster labels. The cluster labels of different constituent clusterings are assumed to be i.i.d..

The number of parameters to be estimated increases with both the number of constituent clusterings as well as with the number of clusters in them. Experiments in Topchy, Jain, and Punch (2004) did not include experiments on data-sets that have more than three clusters. In this chapter we will evaluate the performance of this consensus function on more complex real-life data-sets.

One advantage of this approach is that it is easy to model final clusters of different sizes using this method. Graph partitioning methods tend to yield roughly balanced clusters. This is a disadvantage in situations where the data distribution is not uniform. Using the priors in the mixture model the distribution of data can be accommodated conveniently.

4.2.4 Diversity in Cluster Ensembles

Diversity among the classifiers in an ensemble has been shown to improve its accuracy (Hansen and Salamon, 1990; Melville and Mooney, 2003). Here, we recount some research on the impact of diversity on cluster ensembles.

Ghosh, Strehl, and Merugu (2002) examined the problem of combining multiple clusters of varying resolution and showed that it is possible to obtain robust consensus even when the number of clusters in each of the individual clusterings is different. They also described a simple scheme for selecting a 'good' number of clusters k for the consensus clustering by observing the variation in average normalized mutual information with different k. Fern and Brodley (2003) reported on some experiments on diversity of ensembles. They found that the consensus function's accuracy increased as the ensemble is made more diverse. Kuncheva and Hadjitodorov (2004) studied the diversity of ensembles using multiple measures like the Rand Index, Jaccard measure, etc.. Based on this study they proposed a variant of the Evidence Accumulation framework where the number of over-produced clusters is randomly chosen. This randomization in ensemble generation is shown to increase the diversity of the ensembles thereby leading to better consensus clustering. In a recent follow-up work Hadjitodorov, Kuncheva, and Todorova (2006) reported that selecting constituent clusterings based on median diversity leads to better ensembles.

4.3 SOFT CLUSTER ENSEMBLES

In this section we will formally define the soft cluster ensemble problem and provide intuition on why we expect soft cluster ensembles to yield better results than their corresponding hard versions. We will then introduce a new algorithm based on Information Theoretic k-means (Dhillon, Mallela and Kumar, 2003b) to solve ensembles of soft clusterings. In order to evaluate our new approach objectively, we will describe changes to existing techniques mentioned in Section 4.2 to enable them to handle soft ensembles.

4.3.1 The Soft Cluster Ensemble Problem

In order to facilitate the explanation of various algorithms later in this section, we now define the soft cluster ensemble problem formally.

As in the case of hard ensembles, let $X = \{x_1, x_2, \ldots, x_n\}$ denote a set of instances/objects. Also, let $\lambda^{(q)} \in \{1, 2, \ldots k^{(q)}\}^n$ denote the label vector of the qth clustering of X; i.e., $\lambda_i^{(q)}$ is the label of x_i in the qth clustering. This is the hard labeling defined in Section 4.2.1. In cases where the underlying clustering algorithm outputs soft cluster labels, $\lambda_i^{(q)}$ is defined as $argmax_j P(C_j|x_i)$, where $P(C_j|x_i)$ is the posterior probability of instance x_i belonging to cluster C_j. A soft cluster ensemble is shown in Table 4.3, and its corresponding hard version in Table 4.1.

Instead of *hardening* the posterior probabilities into cluster labels, we construct a matrix $S^{(q)}$ representing the solution of the qth soft clustering algorithm. $S^{(q)}$ has a column for each cluster generated in the clustering and the rows denote the instances of data with $S_{ij}^{(q)}$ being the probability of x_i belonging to cluster j of the qth clustering. Hence, the values in each row of $S^{(q)}$ sum up to 1. There are r such clusterings ($S^{(1,\ldots,r)}$) each with $k^{(q)}$ clusters. Just as in the hard ensemble problem, our goal is to find a consensus function Γ that combines these clusterings into a combined labeling, λ, of the data. It should be noted that the cluster ensemble framework does not specify whether the final clusterings should be hard or soft. In this chapter we only work with algorithms that output hard final clusterings.

Table 4.3 Ensemble of soft clusterings.

	$S^{(1)}$			$S^{(2)}$			$S^{(3)}$		
	s_1	s_2	s_3	s_4	s_5	s_6	s_7	s_8	s_9
x_1	0.7	0.2	0.1	0.1	0.7	0.2	0.6	0.3	0.1
x_2	0.9	0.1	0.0	0.0	0.8	0.2	0.8	0.2	0.0
x_3	0.9	0.0	0.1	0.1	0.4	0.5	0.5	0.5	0.0
x_4	0.2	0.6	0.2	0.1	0.2	0.7	0.2	0.7	0.1
x_5	0.1	0.9	0.0	0.0	0.1	0.9	0.0	0.5	0.5
x_6	0.0	0.2	0.8	0.8	0.1	0.1	0.1	0.2	0.7
x_7	0.1	0.2	0.7	0.7	0.1	0.2	0.1	0.3	0.6

4.3.2 Intuition behind Soft Ensembles

It is fairly obvious from the above definition that hardening a soft cluster ensemble entails a loss of information. However, it is not at all obvious that this additional information is useful. The goal of this study is to show empirically that algorithms designed for soft ensembles improve upon the accuracy of those that operate on the hardened versions of the ensembles. Here, we will try to explain intuitively why we expect this.

For the sake of discussion consider a cluster ensemble where individual clusterings are working on vertically partitioned data. In such a scenario, the underlying clustering algorithms have access to different and often incomplete sets of features. Incomplete data could result from distributed computing constraints (Ghosh, Strehl and Merugu, 2002), random projections in order to facilitate high-dimensional clustering (Fern and Brodley, 2003), or multi-view data-sets as used in (Kreiger and Green, 1999). Under such circumstances there is an increased chance that the underlying clustering algorithms will not be able to assign some objects into clusters with much certainty. If the combining procedure were to accept only hard clusterings, these objects would have to be assigned to the cluster they most belong to (one with the highest posterior probability).

Consider the soft ensemble depicted in Table 4.3. The solution $S^{(2)}$ assigns x_3 to clusters s_4, s_5, and s_6 with probabilities 0.1, 0.4, and 0.5, respectively. If the consensus function were to only accept hard clusterings it would be provided with a vector where $\lambda_i^{(2)}$ is s_6. The combining algorithm would have no evidence that the second underlying clustering algorithm was unsure about the assignment of x_3. It would accept this observation with the same amount of certainty as any other observation that assigns a data point x_i to a cluster s_j with 0.9 probability. If, however, the combining function were to accept soft clusterings, it could potentially use this information to make appropriate cluster assignment of x_3 in the combined clustering. Since it is more likely that clustering algorithms are unsure of their assignments while operating with an incomplete set of features, it is important that the combining function has access to the cluster assignment probabilities, and not just the hard assignments themselves.

4.3.3 Solving Soft Ensembles with Information-Theoretic k-means (ITK)

Information-Theoretic k-means was introduced by Dhillon, Mallela, and Kumar (2003b) as a way to cluster words in order to reduce dimensionality in the document clustering problem. This algorithm is very similar to the k-means algorithm, differing only in the fact that as a measure of distance it uses the KL-divergence (Kullback and Leibler, 1951) instead of the Euclidean distance. The reader is referred to the original paper for more details. Here we just describe the mapping of the soft cluster ensemble problem to the Information-Theoretic k-means problem.

Each instance in a soft ensemble is represented by a concatenation of r posterior membership probability distributions obtained from the constituent clustering algorithms (see matrix S in Table 4.3). Hence, we can define a distance measure between two instances using the Kullback–Leibler (KL) divergence (Kullback and Leibler, 1951), which calculates the 'distance' between two probability distributions. The distance between two instances x_a and x_b can be calculated as

$$KL_{x_a,x_b} = \sum_{q=1}^{r} w^{(q)} \sum_{i=1}^{k^{(q)}} S_{x_a i}^{(q)} \log\left(\frac{S_{x_a i}^{(q)}}{S_{x_b i}^{(q)}}\right) \tag{4.1}$$

where, $w^{(q)}$ are clustering specific weights, such that $\sum_{q=1}^{r} w^{(q)} = 1$.

Equation (4.1) computes the pairwise distance by taking an average of the KL divergence between the two instances in individual constituent clusterings. Here we note that this is equivalent to computing the KL divergence between instances represented by a matrix S in which each row adds up to one. This normalization can be performed by multiplying each value in $S^{(q)}$ by $w^{(q)}/\sum_{q=1}^{r} w^{(q)}$. Now that we have a distance measure between instances based on KL-divergence, we can use existing information-theoretic k-means software mentioned above to solve the soft ensemble.

Computing Equation (4.1) with $w^{(q)} = 1/r$ assumes that all the clusterings are equally important. We can, however, imagine a scenario where we have different importance values for the constituent clusterings. These values could, for instance, be our confidence in the accuracy of these clusterings, possibly based on the number of features they access. These confidence values can easily be integrated into the cost function using the weights $w^{(q)}$.

4.3.4 *Soft Version of CSPA (sCSPA)*

The CSPA algorithm proposed by Strehl and Ghosh (2002) works by first creating a co-association matrix of all objects, and then using METIS (Karypis and Kumar, 1998) to partition this similarity space to produce the desired number of clusters. This algorithm is described in Section 4.2.

sCSPA extends CSPA by using values in S to calculate the similarity matrix. If we visualize each object as a point in $\sum_{q=1}^{r} k^{(q)}$-dimensional space, with each dimension corresponding to the probability of its belonging to a cluster, then SS^T is the same as finding the dot product in this new space. Thus the technique first transforms the objects into a *label-space* and then interprets the dot product between the vectors representing the objects as their similarity. In our experiments we use Euclidean distance in the label space to obtain our similarity measure. The dot product is highly co-related with the Euclidean measure, but Euclidean distance provides for cleaner semantics. Euclidean distance between x_a and x_b is calculated as

$$d_{x_a,x_b} = \sqrt{\sum_{q=1}^{r}\sum_{i=1}^{k^{(q)}}\left(S_{x_ai}^{(q)} - S_{x_bi}^{(q)}\right)^2}.$$

This can be interpreted as a measure of the difference in the membership of the objects for each cluster. This dissimilarity metric is converted into a similarity measure using $s_{x_a,x_b} = \mathrm{e}^{-d_{x_a,x_b}^2}$.

Another distance measure can be defined on the instances in a soft ensemble using KL, divergence (Kullback and Leibler, 1951) as in Section 4.3.3. In our results we observed that all versions of the sCSPA (with Euclidean distance, KL divergence and cosine similarity) gave very similar results. The results obtained while using Euclidean distance were sometimes better, so here we will report results based on only that version of the sCSPA. sCSPA (like CSPA) is impractical for large data-sets, and hence we will only report results for data-sets with less than 2000 data points.

Fern and Brodley (2003) proposed a variant of the evidence accumulation framework that accepts soft clusterings. In this scenario, the similarity of two instances is calculated as the average dot product of the probability distributions describing them. Hence,

$$sim(x_a, x_b) = \frac{1}{r}\sum_{i=1}^{k^{(q)}} S_{x_ai}^{(q)} \times S_{x_bi}^{(q)}.$$

The similarity matrix that results is then clustered using a complete-link agglomerative algorithm. The input matrix used by this framework is essentially equivalent to the one used by sCSPA (using cosine similarity). The only difference is in the combining function. Hence, we will not experiment with this technique further in this chapter.

4.3.5 *Soft Version of MCLA (sMCLA)*

In MCLA each cluster is represented by an n-length binary association vector. The idea is to group and collapse related clusters into meta-clusters, and then assign each object to the meta-cluster in which it belongs most strongly. The clusters are grouped by graph partitioning based clustering.

sMCLA extends MCLA by accepting soft clusterings as input. sMCLA's working can be divided into the following steps (similar steps are followed in MCLA too):

Construct soft meta-graph of clusters. All the $\sum_{q=1}^{r} k^{(q)}$ clusters or indicator vectors s_j (with weights), the hyperedges of S, can be viewed as vertices of another regular undirected graph. The edge weights between two clusters s_a and s_b is set as $W_{a,b} = Euclidean\_dist(s_a, s_b)$. The Euclidean distance is a measure of the difference of membership of all objects to these two clusters. As in the sCSPA algorithm, the Euclidean distance is converted into a similarity value.

Group the clusters into meta-clusters. The meta-graph constructed in the previous step is partitioned using METIS to produce k balanced meta-clusters. Since each vertex in the meta-graph represents a distinct cluster label, a meta-cluster represents a group of corresponding cluster labels.

Collapse meta-clusters using weighting. We now collapse all the clusters contained in each meta-cluster to form its association vector. Each meta-cluster's association vector contains a value for every object's association to it. This association vector is computed as the mean of the association vectors for each cluster that is grouped into the meta-cluster. This is a weighted form of the step performed in MCLA.

Compete for objects. Each object is assigned to the meta-cluster to which it is most associated. This can potentially lead to a soft final clustering, since the ratio of the winning meta-cluster's association value to the sum of association values of all final meta-clusters can be the confidence of assignment of an object to the meta-cluster.

There is, however, one problem with this approach. Because we are using soft clusterings as inputs, the co-association graph of the clusters (meta-graph) is almost complete. More specifically, even clusters from the same clusterings have non-zero similarity to each other. This is not the case with MCLA since it uses a binary Jaccard measure, and for hard clusterings Jaccard similarity between clusters in the same clusterings is necessarily zero. We obtain better consensus clustering results after making the co-association matrix r-partite. Hence, sMCLA forces the similarity of hyperedges coming from the same clustering to be zero. This is, however, only done when the number of clusters in all the constituent clusterings is equal to the desired final number of clusters. In ensembles where the number of clusters in each underlying clustering vary the algorithm does not force the co-association matrix to be r-partite.

4.3.6 Soft Version of HBGF (sHBGF)

HBGF represents the ensemble as a bipartite graph with clusters and instances as nodes, and edges between the instances and the clusters they belong to. This approach can be trivially adapted to consider soft ensembles since the graph partitioning algorithm METIS accepts weights on the edges of the graph to be partitioned. In sHBGF, the graph has $n + t$ vertices, where t is the total number of underlying clusters. The weights on the edges are set as follows:

- $W(i,j) = 0$ if i, j are both clusters or both instances;
- $W_{(i,j)} = S_{i,j}$ otherwise, where i is the instance and j is the cluster.

4.4 EXPERIMENTAL SETUP

We empirically evaluate the various algorithms presented in Sections 4.2 and 4.3 on soft cluster ensembles generated from various data-sets. In this section we describe the experimental setup in detail.

Table 4.4 Data-sets used in experiments.

Name	Type of features	#features	#classes	#instances
8D5K	Real	8	5	1000
Vowel	Real	10	11	990
Pendigits	Real	16	10	10 992
Glass	Real	9	6	214
HyperSpectral	Real	30	13	5211
Yeast	Real	8	10	1484
Vehicle	Real	18	4	846

4.4.1 Data-sets Used

We perform the experimental analysis using the six real-life data-sets and one artificial data-set. Some basic properties of these data-sets are summarized in Table 4.4. These data-sets were selected so as to present our algorithms with problems of varying degrees of difficulty – in terms of the number of desired clusters, the number of attributes, and the number of instances. All these data-sets, with the exception of 8D5K and HyperSpectral, are publicly accessible from the UCI data repository (Black and Merz, 1998).

- *8D5K*. This is an artificially generated data-set containing 1000 points. It was generated from five multivariate Gaussian distributions (200 points each) in eight-dimensional space. The clusters all have the same variance but different means. The means were drawn from a uniform distribution within the unit hypercube. This data-set was used in (Strehl and Ghosh, 2002) and can be obtained from http://www.strehl.com.
- *Vowel*. This data-set contains data on the pronunciation of vowels. We removed some nominal features that corresponded to the context like sex, name, etc., and only retained the 10 real valued features. There are 11 classes in the data and an average of 93 instances per class.
- *Pendigits*. This data-set was generated for the problem of pen-based recognition of handwritten digits. It contains 16 spatial features for each of the 10 992 instances. There are 10 classes of roughly equal sizes corresponding to the digits 0 to 9. In order to get better clustering results, we normalized the columns (features) to sum to one.
- *Glass*. The instances in this data-set are samples of glass used for different purposes. Real-valued features corresponding to their chemical and optical properties describe the instances. There are 214 instances categorized into six classes such as tableware, containers, etc. based on nine attributes.
- *Hyper Spectral*. This data-set contains 5211 labeled pixels from a HyperSpectral snapshot of the Kennedy Space Center. Each data point is described by a set of 30 HyperSpectral signatures pruned from an initial set of 176 features. The pruning was performed by a best-basis feature extraction procedure (Kumar, Ghosh, and Crawford, 2001). The data-set has 13 classes describing the geographical features apparent in the pixel.
- *Yeast*. The Yeast data-set contains information about proteins within yeast cells with the class attribute denoting the localization within the cell. This is a fairly hard problem, and this shows in the clustering results we obtain. The 1484 instances are each characterized by eight attributes, and there are 10 classes in the data-set.
- *Vehicle*. This data-set was designed for the purpose of learning to classify a given silhouette as one of the four types of vehicles, using a set of 18 features extracted from the silhouette. The vehicle may be viewed from one of many different angles. The 846 silhouette instances are classified into four vehicle categories: Opel, Saab, bus, and van.

4.4.2 Ensemble Test-set Creation

In order to compare the hard and soft ensemble methods, as well as to evaluate the k-means Information-Theoretic (ITK) based approach, we created soft cluster ensembles of varying degrees of difficulty. Note

Table 4.5 Data-set specific options for creating ensembles.

Name	# attributes	Numatts options	#clusterings/Numatts-option
8D5K	8	3,4,5,6	10
Vowel	10	3,4,5,6,7	10
Pendigits	16	3,4,6,9,12	15
Glass	9	3,4,5,6,7	10
HyperSpectral	30	5,10,15,20,25	15
Yeast	8	2,3,4,5	10
Vehicle	18	4,5,8,11	15

here that for each soft cluster ensemble we also stored its corresponding hardened version to evaluate methods that only accept hard clusterings.

The individual clusterings in our ensembles were created using the EM algorithm (Dempster, Laird, and Rubin, 1977) with a mixture of Gaussian distribution models, but any algorithm that outputs soft probabilities could have been used. Further, each constituent clustering was created using vertically partitioned subsets of the data-sets. This partial view of the data as well as the dependence of EM on initialization resulted in the diversity in the individual clustering solutions in an ensemble.

As mentioned above, we wanted to evaluate our algorithms on ensembles of varying degrees of difficulty. For this purpose we created ensembles by varying two parameters that controlled the degree of difficulty. The first parameter is the number of attributes that the EM algorithm accesses while creating the constituent clusterings. We expected the difficulty of an ensemble containing clusterings created from less attributes to be higher. The second parameter is the number of constituent clusterings in the ensemble. In general, we expected that as the number of constituent clusterings increased the consensus clusterings obtained should be more accurate. For most data-sets the number of clusterings in the ensembles is varied from two to 10, and in some cases to 15. The entire set of options for all the data-sets is listed in Table 4.5. The column labeled 'Numalts options' in Table 4.5 describes the different settings for a number of features used to create clusterings. For instance, for the 8D5K data-set we can obtain ensembles with constituent clusterings created using 3,4,5, or 6 attributes. Also, for each of these settings we can select from 10 clusterings to form an ensemble. Of course, each of these 10 clusterings is created with a randomly selected set of attributes.

Hence, while creating an ensemble we specify three parameters: the data-set name, the number of attributes, and the number of clusterings. For each set of parameter values, we create multiple ensembles by randomly selecting the clusterings to combine. Also, nondeterministic consensus functions are run multiple times in order to average out variations in results due to initialization.

Here we must note that each individual clustering as well as the consensus function is given the true number of clusters to find. The use of ensembles for finding the true number of clusters, or the effect of different k in constituent clusterings on ensemble accuracy, is not investigated in this study.

4.4.3 Evaluation Criteria

In order to evaluate the final consensus clusterings obtained we use two different criteria. Both these criteria compare the obtained clustering to the true labels of the instances. We also use the geometric mean ratio to present an overall score for the performance of each algorithm.

4.4.3.1 Normalized Mutual Information (NMI)

The first criterion we use was introduced by Strehl and Ghosh (2002). and is called normalized mutual information (NMI).

The NMI of two labelings of instances can be measured as

$$NMI(X,Y) = \frac{I(X,Y)}{\sqrt{H(X)H(Y)}} \tag{4.2}$$

where $I(X, Y)$ denotes the mutual information between two random variables X and Y and $H(X)$ denotes the entropy of X. In our evaluation, X will be consensus clustering while Y will be the true labels.

NMI has some nice properties such as $NMI(X, X) = 1$ and if Y has only one cluster label for all instances $NMI(X, Y) = 0$. With these properties NMI is extensively used for evaluating clustering algorithms in literature.

Another measure of clustering accuracy is adjusted RAND (Hubert and Arabie, 1985). The adjusted RAND compares two labelings based on whether pairs of objects are placed in the same or different clusters in them. The maximum value it takes is 1, and its expected value is 0. We computed the adjusted RAND score for each solution and found it to be highly correlated to the NMI score. Hence we will only report the NMI score in this chapter.

4.4.3.2 Classification via Clustering (CVC)

The CVC is a measure of the purity of the clusters obtained w.r.t. the ground truth. The CVC is calculated by the following procedure:

- To each cluster, assign the label that corresponds to the majority of points.
- Each instance is now labeled by its cluster's label.
- CVC is the fraction of misclassified instances in such a classification of instances.

The CVC measure weighs the contribution of a cluster to the average by its size. This ensures that very small pure clusters do not compensate for large impure ones.

There are other issues with this measure, however. The CVC measure is biased toward solutions with a large number of very small pure clusters. This is not an issue in our evaluation since the number of output clusters is kept constant across all the consensus functions being compared. Also, the CVC measure is not very well defined in the case of empty clusters in the clustering solution. Since we ignore the purity of empty clusters in our calculation of CVC, if all the instances were clustered into one cluster, CVC would be the fraction of instances that belong to the class with the largest number of instances. NMI would have been zero in such a case. This is not a problem for most data-sets since many algorithms are based on graph partitioning approaches and output balanced clusters. However, like most existing literature on cluster ensembles, we will use NMI as our principal measure of goodness.

4.4.3.3 Geometric Mean Ratio

Since we are varying the ensemble parameters over a very wide range for each data-set, we end up with a lot of different points of comparison. In order to report some sort of overall score for each algorithm on all the ensembles used, we use the geometric mean ratio (Webb, 2000). The GMR is calculated as follows. Suppose we have n ensembles that we tested our algorithms on, and NMI_A and NMI_B are vectors of the average NMI values w.r.t. to true labels obtained by algorithms A and B on these runs. GMR is calculated as

$$GMR(A, B) = \left(\prod_{i=1}^{n} \frac{NMI_{Bi}}{NMI_{Ai}} \right)^{\frac{1}{n}}. \tag{4.3}$$

In later sections we display the GMR values in tables with rows and columns representing the algorithms being compared. In these tables element (i, j) represents the value $GMR(algo(i), algo(j))$, where $algo(i)$ and $algo(j)$ are the algorithms represented in row i and column j, respectively. Hence, values > 1 along a column mean that the algorithm corresponding to the column performs better than the other algorithms. Similarly, the values < 1 along the rows indicate that the algorithm corresponding to the row scores better than the other algorithms.

4.5 SOFT VS. HARD CLUSTER ENSEMBLES

In this section we present results from our evaluation of the algorithms we described in earlier sections using the experimental setup described Section 4.4. In Section 4.5.1 we will compare the performance of algorithms accepting soft ensembles as input and those that run on hardened versions of the ensembles. After analyzing these experiments we will compare the Information-Theoretic k-means (ITK) approach with the best performing algorithms from Section 4.5.1. Finally, in Section 4.5.3 and Section 4.5.4, we will examine the variation in performance of algorithms on ensembles of varying difficulty.

4.5.1 Soft Versions of Existing Algorithms

In this section we evaluate the performance of CSPA, MCLA, and HBGF, their soft counterparts, and the mixture of multinomials method. The evaluation measure we employ is the geometric mean ratio (GMR), which is calculated over all the ensembles that were created as described in Section 4.4.2. There were, however, some exceptions to the direct application of the GMR formula over all data-sets. HBGF, CSPA, and their soft versions were not run on the HyperSpectral and Pendigits data-sets because these data-sets are too large to expect solutions in a reasonable time. Hence, when we compare one of these algorithms to the others we do not consider ensembles of these large data-sets. Also, in certain cases (for hard ensembles) the consensus functions output clusterings that score 0 on the NMI measure. This would happen, for example, if all the instances were placed in a single cluster. In such cases the GMR either becomes 0 or ∞ depending on where the zero score appears. Hence, we assign a very small nominal value (0.00001) to the NMI score whenever it is zero. The effect of this nominal score vanishes because we normalize by taking the nth root of the product.

Table 4.6 shows the GMR values of the NMI measure comparing the three original algorithms as well as their soft versions. We can see that for each algorithm the soft version performs better than the corresponding hard version. Keep in mind that algorithm with values < 1 on the rows are performing better than the others. Table 4.6 shows that averaged over all the ensembles we created, the soft versions of the algorithms are slightly better than their hard counterparts. This shows that the soft versions of the algorithms are able to use the extra information in the soft ensembles to obtain better consensus clusterings.

We notice that the mixture of multinomials algorithm (MixMns) performs worse than all other algorithms other than MCLA. This may be because many of the data-sets we used had a large number of clusters, causing parameter estimation problems for the mixture model. Topchy, Jain and Punch (2004) only evaluated their algorithm on real data-sets with very low number of clusters.

Another key observation is the dramatic difference in the performance of the sMCLA and MCLA algorithms. The performance improvement of sMCLA over MCLA is much larger than the improvements by other soft versions like sCSPA and sHBGF. This is because MCLA's performance is very bad when the input clusterings are not accurate. This can be seen by its performance values over tough

Table 4.6 Geometric mean ratio of NMI score over all ensembles. The value $table_{i,j}$ indicates ratio of algorithms j/i.

	CSPA	sCSPA	MCLA	sMCLA	HBGF	sHBGF	MixMns
CSPA	1	1.05	0.718	0.999	0.978	1.02	0.802
sCSPA	0.94	1	0.68	0.948	0.928	0.967	0.76
MCLA	1.163	1.22	1	1.17	1.136	1.18	0.913
sMCLA	1.00	1.05	0.56	1	0.978	1.019	0.77
HBGF	1.02	1.076	0.73	1.02	1	1.04	0.82
sHBGF	0.98	1.03	0.705	0.98	0.959	1	0.787
MixMns	1.25	1.31	0.73	1.297	1.219	1.269	1

Table 4.7 Geometric mean ratio of CVC score over all ensembles. The value $table_{i,j}$ indicates ratio of algorithms j/i.

	CSPA	sCSPA	MCLA	sMCLA	HBGF	sHBGF	MixMns
CSPA	1	1.02	0.795	1.17	0.99	1.01	0.964
sCSPA	0.976	1	0.777	1.146	0.97	0.99	0.94
MCLA	1.015	1.039	1	1.197	1.01	1.03	0.99
sMCLA	0.85	0.873	0.53	1	0.85	0.87	0.80
HBGF	1.004	1.029	0.799	1.179	1	1.02	0.97
sHBGF	0.98	1.009	0.78	1.156	0.98	1	0.95
MixMns	1.037	1.06	0.66	1.24	1.03	1.05	1

ensembles (Table 4.8) as well as ensembles with a very low number of attributes in constituent clusterings (Figure 4.1). sMCLA is not misled during the meta-clustering phase because the distances between the clusters are now determined from soft probabilities. Hence, an error in an input clustering that assigns an instance into the wrong cluster could be alleviated in sMCLA's case if the posterior probabilities of the wrong assignment are small. This phenomenon, however, needs to be investigated further since sMCLA performs on a par with the best algorithms shown in Table 4.6.

Table 4.7 shows the GMR value table for the CVC measure. As we can see from the table the GMR values closely correspond to the values in Table 4.6. Since the values in the two tables closely agree we will henceforth only report results using the NMI measure.

In order to evaluate the intuition that the information obtained from soft ensembles is especially useful when dealing with *tough* ensembles, we have populated Table 4.8 with GMR values calculated over only the *tough* ensembles. *Tough* ensembles are defined as those comprising a small number of clusterings, each of which are obtained using very few features. In our experiments, tough ensembles contained only 2–4 clusterings, which were obtained using the minimum Numatts option number of features for each data-set shown in Table 4.5. For example, a tough ensemble for the 8D5K data-set might contain three clusterings, each obtained using only three features. As we can see from Table 4.8, soft versions of algorithms perform better than their hard counterparts and the difference in their performance is slightly higher than those in Table 4.6. The fact that the differences in performances are higher shows that the extra information in soft clusterings is useful in tough situations.

4.5.2 Information-Theoretic k-means (ITK)

We compare the Information-Theoretic k-means algorithm with only two of the best algorithms from the analysis in the previous section. Table 4.9 displays the GMR values for the ITK, sHBGF, and sMCLA algorithm over all the ensembles. As we can see the ITK algorithm performs appreciably better than both

Table 4.8 Geometric mean ratio of NMI score over tough ensembles. The value $table_{i,j}$ indicates ratio of algorithms j/i.

	CSPA	sCSPA	MCLA	sMCLA	HBGF	sHBGF	MixMns
CSPA	1	1.085	0.652	0.997	0.97	1.06	0.655
sCSPA	0.92	1	0.60	0.919	0.897	0.98	0.604
MCLA	1.53	1.665	1	1.47	1.49	1.63	0.922
sMCLA	1.003	1.088	0.46	1	0.976	1.06	0.627
HBGF	1.028	1.113	0.67	1.025	1	1.09	0.673
sHBGF	0.94	1.024	0.62	0.94	0.92	1	0.618
MixMns	1.53	1.656	0.73	1.59	1.485	1.617	1

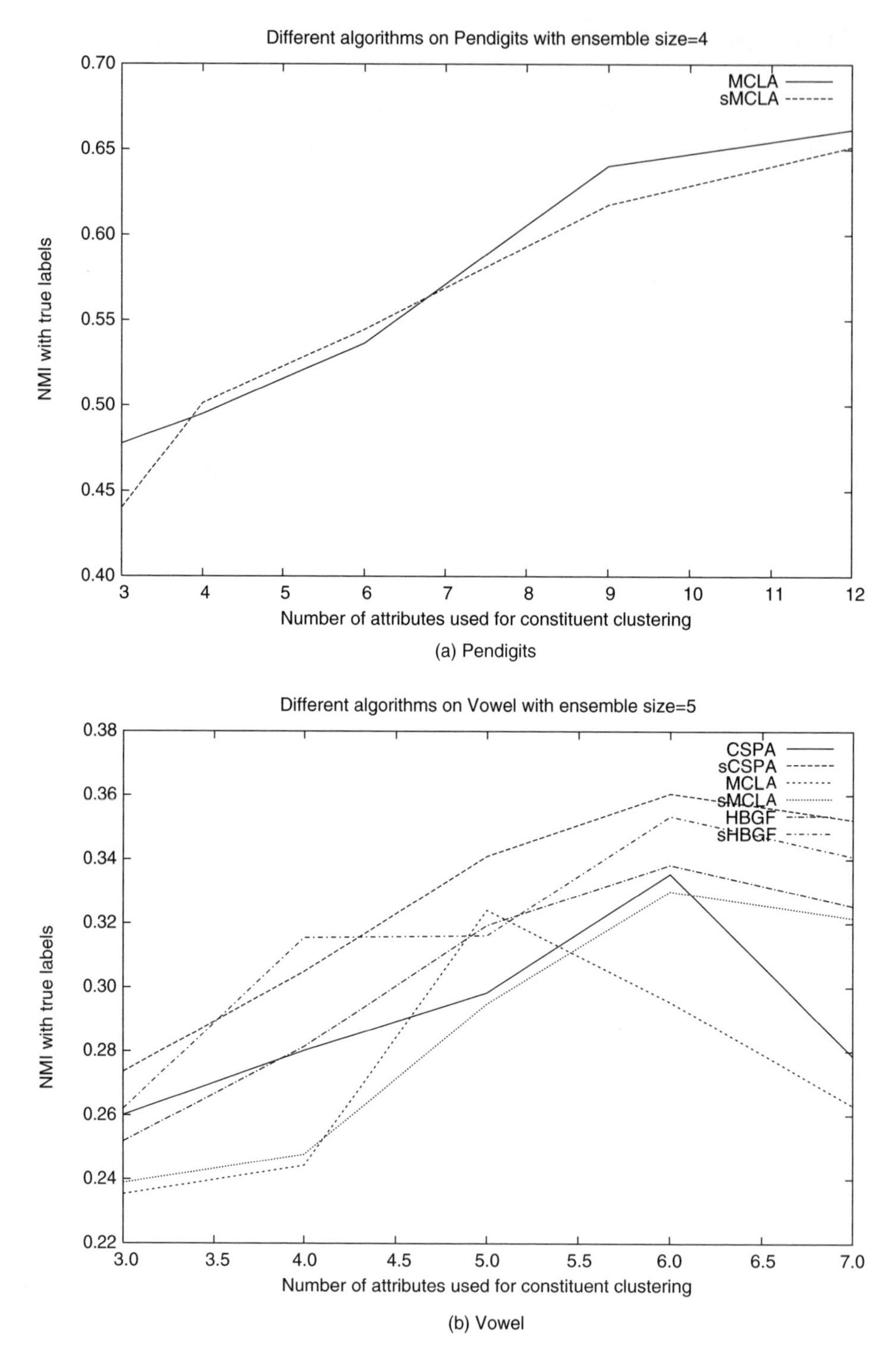

Figure 4.1 Performance of CSPA, MCLA, HBGF, sCSPA, sMCLA, and sHBGF while varying the number of attributes used in constituent clusterings.

Table 4.9 Geometric mean ratio of NMI score over all ensembles. The value $table_{i,j}$ indicates ratio of algorithms j/i.

	ITK 10K	sHBGF	sMCLA
ITK 10K	1	0.856	0.875
sHBGF	1.167	1	0.98
sMCLA	1.142	1.012	1

sHBGF and sMCLA. The sHBGF and sMCLA algorithms are fairly similar to each other in overall performance. The geometric mean ratio matrix for the CVC score is identical to the one for the NMI score, and we do not report those results.

In order to find whether ITK performs better for tougher or simpler ensembles we calculate GMR over only the tough ensembles. Here again the tough ensembles are defined as in Section 4.5.1. The results of this experiment are listed in Table 4.10. As we can see from the two tables the improvement in ITK algorithm's performance over sHBGF/sMCLA is higher for the subset of tougher ensembles.

In the set of data-sets selected for this chapter some present tougher challenges to the clustering algorithms than others. In terms of the NMI score of clusterings 8D5K is the simplest data-set while Yeast is the toughest. We display in Tables 4.11 and 4.12 the GMR value matrix for ensembles of data-sets 8D5K and Yeast, respectively. As we can see from these tables, in the case of the Yeast data-set ITK is by far the best performing algorithm. However, for the 8D5K data-set all algorithms are fairly comparable with sHBGF slightly better than the rest. One reason is that for soft ensembles where most probability values are close to 1 or 0, more complex algorithms like ITK do not perform better than simple graph-theoretic approaches.

Another explanation for ITK's performance on the Yeast data-set can be provided based on the characteristics of the algorithms. The graph partitioning based consensus algorithms are constrained to provide roughly balanced clusters. This can be a problem in cases where the underlying data do not have balanced classes. The 8D5K data-set has perfectly balanced clusters (200 instances each) while the Yeast data-set has classes that range from five instances to 463 instances in size. The ITK algorithm is not constrained to find balanced clusters and hence can adapt the clustering solution better to the natural

Table 4.10 Geometric mean ratio of NMI score over tough ensembles. The value $table_{i,j}$ indicates ratio of algorithms j/i.

	ITK 10K	sHBGF	sMCLA
ITK 10K	1	0.816	0.798
sHBGF	1.226	1	0.94
sMCLA	1.253	1.06	1

Table 4.11 Geometric mean ratio of NMI score for only the 8D5K data-set. The value $table_{i,j}$ indicates ratio of algorithms j/i.

	ITK 10K	sHBGF	sMCLA
ITK 10K	1	1.03	0.97
sHBGF	0.968	1	0.944
sMCLA	1.025	1.05	1

Table 4.12 Geometric mean ratio of NMI score for only the yeast data-set. The value $table_{i,j}$ indicates ratio of algorithms j/i.

	ITK 10K	sHBGF	sMCLA
ITK 10K	1	0.84	0.68
sHBGF	1.18	1	0.817
sMCLA	1.454	1.222	1

distribution of instances in the data. This is why we see the ITK algorithm outperform sHBGF and sMCLA on the Yeast data-set by a large margin.

4.5.3 Performance Variation with Increasing Attributes

In this section we examine how the performances of different consensus functions change as the number of attributes used for the constituent clusterings is changed. The number of attributes is an ad hoc measure of the quality of clustering obtained and hence the difficulty of the ensemble. In general, the fewer the number of attributes in the constituent clusterings the more the confusion in the clustering solutions obtained and, hence, the more the difficulty of obtaining a consensus labeling using these clustering solutions.

Figure 4.1 shows the variation in the performance of the existing ensemble methods and their soft variations on two data-sets. The mixture of multinomial model method is not shown since its performance was much lower than the others. The data-sets selected for these plots are of intermediate difficulty. As we can see, as we increase the number of attributes in the constituent clusterings the accuracy of all algorithms increases in general. For Pendigits, Figure 4.1(a) only has curves for MCLA and sMCLA since we did not run HBGF and CSPA on it.

Figure 4.2 displays curves for the ITK, sHBGF, and sMCLA. As we can see the ITK algorithm outperforms the other algorithms over the whole range of attributes, but as the number of attributes is increased the accuracies of all algorithms tend to saturate.

Fern and Brodley (2003) show experimentally that for high-dimensional domains combining clusterings on subspace projections of the data outperforms clustering on the whole data. They also found that the impact of subspace clustering is more prominent if the number of dimensions is higher (> 60). We have not experimented with data-sets that have very high dimensionality, and hence we did not observe the reduction in accuracy when using the full set of attributes.

4.5.4 Performance Variation with Increasing Ensemble Size

In this section we examine the effect of increasing the number clusterings used in the ensemble on the accuracy of final clustering. Say, we set the number of attributes used to create constituent clusterings to some constant value. We would then expect that as more clusterings are added to the ensemble the combining function would have more information available to create the final clustering. This has been seen previously in the classifier ensemble literature where increasing the size of the ensemble increases the accuracy until a saturation point is reached (Hansen and Salamon, 1990; Melville and Mooney, 2003; Opitz and Maclin, 1999). Hence, the number of clusterings in an ensemble can also be said to be a measure of the difficulty of the task of combining them.

Figure 4.3 shows the variation in accuracy as the number of clusterings is increased in the ensembles. We can see that as the ensembles become easier to solve, the accuracy of all algorithms increases. We can also see that the increasing accuracy of most algorithms reaches a plateau once the number of clusterings grows very large. Figure 4.4 shows the variation in accuracy of the ITK, sMCLA, and sHBGF over the

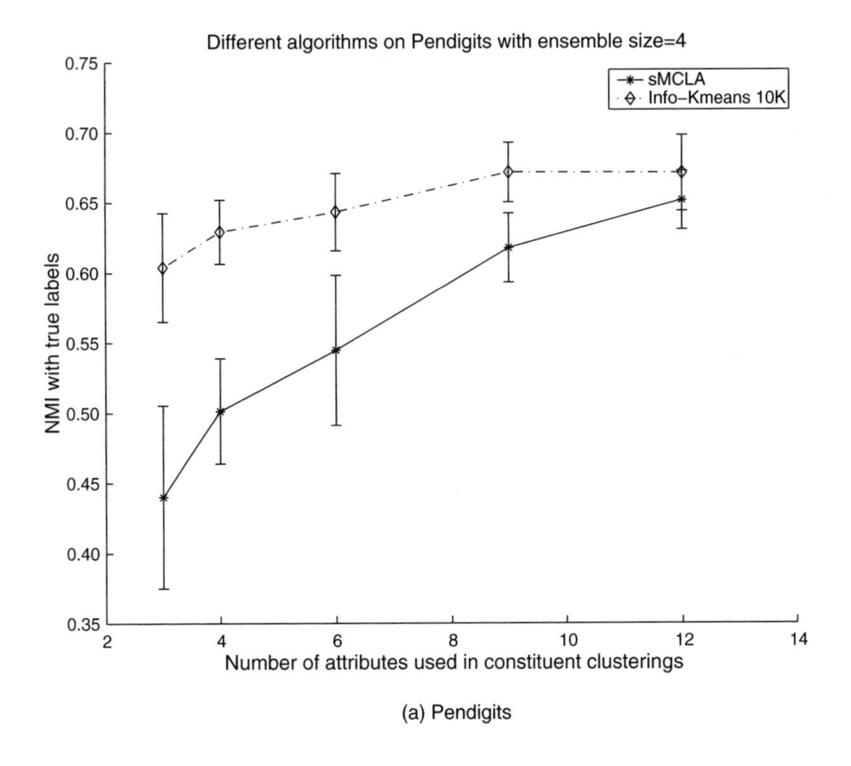

(a) Pendigits

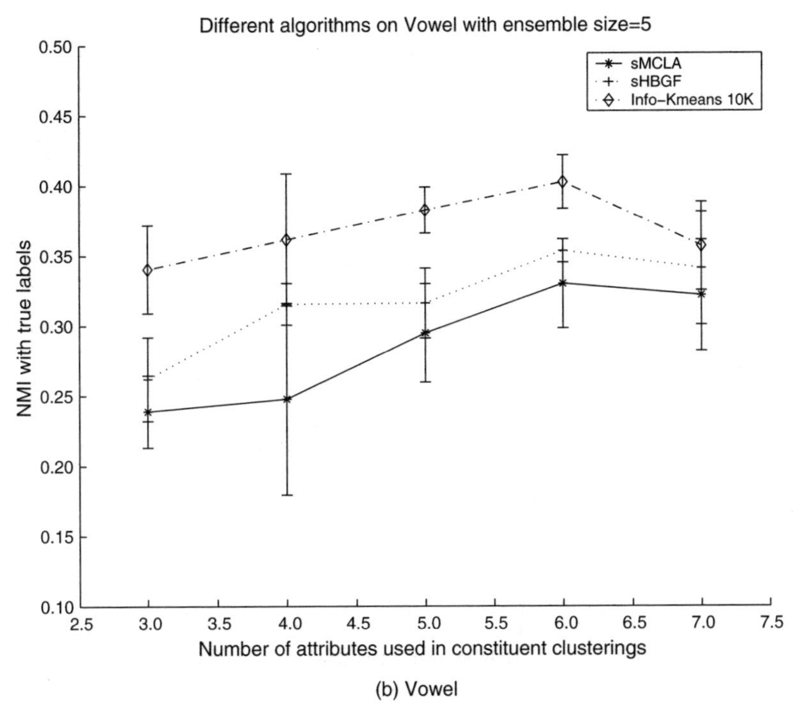

(b) Vowel

Figure 4.2 Performance of ITK, sMCLA, and sHBGF while varying the number of attributes used in constituent clusterings.

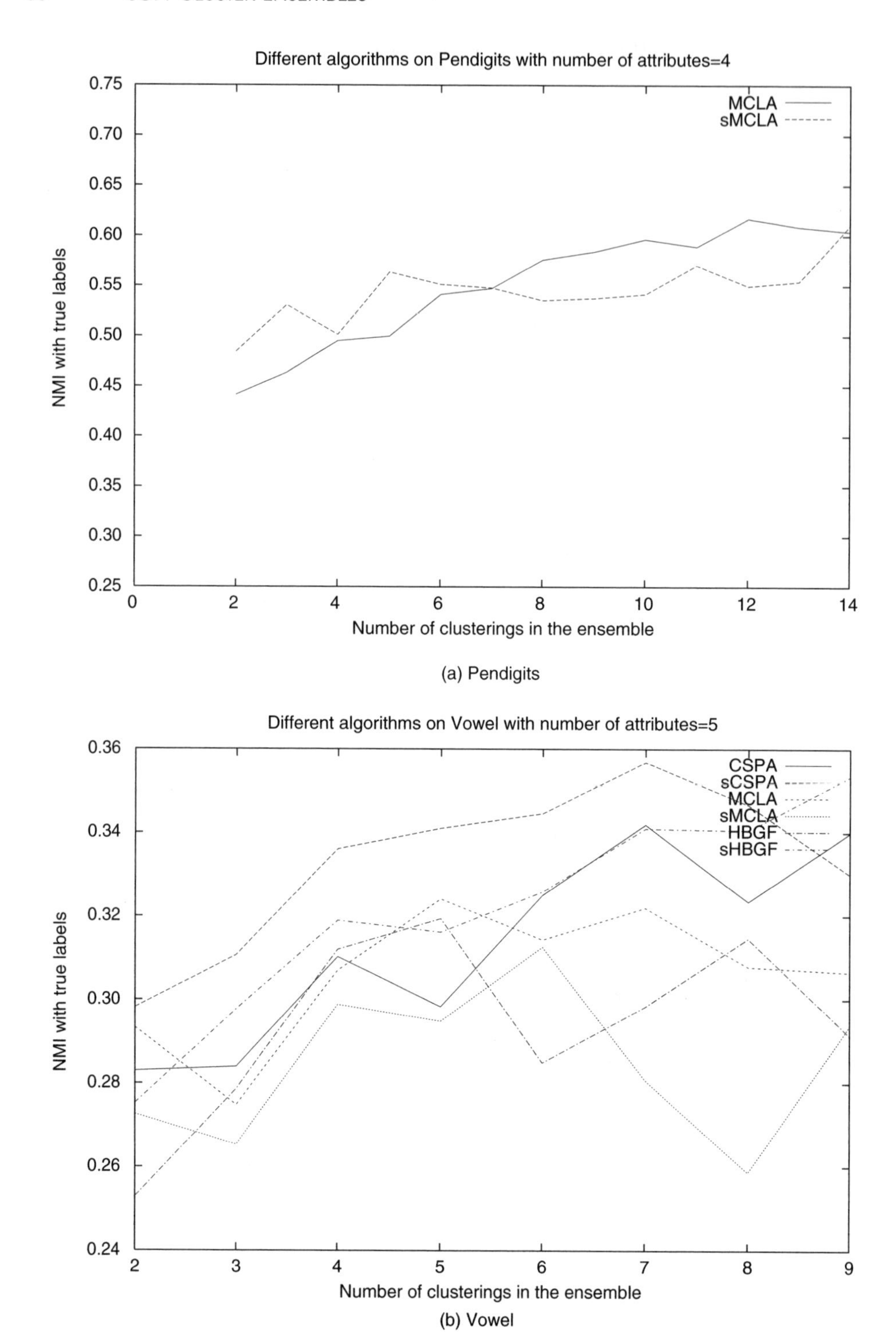

(a) Pendigits

(b) Vowel

Figure 4.3 Performance of CSPA, MCLA, HBGF, sCSPA, sMCLA, and sHBGF while varying the number of constituent clusterings.

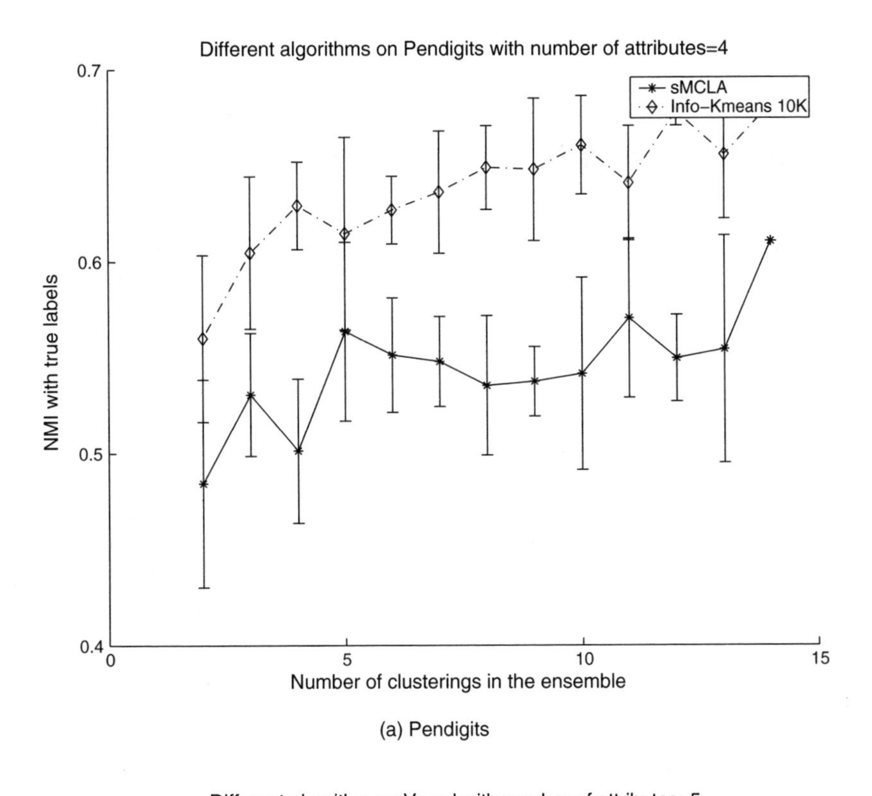

(a) Pendigits

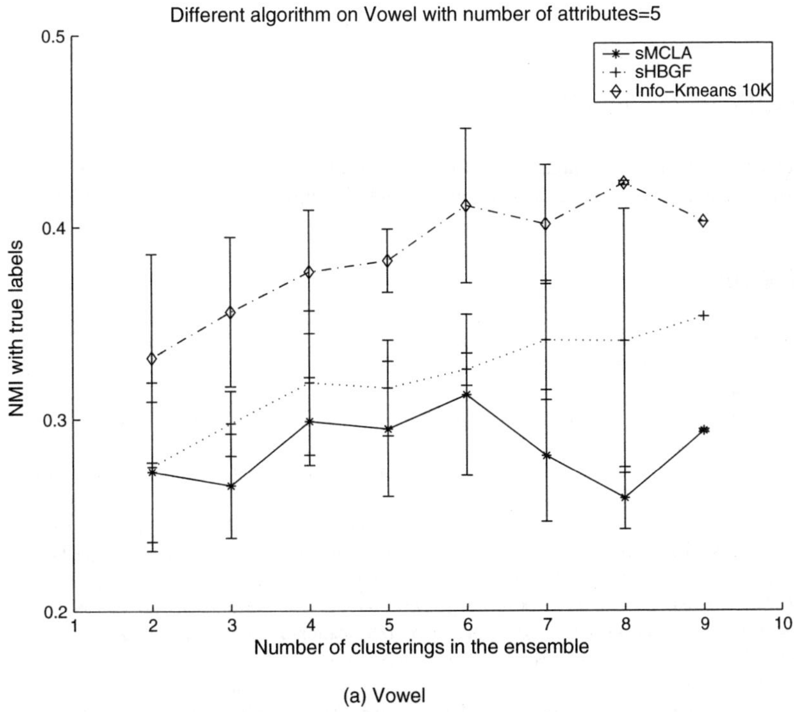

(a) Vowel

Figure 4.4 Performance of ITK, sMCLA, and sHBGF while varying the number of constituent clusterings.

Pendigits and Vowel data-sets as we increase the size of the ensembles. The accuracy of all the algorithms rises but the ITK algorithm performs significantly better than the others.

4.6 CONCLUSIONS AND FUTURE WORK

In this chapter we presented several approaches to solving ensembles of soft clusterings. We introduced a new approach based on Information-Theoretic k-means, and also presented simple extensions of existing approaches for hard ensembles (like sCSPA, sMCLA, and sHBGF). These approaches were extensively evaluated using data-sets and ensembles of varying degrees of difficulty. Some principal conclusions we made were that soft ensembles contain useful information that can be exploited by our algorithms to obtain better consensus clusterings, especially in situations where the constituent clusterings are not very accurate. Also, ITK significantly outperforms existing approaches over most data-sets, with the improvement in performance being especially large when dealing with *tough* ensembles.

Though the experimental results given in this chapter all assume the same number of clusters in each solution, the approaches do allow for varying resolution in the individual solutions. Moreover, the match of the consensus solution at different resolutions with respect to the individual solutions along the lines of Ghosh, Strehl, and Merugu, (2002) provides a good method of model selection. A challenge to the readers of this book is to identify scenarios where the use of soft ensembles provides significantly improved performance over hard ensembles, and if needed devise specialized algorithms to deal with these domains.

While partitioning instances we can also imagine a grouping of the clusters into meta-clusters. Algorithms based on MCLA and HBGF already compute these co-clusterings, albeit using graph partitioning based approaches. There is a significant body of research on co-clustering or bi-clustering using other approaches (Dhillon, Mallela, and Modha, 2003a; Madeira and Oliveira, 2004), and it will be worthwhile to investigate specialized co-clustering approaches for obtaining a consensus of soft clusterings.

ACKNOWLEDGEMENTS

The authors wish to thank Arindam Banerjee, Raymond Mooney, Sreangsu Acharyya, Suju Rajan, Srujana Merugu and for helpful discussions and thoughtful suggestions through the course of this work.

REFERENCES

Bezdek, J.C. and Pal, S. (1992) *Fuzzy Models for Pattern Recognition.* IEEE Press, Piscataway, NJ.

Blake, C. and Merz, C. (1998) UCI repository of machine learning databases, http://www.ics.uci.edu/~mlearn/mlrepository.html.

Breiman, L. (1999) 'Combining predictors' In *Combining Artificial Neural Nets* (ed. Sharkey A) pp. 31–50. Springer-Verlag Berlin, Germany.

Dempster, A., Laird, N. and Rubin, D. (1977) 'Maximum likelihood from incomplete data via the em algorithm' *Journal of the Royal Statistical Society,* vol. 39 Series, B., pp. 1–38.

Dhillon, I.S. (2001) 'co-clustering documents and words using bipartite spectral graph partitioning' *Proceedings of The Seventh, A.C.M. SIGKDD International Conference on Knowledge Discovery and Data Mining(KDD),* pp. 269–274.

Dhillon, I., Mallela, S. and Modha, D. (2003a) 'Information-Theoretic co-clustering' *Proceedings of The Ninth, A.C.M. SIGKDD International Conference on Knowledge Discovery and Data Mining(KDD),* pp. 89–98.

Dhillon, I.S., Mallela, S. and Kumar, R. (2003b) 'A divisive Information-Theoretic feature clustering algorithm for text classification'. *Journal of Machine Learning Research* **3**, 1265–1287.

Dimitriadou, E., Weingessel, A. and Hornik, K. (2001) 'Voting-Merging: An ensemble method for clustering' In *Proceedings of the International Conference on Artificial Neural Networks (ICANN 01)* (ed. Dorffner, G., Bischof, H. and Hornik K), vol. LNCS 2130, pp. 217–224, Vienna, Austria.

Dunn, J.C. (1973) 'A fuzzy relative of the isodata process and its use in detecting compact well-separated clusters'. *Journal of Cybernetics* **3**, 32–57.

Fern, X.Z. and Brodley, C.E. (2003) 'Random projection for high dimensional clustering: A cluster ensemble approach' *Proceedings of the Twentieth International Conference on Machine Learning*. ACM Press.

Fern, X.Z. and Brodley, C.E. (2004) 'Solving cluster ensemble problems by bipartite graph partitioning' *Proceedings of the Twenty-first International Conference on Machine Learning*. ACM Press.

Fred, A. and Jain, A.K. (2001) 'Finding consistent clusters in data partitions' In *Proceedings of the Third International Workshop on Multiple Classifier Systems* (ed. F. Roli, J.K.), vol. LNCS 2364, pp. 309–318.

Fred, A. and Jain, A.K. (2002) 'Data clustering using evidence accumulation' *Proceedings of the Sixteenth International Conference on Pattern Recognition (ICPR)*, pp. 276–280.

Freund, Y. and Schapire, R. (1996) 'Experiments with a new boosting algorithm' *Proceedings of the Thirteenth International Conference on Machine Learning*, pp. 148–156. Morgan Kaufmann.

Frossyniotis, D.S., Pertselakis, M. and Stafylopatis, A. (2002) 'A multi-clustering fusion algorithm' *Proceedings of the Second Hellenic Conference on AI*, pp. 225–236. Springer-Verlag.

Gablentz, W., Koppen, M. and Dimitriadou, E. (2000) 'Robust clustering by evolutionary computation' *Proc. 5th Online World Conference on Soft Computing in Industrial Applications*.

Ghosh, J. (2002) 'Multiclassifier systems: Back to the future (invited paper)' In *Multiple Classifier Systems* (ed. Roli F and Kittler J) LNCS Vol. 2364, Springer pp. 1–15.

Ghosh, J., Strehl, A. and Merugu, S. (2002) 'A consensus framework for integrating distributed clusterings under limited knowledge sharing' *Proceedings of NSF Workshop on Next Generation Data Mining*, pp. 99–108.

Hadjitodorov, S., Kuncheva, L. and Todorova, L. (2006) 'Moderate diversity for better cluster ensembles'. *Information Fusion* **7(3)**, 264–275.

Hansen, L. and Salamon, P. (1990) 'Neural network ensembles'. *IEEE Transactions on Pattern Analysis and Machine Intelligence* **12**, 993–1001.

Hubert, L. and Arabie, P. (1985) 'Comparing partitions'. *Journal of Classification* **2**, 193–218.

Karypis, G. and Kumar, V. (1998) 'A fast and high quality multilevel scheme for partitioning irregular graphs'. *SIAM Journal on Scientific Computing* **20(1)**, 359–392.

Karypis, G., Aggarwal, R., Kumar, V. and Shekhar, S. (1997) 'Multilevel hypergraph partitioning: application in VLSI domain' *Proceedings of the Thirty-fourth Annual Conference on Design Automation*, pp. 526–529.

Kreiger, A.M. and Green, P. (1999) 'A generalized rand-index method for consensus clustering of separate partitions of the same data base'. *Journal of Classification* **16**, 63–89.

Kullback, S. and Leibler, R.A. (1951) 'On information and sufficiency'. *Annals of Mathematical Statistics* **22**, 79–86.

Kumar, S., Ghosh, J. and Crawford, M.M. (2001) 'Best basis feature extraction algorithms for classification of hyperspectral data' *IEEE Transactions on Geoscience and Remote Sensing, Special Issue on Analysis of Hyperspectral Data*, vol. 39(7), pp. 1368–1379.

Kuncheva, L. and Hadjitodorov, S. (2004) 'Using diversity in cluster ensembles' *Proceedings of IEEE International Conference on Systems, Man and Cybernetics*, pp. 1214–1219.

Madeira, S.C. and Oliveira, A.L. (2004) 'Biclustering algorithms for biological data analysis: a survey'. *IEEE/ACM Transactions on Computational Biology and Bioinformatics* **1(1)**, 24–45.

Melville, P. and Mooney, R.J. (2003) 'Constructing diverse classifier ensembles using artificial training examples' *Proceedings of the Eighteenth International Joint Conference on Artificial Intelligence (IJCAI)*, pp. 505–510.

Merugu, S. and Ghosh, J. (2003) 'Privacy-preserving distributed clustering using generative models' *Proceedings of the Third IEEE International Conference on Data Mining (ICDM)*, pp. 211–218.

Ng, A., Jordan, M. and Weiss, Y. (2001) 'On spectral clustering: Analysis and an algorithm' *Proceedings of Advances in Neural Information Processing Systems 14*, pp. 849–856.

Opitz, D. and Maclin, R. (1999) 'Popular ensemble methods: An empirical study'. *Journal of Artificial Intelligence Research* **11**, 169–198.

Pedrycz, W. (2002) 'Collaborative fuzzy clustering'. *Pattern Recognition Letters* **23**(14), 1675–86.

Strehl, A. and Ghosh, J. (2002) 'Cluster ensembles – a knowledge reuse framework for combining multiple partitions'. *Journal on Machine Learning Research (JMLR)* **3**, 583–617.

Topchy, A., Jain, A. and Punch, W. (2004) 'Mixture Model for Clustering Ensembles' *Proceedings of The Fourth, S.I.AM Conference on Data Mining (SDM)*, pp. 379–390.

Webb, G.I. (2000) Multiboosting: 'A technique for combining boosting and wagging'. *Machine Learning* **40**(2), 159–196.

Part II
Visualization

5

Aggregation and Visualization of Fuzzy Clusters Based on Fuzzy Similarity Measures

János Abonyi and Balázs Feil

Department of Process Engineering, Pannon University, Hungary

Most fuzzy clustering algorithms attempt to partition a data-set into self-similar groups (clusters) and describe the geometric structure of the clusters using prototypical cluster shapes such as volumetric clouds, hyperspherical shells, regression models, etc.. The main benefit of the application of these clustering algorithms is that the clustering algorithm not only partitions the data, but in some cases the main result of the clustering is the set of the generated cluster prototypes. Hence, *prototype generator clustering* methods are widely used in the initial steps of complex knowledge discovery in databases (KDD) processes. This is because clustering can be used for the segmentation of the data and this can be very useful at the data selection and preprocessing steps, while they can also be used to initialize a data mining model based on the generated prototypes. According to this field of application several new fuzzy clustering algorithms have been developed tailored to handle classification (Pach, Abonyi, Nemeth, and Arva, 2004; Pedrycz and Sosnowski, 2001), regression (Abonyi, Babuska, and Szeifert, 2002; Johansen and Babuska, 2002; Kim, Park, and Kim, 1998), time-series prediction, and segmentation (Abonyi, Feil, Nemeth, and Arva, 2005; Baldwin, Martin, and Rossiter, 1998; Geva, 1999; Wong 1998) problems.

The performance of prototype based fuzzy clustering methods is highly determined by how the selected prototypes are consistent with the data, how the assumed number of the clusters is correct, which justifies the study of cluster validity techniques, which attempt to assess the correctness of a particular set of clusters in a given data-set. As the large number of the applications of fuzzy clustering techniques and the increasing number of the special prototype based algorithms show, the analysis of the performance of these algorithms is a significant practical and theoretical problem.

Clustering algorithms always fit the clusters to the data, even if the cluster structure is not adequate for the problem. To analyze the adequacy of the cluster prototypes and the number of the clusters, cluster validity measures are used. Appendix 5A.1 will give a short review of these measures. It will be shown

Advances in Fuzzy Clustering and its Applications Edited by J. Valente de Oliveira and W. Pedrycz

that conventional cluster validity techniques represent all the validity information by a single number, which in some cases does not provide as much information about results as needed; namely, how correct the cluster prototypes are, what the required number of clusters could be in case of the applied cluster prototype, which clusters could be merged, or which cluster should be splitted into sub-clusters to get more reliable partitioning of the data and cluster prototypes that locally describe the data.

Since validity measures reduce the overall evaluation to a single number, they cannot avoid a certain loss of information. To avoid this problem, this chapter suggests the approach of the visualization of fuzzy clustering results, since the low-dimensional graphical representation of the clusters could be much more informative than such a single value of the cluster validity. Hence, the aim of this chapter is to give a critical overview about the existing cluster visualization techniques, propose new approaches for the visualization of cluster validity, and show how these tools can be applied.

The impact of visualization of fuzzy clustering results has already been realized in Klawonn, Chekhtman, and Janz (2003), when the membership values of the data obtained by the clustering algorithm were simply projected into the input variables, and the resulted plots served for the same purpose as validity measures.

Bezdek and Hathaway (2002) have updated the method of Johnson and Wichern is based in inter-data distances and developed a set of VAT algorithms for the visual assessment of (cluster) tendency. VAT uses a digital intensity image of the reorganized inter-data distance matrix obtained by efficient reorganization schemes, and the number of dark diagonal blocks on the image indicates the number of clusters in the data. The different VAT modifications can be used to detect how the data are clustered since the algorithm used to reorganize the distance matrix is based on the Prim's algorithm applied to graph-theoretic minimal spanning tree based clustering. The original VAT algorithm will be described in Section 5.4.1. To detect information about the performance of prototype generator clustering methods the same authors proposed a visual cluster validity (VCV) algorithm by a minor modification of VAT. Cluster prototype based clustering algorithms minimize the distances between the cluster and the data, where different cluster prototypes define different distance metrics. Since VAT is based on the reorganization and visualization of the inter-data distance matrix, this tool cannot be directly applied since fuzzy clustering algorithms operate only based on the distances between the cluster and the data. The key idea of VCV is to calculate the missing inter-datum distances based on the triangular inequality property (Hathaway and Bezdeck, 2002). The details of this algorithm will be presented in Section 5.4.2 of this chapter. Beside the critical analysis of this method some minor modifications will also be presented based on the similarity analysis of the clusters.

The ideas behind the cluster validity measures, the VAT and the VCV algorithms already illustrated that a good approach for the cluster validity analysis is the analysis of the cluster-prototype data distances. In higher-dimensional problems this analysis is not possible. One of the approaches applied to the visualization of high-dimensional spaces is the distance preserving mapping of the higher-dimensional space into a lower, usually two-dimensional map. Two general approaches for dimensionality reduction are: (i) feature extraction, transforming the existing features into a lower-dimensional space, and (ii) feature selection, selecting a subset of the existing features without a transformation.

Feature extraction means creating a subset of new features by combination of existing features. These methods can be grouped based on linearity (see Figure 5.1). A linear feature extraction or

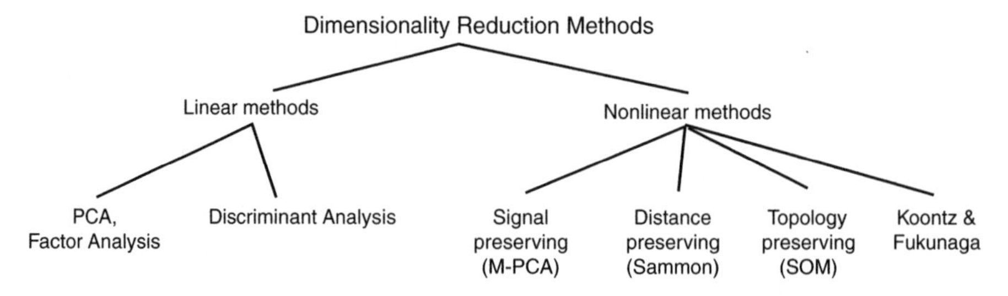

Figure 5.1 Taxonomy of dimensionality reduction methods (Jain and Dubes, 1988).

projection expresses the new features as linear combination of the original variables. The type of linear projection used in practice is influenced by the availability of category information about the patterns in the form of labels on the patterns. If no category information is available, the eigenvector projection (also called principal component analysis (PCA)) is commonly used. Discriminant analysis is a popular linear mapping technique when category labels are available. In many cases, linear projection cannot preserve the data structure because of its complexity. In these cases nonlinear projection methods should be used.

Among the wide range of clustering tools, the self-organizing map (SOM) is often visualized by principal component analysis (PCA) and Sammon mapping to give more insight into the structure of high-dimensional data. Usually, with the use of these tools the cluster centers (the codebook of the SOM) are mapped into a two-dimensional space (Vesanto, 2000). Fuzzy C-means cluster analysis has also been combined with similar mappings and successfully applied to map the distribution of pollutants and to trace their sources to access potential environmental hazard on a soil database from Austria (Hanesch, Scholger and Dekkers, 2001).

While PCA attempts to preserve the variance of the data during the mapping, Sammon's mapping tries to preserve the interpattern distances (Mao and Jain, 1995; Pal and Eluri, 1998). Hence, this chapter focuses on the application of Sammon mapping for the visualization of the results of clustering, as the mapping of the distances is much closer to the task of clustering than preserving the variances. There are two main problems encountered in the application of Sammon mapping to the visualization of fuzzy clustering results:

- The aim of cluster analysis is the classification of objects according to similarities among them, and organizing data into groups. In metric spaces, similarity is often defined by means of distance from a data vector to some prototypical object of the cluster. The prototypes are usually not known beforehand, and are sought by the clustering algorithm simultaneously with the partitioning of the data. The prototypes may be vectors (centers) of the same dimension as the data objects, but they can also be defined as "higher-level" geometrical objects, such as linear or nonlinear subspaces or functions. Hence, classical projection methods based on the variance of the data (PCA) or based on the preservation of the Euclidean interpoint distance of the data (Sammon mapping) are not applicable when the clustering algorithm does not use the Euclidean distance norm.
- As Sammon mapping attempts to preserve the structure of high n-dimensional data by finding N points in a much lower q-dimensional data space, such that the interpoint distances measured in the q-dimensional space approximate the corresponding interpoint distances in the n-dimensional space, the algorithm involves a large number of computations as in every iteration step it requires the computation of $N(N-1)/2$ distances. Hence, the application of Sammon mapping becomes impractical for large N (de Ridder and Duin, 1997).

To avoid these problems this chapter also proposes a new algorithm. By using the basic properties of fuzzy clustering algorithms the proposed tool maps the cluster centers and the data such that the distances between the clusters and the datapoints will be preserved. During the iterative mapping process, the algorithm uses the membership values of the data and minimizes an objective function that is similar to the objective function of the original clustering algorithm.

5.1 PROBLEM DEFINITION

Since clusters can formally be seen as subsets of the data-set, one possible classification of clustering methods can be according to whether the subsets are fuzzy or crisp (hard). Hard clustering methods are based on classical set theory, and require that an object either does or does not belong to a cluster. Hard clustering in a data set $\mathbf{X} = \{\mathbf{x}_k | k = 1, 2, \ldots, N\}$ means partitioning the data into a specified number of mutually exclusive subsets of $\mathbf{X}$. The number of subsets (clusters) is denoted by c. Fuzzy clustering methods allow objects to belong to several clusters simultaneously, with different degrees of membership.

The data-set $\mathbf{X}$ is thus partitioned into c fuzzy subsets. In many real situations, fuzzy clustering is more natural than hard clustering, as objects on the boundaries between several classes are not forced to fully belong to one of the classes, but rather are assigned membership degrees between 0 and 1 indicating their partial memberships. The discrete nature of hard partitioning also causes analytical and algorithmic intractability of algorithms based on analytic functionals, since these functionals are not differentiable.

The objective of clustering is to partition the data set $\mathbf{X}$ into c clusters. For the time being, assume that c is known, based on prior knowledge, for instance. Fuzzy and possibilistic partitions can be seen as a generalization of hard partitioning. Possibilistic partition is similar to fuzzy one because a data point can belong to several clusters simultaneously but it does not require that the sum of memberships of one data point is equal to 1. It is able to deal with outliers better than fuzzy clustering but it raises several problems in the definition and minimization of the objective function. This chapter deals only with fuzzy clustering.

A $c \times N$ matrix $\mathbf{U} = [\mu_{i,k}]$ represents the fuzzy partitions, where $\mu_{i,k}$ denotes the degree of the membership of the $\mathbf{x}_k$th observation belongs to the $1 \leq i \leq c$th cluster, so the ith row of $\mathbf{U}$ contains values of the *membership function* of the ith fuzzy subset of $\mathbf{X}$. The matrix $\mathbf{U}$ is called the fuzzy partition matrix. Conditions for a fuzzy partition matrix are given by:

$$\mu_{i,k} \in [0,1], \ 1 \leq i \leq c, \ 1 \leq k \leq N, \tag{5.1}$$

$$\sum_{i=1}^{c} \mu_{i,k} = 1, \ 1 \leq k \leq N, \tag{5.2}$$

$$0 < \sum_{k=1}^{N} \mu_{i,k} < N, \ 1 \leq i \leq c. \tag{5.3}$$

Fuzzy partitioning space. Let $\mathbf{X} = [\mathbf{x}_1, \mathbf{x}_2, \ldots, \mathbf{x}_N]$ be a finite set and let $2 \leq c < N$ be an integer. The fuzzy partitioning space for $\mathbf{X}$ is the set

$$M_{fc} = \left\{ \mathbf{U} \in \mathbf{R}^{c \times N} | \mu_{i,k} \in [0,1], \forall i,k; \sum_{i=1}^{c} \mu_{i,k} = 1, \forall k; 0 < \sum_{k=1}^{N} \mu_{i,k} < N, \forall i \right\}. \tag{5.4}$$

Equation (5.2)} constrains the sum of each column to 1, and thus the total membership of each $\mathbf{x}_k$ in $\mathbf{X}$ equals 1. The distribution of memberships among the c fuzzy subsets is not constrained.

A large family of fuzzy clustering algorithms is based on minimization of the sum-of-squared error or minimum variance objective function (Duda, Hart, and Stork, 2001) formulated as:

$$J(\mathbf{X}; U, V) = \sum_{i=1}^{c} \sum_{k=1}^{N} (\mu_{i,k})^m d(\mathbf{x}_k, \eta_i)^2 \tag{5.5}$$

where $\mathbf{U} = [\mu_{i,k}]$ is a fuzzy partition matrix of $\mathbf{X}$, η_i is the ith *cluster prototype*, which has to be determined, and $d(\mathbf{x}_k, \eta_i)^2$ is a squared inner-product distance norm between the kth sample and the ith prototype. In the classical fuzzy C-means (FCM) algorithm, the prototypes are centers, $\mathbf{V} = [\mathbf{v}_1, \mathbf{v}_2, \ldots, \mathbf{v}_c]$, $\mathbf{v}_i \in \mathbf{R}^n$; therefore, the distance can be formed in the following way:

$$d(\mathbf{x}_k, \mathbf{v}_i)^2 = \| \mathbf{x}_k - \mathbf{v}_i \|_{\mathbf{A}}^2 = (\mathbf{x}_k - \mathbf{v}_i)^T \mathbf{A} (\mathbf{x}_k - \mathbf{v}_i) \tag{5.6}$$

where $\mathbf{A}$ is the distance measure (if there is no prior knowledge, $\mathbf{A} = \mathbf{I}$), and $m \in \langle 1, \infty)$ is a weighting exponent which determines the fuzziness of the resulting clusters. The measure of dissimilarity in Equation (5.5) is the squared distance between each data point $\mathbf{x}_k$ and the cluster prototype η_i. This distance is weighted by the power of the membership degree of that point $(\mu_{i,k})^m$. The value of the cost function Equation (5.5) is a measure of the total weighted within-group squared error incurred by the representation of the c clusters defined by their prototypes $\mathbf{v}_i$. Statistically, Equation (5.5) can be seen as a measure of the total variance of $\{\mathbf{x_k}$ from η_i. If the prototypes are known, the membership degrees can be determined in the following way:

$$\mu_{i,k} = \frac{1}{\sum_{j=1}^{c} (d(\mathbf{x}_k, \eta_i)/d(\mathbf{x}_k, \eta_j))^{2/(m-1)}}, 1 \leq i \leq c, \ 1 \leq k \leq N. \tag{5.7}$$

As was mentioned above, prototype generator clustering methods always produce clusters. They do it even if there is no (cluster) structure in the data. This chapter deals with other types of problems: how can the user know whether:

- the given partition;
- the number of clusters; or
- the applied type of cluster prototypes are consistent with the analysed data.

The given partition (the partition matrix $\mathbf{U}$) may be inadequate even if the number of clusters and the used prototype are consistent. This is because the clustering methods may be stuck in a local optimum in the search space – or it may happen if the applied method for minimizing the cost function Equation (5.5) is alternating optimization. This problem can be solved by global search (e.g., genetic or evolutionary algorithms). In practice, a simpler approach is used: the clustering algorithm is run several times but from different initial points, and the fuzzy partition with the minimal cost function value (see Equation (5.5)) is chosen as the optimal solution.

5.2 CLASSICAL METHODS FOR CLUSTER VALIDITY AND MERGING

Cluster validity refers to the problem whether a given fuzzy partition fits to all the data, and it is often referred to as the problem of the appropriate number of clusters (especially if the optimal fuzzy partitions are compared). Two main approaches to determine the appropriate number of clusters in data can be distinguished:

- Clustering data for different values of c, and using *validity measures* to assess the goodness of the obtained partitions. This can be done in two ways:
  - The first approach is to define a validity function which evaluates a complete partition. An upper bound for the number of clusters must be estimated ($c_{\max}$), and the algorithms have to be run with each $c \in \{2, 3, \ldots, c_{\max}\}$. For each partition, the validity function provides a value such that the results of the analysis can be compared indirectly.
  - The second approach consists of the definition of a validity function that evaluates individual clusters of a cluster partition. Again, $c_{\max}$ has to be estimated and the cluster analysis has to be carried out for $c_{\max}$. The resulting clusters are compared to each other on the basis of the validity function. Similar clusters are collected in one cluster, very bad clusters are eliminated, so the number of clusters is reduced. The procedure can be repeated until there are no "bad" clusters.
- Starting with a sufficiently large number of clusters, and successively reducing this number by merging clusters that are similar (compatible) with respect to some predefined criteria. This approach is called *compatible cluster merging*.

Appendix 5A.1 gives an overview of the applicable validity measures. Most validation indices proposed during the last decades have focused on two properties: *compactness and separation.* Compactness is used as a measure of the variation or scattering of the data within a cluster, and separation is used to account for inter-cluster structural information. The basic aim of validation indices has been to find the clustering that minimizes the compactness and maximizes the separation. However, the classical indices are limited in their ability to compute these properties because there are several ways to define compactness and separation, and there is no index that can take into account more viewpoints and also other viewpoints besides these ones, e.g., the aim of clustering (partition, regression, classification, etc.). In other words, there is no general validity index and it can be more effective if the results of clustering are visualized because it can give more and detailed information and the user needs can also be taken into account (see also Section 5.4).

The recursive cluster merging technique evaluates the clusters for their compatibility (similarity) and merges the clusters that are found to be compatible. During this merging procedure the number of clusters is gradually reduced. This procedure can be controlled by a *fuzzy decision-making algorithm* based on the similarity of (fuzzy) clusters. Because the compatibility criterion quantifies various aspects of the similarity of the clusters, the overall cluster compatibility should be obtained through an aggregation procedure. A fuzzy decision-making algorithm can be used for this purpose (Babuska, 1998). In this work two criteria were combined and it can be applied to clustering algorithms where the clusters are described by centers and covariance matrices. Let the centers of two clusters be $\mathbf{v}_i$ and $\mathbf{v}_j$. Let the eigenvalues of the covariance matrices of two clusters be $\{\lambda_{i1}, \ldots, \lambda_{in}\}$ and $\{\lambda_{j1}, \ldots, \lambda_{jn}\}$, both arranged in descending order. Let the corresponding eigenvectors be $\{\phi_{i1}, \ldots, \phi_{in}\}$ and $\{\phi_{j1}, \ldots, \phi_{jn}\}$. The following compatibility criteria were defined in Babuska (1998):

$$c_{ij}^1 = |\phi_{in} \cdot \phi_{jn}|, c_{ij}^1 \text{ close to} 1, \tag{5.8}$$

$$c_{ij}^2 = ||\mathbf{v}_i - \mathbf{v}_j\ ||, c_{ij}^2 \text{ close to } 0. \tag{5.9}$$

The first criterion assesses whether the clusters are parallel, and the second criterion measures the distance of the cluster centers. These criteria are evaluated for each pair of clusters. The most similar pair of adjacent clusters has to be merged as long as the value of the corresponding similarity is above a user defined threshold γ. Clusters can be merged in several ways (see, for example, Babuska 1998; Kelly, 1994; Marcelino, Nunes, Lima, and Ribeiro, 2003). A similar compatible cluster merging method was applied to time-series segmentation in Abonyi, Feil, Nemeth, and Arva (2005).

5.3 SIMILARITY OF FUZZY CLUSTERS

The main problem of these approaches is that the similarity of the clusters is measured based on the comparison of certain parameters of the cluster prototypes (e.g., centers and orientation (eigenvectors) of the clusters).

As it will be show in this chapter, for both cluster aggregation and visualization there is a need to determine how similar the resulted clusters are. For that purpose, a fuzzy set similarity measure can be used because fuzzy clusters can be seen as fuzzy sets. The similarity of two sets, A and B can be expressed as follows:

$$\mathrm{S}(A, B) = \frac{|A \bigcap B|}{|A \bigcup B|}. \tag{5.10}$$

For fuzzy sets, instead of conjunction $\bigcap$, several logic operators exist, the so-called *t-norms* (see, for example, Nellas, 2001). Two of them are:

$$\text{Min}: \ A \text{ AND } B = \min(\mu_A, \mu_B), \tag{5.11}$$

$$\text{Product}: \ A \text{ AND } B = \mu_A \mu_B, \tag{5.12}$$

where μ_A and μ_B are the membership functions of set A and B, respectively. For the disjunction, several logic operators also exist, the so-called *t-conorms*. Two of them are:

$$\text{Max}: \ A \text{ OR } B = \max(\mu_A, \mu_B), \tag{5.13}$$

$$\text{Algebraic sum}: \ A \text{ OR } B = \mu_A + \mu_B - \mu_A \mu_B. \tag{5.14}$$

In this way, all i,j pairs of clusters can be compared to each other in the following ways:

$$\mathbf{S}(i,j) = \frac{\sum_{k=1}^{N} \min(\mu_{i,k}, \mu_{j,k})}{\sum_{k=1}^{N} \max(\mu_{i,k}, \mu_{j,k})}, \tag{5.15}$$

or

$$\mathbf{S}(i,j) = \frac{\sum_{k=1}^{N} \mu_{i,k}\mu_{j,k}}{\sum_{k=1}^{N} \mu_{i,k} + \mu_{j,k} - \mu_{i,k}\mu_{j,k}}. \tag{5.16}$$

In this chapter Equation (5.15) is used. Based on the obtained symmetric similarity matrix $\mathbf{S}$, a dendrogram can be drawn to visualize and hierarchically merge the fuzzy clusters based on a hierarchical clustering procedure (in this chapter a single-linkage algorithm was applied). Using this diagram, the human data miner can gain an idea how similar the clusters are in the original space and is able to determine which clusters should be merged if it is needed. This will be illustrated by the three data-sets used below. In all the cases, the classical fuzzy C-means algorithm was used. The parameters of the algorithm: the number of clusters c was 10, the weighting exponent m was equal to 2, and the termination tolerance ϵ was 10^{-4}.

In the following sections, three examples were used to illustrate the detailed methods, one synthetic two-dimensional data set, and the well-known Iris and Wine data-sets from UCI Machine Learning Repository. The *synthetic data-set* can be seen in Figure 5.2. This data-set contains 267 data samples. The *Iris data-set* contains measurements on three classes of Iris flower. The data-set was made by measurements of sepal length and width and petal length and width for a collection of 150 irises (so it is four-dimensional). The problem is to distinguish the three different types (*Iris setosa*, *Iris versicolor*, and *Iris virginica*). These data have been analyzed many times to illustrate various methods. The *Wine data-set* contains the chemical analysis of 178 wines grown in the same region in Italy but derived from three different cultivars. The problem is to distinguish the three different types based on 13 continuous attributes derived from chemical analysis: alcohol, malic acid, ash, alckalinity of ash, magnesium, total phenols, flavanoids, nonflavanoid phenols, proanthocyaninsm, colour intensity, hue, OD280/OD315 of diluted wines and proline.

Example 5.1 Merging similar clusters for the synthetic data-set *In the case of the synthetic data-set, more than 10 clusters are needed to 'cover the data' (based on validity indices, about 15–20). Despite that, only 10 clusters are used because this example wants to illustrate the proposed approach and 10 clusters can easily be overviewed in two dimensions. In Figure 5.3 can be seen the results of the fuzzy C-means algorithm, the contour lines represent equal membership values. As it can be seen on the figure, 3-4-9-6th clusters could be merged, and also 2-7-10-8th clusters represent a group of clusters. This dendrogram corresponds the results of the FCM algorithm. However, a much better solution can be given by another cluster prototype, namely, the Gath–Geva algorithm, which uses adaptive distance norm. However, this clustering method is very sensitive to the initialization, but minimal spanning tree initialization makes it much more robust. These results were published in Vathy-Fogarassy, Feil, and Abonyi (2005).*

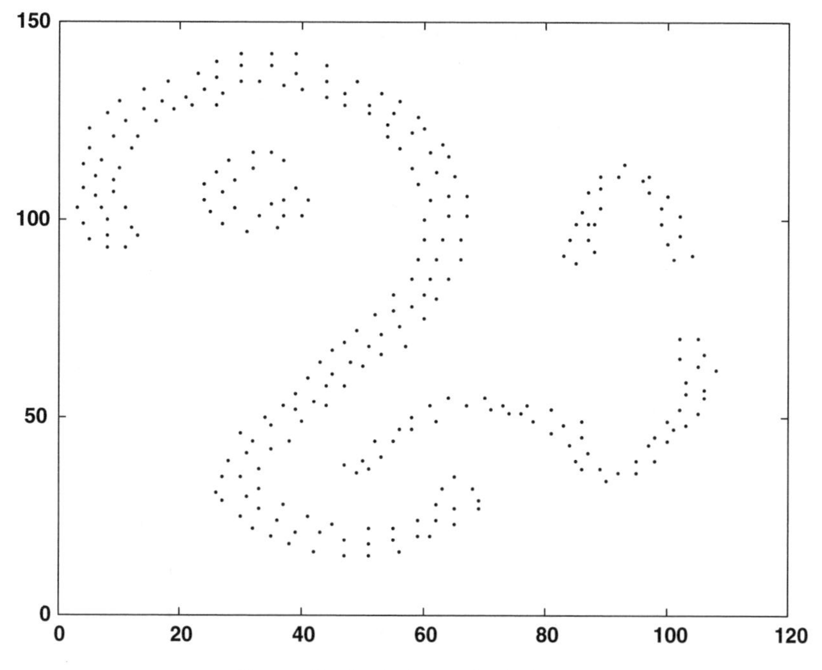

Figure 5.2 Synthetic data for illustration.

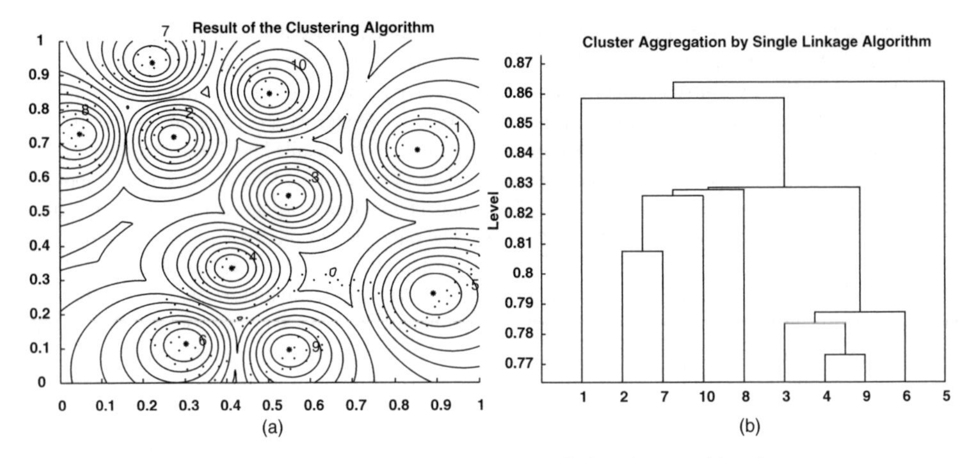

Figure 5.3 Result of the FCM algorithm; similarity of the resulting clusters.

Example 5.2 Merging similar clusters for the Iris data-set *As is known the Iris data contain three labeled classes of Iris flower, but only two well-separated clusters because data of two kinds of Iris are not clearly separable. In the case of the Iris data-set, the number of clusters* $c = 10$ *was clearly greater than the optimal number of clusters. In Figure 5.4 it can be seen that two large groups of clusters can be identified in the case of the Iris data-set.*

Example 5.3 Merging similar clusters for the Wine data-set *In the case of the Wine data-set, the number of clusters* $c = 10$ *was clearly greater than the optimal number of clusters. In Figure 5.5 it can be seen that three well-separated groups of clusters can be identified, and the optimal number of clusters is three. It does not necessarily mean that the real groups of data are well-separated as well. This problem will be discussed in Section 5.4.2 and in Example 5.9.*

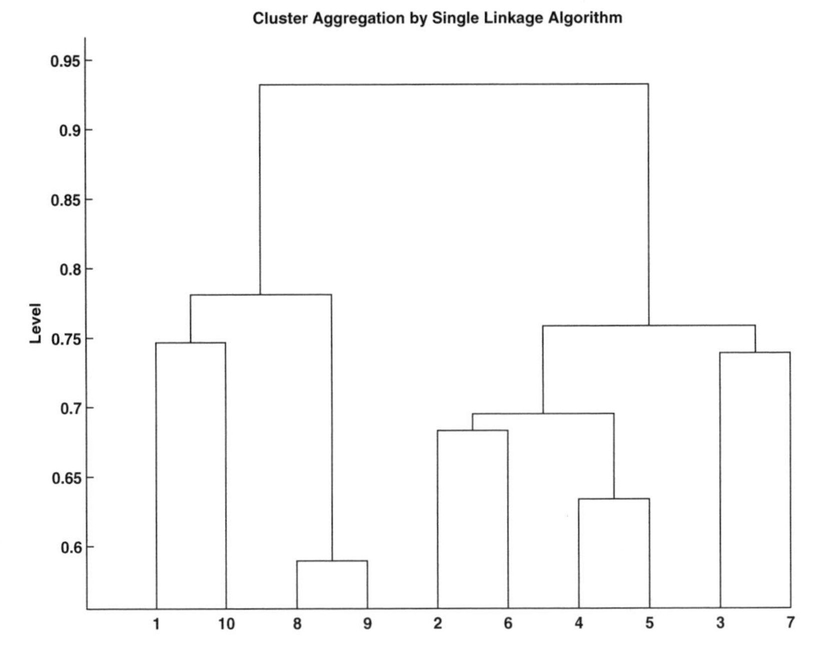

Figure 5.4 Similarity of the resulting clusters for the Iris data-set.

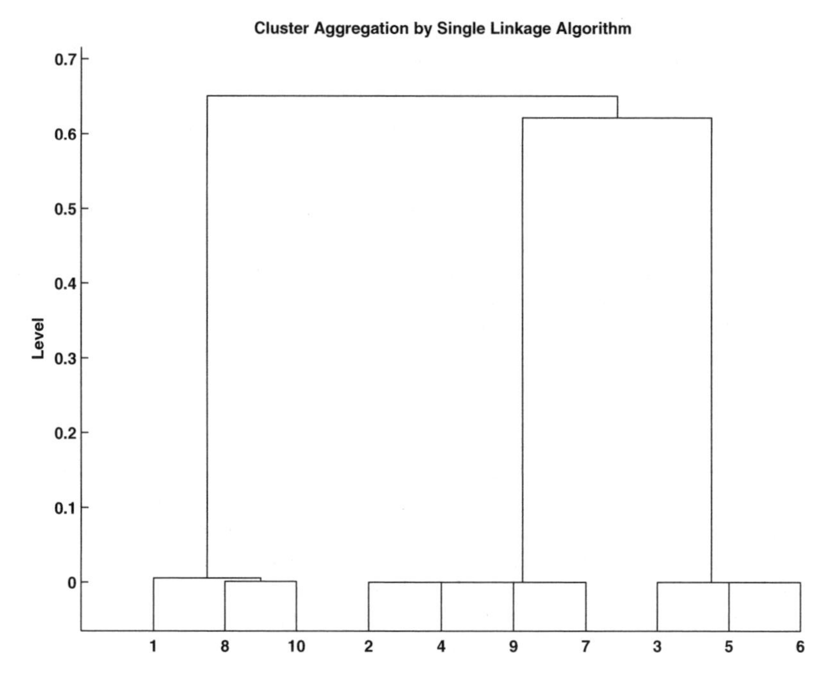

Figure 5.5 Similarity of the resulted clusters for the Wine data-set.

5.4 VISUALIZATION OF CLUSTERING RESULTS

Maybe the most complex problem in cluster validity is to decide whether the applied clarify or re-word cluster prototype fits the data at all. To solve that problem, much more information has to be gathered. Since validity measures reduce the overall evaluation to a single number, they cannot avoid a certain loss of information. A low-dimensional graphical representation of the clusters could be much more informative than such a single value of the cluster validity because one can cluster by eye and qualitatively validate conclusions drawn from clustering algorithms.

In the following sections the most effective visualization techniques will be overviewed and based on their critical analysis new algorithms will be proposed.

5.4.1 Visual Assessment of Cluster Tendency

Visual assessment of cluster tendency (VAT) method was proposed in (Bezdek and Hathaway, 2002), and its variants in Huband, Bezdek, and Hathaway 2004, 2005). Its aim is similar to one of cluster validity indices, but it tries to avoid the 'massive aggregation of information' by scalar validity measures. Instead of a scalar value or a series of scalar values by a different number of clusters, an $N \times N$ intensity image is proposed by Hathaway and Bezdek. It displays the reordered form of the dissimilarity data $\mathbf{D} = [d(\mathbf{x}_i, \mathbf{x}_j)]_{N \times N}$, where $d(\mathbf{x}_i, \mathbf{x}_j)$ is the dissimilarity of the ith and jth samples (not necessarily distance, but in this chapter we use their distance as the dissimilarity measure). The method consists of two steps:

- Step 1 reorder the dissimilarity data and get $\tilde{\mathbf{D}}$, in which the adjacent points are members of a possible cluster;
- Step 2 display the dissimilarity image based on $\tilde{\mathbf{D}}$, where the gray level of a pixel is in connection with the dissimilarity of the actual pair of points.

The key step of this procedure is the *reordering* of $\mathbf{D}$. For that purpose, Bezdek used Prim's algorithm for finding a minimal spanning tree. The undirected, fully connected, and weighted graph analyzed here contains the data points or samples as nodes (vertices) and the edge lengths or weights of the edges are the values in $\mathbf{D}$, the pairwise distances between the samples. There are two differences between Prim's algorithm and VAT: (1) VAT does not need the minimal spanning tree itself (however, it also determines the edges but does not store them), just the order in which the vertices (samples or objects $\mathbf{x}_i$) are added to the tree; (2) it applies special initialization. Minimal spanning tree contains all of the vertices of the fully connected, weighted graph of the samples, therefore any points can be selected as the initial vertex. However, to help ensure the best chance of display success, Bezdek proposed a special initialization: the initial vertex is any of the two samples that are the farthest from each other in the data set ($\mathbf{x}_i$, where i is the row or column index of $\max(\mathbf{D})$). The first row and column of $\tilde{\mathbf{D}}$ will be the ith row and column in $\mathbf{D}$. After the initialization, the two methods are exactly the same. Namely, $\mathbf{D}$ is reordered so that the second row and column correspond to the sample closest to the first sample, the third row and column correspond to the sample closest to either one of the first two samples, and so on.

This procedure is similar to the single-linkage algorithm that corresponds to the Kruskal's minimal spanning tree algorithm and is basically the greedy approach to find a minimal spanning tree. By hierarchical clustering algorithms (such as single-linkage, complete-linkage, or average-linkage methods), the results are displayed as a dendrogram, which is a nested structure of clusters. (Hierarchical clustering methods are not described here, the interested reader can refer, for example, to Jain and Dubes, (1988).) Bezdek and colleagues followed another way and they displayed the results as an intensity image $I(\tilde{\mathbf{D}})$ with the size of $N \times N$. The approach was presented in (Huband, Bezdek, and Hathaway, 2005) as follows. Let $G = \{g_m, \ldots, g_M\}$ be the set of gray levels used for image displays. In the following, $G = \{0, \ldots, 255\}$, so $g_m = 0$ (black) and $g_M = 255$ (white). Calculate

$$(I(\tilde{\mathbf{D}}))_{i,j} = \tilde{\mathbf{D}}_{i,j}\left(\frac{g_M}{\max(\tilde{\mathbf{D}})}\right). \tag{5.17}$$

Convert $(I(\tilde{\mathbf{D}}))_{i,j}$ to its nearest integer. These values will be the intensity displayed for pixel (i,j) of $I(\tilde{\mathbf{D}})$. In this form of display, "white" corresponds to the maximal distance between the data (and always will be two white pixels), and the darker the pixel the closer the two data are. (For large data-sets, the image can easily exceed the resolution of the display. To solve that problem, Huband, Bezdek, and Hathaway (2005) proposed variations of VAT, which are not discussed here.) This image contains information about cluster tendency. Dark blocks along the diagonal indicate possible clusters, and if the image exhibits many variations in gray levels with faint or indistinct dark blocks along the diagonal, then the data set "[...] does not contain distinct clusters; or the clustering scheme implicitly imbedded in the reordering strategy fails to detect the clusters (there are cluster types for which single-linkage fails famously [...])." One of the main advantages of hierarchical clusterings is that they are able to detect non-convex clusters. It is, for example, an "S"-like cluster in two dimensions; and it can be the case that two data points, which clearly belong to the same cluster, are relatively far from each other. (An example will be presented in the following related to that problem.) In this case, the dendrogram generated by single-linkage clearly indicates the distinct clusters, but there will be no dark block in the intensity image by VAT. Certainly, single-linkage does have drawbacks, for example, it suffers from chaining effect, but a question naturally comes up: how much plus information can be given by VAT? It is because it roughly does a hierarchical clustering, but the result is not displayed as a dendrogram but based on the pairwise distance of data samples, and it works well only if the data in the same cluster are relatively close to each other based on the *original distance norm*. (This problem arises not only by clusters with non-convex shape, but very elongated ellipsoids as well.) Therefore, one advantage of hierarchical clustering is lost. There is a need for further study to analyze these problems, and it is not the aim of this chapter. This problem will also be touched on in Section 5.4.2.

Example 5.4 VAT for synthetic data *In Figure 5.6 the results of the single-linkage algorithm and VAT can be seen on the synthetic data. The clusters are well-separated but non-convex, and single-linkage clearly identifies them as can be seen from the dendrogram. However, the VAT image is not as clear as the*

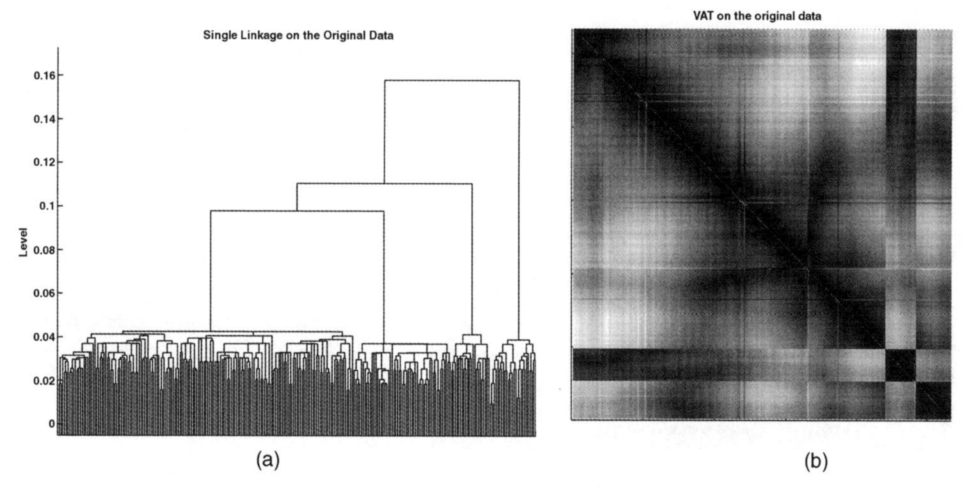

Figure 5.6 Result of the (a) single-linkage algorithm and (b) VAT for synthetic data.

dendrogram in this case because there are data in the "S" shaped cluster that are far from each other based on the Euclidean distance norm (see the top and left corner of the image).

Example 5.5 VAT for Iris data *As was mentioned in Example 5.2, the Iris data contain three labeled classes of Iris flower, but only two well-separated clusters because data of two kinds of Iris are not clearly separable. In Figure 5.7 the results of the single-linkage algorithm and VAT can be seen. Each method clearly identifies two clusters.*

Example 5.6 VAT for Wine data *In Figure 5.8 the results of the single-linkage algorithm and VAT can be seen. In this case, VAT is more informative than the dendrogram. The dendrogram is too "crowded," and all that the user can get to know is that there are no well-separated clusters in this 13-dimensional data-set. VAT provides more information: there are at least three "cores" in the data-set but the clusters are overlapping and not clearly identifiable.*

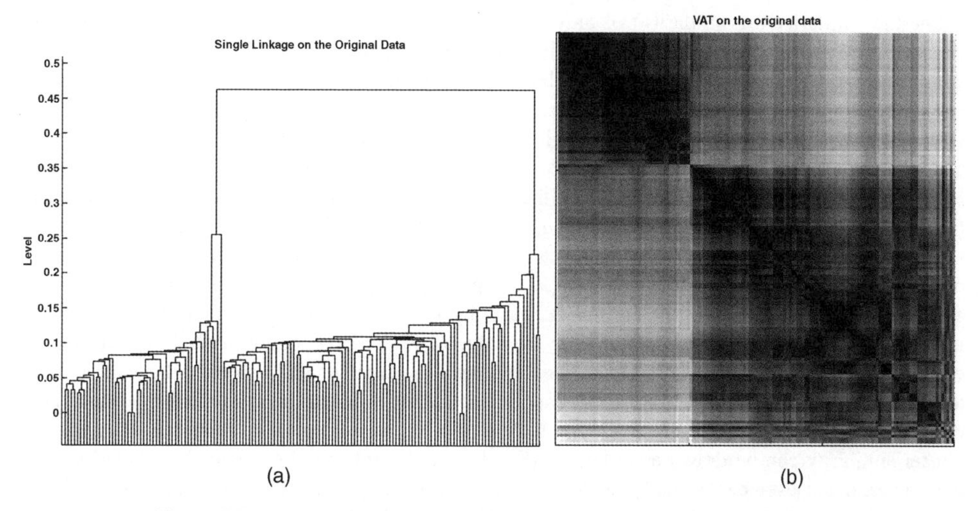

Figure 5.7 Result of the (a) single-linkage algorithm and (b) VAT for Iris data.

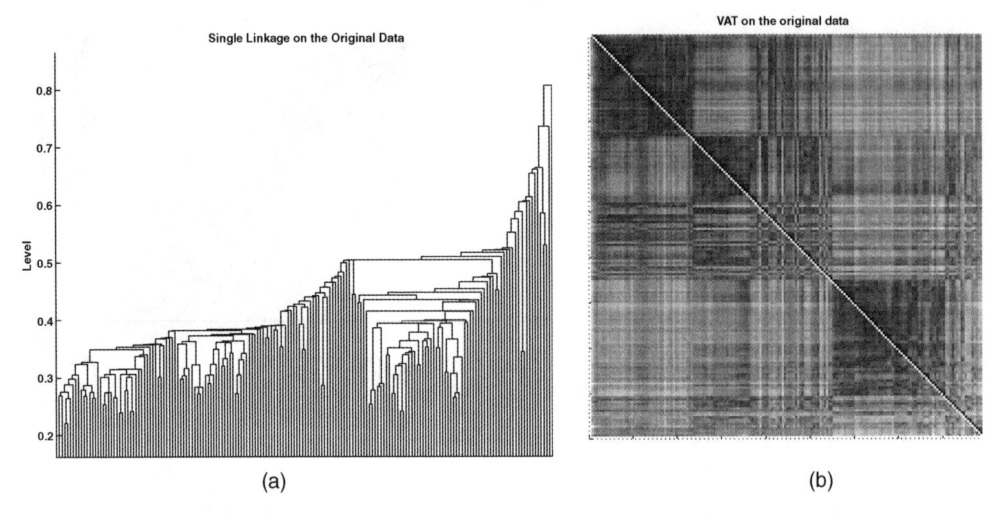

Figure 5.8 Result of the (a) single-linkage algorithm and (b) VAT for Wine data.

5.4.2 Visual Cluster Validity

Visual cluster validity (VCV) technique proposed in Hathaway and Bezdek (2003) is a possible approach to visually validate clustering results. The information about the clusters, which can be hard, fuzzy, or possibilistic, is displayed on a VAT like intensity image described in Section 5.4.1. The differences are the following:

- The order of data is not determined by a minimal spanning tree algorithm but it depends on the clustering result.
- The pairwise distance of data is calculated based on the distances from the cluster prototypes. That is why it is applicable to validate the resulted partition. Thanks to this method, the problems mentioned in Section 5.4.1 related to VAT, *do not occur* in the case of VCV.

Let us briefly overview these differences. The *ordering procedure* is done in two steps. First, the clusters are reordered, and after that the points within the clusters are reordered. There is a need to define the similarity (dissimilarity) of clusters to order them. In Hathaway and Bezdek (2003), the distances between clusters are defined as the Euclidean distance between the parameters defining the cluster prototypes. Because all clusters have the same number of parameters, this method can be applied to any prototype generator clustering approaches. However, this is clearly not the best choice to measure the similarity (dissimilarity) of clusters because prototypes may be very complex and may contain many parameters of different types. (For example, in the Gustafson–Kessel or Gath–Geva algorithms the clusters are parameterized by the mean (center) and the distance norm matrix (related to the covariance matrix of the cluster), see also in Equation (5.6). In the case of fuzzy partition the similarity measure described in Section 5.3 can be used *for all types of fuzzy clustering methods*. Based on the similarity measures in Equation (5.15), pairwise dissimilarity of clusters can be easily calculated as $(1 - S(i,j))$. (In the case of the other t-norm and t-conorm in Equation (5.16), another method has to be used because in this case the value of $(1 - S(i,i))$ is not necessarily zero.) In the original form of VCV, the cluster ordering procedure was a simple greedy approach. Instead of that, in this chapter the single-linkage algorithm is applied to get the order of the cluster based on the cluster distance matrix with the size $c \times c$. (It can have a role by large number of clusters.) After that, the cluster similarity can be drawn as a dendrogram. It contains information about the similarity of the clusters, and the user can visually analyze the clustering result (maybe can determine the proper number of clusters). After the order of the clusters are given, the data are reordered within the clusters.

In the case of fuzzy (or possibilistic) clustering, each datum is assigned to the cluster by which it has the biggest membership. Within each cluster, data are reordered according to their membership in descending order.

After the ordering procedure, there is a need to determine the pairwise dissimilarity between data. In Hathaway and Bezdek (2003), the pairwise dissimilarity is given by

$$d^*(\mathbf{x}_i, \mathbf{x}_j) = \min_{1 \leq k \leq c} \{d(\mathbf{x}_i, \eta_k) + d(\mathbf{x}_j, \eta_k)\}. \tag{5.18}$$

(It is a dissimilarity but not a metric (distance).) This procedure makes it possible to validate the clustering results. It has to be noted that this approach corresponds to the so called *minimax TIBA* (triangle inequality based approximation of missing data) proposed in (Hathaway and Bezdek, 2002). The original TIBA algorithms (minimax, maximin, and maximin/minimax average)were developed to handle missing data problems by relational data. Relational data means that only the pairwise dissimilarity is given (as in **D**) and not the original values of the features (as in **X**). The dissimilarity measure should satisfy the triangle inequality because TIBA algorithms use this property. These methods estimate the missing (i,j) (and (j,i)) pairwise dissimilarity based on the given dissimilarities of the ith and jth data *from other data*. The problem related to the pairwise dissimilarity by clustered data is similar to that: if the distance norms of the clusters are adaptive, it is not known which distance norm should be used to measure the pairwise dissimilarity! However, the pairwise dissimilarities between *the data and cluster prototypes* are known, and using Equation (5.18) is a possible solution to estimate the pairwise dissimilarities between the data. After the order of data and the pairwise dissimilarities are determined, the intensity image is composed in the same way as by VAT (see in Section 5.4.1). Small dissimilarities are represented by dark shades and large dissimilarities are represented by light shades. Darkly shaded diagonal blocks correspond to clusters in the data. This VCV image can be used to determine which clusters should be merged, and what the proper number of clusters could be.

Example 5.7 Visual cluster validity image for the synthetic data-set *In the case of the synthetic data-set the proper number of clusters is probably greater than the applied $c = 10$. However, for the purpose of illustration, it is worth displaying the VCV image in this case as well, and comparing the result with Figure 5.3. In Figure 5.9 similar results can be seen. The vertical lines are the bounds of the "hardened"*

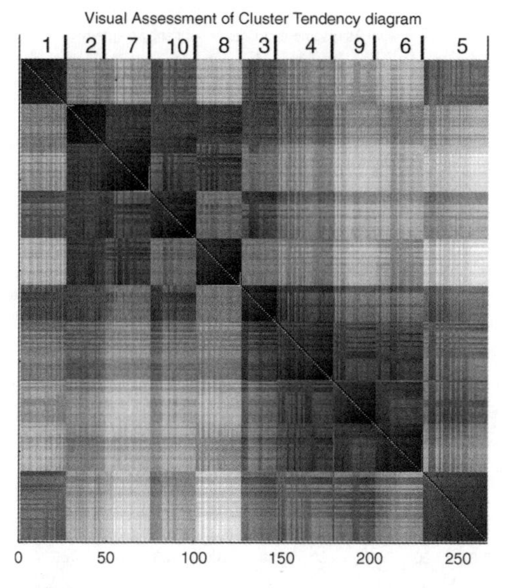

Figure 5.9 VCV image for the synthetic data-set (the numbers mean clusters, see Figure 5.3).

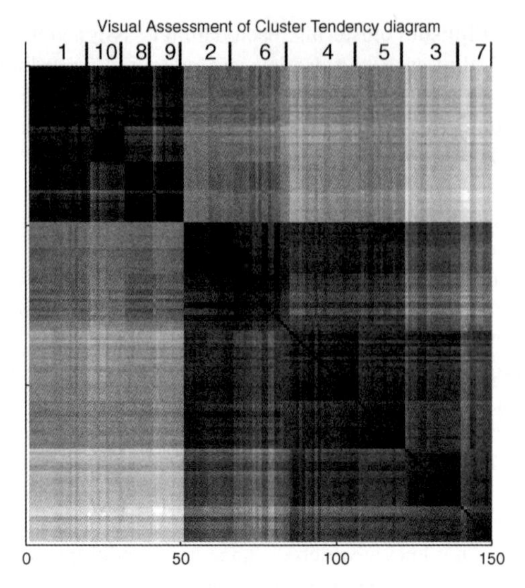

Figure 5.10 VCV image for the Iris data-set.

clusters. Clusters 1 and 5 are clearly well-separated, and at least two groups can be formed from the other clusters. 3-4-9-6th clusters could be merged, they are displayed as a large dark block on the VCV, but 2-7-10-8th clusters are not clearly mergeable. It is the user who has to decide in that type of problem.

Example 5.8 Visual cluster validity image for the Iris data-set *The results plotted in Figure 5.9 are clear. There are only two groups of clusters, as was the case by the data points themselves.*

Example 5.9 Visual cluster validity image for the Wine data-set *The results plotted in Figure 5.9 are not as clear as in Figure 5.5. (In Figure 5.9 the clusters separated by commas came to be so small because of the hardening procedure that their borders could not be displayed. It is because these clusters are very*

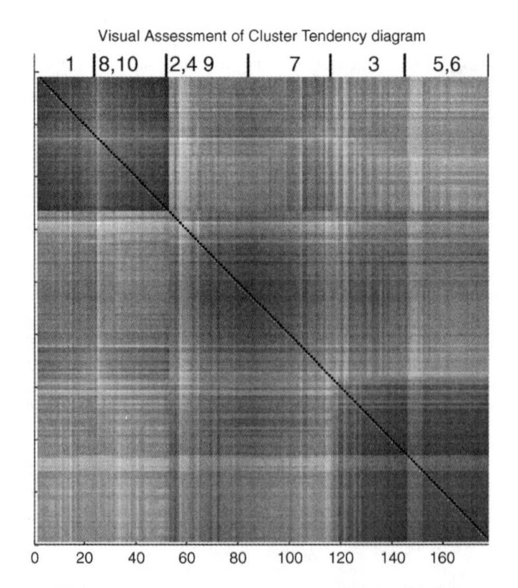

Figure 5.11 VCV image for the Wine data-set.

similar and close to each other.) As was mentioned in Example 5.3, although the optimal number of clusters is three and it is clear from Figure 5.5, the three real clusters in the data-set are not well-separated, because the borders of the middle block in Figure 5.9 are not clear.

5.4.3 Sammon Mapping to Visualize Clustering Results

While PCA attempts to preserve the variance of the data during the mapping, Sammon's mapping tries to preserve the interpattern distances (Mao and Jain, 1995; Pal and Eluri, 1998). For this purpose, Sammon defined the mean-square error between the distances in the high-dimensional space and the distances in the projected low-dimensional space. This square-error formula is similar to the "stress" criterion from multi-dimensional scaling.

The Sammon mapping is a well-known procedure for mapping data from a high n-dimensional space onto a lower q-dimensional space by finding N points in the q-dimensional data space, such that the interpoint distances $d^*_{i,j} = d^*(\mathbf{y}_i, \mathbf{y}_j)$ in the q-dimensional space approximate the corresponding interpoint distances $d_{i,j} = d(\mathbf{x}_i, \mathbf{x}_j)$ in the n-dimensional space. (see Figure 5.12).

This is achieved by minimizing an error criterion, called the Sammon's stress, E:

$$E = \frac{1}{\lambda}\sum_{i=1}^{N-1}\sum_{j=i+1}^{N}\frac{(d_{i,j} - d^*_{i,j})^2}{d_{i,j}} \tag{5.19}$$

where $\lambda = \sum_{i=1}^{N-1}\sum_{j=i+1}^{N} d_{i,j}$.

The minimization of E is an optimization problem in Nq variables $y_{i,l}$, $i = 1, 2, \ldots, N$, $l = 1, \ldots, q$, as $\mathbf{y}_i = [y_{i,1}, \ldots, y_{i,q}]^T$. Sammon applied the method of steepest descent to minimizing this function. Introduce the estimate of $y_{i,l}$ at the tth iteration

$$y_{i,l}(t+1) = y_{i,l}(t) - \alpha\left[\frac{\dfrac{\partial E(t)}{\partial y_{i,l}(t)}}{\dfrac{\partial^2 E(t)}{\partial^2 y_{i,l}(t)}}\right] \tag{5.20}$$

where α is a nonnegative scalar constant (recommended $\alpha \simeq 0.3 - 0.4$), i.e., the step size for gradient search in the direction of

$$\begin{aligned}
\frac{\partial E(t)}{\partial y_{i,l}(t)} &= -\frac{2}{\lambda}\sum_{k=1,k\neq i}^{N}\left[\frac{d_{k,i} - d^*_{k,i}}{d_{k,i}d^*_{k,i}}\right](y_{i,l} - y_{k,l}) \\
\frac{\partial^2 E(t)}{\partial^2 y_{i,l}(t)} &= -\frac{2}{\lambda}\sum_{k=1,k\neq i}^{N}\frac{1}{d_{k,i}d^*_{k,i}}\left[(d_{k,i} - d^*_{k,i}) - \left(\frac{(y_{i,l} - y_{k,l})^2}{d^*_{k,i}}\right)\left(1 + \frac{d_{k,i} - d^*_{k,i}}{d_{k,i}}\right)\right].
\end{aligned} \tag{5.21}$$

It is not necessary to maintain λ for a successful solution of the optimization problem, since the minimization of $\sum_{i=1}^{N-1}\sum_{j=i+1}^{N}(d_{i,j} - d^*_{i,j})^2/d_{i,j}$ gives the same result.

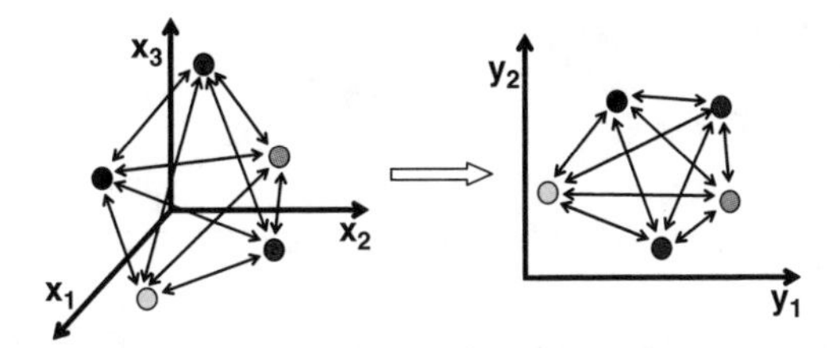

Figure 5.12 Illustration of Sammon mapping.

When the gradient-descent method is applied to search for the minimum of Sammon's stress, a local minimum in the error surface can be reached. Therefore, a significant number of runs with different random initializations may be necessary. Nevertheless, the initialization of **y** can be based on information which is obtained from the data, such as the first and second norms of the feature vectors or the principal axes of the covariance matrix of the data (Mao and Jain, 1995).

The classical Sammon mapping described above can be used to visualize the data themselves because it tries to preserve the interpattern distances. How can it be applied to visualize (fuzzy) clustering results? The following simple idea can be used: the interpattern distances are calculated based on the clustering results and the distances between the cluster prototypes and data, instead of based on the Euclidean distance of the objective data **X**. Exactly the same idea can be applied to obtain these dissimilarities as by VCV in Section 5.4.2. (see Equation (5.18).

Example 5.10 Visualization of synthetic data based on Sammon mapping *The result of the FCM algorithm based Sammon mapping can be seen in Figure 5.13(a) (the number of clusters is 10 as in the previous examples, see Section 5.4.2). These data are very similar to the original two-dimensional data. It is worth comparing the results with another clustering algorithm. The Gath–Geva (GG) method was chosen for that purpose. Its distance norm is adaptive, and it is able to determine ellipsoids with different sizes and shapes (for more information see Gath and Geva (1989)). The results of Sammon mapping based on the Gath–Geva algorithm can be seen in Figure 5.13(b) (the number of clusters was the same by both clustering algorithms). It can be determined that the four clusters are clearly separable and the data are very close to the cluster prototypes. From the two diagrams it can be said that the GG prototype represents best the analyzed data-set than the one of FCM. However, the result of FCM is also acceptable because the possible number of groups of data are known from VAT image or from the dendrogram (see Figure 5.6), and the four clusters are identifiable also from Figure 5.13(a).*

Example 5.11 Visualization of Iris data based on Sammon mapping *The Iris data are displayed in Figure 5.14 based on Sammon mapping using the result of FCM by* $c = 10$. *As it was determined from VAT (Figure 5.7) and VCV (Figure 5.10) as well, this four-dimensional data-set contains two well-separated clusters because two kinds of Iris flower are not clearly separable from these data. Figure 5.14 shows exactly the same. Because in this case the physical label of the data are also known (but of course not used in clustering), the different kinds of flowers are also displayed in the figure by different markers.*

Example 5.12 Visualization of Wine data based on Sammon mapping *In Figure 5.15 the results given by the Sammon mapping can be seen based on two different dissimilarity measures: (a) based on the Euclidean interpattern distances (it is the classical Sammon mapping), and (b) on the VCV based*

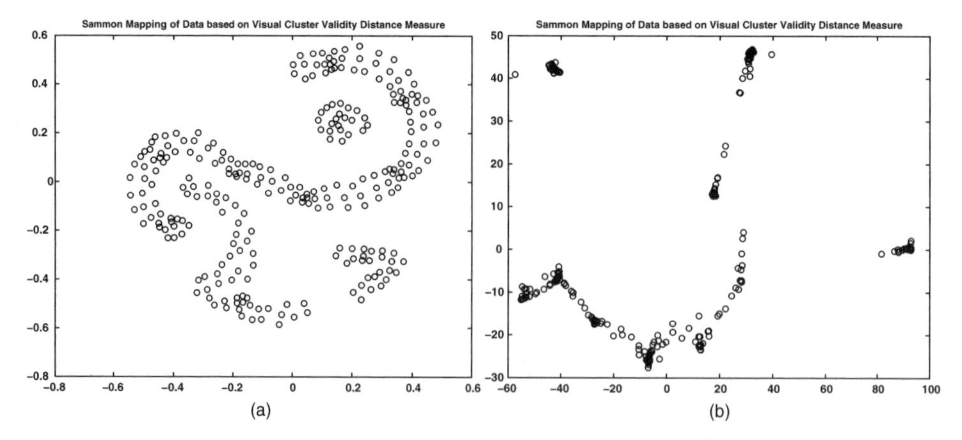

Figure 5.13 Synthetic data visualized by Sammon mapping using VCV dissimilarities based on (a) fuzzy C-means clustering, and (b) Gath–Geva clustering.

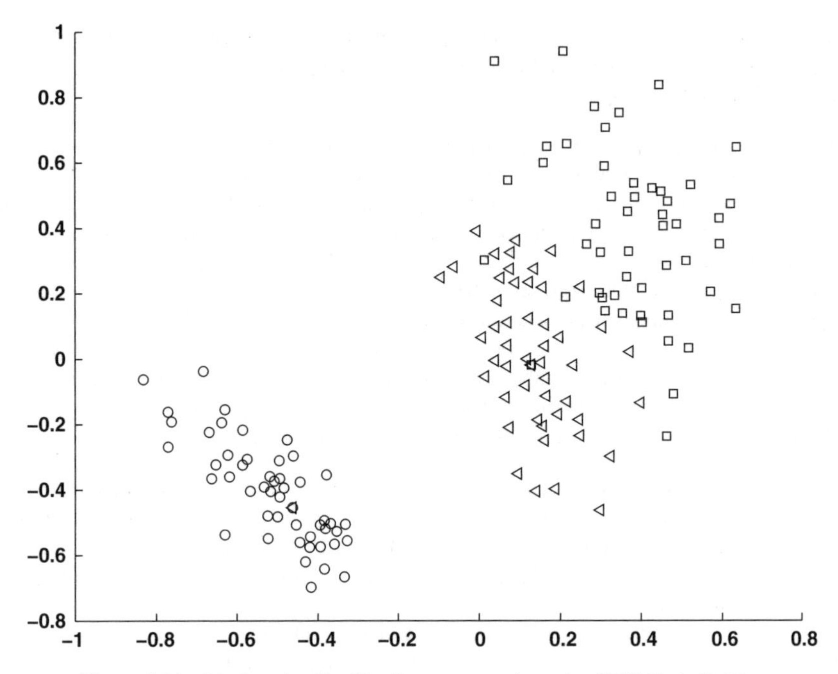

Figure 5.14 Iris data visualized by Sammon mapping using VCV dissimilarities.

dissimilarity (Equation (5.18)) using the FCM algorithm with 10 clusters. As was the case for the Iris data-set as well, the physical labels of the data are known and it is also displayed in the figures. As can be seen, this clustering problem is very complex because the data in the 13-dimensional space likely form a "cloud" and the clusters are not well-separated. (However, the physical classes are not mixed too much but the groups of data are very close to each other.)

5.4.4 Visualization of Fuzzy Clustering Results by Modified Sammon Mapping

This section focuses on the application of Sammon mapping for the visualization of the results of clustering, as the mapping of the distances is much closer to the task of clustering than the preserving the

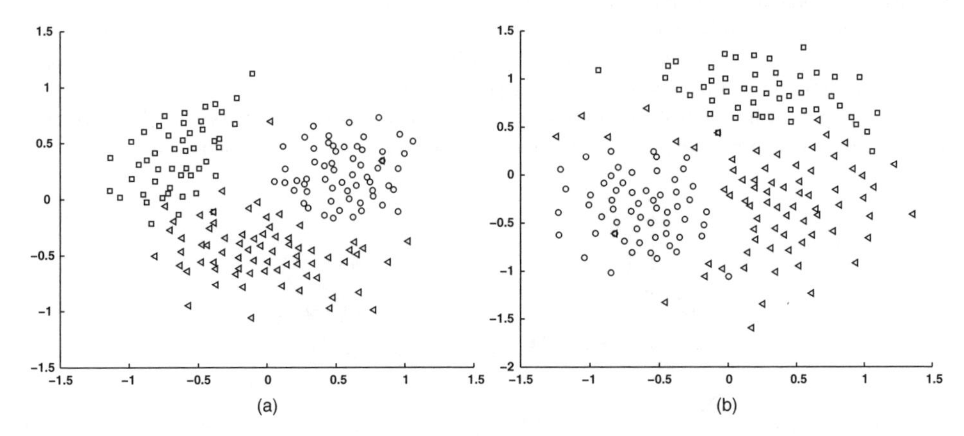

Figure 5.15 Wine data visualized by (a) classical Sammon mapping; (b) VCV dissimilarity based Sammon mapping.

variances. This section is mainly based on a previous work of the authors, for more details see Kovacs and Abonyi (2004). There are two main problems encountered in the application of Sammon mapping to the visualization of fuzzy clustering results:

- The prototypes of clustering algorithms may be vectors (centers) of the same dimension as the data objects, but they can also be defined as "higher-level" geometrical objects, such as linear or nonlinear subspaces or functions. Hence, classical projection methods based on the variance of the data (PCA) or based on the preservation of the Euclidean interpoint distance of the data (Sammon mapping) are not applicable when the clustering algorithm does not use the Euclidean distance norm.
- As Sammon mapping attempts to preserve the structure of high n-dimensional data by finding N points in a much lower q-dimensional data space, such that the interpoint distances measured in the q-dimensional space approximate the corresponding interpoint distances in the n-dimensional space, the algorithm involves a large number of computations as in every iteration step it requires the computation of $N(N-1)/2$ distances. Hence, the application of Sammon mapping becomes impractical for large N (de Ridder and Duin, 1997; Pal and Eluri, 1998).

By using the basic properties of fuzzy clustering algorithms a useful and easily applicable idea is to map the cluster centers and the data such that the distances between the clusters and the data points will be preserved (see Figure 5.16). During the iterative mapping process, the algorithm uses the membership values of the data and minimizes an objective function that is similar to the objective function of the original clustering algorithm.

To avoid the problem mentioned above, in the following we introduce some modifications in order to tailor Sammon mapping for the visualization of fuzzy clustering results. By using the basic properties of fuzzy clustering algorithms where only the distance between the data points and the cluster centers are considered to be important, the modified algorithm takes into account only $N \times c$ distances, where c represents the number of clusters, weighted by the membership values:

$$E_{\text{fuzz}} = \sum_{i=1}^{c} \sum_{k=1}^{N} (\mu_{i,k})^m [d(\mathbf{x}_k, \eta_i) - d^*(\mathbf{y}_k, \mathbf{z}_i)]^2 \tag{5.22}$$

where $d(\mathbf{x}_k, \eta_i)$ represents the distance between the $\mathbf{x}_k$ datapoint and the η_i cluster center measured in the original n-dimensional space, while $d^*(\mathbf{y}_k, \mathbf{z}_i)$ represents the Euclidean distance between the projected cluster center $\mathbf{z}_i$ and the projected data $\mathbf{y}_k$. This means that in the projected space, every cluster is represented by a single point, regardless of the form of the original cluster prototype, η_i. The application of the simple Euclidean distance measure increases the interpretability of the resulting plots (typically in two dimensions, although three-dimensional plots can be used as well). If the type of cluster prototypes is properly selected, the projected data will fall close to the projected cluster center represented by a point resulting in an approximately spherically shaped cluster.

The resulting algorithm is similar to the original Sammon mapping, but in this case in every iteration after the adaptation of the projected data points, the projected cluster centers are recalculated based on the weighted mean formula of the fuzzy clustering algorithms (see Appendix 5A.2).

The resulting two-dimensional plot of the projected data and the cluster centers is easily interpretable since it is based on a normal Euclidean distance measure between the cluster centers and the data points.

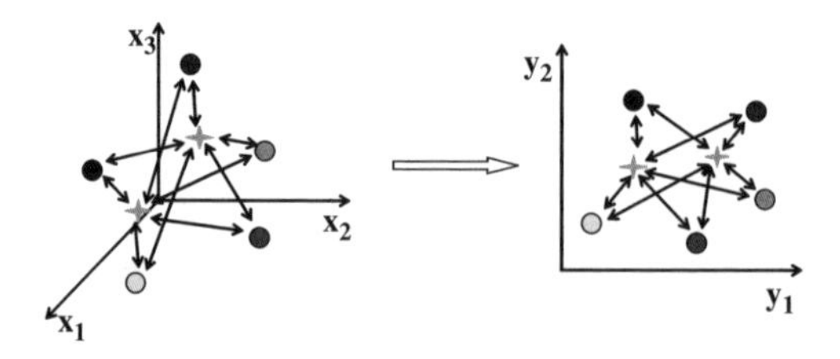

Figure 5.16 Illustration of fuzzy Sammon method.

Based on these mapped distances, the membership values of the projected data can also be plotted based on the classical formula of the calculation of the membership values:

$$\mu^*_{i,k} = 1 / \sum_{j=1}^{c} \left(\frac{d^*(\mathbf{x}_k, \eta_i)}{d^*(\mathbf{x}_k, \eta_j)} \right)^{\frac{2}{m-1}}. \quad (5.23)$$

Of course, the resulting two-dimensional plot will only approximate the original high-dimensional clustering problem. The quality of the approximation can easily be evaluated based on the mean-square error of the original and the recalculated membership values:

$$P = \| \mathbf{U} - \mathbf{U}^* \| \quad (5.24)$$

where $\mathbf{U}^* = [\mu^*_{i,k}]$ represents the matrix of the recalculated memberships.

Of course there are other tools to obtain information about the quality of the mapping of the clusters. For example, the comparison of the cluster validity measures calculated based on the original and mapped membership values can also be used for this purpose.

Several numerical experiments will be given below to demonstrate the applicability of the presented visualization tool. For the sake of comparison, the data and the cluster centers are also projected by principal component analysis (PCA) and the standard Sammon projection. Beside the visual inspection of the results the mean-square error of the recalculated membership values, P, see Equation (5.24), the difference between the original F and the recalculated F^* partition coefficient (Equation (5A.1)) (one of the cluster validity measures described in Appendix 5A.1), and the Sammon stress coefficient (Equation (5.19)) will be analyzed.

Example 5.13 Synthetic data *The aim of the first example is to demonstrate how the resulting plots of the projection should be interpreted and how the distance measure of the cluster prototype is "transformed" into Euclidean distance in the projected two-dimensional space. The visualization performance of the modified Sammon mapping is compared with PCA and the original Sammon mapping.*

This means that in the projected two-dimensional space, each cluster is represented by a single point, regardless of the form of the original cluster prototype, η_i. In this example the result of the Gustafson–Kessel algorithm is visualized, hence the distance norms defined by the inverse of the fuzzy covariance matrices are transferred to Euclidean distances with the presented FUZZSAM mapping. The application of the simple Euclidean distance measure increases the interpretability of the resulted plots. As Figure 5.17 shows in the case of a properly selected cluster prototype the projected data will fall close to the projected cluster center represented by a point resulting in an approximately spherically distributed cluster (compare Figure 5.17(c) and Figure 5.17(d)). The numerical results summarized in Table 5.1 show that the presented FUZZSAM tool outperforms the linear method and the classical Sammon projection tools. The P error of the membership values between are much smaller, and the F and F^* cluster validity measures are similar when the projection is based on the presented FUZZSAM mapping.

5.4.5 Benchmark Examples

The previous example showed that it is possible to obtain a good data structure by the presented mapping algorithm. However, the real advantage of the FUZZSAM algorithm, the visualization of

Table 5.1 Comparison of the performance of the mappings (Example 5.13).

Method	P	F	F^*	E
GK-PCA	0.1217	0.8544	0.7263	0.0000
GK-SAMMON	0.1217	0.8544	0.7263	0.0000
GK-FUZZSAM	0.0204	0.8495	0.8284	0.1177
FCM-FUZZSAM	0.0000	0.7468	0.7468	0.0000

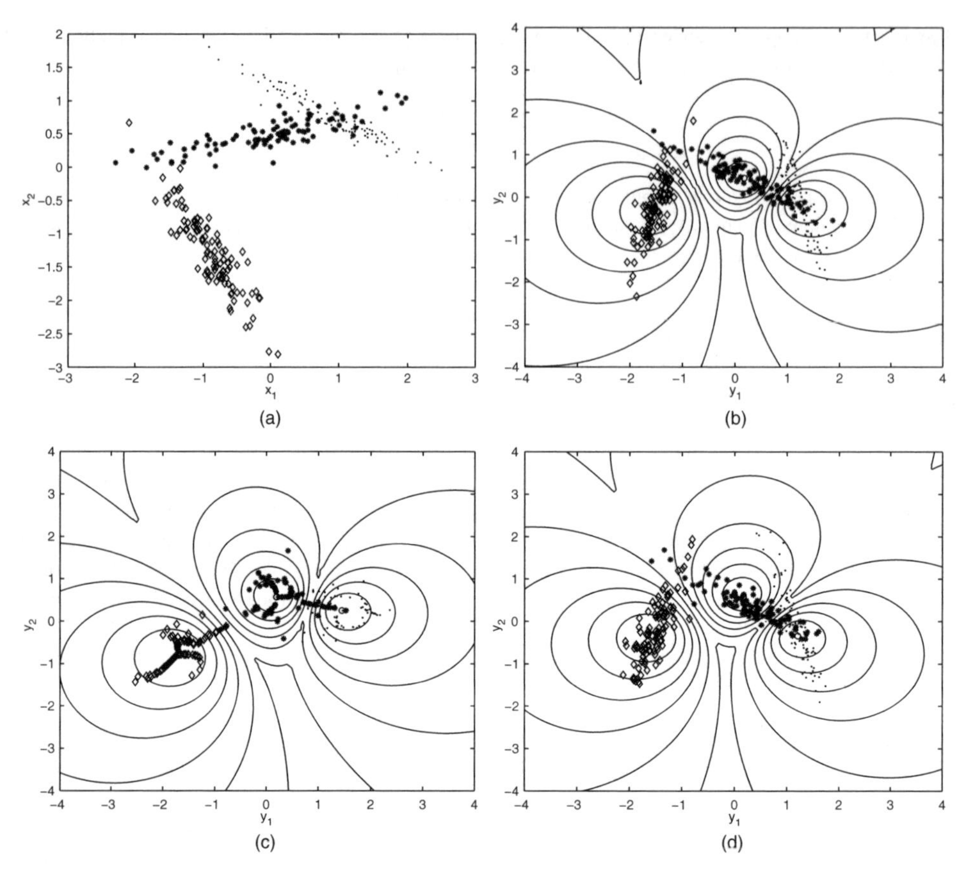

Figure 5.17 (a) Synthetic data in two dimensionals, (b) PCA mapping of the data and the recalculated membership contours, (c) FUZZSAM projection of the results of GK clustering, (d) FUZZSAM projection of the results of FCM clustering.

higher-dimensional spaces was not shown. This will be done by the following real clustering problems. The first example is the visualization of the results of the clustering of the well known Iris data.

Example 5.14 Iris data visualization *To test the presented method the results of the clustering of the Iris data were visualized by principal component analysis (PCA), the original Sammon mapping, and the modified method. The initial conditions in the applied Gustafson–Kessel fuzzy clustering algorithm were the following:* $c = 3$, $m = 2$ *and* $\alpha = 0.4$ *in the Sammon and FUZZSAM mapping algorithms. The results of the projections are shown in Figure 5.18, where the different markers correspond to different types of Iris, and the level curves represent the recalculated membership degrees.*

As Figure 5.18(c) nicely illustrates, the data can be properly clustered by the GK algorithm. One of the clusters is well-separated from the other two clusters. To illustrate how the fuzziness of the clustering can be evaluated from the resulted plot, Figure 5.18(d) shows the result of the clustering when $m = 1.4$. *As can be seen in this plot the data points lie much closer to the center of the cluster and there are many more points in the first iso-membership curves. These observations are confirmed by the numerical data given in Table 5.2.*

Example 5.15 Wine data visualization *This example is used to illustrate how the FUZZSAM algorithm can be used to visualize the clustering of 13-dimensional data, and how this visualization*

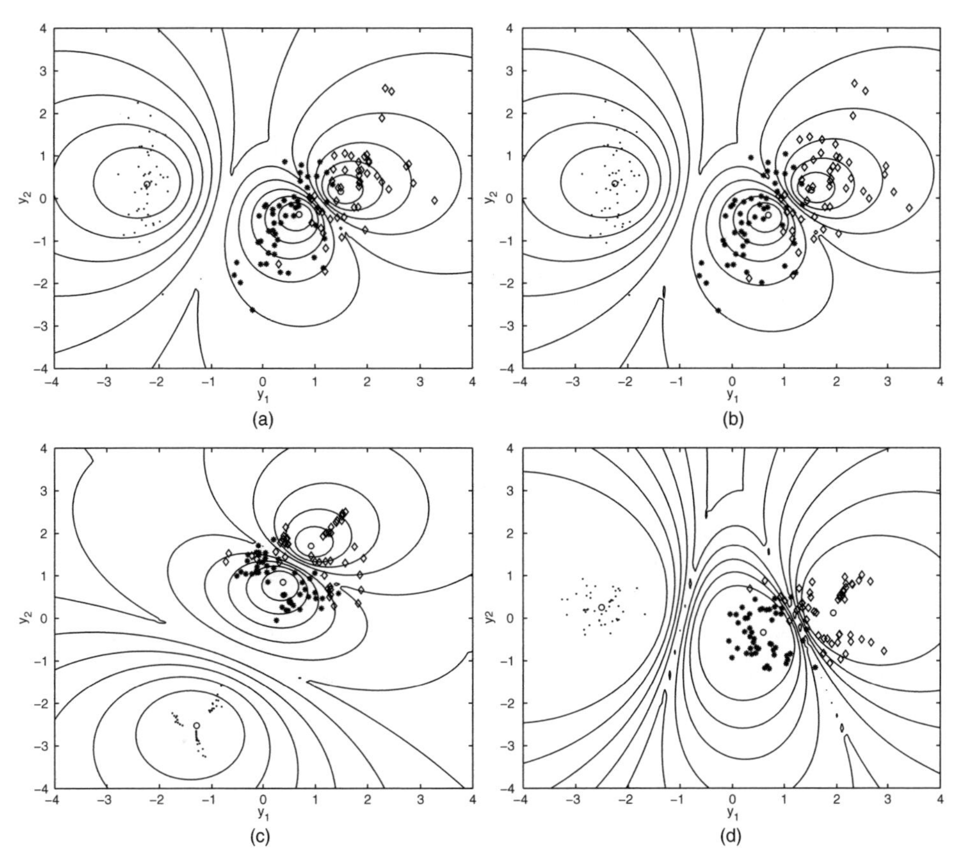

Figure 5.18 (a) PCA projection of the IRIS data and the recalculated membership contours; (b) SAMMON projection of the IRIS data and the recalculated membership contours; (c) FUZZSAM projection of the results of GK clustering of the IRIS data $m = 2$; (d) FUZZSAM projection of the results of FCM clustering of the the IRIS data $m = 1.4$.

can be used to detect the number of clusters (Figure 5.19 and Table 5.3). It can be seen from the values of partition coefficient F that three clusters fit much better to the data than 10. It can also be observed in Figure 5.19(a), (b), and (c) that the clusters are so much overlapping that in these figures only three clusters can be seen. It was also seen by the similarity between clusters (see Example 5.3 and Figure 5.5). There are many points that belong to these three clusters with similar membership values.

Table 5.2 Comparison of the performance of the mappings (Example 5.14).

Method	P	F	F^*	E
GK-PCA	0.1139	0.7262	0.6945	0.0100
GK-SAMMON	0.1153	0.7262	0.6825	0.0064
GK-FUZZSAM	0.0175	0.7262	0.7388	0.1481
GK-PCA-m=1.4	0.1057	0.9440	0.9044	0.0100
GK-SAMMON-m=1.4	0.1044	0.9440	0.8974	0.0064
GK-FUZZSAM-m=1.4	0.0011	0.9440	0.9425	0.0981

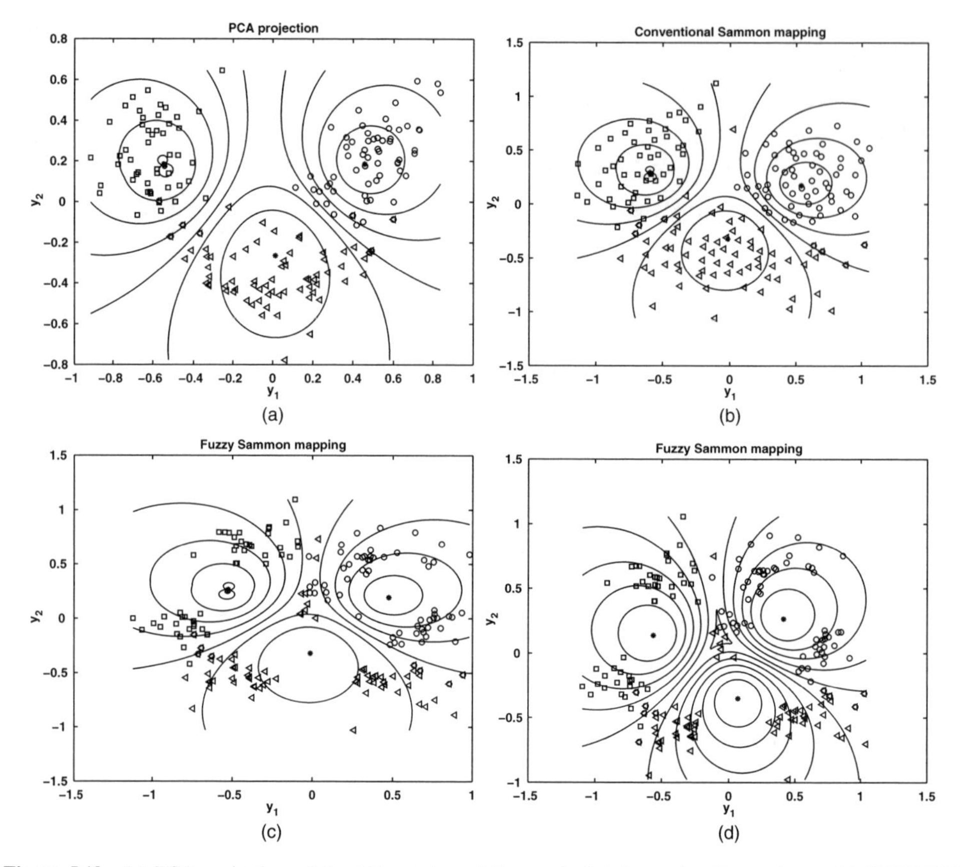

Figure 5.19 (a) PCA projection of the Wine data and the recalculated membership contours; (b) SAMMON projection of the Wine data and the recalculated membership contours; (c) FUZZSAM projection of the Wine data by $c = 10$; (d) FUZZSAM projection of the Wine data by $c = 3$.

Table 5.3 Comparison of the performance of the mappings (Example 6.15).

Method	P	F	F^*	E
FCM-PCA $c = 10$	0.0389	0.1512	0.2252	0.1301
FCM-SAMMON $c = 10$	0.0262	0.1512	0.1984	0.0576
FCM-FUZZSAM $c = 10$	0.0105	0.1512	0.1545	0.0999
FCM-PCA $c = 3$	0.1295	0.5033	0.7424	0.1301
FCM-SAMMON $c = 3$	0.0874	0.5033	0.6574	0.0576
FCM-FUZZSAM $c = 3$	0.0364	0.5033	0.5178	0.1003

5.5 CONCLUSIONS

Visualization of (fuzzy) clustering results may be very important in real-life clustering problems. It is because humans are very good at pattern recognition in two dimensions but are hopeless with more than three dimensions and/or large amounts of data. Nevertheless, it is exactly the case in practice. Cluster validity indices try to solve this problem based on a criterion that measures the "goodness" of the actual results. However, there are many of them and none of them is perfect. It could be much more useful and informative if the clustering result in n dimensions is "projected" into two dimensions in some sense.

Hence, in this chapter the visualization of fuzzy clustering was our focal point. *In the first step* after clustering, there is a need to determine how similar the resulting clusters are. For that purpose, a fuzzy set similarity measure can be used because fuzzy clusters can be seen as fuzzy sets. For crisp sets, the similarity of two sets, A and B can be expressed as follows:

$$\mathrm{sim}(\mathrm{A},\mathrm{B}) = \frac{\mathrm{A} \bigcap \mathrm{B}}{\mathrm{A} \bigcup \mathrm{B}}. \tag{5.25}$$

For fuzzy sets, the *min* and *max* operator should be used instead of $\bigcap$ and $\bigcup$. In this way, all clusters can be compared with each other. Based on this symmetric similarity matrix, hierarchial clustering can be used for cluster merging and for the visualization of the results of the clustering by dendrogram. Sammon mapping can also be used to visualize the cluster prototype in two dimensions. It is important to note that the original prototypes may be really complex but they can be visualized as single points in two dimensions in this way. Using this diagram, the human observer can get an impression of how similar the clusters are in the original space and is able to determine which clusters should be merged if it is needed. *In the next step*, data points should be visualized in this two-dimensional space. It could be done, for example, with triangulation: distances from the two closest cluster prototypes (hence, with the two biggest memberships) can be preserved by every data point. However, other ideas can be used as well.

The distances calculated from the clustering can also be visualized. For this purpose visual assessment of cluster tendency (VAT) plots can be used.

Principal component analysis (PCA) and Sammon mapping can also be used to visualize the n-dimensional data points in two-dimensions. PCA projects the data into a linear subspace, while Sammon mapping tries to preserve the interpoint distances. Both of them are useful but none of them can directly be used to visualize the clustering results, only the data points themselves. If the cluster prototype is not (only) a point in n dimensions, another procedure is needed. A new FUZZSAM algorithm has also been proposed for this purpose. The FUZZSAM method generates two-dimensional informative plots about the quality of the cluster prototypes and the required number of clusters. This tool uses the basic properties of fuzzy clustering algorithms to map the cluster centers and the data such that the distances between the clusters and the data points are preserved. The numerical results show superior performance over principal component analysis and the classical Sammon projection tools.

The proposed tools can be effectively used for interactive and iterative (dynamic) data mining.

Appendices

APPENDIX 5A.1 VALIDITY INDICES

(1) Partition coefficient (PC): measures the amount of "overlapping" between clusters. It is defined by Bezdek (Bezdek, 1981) as follows:

$$PC(c) = \frac{1}{N}\sum_{i=1}^{c}\sum_{k=1}^{N}(\mu_{i,k})^2 \tag{5A.1}$$

where $\mu_{i,k}$ is the membership of data point k in cluster i. The disadvantage of PC is the lack of direct connection to some property of the data themselves. The optimal number of clusters is indicated by the minimum value.

(2) Classification entropy (CE): measures the fuzzyness of the cluster partition only, which is similar to the partition coefficient:

$$CE(c) = -\frac{1}{N}\sum_{i=1}^{c}\sum_{k=1}^{N}\mu_{i,k}\ln(\mu_{i,k}). \tag{5A.2}$$

The optimal number of clusters is indicated by the maximum value.

(3) Partition index (SC): the ratio of the sum of compactness and separation of the clusters. It is a sum of individual cluster validity measures normalized by dividing it by the fuzzy cardinality of each cluster (Bensaid *et al.*, 1996):

$$SC(c) = \sum_{i=1}^{c}\frac{\sum_{k=1}^{N}(\mu_{i,k})^m||\mathbf{x}_k - \mathrm{v}_i||^2}{\sum_{k=1}^{N}\mu_{i,k}\sum_{j=1}^{c}||\mathrm{v}_j - \mathrm{v}_i||^2}. \tag{5A.3}$$

SC is useful when comparing different partitions having an equal number of clusters. A lower value of *SC* indicates a better partition.

(4) ***Separation index (S):*** different from the partition index (SC), the separation index uses a minimum-distance separation for partition validity (Bensaid *et al.*, 1996):

$$S(c) = \frac{\sum_{i=1}^{c}\sum_{k=1}^{N}(\mu_{i,k})^2||\mathbf{x}_k - \mathrm{v}_i||^2}{N\min_{i,j}||\mathrm{v}_j - \mathrm{v}_i||^2}. \tag{5A.4}$$

(5) Xie and Beni's index (XB): indirect indices like the partition coefficient suffer from three drawbacks. First, they are at best indirectly related to any real clusters in **X**; second, they ignore additional parameters (such as **V**); and third, they do not use **X** itself. Xie and Beni defined an index of fuzzy cluster validity that overcomes the second and third problems. It aims to quantify the ratio of the total variation within clusters and the separation of clusters (Xie and Beni, 1991)):

$$XB(c) = \frac{\sum_{i=1}^{c}\sum_{k=1}^{N}(\mu_{i,k})^m||\mathbf{x}_k - \mathrm{v}_i||^2}{N\min_{i,k}||\mathbf{x}_k - \mathrm{v}_i||^2}. \tag{5A.5}$$

The optimal number of clusters should minimize the value of the index.

(6) Dunn's index (DI): this index was originally proposed to be used at the identification of compact and well "separated clusters" (see, for example, in Xie and Beni (1991)). So the result of the clustering has to be recalculated as if it was a hard partition algorithm:

$$DI(c) = \min_{i\in c}\left\{\min_{j\in c, i\neq j}\left\{\frac{\min_{\mathbf{x}\in C_i,\mathbf{y}\in C_j} d(\mathbf{x},\mathbf{y})}{\max_{k\in c}\{\max_{\mathbf{x},\mathbf{y}\in C} d(\mathbf{x},\mathbf{y})\}}.\right\}\right\} \tag{5A.6}$$

The main drawback of Dunn's index is the computational demand since calculating becomes computationally very expensive as c and N increase.

(7) Alternative Dunn index (ADI): the aim of modifying the original Dunn's index was that the calculation becomes more simple, when the dissimilarity function between two clusters ($\min_{\mathbf{x}\in C_i,\mathbf{y}\in C_j} d(\mathbf{x},\mathbf{y})$) is rated in value from beneath by the triangle nonequality:

$$d(\mathbf{x},\mathbf{y}) \geq |d(\mathbf{y},\mathrm{v}_j) - d(\mathbf{x},\mathrm{v}_j)| \tag{5A.7}$$

where $\mathbf{v}_j$ is the cluster center of the jth cluster. Then:

$$ADI(c) = \min_{i \in c}\left\{\min_{j \in c, i \neq j} \frac{\min_{\mathbf{x}_i \in C_i, \mathbf{x}_j \in C_j} |d(\mathbf{y}, \mathrm{v}_j) - d(\mathbf{x}_i, \mathrm{v}_j)|}{max_{k \in c}\{max_{\mathbf{x},\mathbf{y} \in C} d(\mathbf{x}, \mathbf{y})\}}\right\} \tag{5A.8}$$

(8) The fuzzy hyper volume: this index is also widely applied and represents the volume of the clusters:

$$\mathcal{V}(c) = \sum_{i=1}^{c} \det(\mathbf{F}_i), \tag{5A.9}$$

where $\mathbf{F}_i$ is the fuzzy covariance matrix of the ith cluster,

$$\mathbf{F}_i = \frac{\sum_{k=1}^{N} \mu_{i,k}^m (\mathbf{x}_k - \mathrm{v}_i)(\mathbf{x}_k - \mathrm{v}_i)^T}{\sum_{k=1}^{N} \mu_{i,k}^m}. \tag{5A.10}$$

(9) The Fischer interclass separability: this criterion is based on the between-class and within-class scatter or covariance matrices, called $\mathbf{F}_b$ and $\mathbf{F}_w$, respectively, which add up to the total scatter matrix $\mathbf{F}_t$ which is the covariance of the whole training data containing N data pairs (Roubos, Setres, and Abonyi, 2003):

$$\mathbf{F}_t = \frac{1}{N}\sum_{k=1}^{N} (\mathbf{x}_k - \mathbf{v})(\mathbf{x}_k - \mathbf{v})^T, \tag{5A.11}$$

where $\mathbf{v}$ is the mean of the samples:

$$\mathbf{v} = \frac{1}{N}\sum_{k=1}^{N} \mathbf{x}_k = \frac{1}{N}\sum_{i=1}^{c} N_i \mathrm{v}_i, \tag{5A.12}$$

where $N_i = \sum_{k=1}^{N} \mu_{i,k}$ is the "size" of the cluster, i.e., how many data points belong to it in a fuzzy way. The total scatter matrix can be decomposed:

$$\mathbf{F}_t = \mathbf{F}_b + \mathbf{F}_w, \tag{5A.13}$$

where

$$\mathbf{F}_b = \sum_{i=1}^{c} N_i (\mathrm{v}_i - \mathbf{v})(\mathrm{v}_i - \mathbf{v})^T, \tag{5A.14}$$

$$\mathbf{F}_w = \sum_{i=1}^{c} \mathbf{F}_b. \tag{5A.15}$$

The interclass separability criterion is a trade-off between $\mathbf{F}_b$ and $\mathbf{F}_w$:

$$IS(c) = \frac{\det \mathbf{F}_b}{\det \mathbf{F}_w}. \tag{5A.16}$$

It is similar to the so-called invariant criteria for clustering (see Duda, Hart, and Stork (2001)) where the cost function, which has to be maximized in this approach, is not in the form of Equation (5.5) but

$$J_f = \mathrm{trace}[\mathbf{F}_w^{-1} \mathbf{F}_b] = \sum_{i=1}^{n} \lambda_i, \tag{5A.17}$$

where λ_i are the eigenvalues of $\mathbf{F}_w^{-1} \mathbf{F}_b$.

APPENDIX 5A.2 THE MODIFIED SAMMON MAPPING ALGORITHM

- **[Input]** : *Desired dimension of the projection, usually* $q = 2$, *the original data set*, $\mathbf{X}$; *and the results of fuzzy clustering: cluster prototypes*, η_i, *membership values*, $\mathbf{U} = [\mu_{i,k}]$, *and the distances* $D = [d_{k,i} = d(\mathbf{x}_k, \eta_i)]_{N \times c}$.
- **[Initialize]** *the* $\mathbf{y}_k$ *projected data points by PCA based projection of* $\mathbf{x}_k$, *and compute the projected cluster centers by*

$$\mathbf{z}_i = \frac{\sum_{k=1}^{N} (\mu_{i,k})^m \mathbf{y}_k}{\sum_{k=1}^{N} (\mu_{i,k})^m} \tag{5A.18}$$

and compute the distances with the use of these projected points $D^* = [d^*_{k,i} = d(\mathbf{y}_k, \mathbf{z}_i)]_{N \times c}$. *Random initialization can also be used but PCA based projection is a better choice in many cases because it may reduce the number of iterations in the next phase.*

- **[While]** ($E_{\text{fuzz}} > \varepsilon$) and (t$\leq$ maxstep)

$\{for$(i =1:i $\leq$ c: i++)

$\{for(j = 1 : j \leq N : j++)$

$$\{Compute \frac{\partial E(t)}{\partial y_{i,l}(t)}, \frac{\partial^2 E(t)}{\partial^2 y_{i,l}(t)},$$

$$\Delta y_{i,l} = \Delta y_{i,l} + \left[\frac{\frac{\partial E(t)}{\partial y_{i,l}(t)}}{\frac{\partial^2 E(t)}{\partial^2 y_{i,l}(t)}}\right]\}$$

$\}$

$y_{i,l} = y_{i,l} + \Delta y_{i,l}, \forall i = 1, \ldots, N, \; l = 1, \ldots, q$

Compute $\mathbf{z}_i = \sum_{k=1}^{N} (\mu_{i,k})^m \mathbf{y}_k / \sum_{k=1}^{N} (\mu_{i,k})^m$

$D^* = [d^*_{k,i} = d(\mathbf{y}_k, \mathbf{z}_i)]_{N \times c}$

$\}$

Compute E_{fuzz} *by Equation* (5A.22)

where the derivatives are

$$\begin{aligned} \frac{\partial E(t)}{\partial y_{i,l}(t)} &= -\frac{2}{\lambda} \sum_{k=1, k \neq i}^{N} \left[\frac{d_{k,i} - d^*_{k,i}}{d^*_{k,i}} (\mu_{i,k})^m\right] (y_{i,l} - y_{k,l}) \\ \frac{\partial^2 E(t)}{\partial^2 y_{i,l}(t)} &= -\frac{2}{\lambda} \sum_{k=1, k \neq i}^{N} \frac{(\mu_{i,k})^m}{d^*_{k,i}} \left\{(d_{k,i} - d^*_{k,i}) - \frac{(y_{i,l} - y_{k,l})^2}{d^*_{k,i}} [1 + (d_{k,i} - d^*_{k,i})(\mu_{i,k})^m]\right\} \end{aligned} \tag{5A.19}$$

ACKNOWLEDGEMENTS

The authors acknowledge the support of the Cooperative Research Center (VIKKK) (project 2001-II-1), the Hungarian Ministry of Education (FKFP-0073/2001), and the Hungarian Science Foundation (T037600).

REFERENCES

Abonyi, J., Babuska, R. and Szeifert, F. (2002) 'Modified Gath-Geva fuzzy clustering for identification of Takagi Sugeno fuzzy models'. *IEEE Transactions on Systems, Man and Cybernetics* **32(5)**, 612–321.

Abonyi, J., Feil, B., Nemeth, S. and Arva, P. (2005) 'Modified Gath–Geva clustering for fuzzy segmentation of multivariate time-series'. *Fuzzy Sets and Systems – Fuzzy Sets in Knowledge Discovery* **149**(1), 39–56.

Babuska, R. (1998) *Fuzzy Modeling for Control*. Kluwer Academic Publishers, Boston.

Baldwin, J., Martin, T. and Rossiter, J. (1998) 'Time series modelling and prediction using fuzzy trend information' *Proc. of 5th Internat. Conf. on Soft Computing and Information Intelligent Systems*, pp. 499–502.

Bensaid, A. *et al.* (1996) 'Validity-guided (Re)Clustering with Applications to Image Segmentation'. *IEEE Transactions on Fuzzy Systems* **4**, 112–123.

Bezdek, J. (1981) *Pattern Recognition with Fuzzy Objective Function Algorithms*. Plenum Press, New York.

Bezdek, J. and Hathaway, R. (2002) 'VAT: Visual Assessment of (Cluster) Tendency' *Proceedings of IJCNN*, pp. 2225–2230.

de Ridder, D. and Duin, R.P. (1997) 'Sammon's Mapping Using Neural Networks: A Comparison'. *Pattern Recognition Letters* **18**, 1307–1316.

Duda, R., Hart, P. and Stork, D. (2001) *Pattern Classification* 2nd edn. John Wiley & Sons, Inc., New York.

Gath, I. and Geva, A. (1989) 'Unsupervised optimal fuzzy clustering'. *IEEE Transactions on Pattern Analysis and Machine Intelligence* **7**, 773–781.

Geva, A. (1999) 'Hierarchical-fuzzy clustering of temporal-patterns and its application for time-series prediction'. *Pattern Recognition Letters* **20**, 1519–1532.

Hanesch, M., Scholger, R. and Dekkers, M. (2001) 'The application of fuzzy *c*-means cluster analysis and non-linear mapping to a soil data set for the detection of polluted sites'. *Phys. Chem. Earth* **26**, 885–891.

Hathaway, R. and Bezdek, J. (2002) 'Clustering Incomplete Relational Data using the Non-Euclidean Relational Fuzzy *c*-Means Algorithm'. *Pattern Recognition Letters* **23**, 151–160.

Hathaway, R. and Bezdek, J. (2003) 'Visual cluster validity for protoype generator clustering models'. *Pattern Recognition Letters* **24**, 1563–1569.

Huband, J., Bezdek, J. and Hathaway, R. (2004) 'Revised Visual Assessment of (Cluster) Tendency (reVAT)' *Proceedings of the North American Fuzzy Information Processing Society (NAFIPS)*, pp. 101–104.

Huband, J., Bezdek, J. and Hathaway, R. (2005) 'bigVAT: Visual Assessment of Cluster Tendency for Large Data Sets'. *Pattern Recognition* **38**, 1875–1886.

Jain, A. and Dubes, R. (1988) *Algorithms for Clustering Data*. Prentice-Hall, Inc.

Johansen, T. and Babuska, R. (2002) 'On Multi-objective Identification of Takagi–Sugeno Fuzzy Model Parameters' *Preprints 15th IFAC World Congress, Barcelona, Spain*, pp. T–Mo–A04, paper no. 587.

Kelly, P. (1994) 'An algorithm for merging hyperellipsoidal clusters'. *Technical Report, L.A.-UR-94-3306, Los Alamos National Laboratory, Los Alamos, NM.*

Kim, E., Park, M. and Kim, S. (1998) 'A Transformed Input Domain Approach to Fuzzy Modeling'. *IEEE Transactions on Fuzzy Systems* **6**, 596–604.

Klawonn, F., Chekhtman, V. and Janz, E. (2003) 'Visual inspection of fuzzy clustering results' In *Advances in Soft Computing - Engineering, Design and Manufacturing* (ed. Benitez, J., Cordon, O., Hoffmann, F. and Roy R) Springer, London pp. 65–76.

Kovacs, A. and Abonyi, J. (2004) 'Vizualization of Fuzzy Clustering Results by Modified Sammon Mapping' *Proceedings of the 3rd International Symposium of Hungarian Researchers on Computational Intelligence*, pp. 177–188.

Mao, J. and Jain, K. (1995) 'Artificial Neural Networks for Feature Extraction and Multivariate Data Projection'. *IEEE Trans. on Neural Networks* **6**(2), 296–317.

Marcelino, P., Nunes, P., Lima, P. and Ribeiro, M.I. (2003) 'Improving object localization through sensor fusion applied to soccer robots'. *Actas do Encontro Cientifico do Robotica.*

Nelles, O. (2001) *Nonlinear System Identification*. Springer, Berlin, Germany.

Pach, P., Abonyi, J., Nemeth, S. and Arva, P. (2004) 'Supervised Clustering and Fuzzy Decision Tree Induction for the Identification of Compact Classifiers' *5th International Symposium of Hungarian Researchers on Computational Intelligence*, pp. 267–279, Budapest, Hungary.

Pal, N. and Eluri, V. (1998) 'Two Efficient Connectionist Schemes for Structure Preserving Dimensionality Reduction'. *IEEE Transactions on Neural Networks* **9**, 1143–1153.

Pedrycz, W. and Sosnowski, A. (2001) 'The design of decision trees in the framework of granular data and their application to software quality models'. *Fuzzy Sets and Systems* **123**, 271–290.

Roubos, J., Setnes, M. and Abonyi, J. (2003) 'Learning Fuzzy Classification Rules from Labeled Data'. *Information Sciences* **150**(1–2), 77–93.

Vathy-Fogarassy, A., Feil, B. and Abonyi, J. (2005) 'Minimal Spanning Tree based Fuzzy Clustering' In *Transactions on Enformatika, Systems Sciences and Engineering* (ed. Ardil C), vol. 8, pp. 7–12.

Vesanto, J. (2000) 'Neural network tool for data mining: Som toolbox'. *Proceedings of Symposium on Tool Environments and Development Methods for Intelligent Systems (TOOLMET2000)* pp. 184–196.

Wong, J., McDonald, K. and Palazoglu, A. (1998) 'Classification of process trends based on fuzzified symbolic representation and hidden markov models'. *Journal of Process Control* **8**, 395–408.

Xie, X. and Beni, G. (1991) 'Validity measure for fuzzy clustering}'. *IEEE Trans. PAMI* **3**(8), 841–846.

6

Interactive Exploration of Fuzzy Clusters

Bernd Wiswedel[1], David E. Patterson[2], and Michael R. Berthold[1]

[1]*Department of Computer and Information Science, University of Konstanz, Germany*
[2]*Vistamount Consulting, USA*

6.1 INTRODUCTION

Classical learning algorithms create models of data in an uncontrolled, non-interactive manner. Typically the user specifies some (method-dependent) parameters like distance function or number of clusters that he/she likes to identify, followed by the application of the algorithm using these settings. The process of the model generation itself, however, cannot be controlled or influenced by the user. The final outcome is then evaluated by means of some quality measure, for instance the classification error for supervised learning or some cluster validity measure for unsupervised tasks, or it is judged based on the user's impression, provided that the model is interpretable. Depending on the quality of the results, the model generation is either considered successful, which means the model is a good summarization of the data and can be used, for example, for further classification tasks, or it requires further fine-tuning of the parameters and a rerun of the algorithm.

This "learning" scheme is characteristic for most methods in machine learning and data mining. However, in many applications the focus of analysis is not on the optimization of some quality function but rather on the user-controlled generation of interpretable models in order to incorporate background knowledge. This requires tools that allow users to interact with the learning algorithm to inject domain knowledge or help to construct a model manually by proposing good model components, for example, cluster prototypes or classification rules, which can then be accepted, discarded, or fine-tuned. There are generally two different extrema of cooperation between user and system: either the user or the system has all the control and carries out the model generation; however, in general it requires a balance of both.

Several approaches to build interpretable models interactively for classification have been proposed in the past. Ankerst, Elsen, Ester, and Kriegel (1999) propose an interactive decision tree learner that is based on a pixel-oriented visualization of the data. Figure 6.1 shows a screenshot of such an interactive tool. The visualization technique is similar to the one proposed by Ankerst, Elsen, Ester, and Kriegel (1999) but uses a box instead of a circle view. The top half of the figure displays the class distribution for one of the input dimensions with each pixel representing one of the training objects (there are about 15000 training objects

Advances in Fuzzy Clustering and its Applications Edited by J. Valente de Oliveira and W. Pedrycz

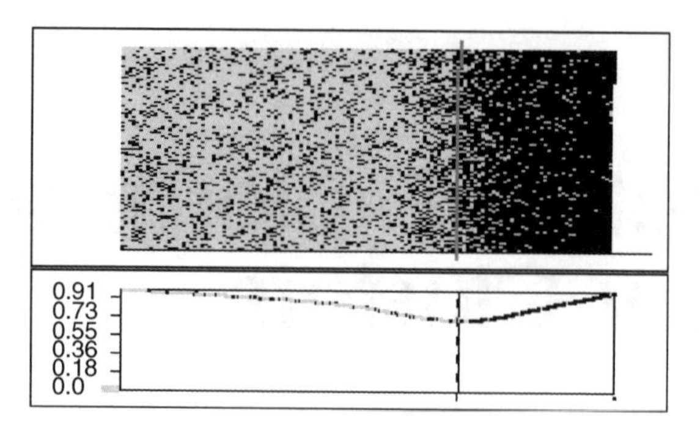

Figure 6.1 Interactive construction of decision trees.

in this data-set). The objects are aligned column by column according to the order of their values in the current dimension, whereby the object with the smallest value is plotted at the top left corner of the pixel display and the others are lined using a snake like alignment. Different colors represent different classes, for instance there are two classes in the example in Figure 6.1. The bottom half of the figure shows the entropy curve for the current dimension, i.e., a classical (automatic) decision tree learner would perform a split at the value with the smallest entropy over all dimensions. This visualization technique allows the class distribution to be inspected immediately for each individual dimension, either by plotting different boxes for different classes or – as proposed by Ankers, Elsen, Ester, and Kriegel (1999) – by using a circle view where different segments of the circle are used to plot the distribution in different dimensions. The construction of the decision tree model will take place interactively whereby the system supports the user by proposing good split points, allowing for look-aheads (such as what would be a resulting sub tree if a split was performed), or automatic construction of sub-trees, for example, when there are only small amounts of data left in a decision branch. The user, on the other hand, can adjust split points or also perform the split on another attribute, which may not lead to such a high information gain as the numerical optimal split point but is, from the user's perspective, more meaningful.

This system for decision tree construction is a typical example for an interactive model generation tool. By using an appropriate visualization technique (in the above example a pixel display), the user can guide the model generation.

Decision trees always partition the instance space, i.e., each point in the input space is assigned a class label. This is often not desirable as some regions of the input space may not contain (training-) data at all and therefore there is no obvious evidence to prefer one class over the other. Rule learning algorithms, on the other hand, generate a set of rules whereby each rule covers only a relatively small region of the input data. Typically, there may be regions in the input space for which more than one rule fires (posing the challenge of conflict handling when the firing rules are of contradicting classes) or for which none of the rules is active. In the latter case, an outcome is often determined using the majority class (the one with highest a priori probability) or – and which is often preferable – a "don't know" answer. Many rule learning algorithms also use the notion of fuzzy membership functions to model regions of high and low confidence (Berthold, 2003; Chiu, 1997).

Fuzzy clusters, similar to fuzzy rules, are well suited for presentation of the resulting classification model to the user. Although the traditional cluster algorithm works on unsupervised data-sets, extensions also allow cluster models to be built that distinguish between areas of different classes. This is an intriguing approach especially for cases where one expects to find various, distributed areas that belong to the same class. Often these clusters are then used directly as fuzzy rules or serve to initialize a fuzzy rule system, which is then optimized. A typical example of this sort of algorithm has been proposed by Chiu (1997): it first finds a set of clusters for each class using subtractive clustering (Chiu, 1994), an algorithm that builds upon the well-known mountain method by Yager and Filev (1994), and then derives classification rules from them.

In this chapter we focus on a supervised approach to construct a set of fuzzy clusters for classification. The algorithm does not use a two stage learning such as in (Chiu, 1997) but rather generates potentially discriminative fuzzy clusters from the beginning. It initially constructs a so-called *Neighborgram* for each object of interest. A Neighborgram is a summarization of the neighborhood of an object, which allows an interpretable view on the underlying data. Such a complete and hence computationally expensive approach obviously only works for all classes of a medium size data-set or – in the case of very large data-sets – to model a minority class of interest. However, in many applications the focus of analysis is on a class with few objects only, a minority class. Such data can be found, for instance, in drug discovery research. Here, huge amounts of data are generated in high throughput screening, but only very few compounds really are of interest to the biochemist. Therefore, it is of prime interest to find clusters that model a small but interesting subset of data extremely well.

The algorithm finds clusters in a set of such Neighborgrams based on an optimality criterion. Since Neighborgrams are easy to interpret, the algorithm can also be used to suggest clusters visually to the user, who is able to interact with the clustering process in order to inject expert knowledge. Therefore, the clustering can be performed fully automatically, interactively, or even completely manually. Furthermore, constructing Neighborgrams only requires a distance matrix, which makes them applicable to data-sets where only distances between objects are known. For many similarity metrics in molecular biology no underlying feature values are known since those similarity (and hence also the distance) values are computed directly. In contrast, methods that compute cluster representatives as mixtures of training objects (like fuzzy c-means by Bezdek (1981)) do require the availability of an underlying feature representation in order to continuously compute and update the cluster centers.

Neighborgrams can naturally be applied to problem settings where there are multiple descriptors for the data available, known as *parallel universes*. One such application is biological data analysis where different descriptors for molecules exist but none of them by itself shows global satisfactory prediction results. We will demonstrate how the Neighborgram clustering algorithm can be used to exploit the information of having different descriptors and how it finds clusters spread out of different universes, each modeling a small subset of the data.

This chapter is organized as follows. We first introduce the concept of Neighborgrams and describe the basic clustering algorithm. We then extend the algorithm to also handle fuzzy clusters before Section 6.3 discusses some aspects of the visual exploration and demonstrates the usefulness of the visual clustering procedure. In Section 6.4 we focus on learning in parallel universes and give an example application using the Neighborgram algorithm.

6.2 NEIGHBORGRAM CLUSTERING

This section introduces Neighborgrams as an underlying data structure of the presented algorithm. We will formalize this structure and derive some properties which help us to judge the quality of a Neighborgram later on. We will refer to one of our previous articles, which discusses the automatic classifier using Neighborgrams more extensively (Berthold, Wiswedel, and Patterson, 2005).

We will assume a set of training objects T with $|T| = M$ instances for which distances, $d(x_i, x_j), i,\ j \in \{1, \ldots, M\}$, are given[1]. Each example is assigned to one of C classes, $c(x_i) = k, 1 \leq k \leq C$.

6.2.1 Neighborgrams

A Neighborgram is a one-dimensional model of the neighborhood of a chosen object, which we will call the *centroid*. Other objects are mapped into the Neighborgram depending on their distance to this centroid. Essentially, a Neighborgram summarizes the neighborhood of the centroid through a ray on

[1]Note that it is not necessary to know the feature values for an instance. It is sufficient to provide the algorithm with distances between objects.

to which the closest neighbors of the centroid are plotted. Obviously, mapping all objects on to the ray would be complicated and the visualization would lose its clarity. Therefore, we introduce a parameter R that determines the maximum number of objects stored in a Neighborgram. Those R stored items represent the R-closest neighbors to the centroid. Hence, a Neighborgram for a certain centroid x_i can also be seen as an ordered list of length R:

$$\mathrm{NG}_i = [x_{l_1}, \ldots, x_{l_R}].$$

The list NG_i is sorted according to the distance of object x_{l_r} to the center vector x_i:

$$\forall r \; : \; 2 \leq r \leq R \wedge d(x_i, x_{l_{(r-1)}}) \leq d(x_i, x_{l_r}),$$

and the objects in the Neighborgram are the closest neighbors of the centroid:

$$\neg \exists r \; : \; r > R \wedge d(x_i, x_{l_r}) < d(x_i, x_{l_R}).$$

Note that $l_1 = i$, because $d(x_i, x_i) = 0$ for all i, that is, each object is closest to itself. Note also that this list is not necessarily a unique representation since it is possible that two entries in the list have exactly the same distance to the centroid. The order of those items would then not be uniquely defined. However, as we will see later, this does not affect the outcome of the clustering algorithm that we are discussing here.

Obviously in the case of large data-sets the computation of Neighborgrams for each training object is excessively time and memory consuming. However, as noted earlier, the main target of the algorithm discussed here are problems where one (or several) minority class(es) are of prime interest. The computation of Neighborgrams for all these objects is then of complexity $O(R \cdot M \cdot M')$, where M' indicates the number of examples of the minority class(es), i.e., $M' \ll M$ in the case of large data-sets. This complexity estimate is derived as follows: for each object $(O(M))$ and for each Neighborgram $(O(M'))$ do an insertion sort into a list of R objects $(O(R))$. If the size R of the Neighborgrams is closer to the overall number of objects M it might make more sense to use a more efficient sorting scheme but for large data-sets usually $R \ll M$ holds and an insertion sort is sufficiently efficient. For large data-sets, M will dominate the above estimate and result in a roughly linear complexity.

Figure 6.2 shows an example of two Neighborgrams for the famous Iris data (Fisher, 1936), a data-set containing four-dimensional descriptions of three different types of Iris plant. For the sake of simplicity, in the example we use only two of the original four dimensions in order to be able to display them in a scatterplot (left-hand side). Different colors are used here to show the different class memberships of the overall 150 objects. Two sample Neighborgrams are shown on the right. They are constructed for the two centroids of the gray class (Iris-Setosa), indicated by the arrows in the scatterplot. Each Neighborgram summarizes the neighborhood of its centroid. The centroid of the top Neighborgram, for instance, has many objects from the same (gray) class in its close vicinity as can be immediately seen when looking at the Neighborgram. The visualization technique uses a simple ray to plot neighbors; however, whenever two objects are too close to each other and might overlap, we stack them. Stacking allows the user to individually select certain objects in one Neighborgram, which are then also highlighted in the other Neighborgrams or another system.

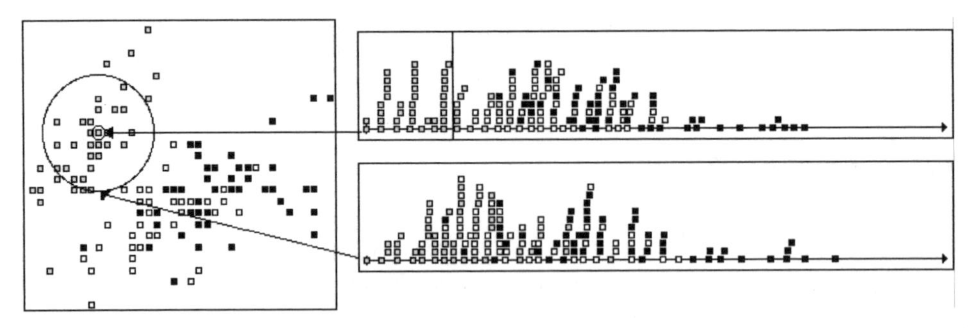

Figure 6.2 Two Neighborgrams for the Iris data is shown on the left. The two-dimensional input space, on the right there are two different Neighborgrams for two selected objects displayed.

Neighborgrams are constructed for all objects that are of interest to the user, for instance all objects of one class. Just from looking at the example in Figure 6.2, we can qualitatively rank Neighborgrams: the centroid of the top Neighborgram has many objects of its class in the close neighborhood whereas the centroid of the bottom Neighborgram is surrounded by some gray but also white objects. The top Neighborgram therefore suggests a better cluster candidate since new, unseen objects that have a small distance to the centroid are likely to be of the same class.

Let us briefly sketch the underlying algorithm to identify a set of good Neighborgrams in the next section before we derive some underlying properties of a Neighborgram using the notation introduced above.

6.2.2 The Basic Clustering Algorithm

The key idea underlying the clustering algorithm is that each object, for which a Neighborgram has been built, is regarded as a potential cluster center. The objective of the algorithm is to rank Neighborgrams in order to greedily find the "best" cluster at each step. The result is a subset of all possible clusters that covers a sufficient number of objects.
The algorithm can be summarized as follows:

1. determine a cluster candidate for each Neighborgram;
2. rank cluster candidates and add the best one as a cluster;
3. remove all objects covered by this cluster;
4. start over at Step 1, unless certain stopping criteria are fulfilled.

Obviously, it needs to be defined, what a cluster candidate is how these candidates can be ranked it, and what removing covered objects really means. In addition, the termination criterion has to be specified. In order to do this, let us first define a few properties of Neighborgrams.

6.2.3 Neighborgram Properties

In Section 6.2.1 we used an ordered list as representation for a Neighborgram. This list contains objects, which are ordered according to their distance to the centroid. The length of the list is determined by the parameter R:

$$\mathrm{NG}_i = [x_{l_1}, \ldots,\ x_{l_r}, \ldots, x_{l_R}].$$

The main parameters to describe a cluster candidate are the following:

- *Coverage* Γ. The default coverage of a cluster with a certain depth $r \leq R$ determines how many positive objects it "explains," that is, the number of objects of the same class as the centroid that fall within its radius:

$$\Gamma_i(r) = \left|\{x_{l_{r'}} \in \mathrm{NG}_i | 1 \leq r' \leq r \wedge c(x_{l_{r'}}) = c(x_i)\}\right|.$$

- *Purity* Π. The purity of a Neighborgram is the ratio of the number of objects belonging to the same class as the centroid to the number of objects encountered up to a certain depth $r \leq R$. The purity is a measure of how many positive vs. negative objects a certain neighborhood around the centroid contains. Positive objects belong to the same class as the centroid, whereas negative objects belong to a different class:

$$\Pi_i(r) = \frac{\Gamma_i(r)}{r}.$$

- *Optimal depth* Ω. The optimal depth is the maximum depth where for all depths r less than or equal to Ω the purity is greater than or equal to a given threshold $p_{\min}$. The optimal depth defines the maximum size of a potential cluster with a certain minimum purity. Note that it is straightforward to derive the corresponding radius from a given depth, that is, $d(x_i, x_{l_r})$:

$$\Omega_i(p_{\min}) = \max\{r | 1 \leq r' \leq r \wedge \Pi_i(r') \geq p_{\min}\}.$$

Furthermore, we introduce a final parameter Ψ for the overall coverage, which is part of the termination criterion for the algorithm. It represents the sum of all coverages of the chosen clusters. Once this threshold is reached, the algorithm stops.

6.2.4 Cluster Candidates and the Clustering Algorithm

Using the above properties we can already clarify the (automatic) clustering procedure. Starting from a user-defined value for parameter purity $\Pi = p_{\min}$ and the stopping criterion Ψ, we can compute values for parameters optimal depth Ω and coverage Γ for each potential cluster. The best cluster is identified as the one with the highest coverage. This cluster then "covers" all objects that are within its radius. These objects are then discarded from the data-set and the cluster-selection process can start again in a sequential covering manner, based on the reduced set of remaining objects. The termination criterion of the algorithm is based on the accumulated coverage of identified clusters: once it exceeds a certain threshold given by the user-defined overall coverage Ψ, the algorithm stops. Thus, the numbers of clusters is implicitly determined by the algorithm as new clusters are being added as long as the coverage is below Ψ. The basic algorithm is outlined in Table 6.1.

Although the clustering scheme as listed in Table 6.1 does not incorporate user interaction, it is fairly easy to integrate: the algorithm determines the (numerically) best Neighborgram (line (6)) and adds it to the set of clusters. However, instead of simply adding the Neighborgram, the ranking (according to $\Gamma_i(\Omega_i)$) can be used to suggest discriminative Neighborgrams to the user, who might be interested in picking another (second choice) Neighborgram or changing the cluster boundaries. Please note that the identification of good Neighborgrams is always bound to an appropriate visualization of the underlying objects in the cluster. For example, as we will also see later in Section 6.3, if the objects represent molecular drug candidates, an accompanying visualization of the molecular structures of the objects in one cluster can help the user to judge if these objects do indeed have something in common or if they are just artifacts in the data.

The basic clustering algorithm in Table 6.1 removes objects once they are covered by a cluster. This effect might be desirable for objects lying close to the center of the new cluster but it will reduce accuracy in areas further away from the cluster center. We therefore introduce the notion of *partial coverage* using fuzzy membership functions, which allows us to model a degree of membership of a particular object to a cluster. The next section will present the membership functions used.

Table 6.1 The basic Neighborgram clustering algorithm (Berthold, Wiswedel and Patterson, 2005).

(1) $\forall x_i$: $c(x_i)$ is class of interest $\Rightarrow$ compute NG_i
(2) $\forall NG_i$: compute $\Omega_i(p_{\min})$
(3) $\forall NG_i$: compute $\Gamma_i(\Omega_i)$
(4) $s := 0$
(5) while $s < \Psi$
(6) $i_{\text{best}} = \arg\max_i\{\Gamma_i(\Omega_i)\}$
(7) add $NG_{i_{\text{best}}}$ to list of clusters,
    add $\Gamma_{i_{\text{best}}}(\Omega_{i_{\text{best}}})$ to s
(8) determine list of covered objects
(9) remove them from all Neighborgrams NG_i
(10) $\forall NG_i$: recompute $\Gamma_i(\Omega_i)$
(11) end while

6.2.5 Membership Functions

The idea underlying the partial coverage is that each cluster is modeled by a fuzzy membership function. This function has its maximum at the centroid and declines toward the cluster boundaries. The coverage is then determined using the corresponding degrees of membership. Objects are removed to a higher degree towards the inner areas of a cluster and to a lesser degree toward the outer bounds. Figures 6.3 to 6.6 show the four membership functions we used. Note that the rectangular membership function corresponds to

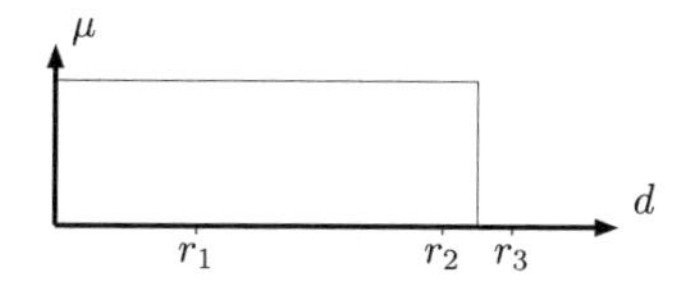

$$\mu(d) = \begin{cases} 1, & \text{if } 0 \leq d \leq \frac{r_2+r_3}{2} \\ 0, & \text{otherwise.} \end{cases}$$

Figure 6.3 The *rectangular* membership function.

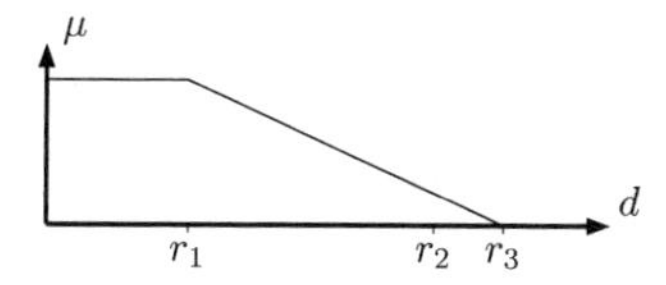

$$\mu(d) = \begin{cases} 1, & \text{if } 0 \leq d \leq r_1 \\ \frac{r_3-d}{r_3-r_1}, & \text{if } r_1 < d \leq r_3 \\ 0, & \text{otherwise.} \end{cases}$$

Figure 6.4 The *trapezoidal* membership function.

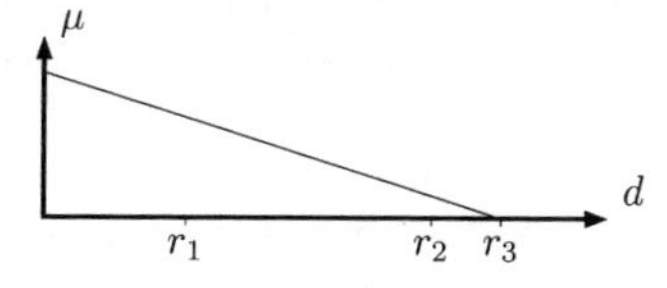

$$\mu(d) = \begin{cases} \frac{r_3-d}{r_3}, & \text{if } 0 \leq d \leq r_3 \\ 0, & \text{otherwise.} \end{cases}$$

Figure 6.5 The *triangular* membership function.

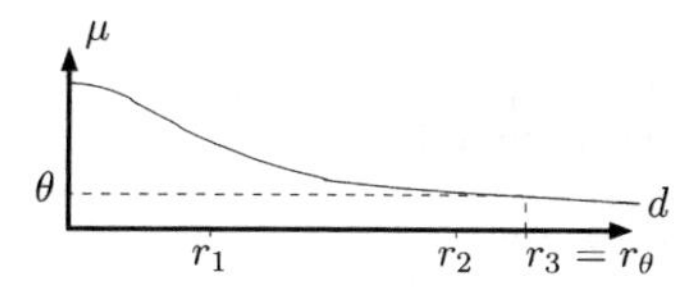

$$\mu(d) = \exp\left(-\frac{d^2}{\sigma^2}\right) \quad \text{with } \sigma^2 = -\frac{r_3^2}{\ln(\theta)}$$

Figure 6.6 The *gaussian* membership function.

the basic algorithm discussed above: objects are covered with degrees of 0 or 1 only, and are therefore removed completely when covered.

In order to describe a cluster by means of a membership function we first need to introduce three radii, which will help to specify different regions of the neighborhood:

- r_1 represents the radius of the last object with $\Pi = 1$ (last known perfect): $r_1 = \max\{r \mid \Pi_i(r) = 1\}$.
- r_2 is the last object with $\Pi \geq p_{\min}$ (last known good), that is, $r_2 = \max\{r \mid 1 \leq r' \leq r \wedge \Pi_i(r') \geq p_{\min}\}$.
- r_3 describes the first object for which $\Pi < p_{\min}$ (first known bad), that is, $r_3 = \max\{r \mid 1 \leq r' \leq r - 1 \wedge \Pi_i(r') \geq p_{\min}\}$.

These radii are sufficient to describe as shown in Figure 6.3 to 6.6 commonly used membership functions.

While the shape of the rectangular, trapezoidal and triangular membership functions are determined by the three radii, the Gaussian membership function is specified using the additional parameter θ. The inverse value of θ, r_θ, determines radius r_3. For a minimum required purity $p_{\min}$ equal to 1, the parameter θ determines the maximum degree of membership of an incorrect class for other objects in the training data (see Berthold and Diamond (1998) for details).

Using these fuzzy membership functions the clustering algorithm changes slightly. First, a degree of exposure (measuring how much is still uncovered), $\eta \in [0, 1]$, needs to be assigned to each object of the classes of interest. At the beginning this value is initialized to 1.0 for each object, that is, each object still needs to be covered completely. Subsequently this value will be decreased during clustering. A new cluster which (partly) covers an object will reduce this value accordingly. Obviously an object can only be covered until $\eta = 0$. Let $\eta(x)$ be an object's degree of exposure and $\mu_{Cluster}(d(x_i, x))$ the degree of membership to the cluster. Then the partial coverage Φ of a cluster is defined as:

$$\Phi_i(\Omega_i) = \sum_{\substack{x_{l_{r'}} \in \text{NG}_i \mid 1 \leq r' \leq \Omega_i \\ \wedge c(x_{l_{r'}}) = c(x_i)}} \min\{\eta(x_{l_{r'}}),\ \mu_{Cluster}(d(x_i, x_{l_{r'}}))\}.$$

The new fuzzy version of the algorithm is shown in Table 6.2. A list of objects for the class of interest needs to be created in conjunction with their degrees of coverage (step (2)). Steps (8) and (9) of the basic algorithm are modified to incorporate the notion of partial coverage. Note that we do not need to remove covered objects from other Neighborgrams anymore, since the added degree of exposure does this implicitly.

The introduction of this concept of partial coverage improves the accuracy substantially. Experiments on publicly available data-sets from the StatLog project (Michie, Spiegelhalter, and Taylor, 1994) demonstrate that the performance of such an automatic classifier is comparable to state-of-the-art

Table 6.2 The fuzzy Neighborgram clustering algorithm (Berthold, Wiswedel, and Patterson, 2005).

(1)	$\forall x_i : c(x_i)$ is class of interest $\Rightarrow$ compute NG_i
(2)	$\forall x_i : c(x_i)$ is class of interest $\Rightarrow$ store $\eta(x_i) = 1$
(3)	$\forall \text{NG}_i$: compute Ω_i
(4)	$\forall \text{NG}_i$: compute $\Phi_i(\Omega_i)$
(5)	$s := 0$
(6)	while $s < \Psi$
(7)	$i_{\text{best}} = \arg\max_i\{\Phi_i(\Omega_i)\}$
(8)	add $\text{NG}_{i_{\text{best}}}$ to list of clusters, add $\Phi_{i_{\text{best}}}(\Omega_{i_{\text{best}}})$ to s
(9)	recompute η for each object and
(10)	$\forall \text{NG}_i$: recompute $\Phi_i(\Omega_i)$
(11)	end while

techniques (among others c4.5, k nearest neighbor, and a multi-layer perceptron). We do not discuss these experiments here but rather refer to (Berthold, Wiswedel, and Patterson, 2005). The use of fuzzy membership functions as cluster description increased the generalization ability of the classifier significantly. The best performance was always achieved using the Gaussian membership function, which, however, has one important drawback: it always produces an output since the membership value is always greater than 0. In most cases a classifier that also produces a "do not know" answer is preferable, as it allows an obviously uncertain classification to be deferred to an expert or to another system. In the following we will therefore concentrate on the other membership functions instead, also because they allow for a more intuitive visualization.

6.3 INTERACTIVE EXPLORATION

As already outlined in the introduction, the main focus of this chapter lies on the interactive exploration of such supervised fuzzy clusters. The quality measures as introduced in the previous sections allow Neighborgrams to be ranked based on their coverage given a user-defined threshold value for the purity $p_{\min}$. The system uses these values to suggest potentially interesting Neighborgrams to the user who is then able to evaluate how interesting they are. This section demonstrates the usefulness of this approach by means of some examples. We will first briefly explain the visualization technique of a Neighborgram and its cluster candidate before we show its applicability in practice on a real-world data-set from the National Cancer Institute.

6.3.1 Neighborgram Visualization

Since a Neighborgram is a one-dimensional representation of the neighborhood of an object, it is straightforward to visualize. Figure 6.7 shows the Neighborgrams for two objects of the Iris data-set (Fisher, 1936). In comparison to Figure 6.2 the figure also contains a visualization of the (trapezoidal) fuzzy membership function as proposed by the system. While the vertical axis does not have a meaning for the visualization of points (as noted earlier, we use it only to avoid overlaps), it is used to display the membership function.

Both clusters in Figure 6.7 are built for the Iris-Virginica class (points shown in black); however, the top Neighborgram suggests better clustering behavior as it covers almost all of the objects of Virginica class (the same class as the centroid), whereas the bottom Neighborgram also has objects of Iris-Versicolor class (white points) in its close neighborhood. Note also how objects of Iris-Setosa class (gray) form a nice separate cluster far away in both Neighborgrams, a fact well-known from the literature. In this case automatic ranking is likely to be a good choice; however, in a less obvious case, the user could

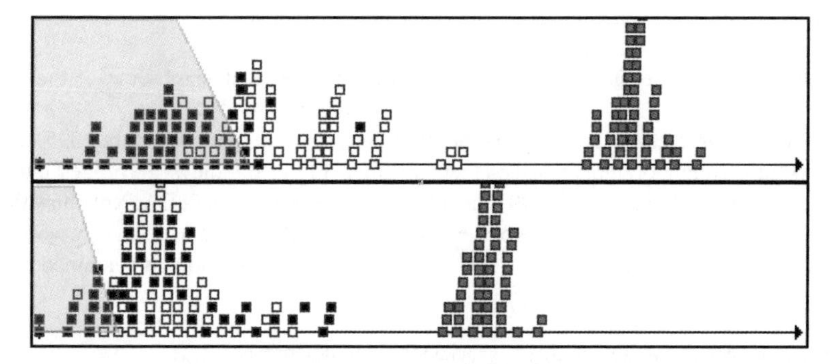

Figure 6.7 Two Neighborgrams built for the Iris data.

overrule the algorithm's choice, select individual clusters, and also modify their membership functions if so desired.

6.3.2 NCI's HIV Data

To show the usefulness of the proposed visual clustering algorithm in a real-world scenario, let us look at the application of Neighborgrams on a well-known data-set from the National Cancer Institute, the DTP AIDS Antiviral Screen data-set. The screen utilized a soluble formazan assay to measure protection of human CEM cells from HIV-1 infection. All compounds in the data-set were tested on their protection of the CEM cell; those that did not provide at least 50 % protection were labeled as confirmed inactive (**CI**). All others were tested in a second screening. Compounds that provided protection in this screening, too, were labeled as confirmed active (**CA**), the remaining ones as moderately active (**CM**). Available online (National Cancer Institute, 2005) are screening results and chemical structural data on compounds that are not protected by a confidentiality agreement. Available are 41 316 compounds of which we have used 36 045[2]. A total of 325 belongs to class **CA**, 877 are of class **CM** and the remaining 34 843 are of class **CI**. Note the class distribution for this data-set is very unbalanced, there are about 100 times as many inactive compounds (**CI**) as there are active ones (**CA**). The focus of analysis is on identifying internal structures in the set of active compounds as they showed protection to the CEM cells from an HIV-1 infection.

This data-set is a very typical application example of the Neighborgram classifiers: although it is a relatively large data-set, it has an unbalanced class distribution with the main focus on a minority class.

In order to generate Neighborgrams for this data-set, a distance measure needs to be defined. We initially computed Fingerprint descriptors (Clark, 2001), which represent each compound through a 990-dimensional bit string. Each bit represents a (hashed) specific chemical substructure of interest. The used distance metric was a Tanimoto distance, which computes the number of bits that are different between two vectors normalized over the number of bits that are turned on in the union of the two vectors. The Tanimoto distance is often used in cases like this, where the used bit vectors are only sparsely occupied with 1s.

NCI lists a small number (75) of known active compounds, which are grouped into seven chemical classes:

- azido pyrimidines;
- pyrimidine nucleosides;
- heavy metal compounds;
- natural products or antibiotics;
- dyes and polyanions;
- purine nucleosides;
- benzodiazepines, thiazolobenzimidazoles, and related compounds.

One would expect that a decent clustering method would retrieve at least some of these classes of compounds.

Therefore, we constructed Neighborgrams for all of the active compounds (overall 325) and used the system to rank these Neighborgrams based on their coverage. Figure 6.8 shows the biggest cluster that the algorithm encountered using a purity of 90 % and building clusters for class **CA**. Note how the next two rows show Neighborgrams for compounds of the same cluster, both of them with slightly worse computed *purity* and *coverage*. At first we were surprised to see that none of the compounds contained in this cluster

[2]For the missing compounds we were unable to generate the used descriptors.

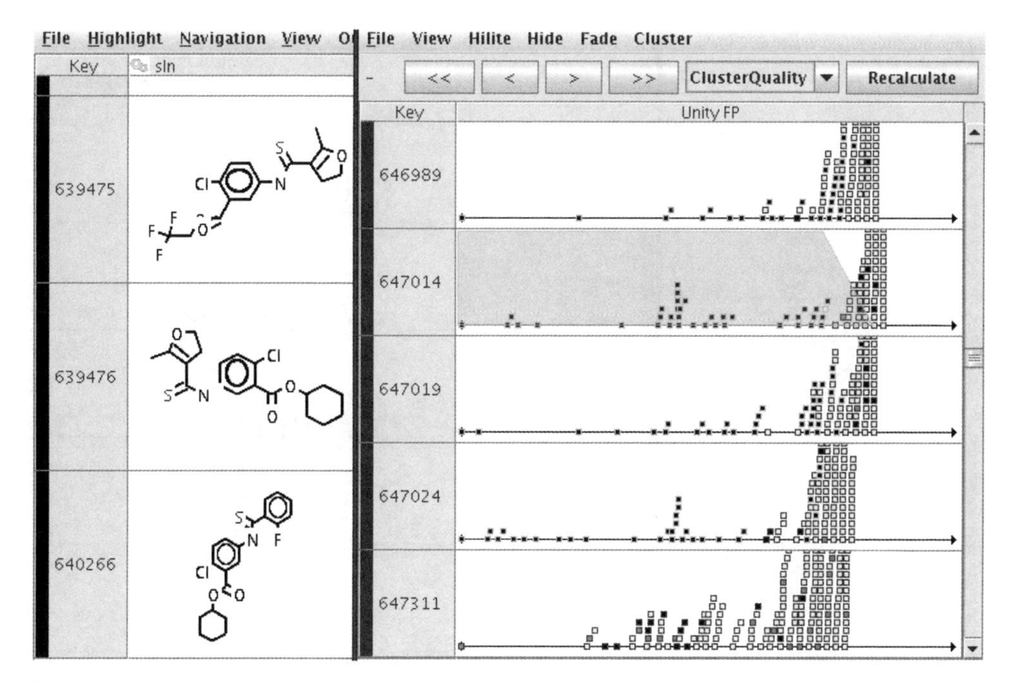

Figure 6.8 The first cluster of the NIH-Aids data centered around compound #647014. On the right the Neighborgrams in unity-fingerprint space (black=**CA**, gray=**CM**, white=**CI**), on the left a view showing some of the structures contained in this cluster.

fall in any of the classes of active compounds listed on NIH's Web site (National Cancer Institute, 2005). As it turns out when looking at the corresponding structures, this cluster covers m-acylaminobenzamides which probably all inhibit folic acid synthesis, but are likely to be too toxic and hence not very interesting as active compounds to fight HIV. This is therefore a nice example of a cluster that a chemist might discard as 'useful but not very interesting for the current task at hand." The clustering algorithm has no insights other than numerical cluster measures and therefore would assign a high ranking value to this cluster without any expert interaction.

Subsequent clusters reveal groupings very much in line with the classes listed above, one particular example is shown in Figure 6.9. Here the group of "dyes and polyanions" are grouped together in a nice cluster with almost perfect purity (two inactive compounds are covered as well). Figure 6.10 shows another example, this time grouping together parts of the group of "azido pyrimidines," probably one of the best-known classes of active compounds for HIV.

Experiments with this (and other similar) data-sets showed nicely how the interactive clustering using Neighborgrams helps to inject domain knowledge in the clustering process and how Neighborgrams help promising cluster candidates to be quickly inspected visually. Without the additional display of chemical structure this would not have worked as convincingly. It is important to display the discovered knowledge in a "language" the expert understands.

In our experiments we were confronted with the problem of which descriptor to use. In the previous experiments we always used unity fingerprint descriptors in conjunction with the Tanimoto distance. We have ignored that there may be more than just this single description available. However, particularly in the field of molecular data analysis there are numerous ways to describe molecules and it is hardly ever known which descriptor is best. This problem relates to the learning in parallel universes, which will be addressed in the next section.

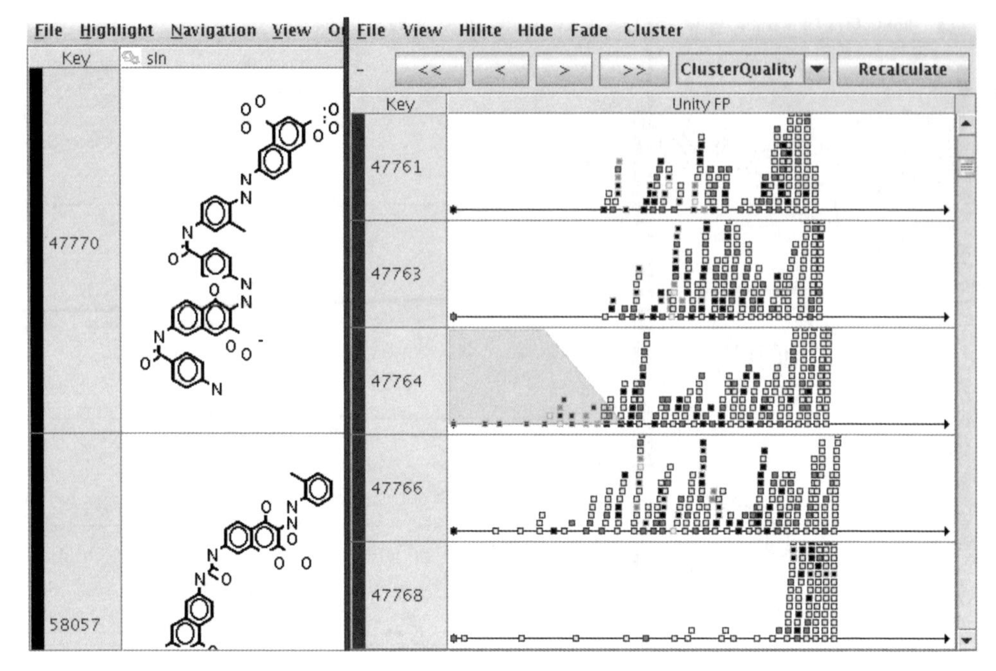

Figure 6.9 Another cluster of the NIH-Aids data centered around compound #47764 (right: Neighborgrams (black=**CA**, gray=**CM**, white=**CI**), left: structures). This cluster nicely covers one of the classes of active compounds: dyes and polyanions.

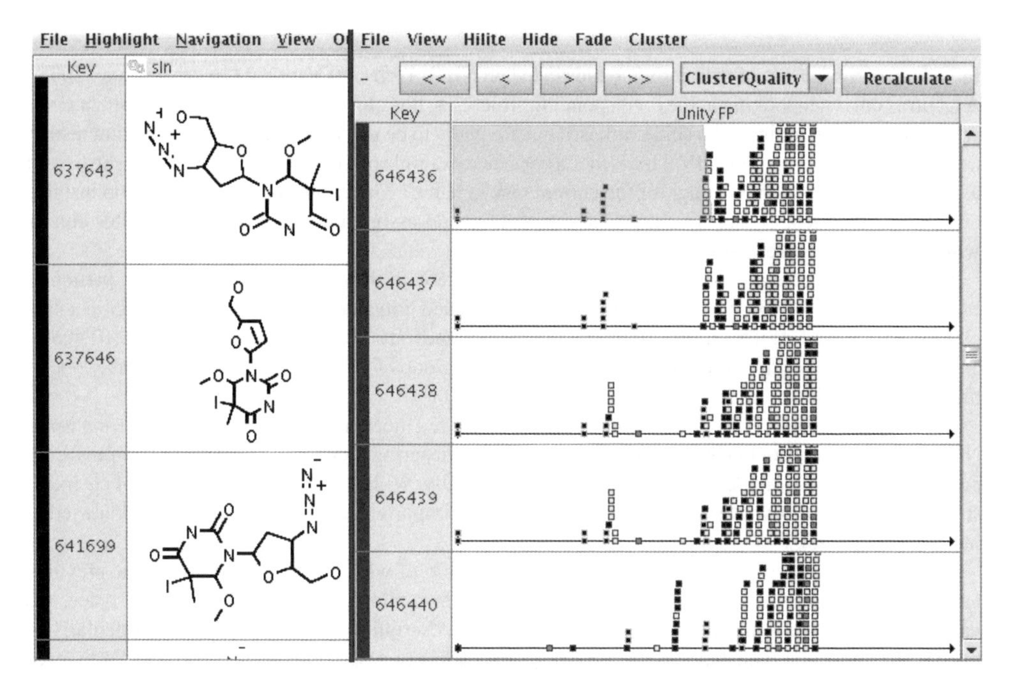

Figure 6.10 Another cluster of the NIH-Aids data centered around compound #646436 (right: Neighborgrams (black=**CA**, gray=**CM**, white=**CI**), left: structures). This cluster nicely covers part of one of the most well-known classes of active compounds: azido pyrimidines.

6.4 PARALLEL UNIVERSES

In the foregoing experiment we assumed that there is an adequate description of the data available. This descriptor was expected to comprise all necessary information to classify an object. However, in many real-world applications, the generation of an appropriate descriptor is far from trivial as the underlying objects are complex and can be described in various ways, focusing on different aspects of the object's nature. An example, other than molecules, are musical songs, i.e., audio streams, which can be represented based on dynamics, melody and key or – as a different representation – based on rhythm and harmony. A third representation may be more descriptive, such as interpreter, position in music charts, length, and so on. Further examples of such complex objects are images, and three-dimensional objects. For the learning it is often unclear, which of the available descriptors are optimal for any given task. This leads to the notion of *learning in parallel universes*, where we can find interesting patterns, e.g., clusters, in different descriptor spaces in parallel. Wiswedel and Berthold (2006), for instance, applied a fuzzy *c*-means algorithm to data described in parallel universes. However, this is an unsupervised approach. When looking at the algorithm in Table 6.1, one notes that the Neighborgram methodology lends itself naturally to handling different descriptor spaces: the clusters do not interact with each other based on any universe-specific information. Besides the fact that a chosen cluster removes covered objects from consideration there is no obvious need for two clusters to originate from the same universe. Instead of constructing just one Neighborgram for each object of interest, we can easily create Neighborgrams for each available descriptor space and consider these as potential cluster candidates. We can then modify the clustering algorithm to investigate all Neighborgrams in all feature spaces in parallel and choose the best cluster among all universes. Covered objects will subsequently be removed from all universes and the result is a set of clusters, spread out over different feature spaces.

Figure 6.11 shows an example for the HIV data from the previous section. However, rather than just having one fingerprint descriptor as before, we computed two other descriptors of the underlying compounds. Firstly, we generated an AtomPair fingerprint descriptor, a 1200-dimensional bit vector, which encodes the presence or absence of certain pairs of atoms in the molecule. The second one is a VolSurf descriptor (Cruciani, Crivori, Carrupt and Testa, 2000), i.e., a 56-dimensional numerical feature vector encoding molecular properties such as molecular weight and two-dimensional numerical descriptions describing properties of the three-dimensional structure. Distances were calculated using the Euclidean distance in VolSurf space and Tanimoto distance for both fingerprints (Unity and AtomPair). The left column in the figure shows Neighborgrams in AtomPair space, the middle column Neighborgrams in Unity space, whereas the Neighborgrams in the right column are constructed using the VolSurf descriptor.

Note how the Neighborgrams in the first row differ substantially although they represent the neighborhood of the same object (#662427). The Unity fingerprint descriptor (middle) suggests for this compound

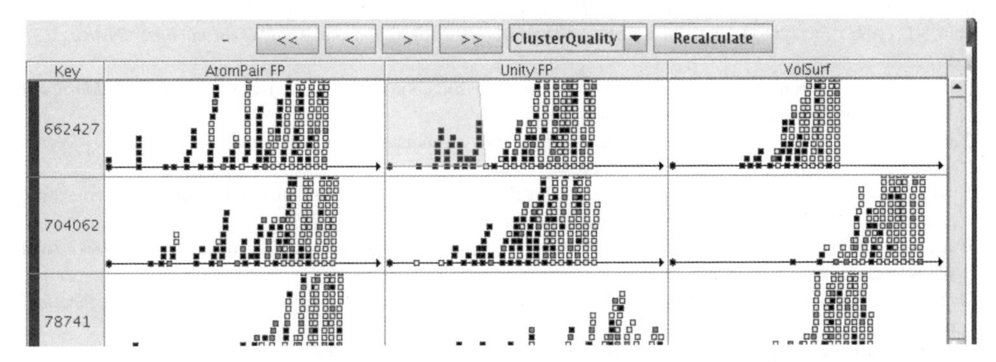

Figure 6.11 Neighborgrams for the NIH-Aids data in parallel universes. The left column in a AtomPair fingerprint space (1200-dimensional bit vectors), the middle column Unity fingerprint space (990-dimensional bit vectors), and the right column 56-dimensional Euclidean space, the VolSurf descriptor. Note how the Neighborgrams in the top row differ, although they represent the neighborhood of the same compound, however in different universes.

the best cluster as it covers about 25 active molecules (black points). However, the neighborhood of this compound in the VolSurf space (right) contains only few actives (less than 10) and in the AtomPair fingerprint space about 15. This example demonstrates that the definition of a cluster depends on the underlying object description. The user can consequently inspect the cluster and potentially gain new insights as to why objects group in one universe but not in another. Especially for data-sets that involve structural descriptions of molecules it is hardly ever known which descriptor is optimal for a particular problem. The final outcome of the clustering algorithm itself is a set of clusters originating from different feature spaces.

6.5 DISCUSSION

In this chapter we discussed a supervised approach to identify and visually explore a set of fuzzy clusters. We used a one-dimensional data structure, a so-called Neighborgram, to depict local neighborhoods of each object. Constructing Neighborgrams for all objects of interest, for example, all objects of a particular class, and deriving potential cluster candidates from them, allowed us to rank these Neighborgrams. An automatic clustering algorithm sequentially accepts the top-ranked cluster and removes all objects covered by this cluster from consideration. More important, however, is that the accompanying visualization of a Neighborgram provides a powerful way to explore the proposed cluster selection and enables the user to inject domain knowledge into the clustering process by accepting, discarding, or fine-tuning potential cluster candidates. Using a real-world data-set from a bioinformatics application we demonstrated how this method of visual exploration supports the user in finding potentially interesting groupings in the data. The described technique provides a tool for interactive exploration of large data-sets, allowing for truly intelligent data analysis.

REFERENCES

Ankerst, M., Elsen, C., Ester, M. and Kriegel, H.P. (1999) 'Visual classification: An interactive approach to decision tree construction'. *Proceedings of the Fifth, A.C.M. SIGKDD International Conference on Knowledge Discovery and Data Mining*, pp. 392–396.

Berthold, M.R. (2003) 'Mixed fuzzy rule formation'. *International Journal of Approximate Reasoning (IJAR)* **32**, 67–84.

Berthold, M.R. and Diamond, J. (1998) 'Constructive training of probabilistic neural networks'. *Neurocomputing* **19**, 167–183.

Berthold, M.R., Wiswedel, B. and Patterson, D.E. (2005) 'Interactive exploration of fuzzy clusters using neighborgrams'. *Fuzzy Sets and Systems* **149**(1), 21–37.

Bezdek, J.C. (1981) *Pattern Recognition with Fuzzy Objective Function Algorithms*. Plenum Press, New York.

Chiu, S.L. (1994) 'Fuzzy model identification based on cluster estimation'. *Journal of Intelligent and Fuzzy Systems* **2**(3), 267–278.

Chiu, S.L. (1997) 'An efficient method for extracting fuzzy classification rules from high dimensional data'. *Journal of Advanced Computational Intelligence* **1**(1), 31–36.

Clark, R.D. (2001) 'Relative and absolute diversity analysis of combinatorial libraries' *Combinatorial Library Design and Evaluation* Marcel Dekker New York pp. 337–362.

Cruciani, G., Crivori, P., Carrupt, P.A. and Testa, B. (2000) 'Molecular fields in quantitative structure-permeation relationships: the VolSurf approach'. *Journal of Molecular Structure* **503**, 17–30.

Fisher, R.A. (1936) 'The use of multiple measurements in taxonomic problems' *Annual Eugenics, II*, pp. 179–1887. John Wiley & Sons, Inc., New York.

Michie, D., Spiegelhalter, D.J. and Taylor, C.C. (eds) (1994) *Machine Learning, Neural and Statistical Classification*. Ellis Horwood Limited Chichester, UK.

National Cancer Institute (2005) http://dtp.nci.nih.gov/docs/aids/aids_data.html.

Wiswedel, B. and Berthold, M.R. (2007) 'Fuzzy clustering in parallel universes'. *International Journal of Approximate Reasoning* (in press).

Yager, R.R. and Filev, D.P. (1994) 'Approximate clustering via the mountain method'. *IEEE Transaction on Systems, Man, and Cybernetics* **24**(8), 1279–1284.

Part III

Algorithms and Computational Aspects

7 Fuzzy Clustering with Participatory Learning and Applications

Leila Roling Scariot da Silva[1], Fernando Gomide[1], and Ronald Yager[2]

[1]*State University of Campinas – FEEC – DCA, Campinas, SP – Brazil*
[2]*Iona College, New Rochelle, New York, USA*

7.1 INTRODUCTION

Clustering is an essential task in information processing, engineering, and machine learning domains. Applications include man–machine communication, pattern recognition, decision-making, data mining, system modeling, forecasting, and classification (Bezdek and Pal, 1992).

A recurring problem in clustering concerns the estimation of the number of clusters in a data-set. Most clustering algorithms are supervised in the sense that they assume that the number of clusters is known a priori. If the algorithm assumes that the number of clusters is unknown, then it is unsupervised (Gath and Geva, 1989). When clustering is performed by optimizing a performance criterion, a common approach to find an appropriate number of clusters is to repeat the clustering algorithm for distinct values of c, the number of clusters, and observe how the performance changes. Unsupervised algorithms attempt to find the clusters based on information contained in the data itself. Often, however, validation procedures are used to find the number of clusters. Many validity criteria have been proposed in the literature, but currently there is no consensus on which one is the best since the results depend heavily on the data-set and clustering algorithms (Geva, Steinberg, Bruckmair and Nahum, 2000). It is well known that to find the optimal number of clusters is a complex issue (Bezdek and Pal, 1992; Duda and Hart, 1973; Duda, Hart, and Stork, 2001) and in practice users must validate cluster results using perception and knowledge of the intended application. The user examines the cluster structure directly from data and relative criteria are built comparing different cluster structures to find a reference and decide which one best reveals data characteristics. In these circumstances, unsupervised clustering algorithms are significant once they provide useful information without polarization of the user's perception and knowledge.

One way of clustering a p-dimensional data space into c clusters is to assume a performance criterion P and initially set the c cluster centers randomly in the data space (Pedrycz, 2005). Then, a cluster assignment is performed *globally* by either assigning each data point to one cluster, or assigning a

Advances in Fuzzy Clustering and its Applications Edited by J. Valente de Oliveira and W. Pedrycz

membership grade, the degree with which the data point is compatible with each of the c clusters. Cluster updating and assignment continues during several iterations until convergence, when no significant difference in the value of P or in the cluster assignments is observed between consecutive iterations. An alternative way of performing clustering is to update cluster centers and perform cluster assignment *sequentially*. Often, sequential updating either uses online gradient of P or a learning law. At each iteration one data point is presented and assigned to a cluster. Next, the respective cluster center is updated using the learning law. This procedure is a form of competitive learning similar to that found in the neural networks literature. Neural network-based clustering has been dominated by the self-organizing maps, learning vector quantization, and adaptive resonance theory (Xu and Wunsch, 2005).

This chapter introduces a fuzzy clustering algorithm in which cluster centers are updated using the participatory learning law of Yager (1990). The algorithm can be implemented either globally or sequentially. The participatory learning clustering algorithm is an unsupervised procedure in which the number of clusters depends on the cluster structure developed by the algorithm at each iteration. This accounts for the participatory nature of the clustering algorithm once the current cluster structure affects acceptance and processing of new data. In participatory learning clustering, cluster structures play the role of belief and data operate as information.

The organization of the chapter is as follows. Section 7.2 overviews the main idea and conceptual structure of participatory learning. Section 7.3 shows how participatory learning (PL) is used to cluster data and details the fuzzy clustering procedures. Section 7.4 compares the PL algorithms with the Gustafson–Kessel (GK) and modified fuzzy k-means (MFKM) since they are amongst the most efficient clustering algorithms reported in the literature. Section 7.5 addresses applications of PL clustering algorithm in evolutionary optimization of complex systems, and in system modeling for time series forecasting. The chapter concludes by summarizing issues that deserve further investigation.

7.2 PARTICIPATORY LEARNING

In many environments, learning is a bootstrap process in the sense that we learn and revise our beliefs in the framework of what we already know or believe. Such an environment is called a participatory learning environment. A prototypical example of this environment is that of trying to convince a scientist to discard an old theory for a new one. In this situation, it is worth relating and explaining the new theory in terms of the old, and the faults of the old theory must lie within itself. The old theory must participate in the learning and believing of the new theory (Yager, 1990). Thus, participatory learning assumes that learning and beliefs about an environment depend on what the system already knows about the environment. The current knowledge is part of the learning process itself and influences the way in which new observations are used for learning. An essential characteristic of participatory learning is that an observation impact in causing learning or belief revision depends on its compatibility with the current system belief.

Let $\boldsymbol{v} \in [0, 1]^p$ be a vector that encodes the belief of a system. Our aim is to learn the values of this variable $\boldsymbol{v}$. We will assume that our knowledge about the values of the variable comes in a sequence of observations $\boldsymbol{x}_k \in [0, 1]^p$, where $\boldsymbol{x}_k$ is a manifestation of a value of $\boldsymbol{v}$ in the kth observation. Thus we use vector $\boldsymbol{x}$ as a means to learn valuations of $\boldsymbol{v}$. The learning process is participatory if the usefulness of each observation $\boldsymbol{x}_k$ in contributing to the learning process depends upon its acceptance by the current estimate of the value of $\boldsymbol{v}$ as being valid observation. Implicit in this idea is that, to be useful and to contribute to the learning of values of $\boldsymbol{v}$, observations $\boldsymbol{x}_k$ must somehow be compatible with the current estimates of $\boldsymbol{v}$. Let $\boldsymbol{v}_k$ be the estimate of $\boldsymbol{v}$ after k observations. Participatory learning means that, to be relevant for the learning process, $\boldsymbol{x}_k$ must be close to $\boldsymbol{v}_k$. Intuitively, participatory learning is saying that the system is willing to learn from information that is not too different from the current beliefs. A mechanism to update the estimate, or belief, of $\boldsymbol{v}$ is a smoothing-like algorithm:

$$\boldsymbol{v}_{k+1} = \boldsymbol{v}_k + \alpha \rho_k (\boldsymbol{x}_k - \boldsymbol{v}_k) \tag{7.1}$$

where $k = 1, \ldots, n$, and n is the number of observations, $\boldsymbol{v}_{k+1}$ is the new system belief, $\boldsymbol{v}_k \in [0, 1]^p$ and $\boldsymbol{x}_k \in [0, 1]^p$ are defined above, $\alpha \in [0, 1]$ is the learning rate, and $\rho_k \in [0, 1]$ is the compatibility degree between $\boldsymbol{x}_k$ and $\boldsymbol{v}_k$, given by:

$$\rho_k = F(S_{k1}, S_{k2}, \ldots, S_{kp}),$$

where S_{kj} is a similarity measure,

$$S_{kj} = G_{kj}(v_{kj}, x_{kj}),$$

$S_{kj} \in [0, 1]$, $j = 1, \ldots, p$, and F is an aggregation operator (Pedrycz and Gomide, 1998). $S_{kj} = 1$ indicates full similarity whereas $S_{kj} = 0$ means no similarity. Notice that G_{kj} maps pairs (v_{kj}, x_{kj}) into a similarity degree and this frees the values v_{kj} and x_{kj}, $j = 1, \ldots, p, k = 1, \ldots, n$, from being in the unit interval. Moreover, G_{kj} allows that two vectors $\boldsymbol{v}_k$ and $\boldsymbol{x}_k$ to have $S_{kj} = 1$ even if they are not exactly equal. This formulation also allows for different perceptions of similarity for different components of the vectors, that is, for different js. A possible formulation is:

$$\rho_k = 1 - \frac{1}{p}\sum_{j=1}^{p} d_{kj} \tag{7.2}$$

where $d_{kj} = |x_{kj} - v_{kj}|$. Clearly, ρ_k provides a compatibility measure between observation x_{kj} and the current belief v_{kj}. Notice that, in this case, ρ_k is the complement of the average absolute difference between each observation and the corresponding current belief, that is, between x_{kj} and v_{kj}. We note that Equation (7.2) is a special case of the Hamming distance, more precisely a complement of the normalized Hamming distance. In some instances, especially in machine learning, it is common to adopt the Euclidean distance as an alternative. In general, the compatibility measure ρ_k can be defined as:

$$\rho_k = 1 - \frac{1}{p} d_k \tag{7.3}$$

where $d_k = \| \boldsymbol{x}_k - \boldsymbol{v}_k \|^2$ and $\| \cdot \|$ is a distance function.

One concern about this is that the above participatory learning environment ignores the situation where a stream of conflicting observations arises during a certain period of time. In this circumstance, the system sees a sequence of low values of ρ_k, that is, incompatibility of belief and observations. While in the short term low values of ρ_k cause an aversion to learning, actually it should make the system more susceptible to learning because it may be the case that the current belief structure is wrong. Yager (1990) identified this situation with a type of arousal, a mechanism that monitors the compatibility of the current beliefs with the observations. This information is translated into an arousal index used to influence the learning process, as Figure 7.1 suggests. The higher the arousal rate, the less confident is the system with the current belief, and conflicting observations become important to update the beliefs.

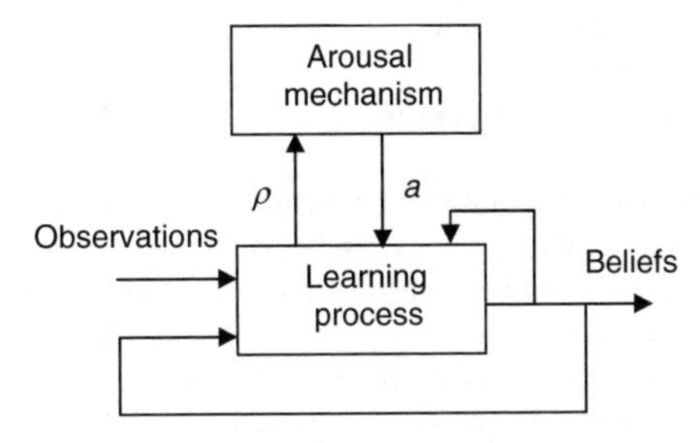

Figure 7.1 An overall scheme of participatory learning.

Let us denote the arousal index by $a_k \in [0, 1]$. The higher the values of a_k, the more aroused the system is. The arousal index is updated as follows:

$$a_{k+1} = a_k + \beta[(1 - \rho_{k+1}) - a_k]. \tag{7.4}$$

The value of $\beta \in [0, 1]$ controls the rate of change of arousal, the closer β is to one, the faster the system is to sense compatibility variations. The arousal index can be viewed as the complement of the confidence in the belief structure currently held.

One way for the participatory learning procedure to consider the arousal mechanism appropriately is to incorporate the arousal index in the basic procedure (7.1) as follows:

$$\boldsymbol{v}_{k+1} = \boldsymbol{v}_k + \alpha(\rho_k)^{1-a_k}(\boldsymbol{x}_k - \boldsymbol{v}_k). \tag{7.5}$$

The form of Equation (7.5) introduces a self-equilibrium mechanism in the participatory learning. While ρ_k measure how much the system changes its credibility in its own beliefs, the arousal index a_k acts as a critic to remind us when the current belief should be modified in the light of new evidence.

Figure 7.1 highlights the main components of participatory learning. The current beliefs, in addition to providing a standard against which observations are compared via the lower feedback loop, directly affect the process used for learning via the upper feedback loop. This upper feedback corresponds to the participatory nature of the model. In Equation (7.5) the upper feedback appears in the form of ρ_k, the compatibility degree between $\boldsymbol{x}_k$ and $\boldsymbol{v}_{k,}$. The arousal mechanism monitors the performance of the lower system by observing the compatibility of the current model with the observations. Therefore learning is dynamic in the sense that Equation (7.5) can be viewed as a belief revision strategy whose effective learning rate

$$\alpha' = \alpha(\rho_k)^{1-a_k}$$

depends on the compatibility between new observations and current beliefs, and on model confidence as well. The primary learning rate is modulated by the compatibility. In models such as competitive learning and gradient based models there are no participatory considerations and the learning rate is usually set small to avoid undesirable swings due to spurious values of $\boldsymbol{x}$ that are far from $\boldsymbol{v}$. Small values of the primary learning rate while protecting against the influence of bad observations, slow down learning. Participatory learning allows the use of higher values of the primary learning rate α once ρ_k, acts to lower the effective learning rate when large deviations occur. Conversely, when the compatibility is large, ρ_k is such that it increases the effective rate that means speeding up the learning process. Next we show how the participatory learning paradigm naturally induces an unsupervised clustering algorithm.

7.3 PARTICIPATORY LEARNING IN FUZZY CLUSTERING

This section introduces participatory learning (PL) as a fuzzy clustering algorithm. One of the main characteristics of the PL clustering algorithm is to naturally partition a data-set X into a suitable number of clusters. Participatory learning clustering is an instance of unsupervised fuzzy clustering algorithm.

First, it is worth noting that there is a close relationship between the participatory learning paradigm and data clustering if we associate data points $\boldsymbol{x}_k$ and cluster centers $\boldsymbol{v}_i$ with observations and beliefs, respectively. The compatibility ρ_{ki} between observation k and the ith belief of the system is viewed as the compatibility degree between $\boldsymbol{x}_k$ and the cluster center $\boldsymbol{v}_i$. The same happens with the arousal index a_{ki}, since it gives an incompatibility degree between current beliefs and observations, that is, it provides an evaluation on how far are observations $\boldsymbol{x}_k, k = 1, \ldots, n$, from the current centers $\boldsymbol{v}_i, i = 1, \ldots, c$.

To partition a data-set X into clusters, the participatory learning fuzzy clustering algorithm uses a parameter τ. The parameter τ is a threshold whose purpose is to advise when an observation should be declared incompatible with the current system belief, the current group structure. In this circumstance belief must be reviewed to accommodate new knowledge. In data clustering this means that if a data point $\mathbf{x}_k$ is far enough from *all* cluster centers, then there is enough motivation to create a new cluster and to declare the discovery of a new cluster structure. An alternative, the one adopted in this chapter, is to set $\mathbf{x}_k$ itself as the new cluster representative, that is, the new cluster center.

More formally, participatory learning clustering partitions a set of data $X = \{\mathbf{x}_1, \ldots, \mathbf{x}_n\}$, $\mathbf{x}_k \in [0,1]^p, k = 1, 2, \ldots, n$, into $c, 2 \leq c \leq n$, fuzzy sets of X. Many clustering models assume spherical clusters of equal size. In this case, an appropriate measure is the Euclidean distance. When spherical clusters are not justifiable, elliptical clustering using the Mahalanobis distance is a more appropriate choice. Therefore, we may, without loss of generality, adopt the Mahalanobis distance in Equation (7.6) to compute the similarity measure, due to its practical usefulness (Gustafson and Kessel, 1979). Application and real-time constraints may require computationally simpler distance measures or recursive computation of the covariance matrix and its inverse:

$$d_{ki}^2 = (\mathbf{x}_k - \mathbf{v}_i)^T \{[(\det(F_i)^{1/n+1})F_i^{-1}]\}(\mathbf{x}_k - \mathbf{v}_i) \tag{7.6}$$

where F_i is the covariance matrix (7.7) associated with the ith cluster, that is,

$$F_i = \frac{\sum_{j=1}^{n} [u_{ji}]^m (\mathbf{x}_j - \mathbf{v}_i)(\mathbf{x}_j - \mathbf{v}_i)^T}{\sum_{j=1}^{n} [u_{ji}]^m}. \tag{7.7}$$

In fuzzy clustering, the fuzzy partition can be represented by a membership matrix U $(n \times c)$ whose element $u_{ki} \in [0,1]$ $i = 1, 2, \ldots, c$ is the membership degree of the kth data point $\mathbf{x}_k$ to the ith cluster, the one with center $\mathbf{v}_i \in [0,1]^p$. For instance, membership degrees, in the same way as in the fuzzy c-means algorithm (Bezdek, 1981; Pedrycz, 2005), can be found using the membership assignment, where $m > 1$:

$$u_{ki}^l = \frac{1}{\sum_{j=1}^{c} (d_{ki}/d_{kj})^{(m-1)}}. \tag{7.8}$$

The participatory fuzzy clustering algorithm can be summarized as follows. Given initial values for α, β, and τ, two random points of X are chosen to assemble the set V^0 of the initial cluster centers. Next, the compatibility ρ_{ki} and arousal a_{ki} indexes are computed to verify if, for all $\mathbf{v}_i$, the arousal index of $\mathbf{x}_k$ is greater than the threshold τ. If the arousal index is greater than the threshold τ, then $\mathbf{x}_k$ is declared as the center of a new cluster. Otherwise, the center closest to $\mathbf{x}_k$ is updated, that is, the center $\mathbf{v}_s$ that has the greatest compatibility index with $\mathbf{x}_k$ is adjusted.

Notice that whenever a cluster center is updated or a new cluster is added, it is necessary to verify if redundant clusters are being formed. This is because updating a cluster center may push it closer to a different cluster center and redundant knowledge may be formed. Therefore a mechanism to exclude close cluster centers is needed since redundancy does not add new information from the point of view of participatory learning once redundancy means the same belief.

A mechanism to exclude redundant cluster centers adopts a compatibility index between cluster centers using the Euclidean distance. A center is excluded when the compatibility index is greater than a threshold that depends on a parameter λ whose value, found experimentally, is $\lambda = 0.95\tau$. Thus, a cluster center i is excluded whenever its compatibility λ_{vi} with another center is less than or equal to λ, or when the compatibility index between the two centers is high. The value of λ_{vi} is inspired in the arousal index idea Equation (7.4).

When a finite set of data is clustered, the algorithm stops when either the maximum number $l_{\max}$ of iterations is reached or no significant variation in the location of the cluster centers has been noted. It is worth noting that, since the cluster centers are updated whenever a data point is provided, the fuzzy partition matrix U must be updated accordingly. Note that steps 2 and 3 are relevant for finite data sets only, that is, when n is fixed, as in the global case. In its sequential version $k = 1, \ldots$ and steps 2 and 3 must be skipped. The detailed steps of the global algorithm (PL-A) are as follows (Silva, Gomide, and Yager, 2005).

7.3.1 Participatory Learning Fuzzy Clustering Algorithm

Input: $\boldsymbol{x}_k \in [0,1]^p$, $k = 1, \ldots, n$, $\alpha \in [0,1]$, $\beta \in [0,1]$, $\tau \in [0,1]$, $\varepsilon > 0$ e $m > 0$.
Choose $l_{\max}$ maximum number of iterations. Set $c = 2$, choose $V^0 = \{\boldsymbol{v}_1, \boldsymbol{v}_2\}$ randomly.
Compute U^0 from V^0, set $l = 1$, $a_{ki}^0 = 0$, $k = 1, 2, \ldots, n$, and $i = 1, \ldots, c$.
Output: c, $\boldsymbol{v}_i \in [0,1]^p$, $i = 1, \ldots, c$ and U.

1. For $k = 1, \ldots, n$;
   1.1 For $i = 1, \ldots, c$;
   Compute covariance matrix using Equation (7.7).
   Compute $d_{ki}(\boldsymbol{x}_k, \boldsymbol{v}_i)$, using, for instance, Equation (7.6).
   Determine ρ_{ik}^l

$$\rho_{ki}^l = 1 - \frac{1}{p} d_{ki}.$$

   1.2 For $i = 1, \ldots, c$;
   Compute a_{ki}^l

$$a_{ki}^l = a_{ki}^{l-1} + \beta\lfloor(1 - \rho_{ki}^l) - a_{ki}^{l-1}\rfloor.$$

   1.3 If $a_{ki}^l \geq \tau$, $\forall i \in \{1, \ldots, c\}$ then
   create new center
   else update $\boldsymbol{v}_s$, $\boldsymbol{v}_s^l = \boldsymbol{v}_s^{l-1} + \alpha \rho_{ks}^{1-a_{ks}^l}(\boldsymbol{x}_k - \boldsymbol{v}_s^{l-1})$, $s = \arg\max_i\{\rho_{ki}\}$.
   1.4 Update the number of cluster centers: compute the compatibility index among cluster centers
   for $i = 1, \ldots, c-1$;
   for $j = i+1, \ldots, c$;

$$\rho_{v_i}^l = 1 - \frac{1}{p}\sum_{h=1}^{p} |v_{ih}^l - v_{jh}^l|^2$$
$$\lambda_{v_i}^l = \beta(1 - \rho_{v_i}^l)$$

if $\lambda_{v_i} \leq 0.95\tau$ then eliminate $\boldsymbol{v}_i$
   Update U excluding the ith cluster.
2. Compute *error*

$$error = \| V^l - V^{l-1} \| = \max_{ij} |v_{ij}^l - v_{ij}^{l-1}|.$$

3. If $error > \varepsilon$ and $l < l_{\max}$, $l = l + 1$ then return to step 1; else stop.
4. Update fuzzy partition matrix U.

End

When there is no requirement on the shape of clusters, a simpler version of the participatory learning algorithm, PL-B, can be implemented computing ρ_{ik}^l in step 1.1 as follows

$$\rho_{ki}^l = 1 - \frac{1}{p}\sum_{j=1}^{p} d_{kj}, \quad d_{kj} = |x_{kj} - v_{kj}| \tag{7.9}$$

instead of using Equations (7.6) and (7.7).

The complexity of the PL clustering algorithm is affected mainly by how ρ_{ik}^l is computed. Therefore, for spherical and ellipsoidal clusters, PL is as complex as FCM and GK algorithms. When PL uses Equation (7.9) to compute ρ_{ik}^l its complexity is less than FCM and GK. Recall, however, that FCM and GK are supervised while PL is not. Moreover, PL clustering can be done sequentially and can be performed in real time while FCM and GK cannot.

Notice that the update scheme given by (7.5) is similar to the Widrow–Hoff learning rule, but results in a fundamentally different convergence behavior: while in the classic rule the learning rate α is kept small to maintain learning responding smoothly to outlier observations, the PL rule can keep the learning rate higher because of the effect of any observation incompatible with current cluster structure is modulated by the arousal term. Similarly to FCM, the PL rule of Equation (7.5) can also be viewed as a gradient-based updating rule relative to an implicit quadratic objective function. Also, the PL update rule is very close to the fuzzy competitive learning (FCL) rule (Backer and Sheunders, 1999), but the FCL learning rate must decrease along iterations to assure convergence. In FCL, cluster information is inserted via membership degrees.

7.4 EXPERIMENTAL RESULTS

In this section we address clustering examples discussed in the literature using participatory learning fuzzy clustering (PL) and two representative fuzzy clustering algorithms, namely the Gustafson–Kessel or GK (Gustafson and Kessel, 1979) and the modified fuzzy k-means or MFKM (Gath, Iskoz, Cutsem, and Van, 1997). The GK is as a generalization of the fuzzy C-means in which the Mahalanobis distance is used in the objective function. GK is a partition-based algorithm. The GK has shown to be particularly effective to find spherical, ellipsoidal, and convex clusters. Differently from the partition-based methods, the MFKM uses a data induced metric with the Dijkstra shortest path procedure in a graph-based representation. The outstanding feature of the MFKM worth mentioning is its ability to find non-convex clusters.

Bezdek and Pal (1992) suggested the following items to evaluate clustering algorithms: (1) need to choose the number of clusters; (2) initialization of cluster centers; (3) order in which data are input; (4) geometric properties of data; (5) diversity of geometric forms of the clusters; (6) variation of data density among groups; (7) separation degree of clusters.

The experiments reported here adopt the following parameters: $error = 0.001$, $l_{\max} = 8$, and $m = 2$. Data have been normalized in [0,1], and the order of data input and cluster centers initialization were the same for all algorithms. When cluster center initializations were kept fixed, data presentation order was random. Also, we consider a finite set of data only, once there is no substantial challenge when using the PL clustering algorithm in its sequential, real-time form.

The parameters chosen for the participatory learning fuzzy clustering algorithm are $\alpha = 0.01$ and $\beta = 0.9$, and τ varies depending on the data-set. They are shown at the bottom of the figures. Similarly, for the MFKM the grid parameter η is depicted at the bottom of the figures.

First we recall that participatory learning fuzzy clustering and MFKM are unsupervised algorithms whereas GK is not. Both PL and GK may fail when clustering complex data-sets, especially when they are not convex. A classic example is shown in Figure 7.2. On the other hand, the MFKM fails when clusters overlap while PL and GK successfully find the clusters, as Figures 7.3 and 7.4 show. Figure 7.5 concerns the Iris data. Since PL and GK use the same distance measure, they behave similarly as Figures 7.3, 7.4, and 7.5 suggest.

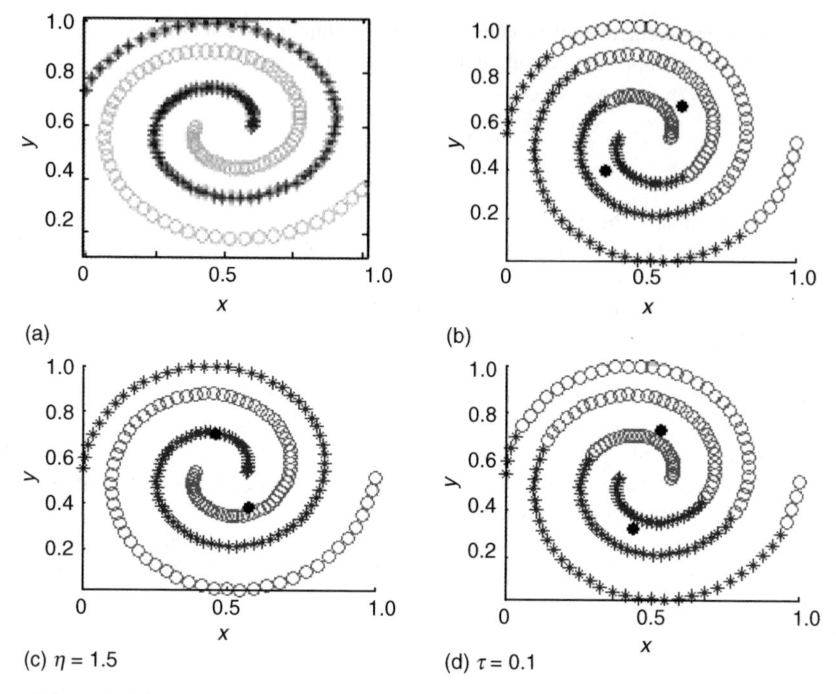

Figure 7.2 (a) Original data and clusters, (b) GK, (c) MFKM, (d) PL. Cluster centers are marked with "•".

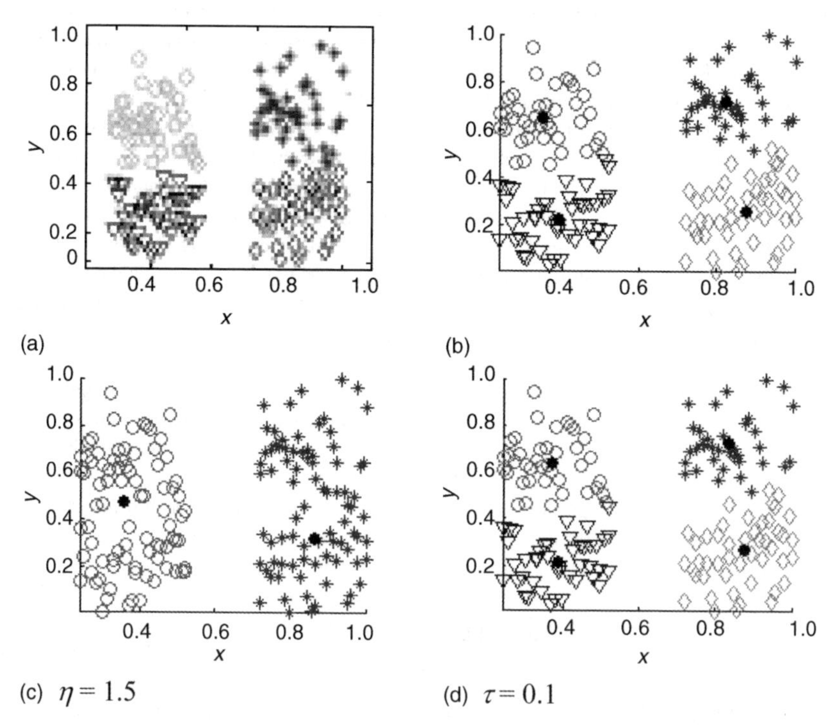

Figure 7.3 (a) Original data and clusters, (b) GK, (c) MFKM, (d) PL. Cluster centers are marked with "•".

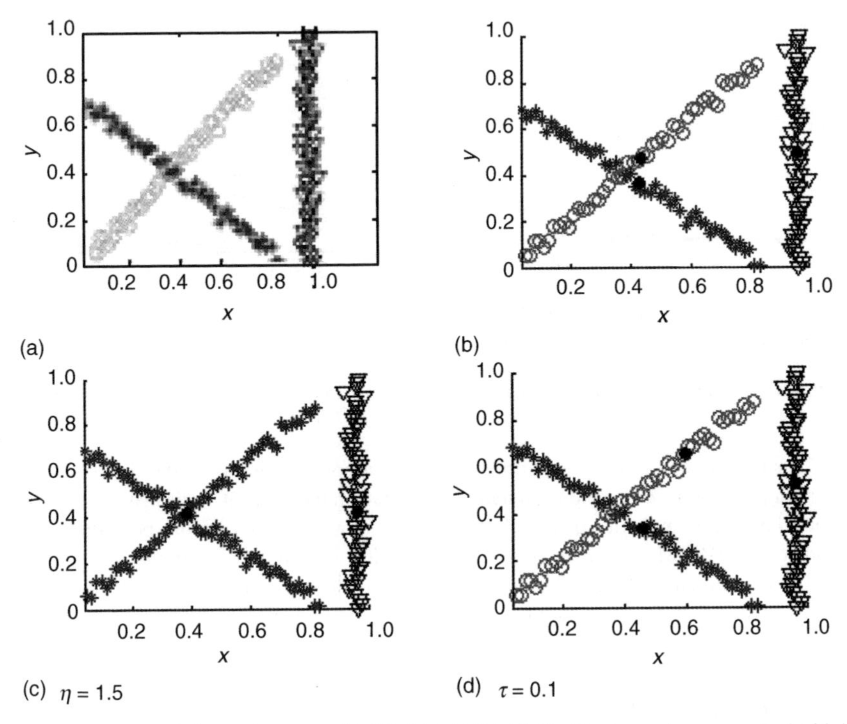

Figure 7.4 (a) Original data and clusters, (b) GK, (c) MFKM, (d) PL. Cluster centers are marked with "•".

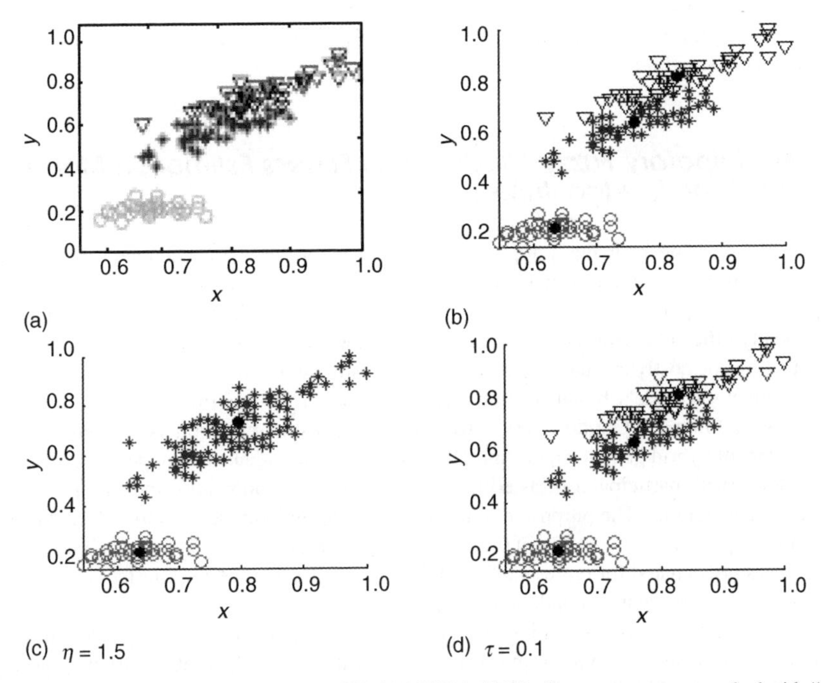

Figure 7.5 (a) Iris dada and clusters, (b) GK, (c) MFKM, (d) PL. Cluster centers are marked with "•".

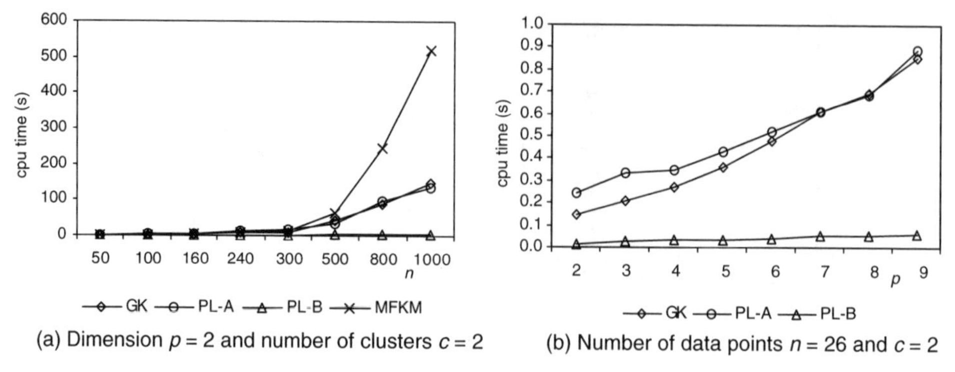

Figure 7.6 Time performance as (a) data dimension and (b) number of points in data-set changes.

Figure 7.6 summarizes a series of experiments and shows how the runtime of the algorithms changes as two important parameters, dimension and number of points in the data-set, varies. We note, as discussed above, that the time complexity of PL clustering algorithms depends on the procedure to compute compatibility. The MFKM runtime increases exponentially as the number of points and data dimension increase. The experiment of Figure 7.6 (b) does not include results for MFKM because runtimes were substantially greater than the ones required for GK, PL-A, and PL-B.

7.5 APPLICATIONS

In this section we address two application examples to illustrate the usefulness of participatory learning clustering in practice. The first concerns a class of hybrid genetic algorithm in which clustering is used as a strategy to improve computational performance through fitness estimation. The second deals with adaptive fuzzy system modeling for time series forecasting. In this case, clustering is used to learn the forecasting model structure.

7.5.1 Participatory Fuzzy Clustering in Fitness Estimation Models for Genetic Algorithms

Despite the success achieved in many applications, genetic algorithms still encounters challenges. Often, genetic algorithms need numerous fitness evaluations before acceptable solutions are found. Most real-world applications require complex and nontrivial fitness evaluation. Fitness evaluation can be costly and computationally efficient performance estimation models must be adopted in these circumstances. One approach to alleviate costly evaluations is to use fitness estimation models.

Fitness estimation models based on fuzzy clustering are a way of improving runtime of genetic algorithms (Mota Filho and Gomide, 2006). Together, fitness estimation models and genetic algorithms assemble a class of hybrid genetic algorithm (HGA) in which individuals of a population are genetically related. In HGA, fuzzy participatory clustering can be used to cluster population into groups during fitness evaluations in generations. The purpose of clustering population individuals is to reduce direct evaluations, to improve processing speed and, at the same, to keep time population diversity and solution quality. In HGA, fitness is evaluated for representative individuals of the population. Cluster centers are natural candidates to act as representative individuals.

The main idea can be grasped looking at Figure 7.7. As the population evolves, individuals tend to concentrate around an optimal solution and become genetically similar. This observation suggests that the number of clusters should reduce over the generations. Also, in Figure 7.6 we note that unsupervised clustering algorithms such as FCM always groups the population in a fixed number of clusters and

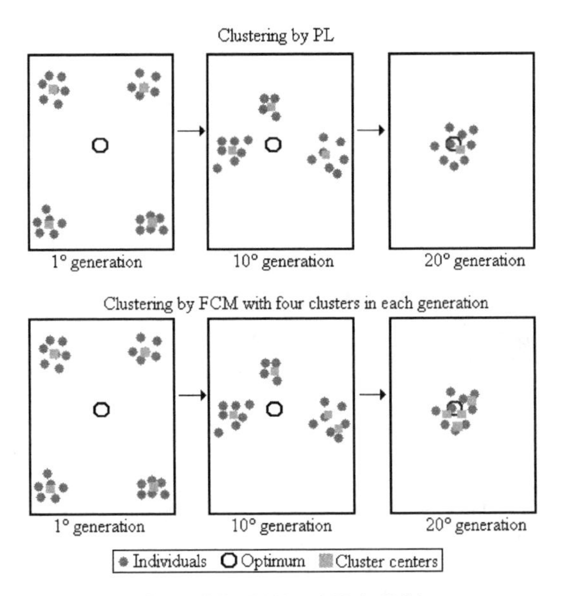

Figure 7.7 FCM and PL in HGA.

generates genetically redundant cluster centers. In the example of Figure 7.6 this is the case in the 10th and 20th generation. Unsupervised procedures, such as the PL fuzzy clustering introduced in this chapter, recognize the distribution of the population over the search space and cluster the population in a smaller number of groups along generations. This avoids the creation of genetically redundant clusters. In general, PL tends to perform better than supervised clustering once it naturally adapts and evolves together with the population. The hybrid genetic algorithm is summarized in Figure 7.8. As an illustration, consider the classic Schwefel function

$$f(\boldsymbol{x}) = 418.9829p + \sum_{i=1}^{p} x_i \cdot \sin(\sqrt{|x_i|}), \ x \in R^p$$

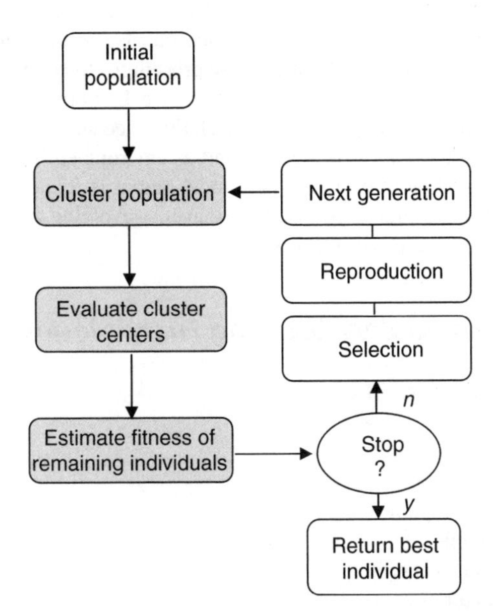

Figure 7.8 HGA based on fitness estimation.

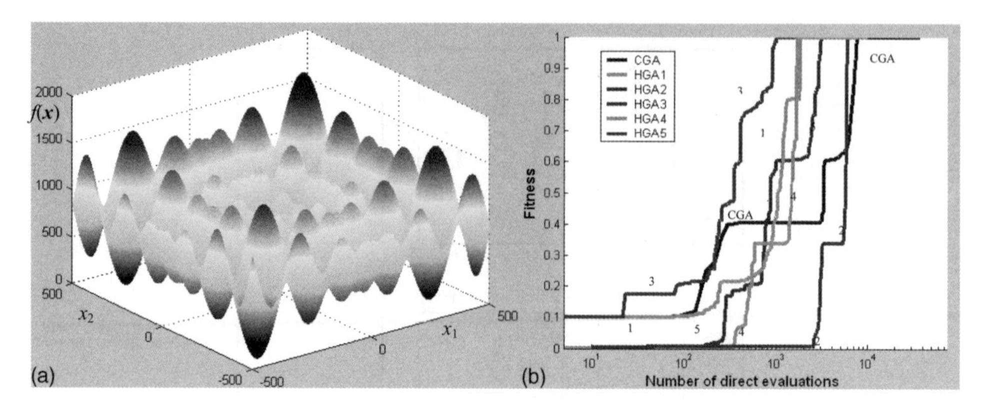

Figure 7.9 HGAs behavior for the Schwefel function.

depicted in Figure 7.9 (a). The global optimum is at $\boldsymbol{x} = (-420.9687, -420.9687)$. Despite several local minima, the HGA and its variations successfully converged to the global optimum faster than the classic genetic algorithm (CGA), Figure 7.8 (right).

Note that the HGA with PL fuzzy clustering (HGA3) requires less direct evaluations of the individuals than the remaining algorithms, especially the conventional genetic algorithm (CGA). They are based on the fuzzy C-means and on fitness imitation schemes (see Mota Filho and Gomide (2006) for further details and a description of a real-world application concerning real-time train scheduling).

7.5.2 Evolving Participatory Learning Fuzzy Modeling

When learning models online, data are collected and processed continuously. New data may either reinforce and confirm the current model or suggest model revision. This is the case when operating conditions of a system modify, faults occur or parameters of a dynamic process change. In adaptive system modeling, a key question is how modify the current model structure using the newest information. Fuzzy functional model identification, Takagi–Sugeno (TS) models in particular, considers a set of rule-based models with fuzzy antecedents and functional consequents. Online learning of TS models needs online clustering to find cluster centers and least square procedures to compute consequent parameters. Each cluster defines a rule and the cluster structure the rule base. Participatory fuzzy clustering modeling adopts this same scheme. Clustering is performed at each time step and a new cluster can be created, an old cluster modified, and redundant clusters eliminated as environment information is updated. Each cluster center defines the focal point of a rule and model output is found as the weighted average of individual rules output. This constitutes a form of participatory evolving system modeling suggested in (Lima et al, 2006) whose procedure is as follows.

7.5.3 Evolving Participatory Learning Fuzzy Modeling Algorithm

Input: *data samples* $\mathbf{x}_k \in [0, 1]^p$, $k = 1, \ldots$

Output: *model output*

begin{enumerate}

1. Initialize the rule base structure
2. Read the next data sample
3. Compute cluster centers using PL
4. Update rule base structure

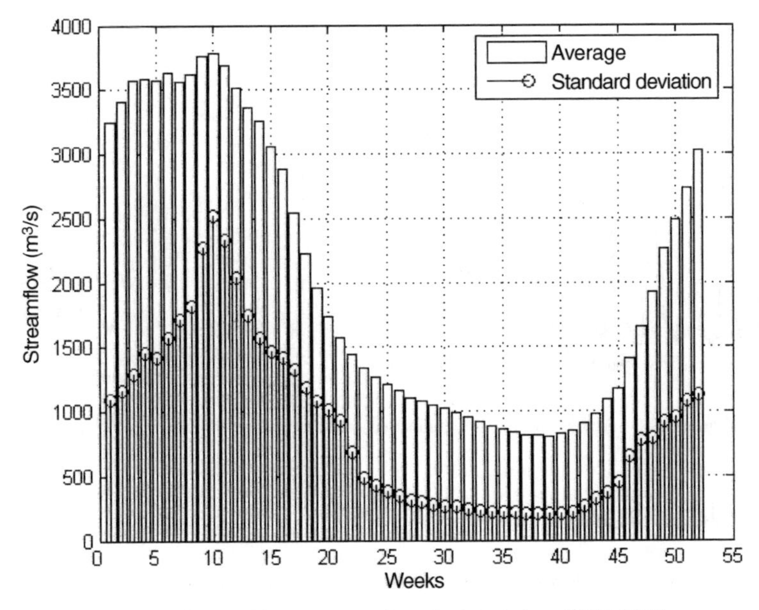

Figure 7.10 Weekly averages and standard deviations (1931–2000).

5. Compute the consequent parameters
6. Compute model output

end

The evolving participatory learning (ePL) modeling approach was adopted to forecast average weekly inflows of a large hydroelectric plant. Hydrological data covering the period of 1931–1990 was used. The analysis and forecast of inflow are of utmost importance to operate water resource-based systems. One of the greatest difficulties in forecasting is the nonstationary nature of inflows due to wet and dry periods of the year as Figure 7.10 shows via weekly averages and standard deviations for the 1931–2000 period.

The performance of the ePL was compared with eTS, an evolving modeling technique introduced by Angelov and Filev (2004) from which ePL was derived. Both eTS and ePL use an online clustering phase followed by a least squares phase to estimate the parameter of linear TS rule conequents.

The performances of ePL and eTS forecasting models were evaluated using the root mean square error (RMSE), mean absolute error (MAD), mean relative error (MRE), and maximum relative error (REmax). Their values are summarized in Table 7.1.

Figure 7.11 shows the actual inflows and forecasted values for ePL and eTS. Both models developed two rules with linear consequents. Further details are given in Lima, Ballini, and Gomide (2006).

Table 7.1 Performance of weekly inflow forecasts using ePL and eTS.

Evaluation method	ePL	eTS
RME (m^3/s)	378.71	545.28
MAE (m^3/s)	240.55	356.85
MRE (%)	12.54	18.42
REmax (%)	75.51	111.22

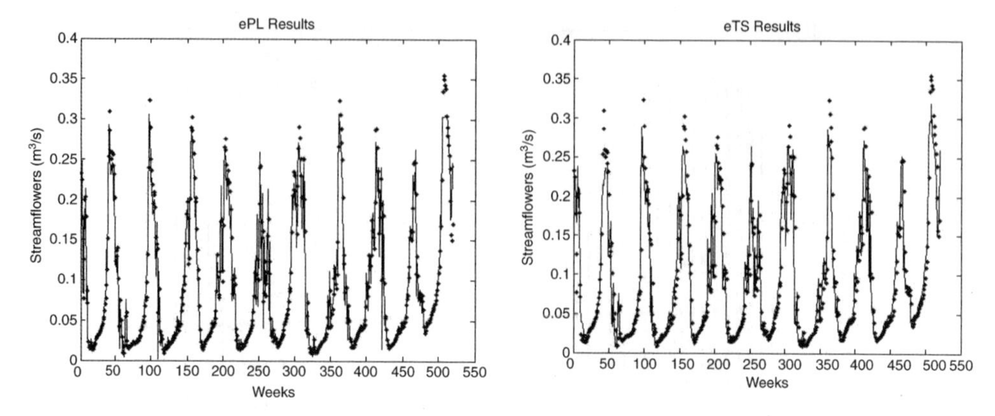

Figure 7.11 Actual (dotted line) and forecasted (solid line) inflows for ePL and eTS.

7.6 CONCLUSIONS

In this chapter, we have introduced a fuzzy clustering algorithm as a participatory learning mechanism whose purpose is to learn cluster structure embedded in data. Participatory learning is a model in which the representation of current knowledge is part of the learning process itself and influences the way in which new observations are used for learning. In clustering this means that current cluster structure is part of and influences the way data is processed to find the cluster structure itself.

Computational experiments suggest that the participatory learning fuzzy clustering algorithm is an attractive alternative for unsupervised fuzzy clustering. PL clustering is as efficient as GK and MFKM, two other major fuzzy clustering algorithms. Its computational complexity depends on the distance measure adopted to express compatibility.

The use of the participatory learning fuzzy clustering algorithm in applications such as in hybrid genetic algorithms enhances computational performance and helps to solve complex optimization problems. PL also improves the effectiveness of evolving system modeling and adaptive systems applications.

Participatory learning fuzzy clustering algorithms, however, still need further improvement. For instance, mechanisms to find values for the threshold τ still need further investigation because, indirectly, τ determines the cluster structure. Here clustering evaluation functions and validation indexes could be of value when combined with genetic algorithms. The effectiveness of this approach will, however, depend on the availability of universal clustering validation functions, an issue still open. These are the questions to be addressed in future research.

ACKNOWLEDGEMENTS

The first two authors are grateful to CNPq, the Brazilian National Research Council, for its support via fellowships 133363/2000-7 and 304299/2003-0, respectively. The second author also acknowledges FAPESP, the Research Foundation of the State of São Paulo, for grant 03/10019-9.

REFERENCES

Angelov, P. and Filev, D. (2004) 'An approach to online identification of Takagi-Sugeno fuzzy models'. *IEEE Trans. Systems Man, and Cybernetics, Part B*, **34**, 484–498.

Backer, S. and Scheunders, P. (1999) 'A competitive elliptical clustering algorithm'. *Pattern Recognition Letters*, **20**, 1141–1147.

Bezdek, J. C. and Pal, S. K. (1992) *Fuzzy Models for Pattern Recognition: Methods that Search for Structures in Data.* IEEE Press, New York, USA.

Duda, R. O. and Hart, P. E. (1973) *Pattern Classification and Scene Analysis.* John Wiley & Sons, Inc., New York, USA.

Duda, R. O., Hart, P. E. and Stork G. D. (2001) *Pattern Classification.* John Wiley & Sons, Inc., New York, USA.

Gath, I. and Geva, A. B. (1989) 'Unsupervised optimal fuzzy clustering'. *IEEE Trans. Pattering Analysis and Machine Intelligence*, **11**, 773–781.

Gath, I., Iskoz A. S., Cutsem, B. and Van M. (1997) 'Data induced metric and fuzzy clustering of non-convex patterns of arbitrary shape'. *Pattern Recognition Letters*, **18**, 541–553.

Geva, A., Steinberg, Y., Bruckmair, S. and Nahum, G. (2000) 'A comparison of cluster validity criteria for a mixture of normal distributed data'. *Pattern Recognition Letters*, **18**, 511–529.

Gustafson, D. and Kessel, W. (1979) 'Fuzzy clustering with a fuzzy covariance matrix'. *Proc. IEEE Conference on Decision and Contro*l, San Diego, USA, pp. 761–766.

Lima, E., Ballini, R. and Gomide, F. (2006) 'Evolving participatory learning modeling'. *Proc. 2nd Int. Symposium on Evolving Fuzzy Systems*, Lake District, UK.

Mota, Filho, F. and Gomide F. (2006) 'Fuzzy clustering in fitness estimation models for genetic algorithms and applications'. *Proc. 15th IEEE Int. Conf. on Fuzzy Systems*, Vancouver, Canada.

Pedrycz, W. and Gomide, F. (1998) *An Introduction to Fuzzy Sets: Analysis and Design.* MIT Press, Cambridge, MA, USA.

Pedrycz, W. (2005) *Knowledge-based Clustering: from Data to Information Granules.* John Wiley & Sons, Inc., Hoboken, NJ, USA.

Silva, L. Gomide, F. and Yager R. (2005) 'Participatory learning in fuzzy clustering'. *Proc. 14th IEEE Int. Conf. on Fuzzy Systems*, Reno, USA, pp. 857–861.

Xu, R. and Wunsch, D. (2005) 'Survey of clustering algorithms'. *IEEE Trans. on Neural Networks*, **6**, 645–678.

Yager, R. (1990) 'A model of participatory learning'. *IEEE Trans. on Systems, Man and Cybernetics*, **20**, 1229–1234.

8

Fuzzy Clustering of Fuzzy Data

Pierpaolo D'Urso

Dipartimento di Scienze Economiche, Università degli Studi del Molise, Campobasso, Italy

8.1 INTRODUCTION

Exploratory data analysis represents a particular category of knowledge acquisition, since it defines a class of statistical methods referring to a specific type of information element and to the associated processing procedures, i.e., the data and the models. Thus the exploratory data analysis can be described as a cognitive process based on the so-called *informational paradigm* constituted by the data and the models (Coppi, D'Urso, and Giordani, 2007). Since, in the real world, this paradigm is often inherently associated with the factor of fuzziness, it happens that the available information is completely or partially fuzzy (i.e., we can have fuzzy data and crisp (nonfuzzy) model, crisp data and fuzzy model, fuzzy data and fuzzy model). Likewise in different exploratory procedures, the cluster analysis can be based on the informational paradigm in which the two informational components can be separately or simultaneously fuzzy. In this chapter, we analyze only the case in which the information is completely fuzzy.

In the last few years a great deal of attention has been paid to the classification of imprecise (vague or fuzzy) data and, in particular, to the fuzzy clustering of fuzzy data (see, for example, Sato and Sato, 1995; Hathaway, Bezdek, and Pedrycz, 1996; Pedrycz, Bezdek, Hathaway, and Rogers, 1998; Yang and Ko, 1996; Yang and Liu, 1999; Yang, Hwang, and Chen, 2004; Hung and Yang, 2005; Alvo and Théberge, 2005; Colubi, Gonzales Rodriguez, Montenegro, and D'Urso, 2006; D'Urso, Giordani, 2006a; see in the following Sections 8.4 and 8.5).

The aim of this chapter is to review and compare various fuzzy clustering models for fuzzy data.

The study is structured as follows. In Section 8.2, we explain the informational paradigm and discuss the fuzziness in the clustering processes by analyzing the different informational situations. In Section 8.3, we define fuzzy data and analyze the different features connected to mathematical (algebraic and geometric formalization, mathematical transformations, metrics) and conceptual (elicitation and specification of the membership functions) aspects. An organic and systematic overview and a comparative assessment of the different fuzzy clustering models for fuzzy univariate and multivariate data are shown in Section 8.4. In Section 8.5, we analyze some extensions of the fuzzy clustering models for complex

Advances in Fuzzy Clustering and its Applications Edited by J. Valente de Oliveira and W. Pedrycz

structures of fuzzy data, the so-called three-way fuzzy data. In particular, we formalize mathematically and geometrically the fuzzy data time array and define suitable distance measures between the so-called fuzzy time trajectories. Then, we show an example of dynamic fuzzy clustering model. In Section 8.6, for evaluating the empirical capabilities and the different performances of the illustrated clustering models, several applicative examples are shown. Final remarks and future perspectives in this methodological domain are provided in Section 8.7.

8.2 INFORMATIONAL PARADIGM, FUZZINESS, AND COMPLEXITY IN CLUSTERING PROCESSES

8.2.1 Informational Paradigm, Fuzziness and Complexity

In decision making and in the more systematic processes of knowledge acquisition in the various scientific domains, the important role of *vagueness* has been widely recognized (Ruspini, Bonissone, and Pedrycz, 1998). In general, a "corpus of knowledge" is a set of "information elements." Each information element is represented by the following quadruple (attribute, object, value, confidence), in which attribute is a function mapping of an object to a value, in the framework of a reference universe; value is a predicate of the object, associated to a subset of a reference universe, and confidence indicates the reliability of the information elements (Coppi, 2003). In the real world, an element of information is generally characterized by imprecision (with regards to value) and uncertainty (expressed through the notion of confidence). Imprecision and/or uncertainty define what we may call *imperfect information* (here the term imperfect indicates that the information presents one or more of the following features: vagueness, roughness, imprecision, ambiguity, and uncertainty). When the phenomena and the situation under investigation are *complex* (here the term complex refers to the presence, in the phenomenon under investigation, of at least some of the following features: dynamic evolution, many variables of possibly different nature, feedback loops and so on), then the "Incompatibility Principle" comes into play. Let us recall its essence as originally formulated by Zadeh. As the complexity of a system increases, our ability to make precise and yet significant statements about its behavior diminishes until a threshold is reached beyond which precision and significance (or relevance) become almost mutually exclusive characteristics (Zadeh, 1973). This principle justifies the development and application of logics allowing the utilization of *imperfect* information, in order to draw *relevant* conclusions when faced with complex situations (Coppi, 2003; Coppi, D'Urso, and Giordani, 2007).

Data analysis represents, in the statistical framework, a particular category of knowledge acquisition, since it defines a class of statistical techniques referring to a specific type of information element and to the associated processing procedures. This class is characterized by two entities: the empirical *data* and the *models* for data. Then, an exploratory data analysis can be described as a cognitive process which, starting from *initial information* (both empirical and theoretical), through some computational procedures (algorithms), gets *additional information* (information gain) having a cognition and/or operational nature (Coppi, 2003; D'Urso and Giordani, 2006a). This description defines the so-called *informational paradigm* (*IP*) represented by the following pair $(\Im_E, \Im_T)$, where $\Im_E$ is the *empirical information* (data) and $\Im_T$ is the *theoretical information* (e.g., models). Both $\Im_E$ and $\Im_T$, to the extent to which they are "informational" entities (corpus of knowledge), are constituted by informational elements that can be imprecise and uncertain (Coppi, 2003). Then, we have that the statistical data analysis can be based on the *informational paradigm.*

In Figure 8.1, we summarize, schematically, the informational paradigm. Notice that, when the information is not affected by fuzziness (in this case we have the *crisp informational paradigm* (*CIP*)) the information is *perfect*; vice versa, if the information is affected by fuzziness, then the information becomes *imperfect*. In particular, if the fuzziness concerns only a part of the information (data or model), we have the *partial fuzzy informational paradigm* (*PFIP*); if the fuzziness regards all the information (data and model), we have the *complete fuzzy informational paradigm* (*CFIP*).

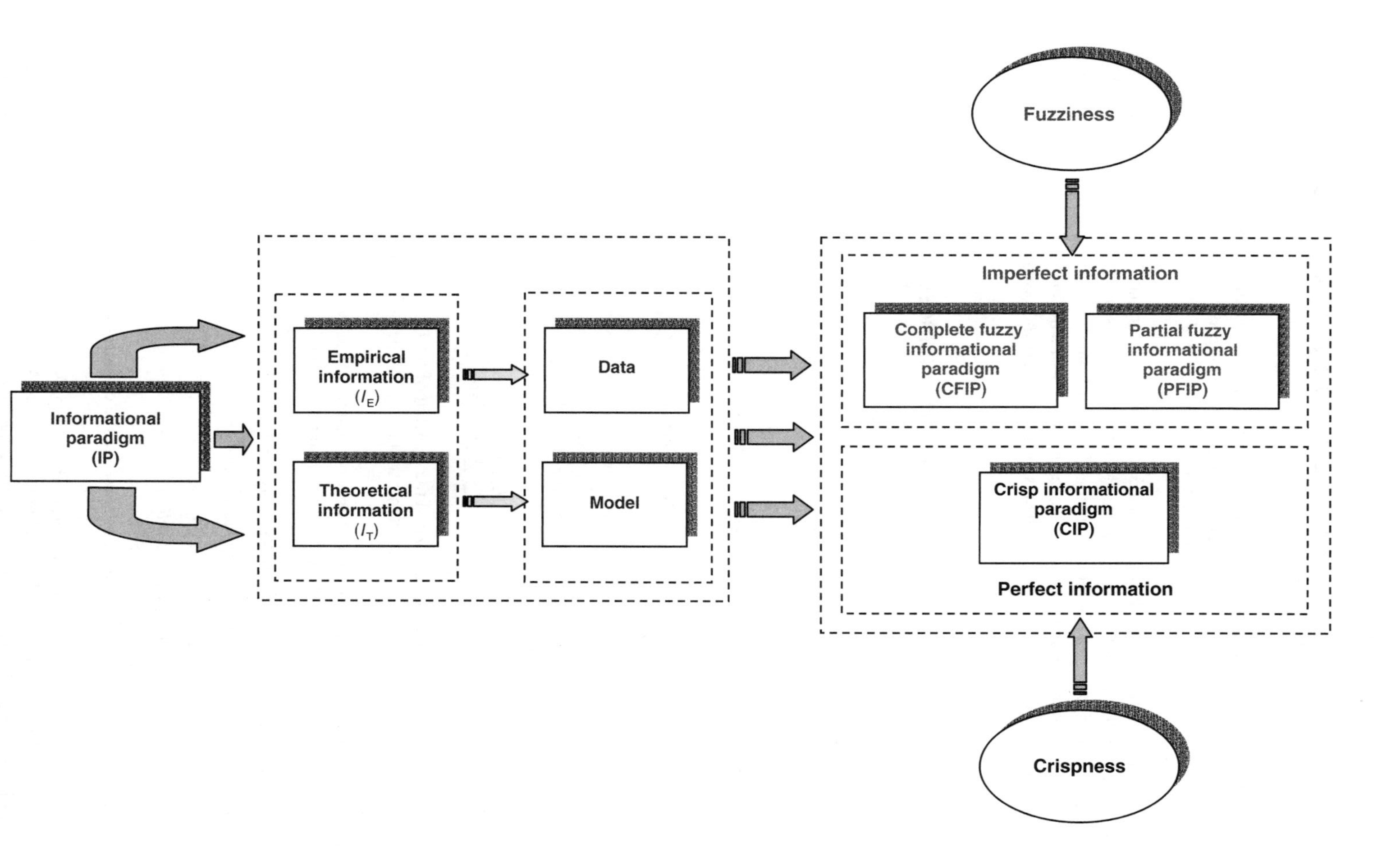

Figure 8.1 Informational paradigm scheme.

8.2.2 Theoretical Information: Methodological Approaches, Models, and Fuzziness

Analogously to various data analysis procedures, cluster analysis can also be based on the informational paradigm. In this way, we recognize the informational nature of the ingredients of the clustering procedures: the observed data (empirical information, $\Im_E$) and the clustering models (theoretical information, $\Im_T$).

By considering the informational paradigm ($\Im_E, \Im_T$), it can occur that $\Im_E$ and $\Im_T$ are characterized by *empirical* and/or *theoretical fuzziness* (Zadeh, 1965). The theoretical fuzziness is embodied in the clustering model, in particular in the assignment process of objects to clusters. In order to incorporate the theoretical fuzziness in the clustering procedure, the so-called *membership degree* of each unit to different groups can be considered that evaluates the fuzziness in the assignment procedure (D'Urso and Giordani, 2006a).

8.2.3 Empirical Information: Data and Fuzziness

Empirical fuzziness is connected to the *imprecision* embodied, for instance, in the *human perception* expressed in judgments on certain observational situations (hot, cold; excellent, very good, good, fair, poor) *interval valued data* (temperature, pulse rate, systolic, and diastolic pressure over a given time period, etc.), *vague measurements* (*granularity*) (D'Urso and Giordani, 2006a). In these cases, the data are contained in the so-called *fuzzy data matrix*, where each fuzzy datum is represented by the central value, called *center* (or *mode*), and the *spreads* that represent the uncertainty around the center (cf. Section 8.3).

8.2.4 Different Informational Typologies in Clustering Processes

By considering the K-means clustering (MacQueen, 1967) and using the notion of the nonfuzzy (crisp) and the partial or complete fuzzy informational paradigm, as we described so far, we may define various typologies of K-means clustering models, where the following objective function is to be minimized (D'Urso and Giordani, 2006a):

$$\sum_{i=1}^{I} \sum_{k=1}^{K} u_{ik}^{m} d_{ik}^{2}, \tag{8.1}$$

where d_{ik} is a distance between the unit i and the centroid of the cluster k, u_{ik} indicates the membership degree of the unit i to the cluster k, m is a suitable weighting exponent that controls the fuzziness/crispness of the clustering.

By taking into account the possible (crisp or fuzzy) nature of the two informational components, we get four informational cases (see Figure 8.2) (D'Urso and Giordani, 2006a):

(1) $\Im_E$ and $\Im_T$ crisp (*complete crisp paradigm*);

(2) $\Im_E$ fuzzy and $\Im_T$ crisp (*partial fuzzy paradigm*);

(3) $\Im_E$ crisp and $\Im_T$ fuzzy (*partial fuzzy paradigm*);

(4) $\Im_E$ and $\Im_T$ fuzzy (*complete fuzzy paradigm*).

In Figure 8.2, we show the four information cases in the clustering process:

- In case (1) ($\Im_E$ and $\Im_T$ crisp) the clustering model represents the traditional (crisp) K-means model (MacQueen, 1967). We observe that (cf. Figure 8.2, case (1)), in this situation, there is not empirical and theoretical fuzziness; in fact, the data are crisp and the clusters are *well-separated* (there is not fuzziness in the assignment process of units to clusters and, thus, we can utilize a crisp clustering model). So, for

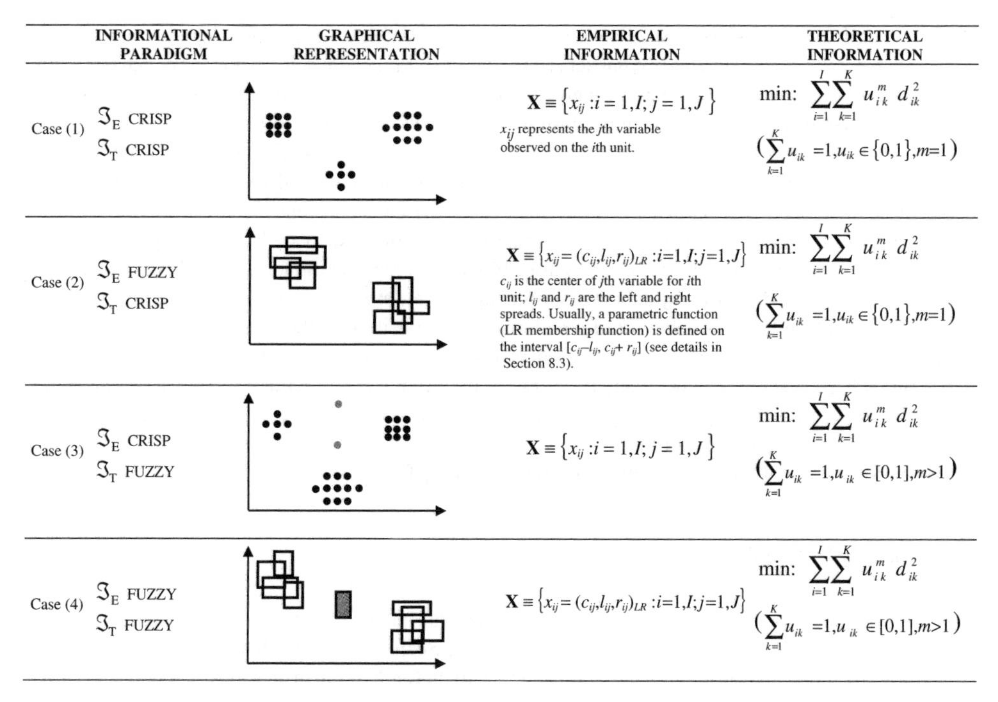

Figure 8.2 Different informational typologies in clustering processes.

partitioning the units we can consider a non-overlapping clustering model (crisp K-means clustering model) by assigning each unit to *exactly* one cluster (in fact, we have $\sum_{k=1}^{K} u_{ik} = 1, u_{ik} \in \{0, 1\}$, $m = 1$) and by analyzing, for instance, the data matrix $\mathbf{X} \equiv \{\mathbf{x}_i : i = 1, 25\} = \{x_{ij} : i = 1, 25;$ $j = 1, 2\}$, where $\mathbf{x}_i = (x_{i1}, x_{i2})$ and, for example, x_{i1} and x_{i2} representing, respectively, the (crisp) gross domestic product per capita and gross domestic investment observed on the ith country, $i = 1, 25$. Then, in this example, according to the data configuration, the crisp K-means clustering determines an optimal crisp partition of the 25 countries in three well-separated and compacted clusters without fuzziness in the assignment procedure (D'Urso and Giordani, 2006a).

- In case (2) ($\Im_E$ fuzzy and $\Im_T$ crisp), we have fuzzy data (empirical fuzziness), but the assignment process of the units to the clusters is not fuzzy; then, in this situation, we have a crisp partitioning model with fuzzy data. For example, suppose that the fuzzy scatter shown in Figure 8.2 (case (2)) represents the graphical configuration of eight vehicles obtained by observing a feature component (a linguistic assessment that can be modeled by real fuzzy sets) (e.g. comfort) and a fuzzy measurement (e.g., speed) under the assumption that measurements cannot be infinitely precise since the measurement values are inherently fuzzy (i.e., always approximate), the source of the measurement is inherently unreliable and the range of possible measurement values is limited by physical realities (see, for example, Ramot, Milo, Friedman, and Kandel, 2001). Then, we can utilize the crisp K-means clustering model for partitioning the set of cars and obtain a classification of the vehicles into two classes (D'Urso and Giordani, 2006a).
- In case (3) ($\Im_E$ crisp and $\Im_T$ fuzzy), we have crisp data and fuzziness in the assignment process of units to clusters, then, to take into account the theoretical fuzziness, we utilize a fuzzy clustering model for classifying the units (Dunn, 1974; Bezdek, 1974a, 1981; Pedrycz, 2005). For considering the particular configuration of the data in the space $\Re^2$, we can classify the crisp data (without fuzziness) by considering a fuzzy clustering technique where each unit can be simultaneously assigned to more clusters (in fact, in this case, the constraints are $\sum_{k=1}^{K} u_{ik} = 1, u_{ik} \in [0, 1], m > 1$). For example, Figure 8.2 (case (3)) can represent the values of two (crisp) variables (e.g., gross domestic product from agriculture and labour

force in agriculture) observed on 27 countries. Then, by applying a fuzzy clustering algorithm, we can classify the 27 countries into three compact classes, taking into account suitably the fuzzy behavior of the two countries indicated in the figure with grey points (i.e., assigning, objectively, for these two countries, low membership degrees to all the three clusters) (D'Urso and Giordani, 2006a).

- In case (4) ($\Im_E$ fuzzy and $\Im_T$ fuzzy), we have, simultaneously, fuzziness in the empirical and theoretical information. For this reason, in this case, for partitioning the units, we utilize a fuzzy clustering model for fuzzy data. The plot shown in Figure 8.2 (case (4)) can represent the bivariate configuration corresponding to the values, observed on a group of nine patients who are hospitalized, for two fuzzy variables: the range of both diastolic and systolic blood pressure over the same day. Then, by applying a fuzzy clustering model for fuzzy data, we can consider the empirical fuzziness (characterizing the data) and capture the ambiguous behavior of a patient (see the position of the grey rectangle in the space), obtaining two classes (each class composed of four patients with high membership degrees and a patient with membership degree "spread" between the two clusters). In this case, we have a clustering with a double type of fuzziness (empirical and theoretical fuzziness), i.e., fuzzy clustering for fuzzy data (D'Urso and Giordani, 2006a).

In this chapter, we shall deal only with case (4) ($\Im_E$ and $\Im_T$ fuzzy).

8.3 FUZZY DATA

Models based on imprecise (fuzzy) data are used in several fields. Sometimes such models are used as simpler alternatives to probabilistic models (Laviolette, Seaman, Barrett, and Woodall, 1995). Other times they are, more appropriately, used to study data which, for their intrinsic nature, cannot be known or quantified exactly and, hence, are correctly regarded as vague or fuzzy. A typical example of fuzzy data is a human judgment or a linguistic term. The concept of fuzzy number can be effectively used to describe formally this concept of vagueness associated with a subjective evaluation. Every time we are asked to quantify our sensations or our perceptions, we feel that our quantification has a degree of arbitrariness. However, when our information is analyzed through nonfuzzy techniques, it is regarded as if it were exact, and the original fuzziness is not taken into account in the analysis. The aim of fuzzy techniques is to incorporate all the original vagueness of the data. Therefore, models based on fuzzy data use more information than models where the original vagueness of the data is ignored or arbitrarily canceled. Further, models based on fuzzy data are more general because a crisp number can be regarded as a special fuzzy number having no fuzziness associated with it (D'Urso and Gastaldi, 2002).

The need for fuzzy data arises in the attempt to represent vagueness in everyday life (Coppi, D'Urso, and Giordani, 2007). To understand this concept: suppose, as you approach a red light, you must advise a driving student when to apply the brakes. Would you say "Begin braking *74 feet* from the crosswalk?" Or would your advice be more like. "Apply the brakes *pretty soon*?" The latter, of course; the former instruction is too precise to be implemented. This illustrates that precision may be quite useless, while vague directions can be interpreted and acted upon. Everyday language is one example of the ways in which vagueness is used and propagated. Children quickly learn how to interpret and implement fuzzy instructions ("go to bed *about 10*"). We all assimilate and use (act on) fuzzy data, vague rules, and imprecise information, just as we are able to make decisions about situations that seem to be governed by an element of chance. Accordingly, computational models of real systems should also be able to recognize, represent, manipulate, and use (act on) fuzzy uncertainty (Bezdek, 1993).

In order to understand the fuzziness in the data, other explicative examples can be considered. Coppi (2003) considers an example of multivariate fuzzy empirical information in which a set of individuals (e.g., a population living in a given area) are taken into account. Each individual, from a clinical viewpoint, can be characterized according to her/his "health state." This refers to the "normal" functioning of the various "aspects" of her/his organism. Generally, any "aspect" works correctly to a certain extent. We often use the notion of "insufficiency," related to different relevant functions of parts of the body (e.g., renal or hepatic insufficiency, aortic incompetence, etc.). Insufficiency (as referring to the various relevant aspects mentioned above) is a concept that applies to a certain degree to any

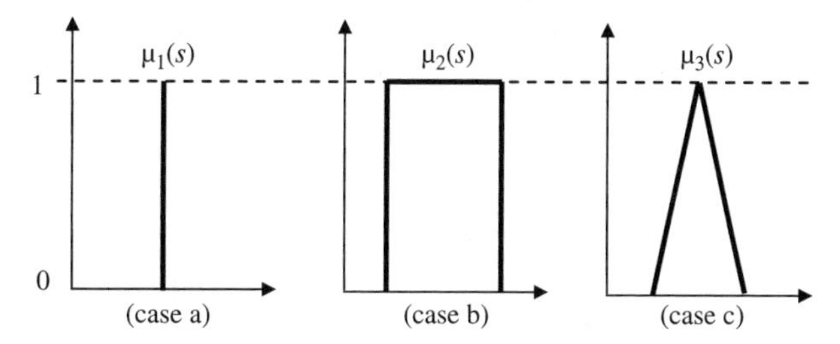

Figure 8.3 Examples of membership function (cf. Pedrycz, Bezdek, Hathaway and Rogers, 1998).

individual (depending on several factors such as age, sex, previous illnesses, and so on). This may be expressed by means of a fuzzy number on a standard scale (say, from 0≡perfect functioning to 10≡complete insufficiency). Consequently, each individual can be more realistically characterized by a vector of fuzzy variables concerning "insufficiency" of various relevant aspects of her/his organism (Coppi, 2003).

Pedrycz, Bezdek, Hathaway, and Rogers (1998) remarked that the features utilized for describing and classifying vehicle speed can have different representations. Consider the speed s of a vehicle. If measured precisely at some time instant, speed s is a real number, say $s = 100$. Figure 8.3 (case a) shows the membership function $\mu_1(s) = 1$ for this case, $\mu_1(s) = 1 \Leftrightarrow s = 100$; otherwise, $\mu_1(s) = 0$. This piece of data could be collected by one observation of a radar gun. Next, suppose that two radar guns held by observers at different locations both measure s at the same instant. One sensor might suggest that $s = 80$, while the second measurement might be $s = 110$. Uncalibrated instruments could lead to this situation. In this case, several representations of the collected data offer themselves. One way to represent these temporally collocated data points is by the single interval [80, 110], as shown by the membership function $\mu_2(s)$ in Figure 8.3 (case b), $\mu_2(s) = 1 \Leftrightarrow 80 < s < 110$, otherwise, $\mu_2(s) = 0$. Finally, it may happen that vehicle speed is evaluated nonnumerically by a human observer who might state simply that "s is very high." In this case, the observation can be naturaly modeled by a real fuzzy set. The membership function $\mu_3(s)$ shown in Figure 8.3 (case c) is one (of infinitely many) possible representations of the linguistic term "very high" $\mu_3(s) = \max\{0, 1\text{-}0.1|100\text{-}s|\}$, $s \in \Re$ (Pedrycz, Bezdek, Hathaway, and Rogers, 1998).

In conclusion, we remark some fields in which fuzzy data have been widely analyzed. For instance, empirical studies with fuzzy data regard the following areas: ballistics (Celmins, 1991), event tree analysis (Huang, Chen, and Wang, 2001), food chemistry (Kallithraka *et al.*, 2001), group consensus opinion and multicriteria decision making (Raj and Kumar, 1999), human errors rate assessment (Richei, Hauptmanns, and Urger, 2001), machine learning (Chung and Chan, 2003), management talent assessment (Chang, Huang, and Lin, 2000), maritime safety (Sii, Ruxton, and Wang, 2001), material selection analysis (Chen, 1997), medical diagnosis (Di Lascio *et al.*, 2002), military application (Cheng and Lin, 2002), nuclear energy (Moon and Kang, 1999), public opinion analysis (Coppi, D'Urso, and Gordani, 2006a), intelligent manufacturing (Shen, Tan, and Xie, 2001), petrophysics (Finol, Guo, and Jing, 2001), risk analysis (Lee, 1996), software reliability (D'Urso and Gastaldi, 2002), tea evaluation (Hung and Yang, 2005), technical efficiency (Hougaard, 1999), thermal sensation analysis (Hamdi, Lachiver, and Michand, 1999), and VDT legibility (Chang, Lee and Konz, 1996).

8.3.1 Mathematical and Geometrical Representation

The so-called *LR fuzzy data* represent a general class of fuzzy data. This type of data can be collected in a matrix called *LR fuzzy data matrix* (units × (fuzzy) variables):

$$\mathbf{X} \equiv \{x_{ij} = (c_{ij}, l_{ij}, r_{ij})_{LR} : i = 1,I; j = 1,J\}, \tag{8.2}$$

where i and j denote the units and fuzzy variables, respectively; $x_{ij} = (c_{ij}, l_{ij}, r_{ij})_{LR}$ represents the LR fuzzy variable j observed on the ith unit, where c_{ij} denotes the center and l_{ij} and r_{ij} indicate, respectively,

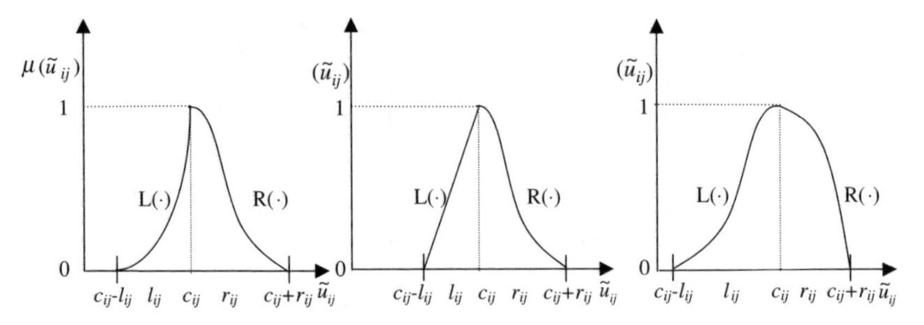

Figure 8.4 Examples of membership functions of LR fuzzy data.

the left and right spread, with the following *membership function*:

$$\mu(\tilde{u}_{ij}) = \begin{cases} L\left(\dfrac{c_{ij} - \tilde{u}_{ij}}{l_{ij}}\right) & \tilde{u}_{ij} \leq c_{ij} \quad (l_{ij} > 0) \\ R\left(\dfrac{\tilde{u}_{ij} - c_{ij}}{r_{ij}}\right) & \tilde{u}_{ij} \geq c_{ij} \quad (r_{ij} > 0), \end{cases} \tag{8.3}$$

where $L(z_{ij})$ (and $R(z_{ij})$) is a decreasing "shape" function from $\Re^+$ to [0,1] with $L(0) = 1$; $L(z_{ij}) < 1$ for all $z_{ij} > 0, \forall i,j$; $L(z_{ij}) > 0$ for all $z_{ij} < 1, \forall i,j$; $L(1) = 0$ (or $L(z_{ij}) > 0$ for all $z_{ij}, \forall i,j$, and $L(+\infty) = 0$) (Zimmermann, 2001). Some examples of geometric representations of membership functions for *LR fuzzy data* are shown in Figure 8.4.

However, in several real applications, the most utilized type of fuzzy data is a particular class of the LR family: the so-called *L fuzzy data* or *symmetric fuzzy data*. In fact, we get $L = R$ and $l = r$ (the left and the right spreads are the same) and then the fuzzy data and their membership functions are symmetric.

A *symmetric fuzzy data matrix* is formalized as follows:

$$\mathbf{X} \equiv \{x_{ij} = (c_{ij}, l_{ij})_L : i = 1,I; j = 1,J\} \tag{8.4}$$

with the following membership function:

$$\mu(\tilde{u}_{ij}) = L\left(\frac{c_{ij} - \tilde{u}_{ij}}{l_{ij}}\right) \quad \tilde{u}_{ij} \leq c_{ij} \quad (l_{ij} > 0). \tag{8.5}$$

Particular cases of the family of the membership function (8.5) are shown in Figure 8.5.

A very interesting class of symmetric fuzzy data is the *symmetric triangular fuzzy data* characterized by the following family of membership functions (cf. Figure 8.5, case (b)):

$$\mu(\tilde{u}_{ij}) = \begin{cases} 1 - \dfrac{c_{ij} - \tilde{u}_{ij}}{l_{ij}} & \tilde{u}_{ij} \leq c_{ij} \quad (l_{ij} > 0) \\ 1 - \dfrac{\tilde{u}_{ij} - c_{ij}}{r_{ij}} & \tilde{u}_{ij} \geq c_{ij} \quad (r_{ij} > 0). \end{cases} \tag{8.6}$$

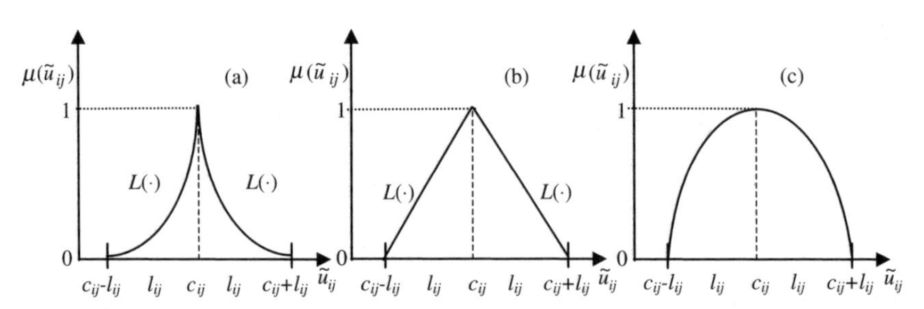

Figure 8.5 Examples of membership functions of *L* fuzzy data.

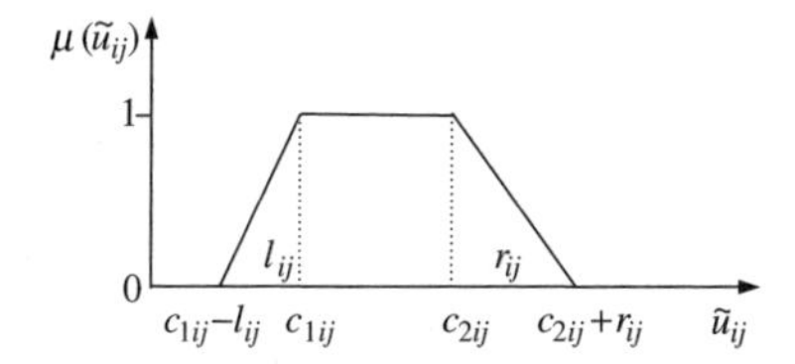

Figure 8.6 Trapezoidal membership function.

In literature, we have another type of LR fuzzy data. By indicating with *LR-I type fuzzy data* the previous LR fuzzy data (8.2), we can denote the new type of LR fuzzy data with *LR-II type fuzzy data*. An *LR-II type fuzzy data matrix* is defined as follows:

$$\mathbf{X} \equiv \{x_{ij} = (c_{1\,ij}, c_{2\,ij}, l_{ij}, r_{ij})_{LR} : i = 1,I; j = 1,J\}, \tag{8.7}$$

where $x_{ij} = (c_{1\,ij}, c_{2\,ij}, l_{ij}, r_{ij})_{LR}$ represents the *LR-II type* fuzzy variable j observed on the ith unit, c_{1ij} and c_{2ij} denote, respectively, the left and right "center" and l_{ij} and r_{ij} the left and right spread, respectively, with the following family of membership functions:

$$\mu(\tilde{u}_{ij}) = \begin{cases} L\left(\dfrac{c_{1ij} - \tilde{u}_{ij}}{l_{ij}}\right) & \tilde{u}_{ij} \leq c_{1ij} \quad (l_{ij} > 0) \\ 1 & c_{1ij} \leq \tilde{u}_{ij} \leq c_{2ij} \\ R\left(\dfrac{\tilde{u}_{ij} - c_{2ij}}{r_{ij}}\right) & \tilde{u}_{ij} \geq c_{2ij} \quad (r_{ij} > 0). \end{cases} \tag{8.8}$$

A particular case of *LR-II type* fuzzy data is the trapezoidal one, i.e., the *LR-II type* fuzzy data with the following family membership functions (see Figure 8.6):

$$\mu(\tilde{u}_{ij}) = \begin{cases} 1 - \dfrac{c_{1ij} - \tilde{u}_{ij}}{l_{ij}} & \tilde{u}_{ij} \leq c_{1ij} \quad (l_{ij} > 0) \\ 1 & c_{1ij} \leq \tilde{u}_{ij} \leq c_{2ij} \\ 1 - \dfrac{\tilde{u}_{ij} - c_{2ij}}{r_{ij}} & \tilde{u}_{ij} \geq c_{2ij} \quad (r_{ij} > 0). \end{cases} \tag{8.9}$$

8.3.2 Membership Function: Elicitation and Specification

As for the "subjectivistic" approach to probability, also the choice of the membership functions is subjective. In general, these are determined by experts in the problem area. In fact, the membership functions are context-sensitive. Furthermore, the functions are not determined in an arbitrary way, but are based on a sound psychological/linguistic foundation. It follows that the choice of the membership function should be made in such a way that the function captures the approximate reasoning of the person involved. In this respect, the *elicitation* of a membership function requires a deep psychological understanding. Suppose that an interviewer asks how a person judges her (his) health. The concept of health refers to the functioning of the various "aspects" of the organism. Generally, any "aspect" works correctly to a certain extent. If a person is optimistic and has never had considerable health diseases, it is plausible that she (he) feels "very well." The term "very well" can be fuzzified as a number in [0.85, 1.00] in the scale [0,1]. Conversely, another optimistic person who recently had a critical surgery operation may still answer "very well" but now the term could be fuzzified as a fuzzy number in [0.60, 0.75]. Similarly, if a person says "wait for me about 10 minutes," the fuzzification of "about 10" may depend on the nationality. Specifically, this usually means from 5 to 15 minutes but, for instance, if she (he) is Italian, the appropriate fuzzy coding could be from 10 minutes to half an hour or more. Therefore, if possible, the process of fuzzification should be constructed ad hoc for each person to be analyzed. Unfortunately, it is sometimes hard to adopt an ad hoc fuzzification procedure. In these cases, one has to choose fuzzy numbers such that they capture the approximate reasoning of all of the persons involved (Coppi, D'Urso, and Giordani, 2007).

Notice that, when we speak of fuzzy data related to a single variable we think of a *vector* of fuzzy number. In the more general case of multivariate analysis, if all the variables are fuzzy, we have a matrix of fuzzy numbers (see, for example, (8.2)). In this case, particular attention must be paid to the *specification* of the membership functions when we deal simultaneously with J variables. To this purpose, we can distinguish two approaches: the *conjunctive* approach and the *disjunctive* approach (Coppi, 2003).

In the *conjunctive* approach, we take into account the fuzzy relationship defined on the Cartesian product of the reference universes of the J variables. In this perspective, we can distinguish *noninteractive* and *interactive* variables. From the statistical point of view, the adoption of the conjunctive approach to the multi-dimensional fuzzy variables involves a specific interest in studying the fuzzy relationship looked at as a "variable" in itself, which could be observed on the I units. Conversely, in the *disjunctive* approach, we are not interested in studying a fuzzy variable which constitutes the resultant of the J original variables. Instead, our interest focuses upon the set of the J "juxtaposed" variables, observed as a whole in the group of I units. In this case, we have J membership functions and the investigation of the links among the J fuzzy variables is carried out directly on the matrix of fuzzy data concerning the IJ-variate observations (Coppi, 2003).

8.3.3 Fuzzy Data Preprocessing

In data analysis processes, in order to take into account the heterogeneity problems, with particular reference to the variables (different variances and/or units of measurement), an appropriate preprocessing of the data may be required, such as *centering*, *normalization*, and *standardization*. In this connection, we can consider different types of preprocessing procedures for centers and (left and right) spreads of the fuzzy data (Coppi and D'Urso, 2003):

- *Centering of the centers*, by taking into account the average of the centers. For instance, we can utilize the following transforms $\tilde{c}_{ij} = c_{ij} - \bar{c}_{.j}$, where the subscript dot is used to indicate the mean across $i = 1, I(\bar{c}_{.j})$.
- *Normalization of the centers*, by dividing the centers, for instance c_{ij}, by the normalization factor $\bar{c}_j$. In this case, we obtain $\tilde{\tilde{c}}_{ij} = c_{ij}/\bar{c}_{.j}$.
- *Standardization of the centers*, by using, for instance, $c^*_{ij} = \tilde{c}_{ij}/\left(\frac{1}{I}\sqrt{\sum_{i=1}^{I} \tilde{c}^2_{ij}}\right)$.
- *Normalization of the spreads*, by setting, for example $\tilde{l}_{ij} = l_{ij}/\bar{l}_{.j}$, $\tilde{r}_{ij} = r_{ij}/\bar{r}_{.j}$.

Normalization of the centers and spreads, as illustrated, is particularly indicated for coping with problems of heterogeneity of units of measurement and/or of size of the variables. In any case, when choosing a specific transform of the original data, we should consider the particular informational features we would like to keep in or eliminate from the analysis.

8.3.4 Metrics for Fuzzy Data

In literature, several topological measures have been generalized to the fuzzy framework (Rosenfeld, 1979; Goetshel and Voxman, 1983; Diamond and Kloeden, 1994). By focusing on metrics between fuzzy data, we can consider first the *Hausdorff metric*:

$$d_H(A, B) = \max\left\{\sup_{a\in A}\inf_{b\in B} \| a - b \|, \sup_{b\in B}\inf_{a\in A} \| a - b \|\right\}, \tag{8.10}$$

where $A, B \subseteq \Re^d$ denote crisp sets. According to the so-called κ-cuts, the Hausdorff metric d_H can be generalized to fuzzy numbers F, G, where F (or G): $\Re \rightarrow [0,1]$:

$$d_\rho(F,G) = \begin{cases} \left[\int_0^1 (d_H(F_\kappa, G_\kappa))^\rho d\kappa\right]^{\frac{1}{\rho}} & \rho \in [1,\infty) \\ \sup_{\kappa\in[0,1]} d_H(F_\kappa, G_\kappa) & \rho = \infty, \end{cases}$$

where the crisp set $F_\kappa \equiv \{x \in \Re^d : F(x) \geq \kappa\}$, $\kappa \in (0, 1]$, is called the κ-cut of F (Näther, 2000).

Another typology of distance measures can be defined via *support functions* (Diamond and Kloeden, 1994; Näther, 2000). For any compact convex set $F \subset \Re^d$, the support function s_F is defined as $s_F(u) = \sup_{y\in F}\langle u, y\rangle$; $u \in S^{d-1}$, where $\langle\cdot,\cdot\rangle$ is the scalar product in $\Re^d$ and S^{d-1} the $(d-1)$-dimensional unit sphere in $\Re^d$. Notice that, for convex and compact $F \subset \Re^d$, the support function s_F is uniquely determined. A fuzzy set F can be characterized κ-cut-wise by its support function: $s_F(u, \kappa) = s_{A_\kappa}(u)$; $\kappa \in [0, 1]$, $u \in S^{d-1}$. Thus, via support functions, we can define a metric using, for example, a special L_2-metric, i.e.,

$$d_S(F,G) = \sqrt{d\int_0^1 \int_{S^{d-1}} (s_F(u,\kappa) - s_G(u,\kappa))^2 \nu(du) d\kappa}, \tag{8.11}$$

where ν is the Lebesgue measure on S^{d-1}. Different distances for LR fuzzy data can be derived by (8.11) (see Diamond and Kloeden, 1994; Näther, 2000).

Furthermore, a new class of distances between fuzzy numbers is suggested by Bertoluzza, Corral, and Salas (1995).

Most distance measures utilized in the fuzzy clustering for fuzzy data can be considered particular cases of the previous classes of distances. In Section 8.4, when we characterize the different fuzzy clustering models for fuzzy data, we analyze some of these distance measures.

In this section, in general, we point out the proximity measures (dissimilarity, similarity and distance measures) for fuzzy data suggested by Abonyi, Roubos, and Szeifert, 2003; Bloch, 1999; Diamond and Kloeden, 1994; Grzegorzewski, 2004; Hathaway, Bezdek, and Pedrycz, 1996; Hung and Yang, 2004, 2005; Li et al., 2007; Kim and Kim, 2004; Näther, 2000; Yang, Hwang, and Chen, 2005; Yong, Wenkang, Feng, and Qi, 2004; Pappis and Karacapilidis, 1993; Szmidt and Kacprzyk, 2000; Tran and Duckstein, 2002; Zhang and Fu, 2006; Zwich, Carlstein, and Budescu, 1987. In particular, some of these distance measures between fuzzy data are obtained by comparing the respective membership functions. These distances, can be classified according to different approaches (Bloch, 1999; Zwich, Carlstein, and Budescu, 1987): the "functional approach," in which the membership functions are compared by means of Minkowski and Canberra distances extended to the fuzzy case (Dubois and Prade, 1983; Kaufman, 1973; Lowen and Peeters, 1998; Pappis and Karacapilidis, 1993); the "information theoretic approach," based on the definition of fuzzy entropy (De Luca and Termini, 1972) and the "set theoretic approach," based on the concepts of fuzzy union and intersection (Chen, Yeh, and Hsio, 1995; Pappis and Karacapilidis, 1993; Wang, 1997; Wang, De Baets, and Kerre, 1995; Zwich, Carlstein, and Budescu, 1987).

8.4 FUZZY CLUSTERING OF FUZZY DATA

Cluster analysis constitutes the first statistical area that lent itself to a fuzzy treatment. The fundamental justification lies in the recognition of the vague nature of the cluster assignment task. For this reason, in the last 30 years, many fuzzy clustering models for crisp data have been suggested (in a more general way, for the fuzzy approach to statistical analysis, see Coppi, Gil, and Kiers, 2006b). In the literature on fuzzy clustering, the fuzzy K-means clustering model introduced, independently, by Dunn (1974) and Bezdek (1974a) and then extended by Bezdek (1981), is the first model that is computationally efficient and powerful and therefore represents the best-known and used clustering approach. Successively, several models have been set up in this connection. Yet, we can observe that, already Bellman, Kalaba, and Zadeh (1966) and Ruspini (1969, 1970, 1973) proposed pioneering fuzzy clustering. With regard to Ruspini's model, however, the original algorithm due is said to be rather difficult to implement. Its computational efficiency should be weak and its generalization to more than two clusters should be of little success. However, it was the pioneer for a successful development of this approach (Bandemer, 2006). In fact, Ruspini's model opened the door for further research, especially since he first put the idea of fuzzy K-partitions in cluster analysis (Yang, 1993).

In particular, the version proposed by Bezdek in 1981 is the best-known and applied model in the body of literature. It is formalized in the following way:

$$\begin{aligned} &\min: \sum_{i=1}^{I}\sum_{k=1}^{K} u_{ik}^m d_{ik}^2 = \sum_{i=1}^{I}\sum_{k=1}^{K} u_{ik}^m \parallel \mathbf{x}_i - \mathbf{h}_k \parallel^2 \\ &\sum_{k=1}^{K} u_{ik} = 1, u_{ik} \geq 0 \end{aligned} \tag{8.12}$$

where u_{ik} denotes the membership degree of the ith unit to the kth cluster, $d_{ik}^2 = \parallel \mathbf{x}_i - \mathbf{h}_k \parallel^2$, is the Euclidean distance between the ith unit and the centroid which characterizes the kth cluster ($\mathbf{x}_i (i = 1, I)$ and $\mathbf{h}_k$ $(k = 1, K)$ are crisp vectors), $m > 1$ is the fuzzification factor.

The iterative solutions of (8.12) are:

$$u_{ik} = \frac{1}{\sum_{k'=1}^{K} \left[\frac{\parallel \mathbf{x}_i - \mathbf{h}_k \parallel}{\parallel \mathbf{x}_i - \mathbf{h}_{k'} \parallel}\right]^{\frac{2}{m-1}}}, \qquad \mathbf{h}_k = \frac{\sum_{i=1}^{I} u_{ik}^m \mathbf{x}_i}{\sum_{i=1}^{I} u_{ik}^m}. \tag{8.13}$$

This approach has been generalized to fuzzy clustering of fuzzy data. In particular, in the last decades, a great deal of attention has been paid to the fuzzy clustering analysis for fuzzy data. From the "informational" perspective, we are assuming that both the theoretical information (the model) and the empirical information (the data) are fuzzy (case (4)) (see Section 8.2.4).

The fuzzy clustering of fuzzy data has been studied by different authors. Sato and Sato (1995) suggest a fuzzy clustering procedure for interactive fuzzy vectors (Fullér and Majlender, 2004), i.e., fuzzy vector defined as "vectors" where each element cannot be separated from the others. Precisely, an interactive fuzzy vector is one that cannot be realized as the cylindrical closure of its projections. An example of an interactive fuzzy vector is "John is a BIG man" where BIG is characterized by both weight and height (Auephanwiriyakul and Keller, 2002). In particular, these authors analyze, in a clustering framework, fuzzy data that are defined by convex and normal fuzzy sets (CNF sets); then, in order to suitably represent the CNF sets, they define a conical membership function and suggest a fuzzy asymmetrical dissimilarity measure between fuzzy data. Successively, for solving the clustering problem, they utilize an additive fuzzy clustering model based on a multicriteria procedure. Hathaway, Bezdek, and Pedrycz (1996) and Pedrycz, Bezdek, Hathaway, and Rogers (1998) introduce models that convert parametric or nonparametric linguistic variables to generalize coordinates (vectors of numerical numbers) before performing fuzzy c-means clustering. Hathaway, Bezdek, and Pedrycz (1996) analyze the fusing heterogeneous fuzzy data (real numbers, real intervals, and linguistic assessment (real fuzzy sets)) by utilizing fuzzy clustering. By considering the simple case of univariate interval data, the authors consider three schemes for extracting the parameters to represent the interval data, i.e., for each interval datum, they consider the center and the radius, the center and the diameter, and the left and right endpoints. For each representation, by using the Euclidean metric they compare the two parameters of each pair interval data obtaining suitable Euclidean distance measures between interval data. Then, utilizing these distances in a clustering framework, the authors suggest, for each scheme, a fuzzy clustering model. Notice that, for integrating heterogeneous fuzzy data, the authors consider a parametric approach. Nonparametric models for fusing heterogeneous fuzzy data are suggested by Pedrycz, Bezdek, Hathaway, and Rogers (1998). Yang and Ko (1996) propose a class of fuzzy *K*-numbers clustering procedures for LR fuzzy univariate data (i.e., the authors consider LR fuzzy data with a different type of membership function: triangular, normal, trapezoidal, etc.). The same authors apply fuzzy clustering to overcome the heterogeneous problem in the fuzzy regression analysis with fuzzy data (i.e., fuzzy input and fuzzy output; crisp input and fuzzy output) (Yang and Ko, 1997). In particular, they suggest two clusterwise fuzzy regression models (the two-stage weighted fuzzy regression and one-stage generalized fuzzy regression). This is a typical example in which the fuzzy clustering is incorporated in a fuzzy regression framework with complete or partial fuzzy empirical information (i.e., the data are all or in part fuzzy). Furthermore, in this case, the aim of the fuzzy clusterwise regression analysis is twofold: clustering, in a fuzzy manner, of a set of units and interpolation of fuzzy data. Then, the computation problem is the calculation of the membership degrees of each unit to different (e.g., linear) clusters and the estimation of the parameter that characterizes the analytic form of the fuzzy prototypes. Successively, Yang and Liu (1999) extend the clustering procedures proposed by Yang and Ko (1996) to high-dimensional fuzzy vectors (conical fuzzy vectors). Tanaka, Miyamoto, and Unayahara (2001) discuss the fuzzy clustering technique for data with uncertainties using minimum and maximum distances based an on L_1 metric.

Auephanwiriyakul, Keller (2002) develop a linguistic fuzzy *K*-means model that works with a vector of fuzzy numbers. This model is based on the extension principle and the decomposition theorem. In particular, it turns out that using the extension principle to extend the capability of the standard membership update equation to deal with a linguistic vector has a huge computational complexity. To cope with this problem, the authors develop an efficient procedure based on fuzzy arithmetic and optimization. They also prove that the algorithm behaves in a similar way to the fuzzy *K*-means clustering suggested by Bezdek (1981) in the degenerate linguistic case (Auephanwiriyakul and Keller, 2002). Yang, Hwang, and Chen (2004) propose a fuzzy clustering model for mixed data, i.e., fuzzy and symbolic data. Then, if we have in the data-set only fuzzy data, this model can be considered as a fuzzy clustering model for fuzzy data. In particular, to take into account the peculiarity of the symbolic and fuzzy component of the mixed data, the authors define a "composite" dissimilarity measure. For the symbolic component, they consider a modification of the Gowda–Diday dissimilarity; for the fuzzy component, they utilize the parametric approach proposed by Hathaway (1996) and the Yang–Ko dissimilarity (1996). Hung and Yang (2005) suggest the so-called alternative fuzzy *K*-numbers clustering algorithm for LR fuzzy numbers (i.e., LR fuzzy univariate data) based on an exponential-type distance measure. The authors show this distance is claimed to be robust as regards noise and outliers. Hence, the model is more robust than the fuzzy *K*-numbers clustering proposed by Yang and Ko (1996). To explain the applicative performances of the suggested model, the authors consider a nice application with tea evaluation data (see Section 8.6.1.1). D'Urso and Giordani (2006a) propose a fuzzy clustering model for fuzzy data based on a "weighted" dissimilarity for comparing pairs of fuzzy data. This dissimilarity is composed of two distances, the so-called center distance and the spread distance. A peculiarity of the suggested fuzzy clustering is the objective computation, incorporated in the clustering procedure, of weights pertaining to the center distance and spread distance of the fuzzy data. Then, the model automatically tunes the influence of the two components of the fuzzy data for calculating the center and spreads centroids in the fuzzy clustering.

In the class of fuzzy clustering models for fuzzy data, we can also include, as a particular case, the fuzzy clustering model for symbolic data suggested by El-Sonbaty and Ismail (1998) when all symbolic data are interval data. For the fuzzy clustering of symbolic (interval) data, see also de Carvalho (2007).

In the last few years, increasing attention has also been paid to developing fuzzy clustering models for the so-called fuzzy data time arrays. In this respect, Coppi and D'Urso (2002, 2003) propose different fuzzy *K*-means clustering models (*D-DFKMC models*, i.e., *dynamic double fuzzy K-means clustering models*) for fuzzy time trajectories that are a particular geometrical representation of the fuzzy data time array (Coppi and D'Urso, 2000). Coppi, D'Urso, and Giordani (2004, 2006a) propose other dynamic fuzzy clustering models for fuzzy trajectories: the *dynamic double fuzzy clustering with entropy regularization* models (*D-DFCER* models) (Coppi, D'Urso, and Giordani, 2004) and the *dynamic double fuzzy K-medoids clustering* models (*D-DFKMDC* models) (Coppi, D'Urso, and Giordani, 2006a).

In conclusion, by considering the multivariate framework, we observe that examples of fuzzy clustering models based on the *conjunctive* approach are the models suggested by Sato and Sato (1995) and Yang and Liu (1999). Conversely, examples of models based on a *disjunctive* representation of the fuzzy data are the ones proposed by Hathaway, Bezdek, and Pedrycz (1996), Yang, Hwang, and Chen (2004), D'Urso and Giordani (2006a) and the fuzzy clustering models for fuzzy data time arrays (see Section 8.5) (Coppi and D'Urso, 2002, 2003; Coppi, D'Urso, and Giordani, 2004, 2006a).

In this section, we focus our attention on the fuzzy clustering models for fuzzy data proposed by Yang and Ko (1996) (*DFKNC model*, i.e., *double fuzzy K-numbers clustering model*), Yang and Liu (1999) (*FKMCCFV model*, i.e., *fuzzy K-means clustering model for conical fuzzy vectors*), Yang, Hwang, and Chen (2004) (*FKMCMD model*, i.e., *Fuzzy K-means clustering model for mixed data*), Hung, and Yang (2005) (*ADFKNC model*, i.e., *alternative double fuzzy K-numbers clustering model*) and D'Urso and Giordani (2006a) (*DFWKMC model*, i.e., *double fuzzy weighted K-means clustering model*). A fuzzy clustering model for fuzzy data time arrays is analyzed in Section 8.5. In the same section, a concise illustration of the fuzzy time data arrays and metrics is also shown.

8.4.1 Double Fuzzy K-numbers Clustering Model (DFKNC Model)

The clustering model proposed by Yang and Ko (1996), called *double fuzzy K-numbers clustering* model (*DFKNC* model), deals with a single fuzzy variable observed on I units. It is assumed that the membership function of the fuzzy variable belongs to the *LR* family (8.3) and the univariate fuzzy data are represented by $x_i = (c_i, l_i, r_i)_{LR}, i = 1,I$.

Firstly, a *distance measure* between the realizations of such a variable is needed. The authors suggested the following (squared) distance between each pair of fuzzy numbers, say x_i and $x_{i'}$:

$$_{YC}d^2_{ii'}(\lambda,\rho) = (c_i - c_{i'})^2 + [(c_i - \lambda l_i) - (c_{i'} - \lambda l_{i'})]^2 + [(c_i + \rho r_i) - (c_{i'} + \rho r_{i'})]^2, \tag{8.14}$$

where $\lambda = \int_0^1 L^{-1}(\omega)\mathrm{d}\omega$, $\rho = \int_0^1 R^{-1}(\omega)\mathrm{d}\omega$ are parameters that summarize the shape of the left and right tails of the membership function (then, for each value of λ and ρ, we have a particular membership function (see, for example, Figure 8.7)).

Then, the *DFKNC* model is characterized as:

$$\min: \sum_{i=1}^{I}\sum_{k=1}^{K} u_{ik}^m {}_{YC}d_{ik}^2 = \sum_{i=1}^{I}\sum_{k=1}^{K} u_{ik}^m [(c_i - c_k)^2 + [(c_i - \lambda l_i) - (c_k - \lambda l_k)]^2 + [(c_i + \rho r_i) - (c_k + \rho r_k)]^2]$$
$$\sum_{k=1}^{K} u_{ik} = 1, u_{ik} \geq 0, \tag{8.15}$$

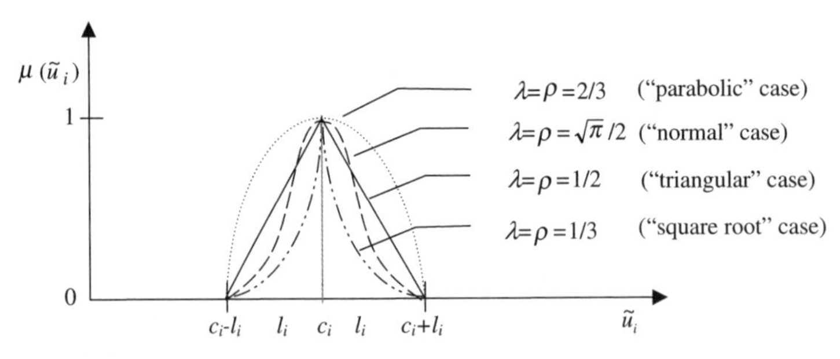

Figure 8.7 Examples of membership functions for particular values of λ and ρ.

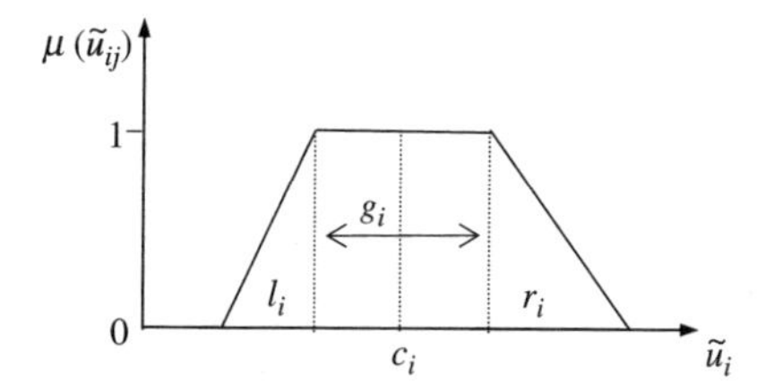

Figure 8.8 Trapezoidal membership function with a new parameterization.

where c_k, l_k and r_k are, respectively, the center and the left and right spreads of the k-th centroid. By solving the optimization problem (8.15) by means of the Lagrangian multiplier method, we obtain the following *iterative solutions*:

$$u_{ik} = \frac{1}{\sum_{k'=1}^{K}\left[\frac{{}_{YC}d_{ik}(\lambda,\rho)}{{}_{YC}d_{ik'}(\lambda,\rho)}\right]^{\frac{2}{m-1}}}, \quad c_k = \frac{\sum_{i=1}^{I} u_{ik}^m[3c_i - \lambda(l_i - l_k) + \rho(r_i - r_k)]}{3\sum_{i=1}^{I} u_{ik}^m}, \quad l_k = \frac{\sum_{i=1}^{I} u_{ik}^m(c_k + \lambda l_i - c_i)}{\lambda\sum_{i=1}^{I} u_{ik}^m},$$

$$r_k = \frac{\sum_{i=1}^{I} u_{ik}^m(c_i + \rho r_i - c_k)}{\rho\sum_{i=1}^{I} u_{ik}^m}. \tag{8.16}$$

For computing the iterative solutions, starting with an initial fuzzy partition, the authors proposed an operative algorithm (for more detail see Yang and Ko, 1996). Notice that, the authors also formalized the clustering model for trapezoidal data (see formula (8.9) and Figure 8.6).

8.4.2 Fuzzy K-means Clustering Model for Conical Fuzzy Vectors (FKMCCFV Model)

Different from the *DFKNC* model, the *FKMCCFV* model proposed by Yang and Liu (1999) is applicable to multi-dimensional fuzzy variables observed on I units. The membership function of the vector valued variable is assumed to be *conical*:

$$\mu(\tilde{\mathbf{u}}) = 1 - \min\{1, \parallel \tilde{\mathbf{u}} - \mathbf{c} \parallel_{\mathbf{A}}\} = \max\{0, 1 - \parallel \tilde{\mathbf{u}} - \mathbf{c} \parallel_{\mathbf{A}}\}, \tag{8.17}$$

where: $\mu(\mathbf{c}) = 1$, $\parallel \tilde{\mathbf{u}} - \mathbf{c} \parallel_{\mathbf{A}} = [(\tilde{\mathbf{u}} - \mathbf{c})'\mathbf{A}^{-1}(\tilde{\mathbf{u}} - \mathbf{c})]^{\frac{1}{2}}$, $\mathbf{c}$ = apex (center), $\mathbf{A}$ = *panderance* matrix (Celmin, 1987a,b). Under the above assumption, the fuzzy data are represented by:

$$\mathbf{x}_i = (\mathbf{c}_i, \mathbf{A}_i), \ i = 1, I. \tag{8.18}$$

Then, the authors introduced a distance measure between fuzzy conical vectors. In particular, given two realizations, say $\mathbf{x}_i = (\mathbf{c}_i, \mathbf{A}_i)$ and $\mathbf{x}_{i'} = (\mathbf{c}_{i'}, \mathbf{A}_{i'})$, $i, i' = 1, I$, Yang and Liu proposed the following (squared) *distance*:

$${}_{YL}d_{ii'}^2 = \parallel \mathbf{c}_i - \mathbf{c}_{i'} \parallel^2 + tr[(\mathbf{A}_i - \mathbf{A}_{i'})'(\mathbf{A}_i - \mathbf{A}_{i'})]. \tag{8.19}$$

By using the distance (8.19) the *FKMCCFV* model is formalized in the following way:

$$\min: \sum_{i=1}^{I}\sum_{k=1}^{K} u_{ik}^m {}_{YL}d_{ik}^2 = \sum_{i=1}^{I}\sum_{k=1}^{K} u_{ik}^m \parallel \mathbf{c}_i - \mathbf{c}_k \parallel^2 + tr[(\mathbf{A}_i - \mathbf{A}_k)'(\mathbf{A}_i - \mathbf{A}_k)]$$
$$\sum_{k=1}^{K} u_{ik} = 1, u_{ik} \geq 0. \tag{8.20}$$

By utilizing the Lagrangian multiplier method, we arrive at the following solutions being produced in an iterative fashion,

$$u_{ik} = \frac{1}{\sum_{k'=1}^{K} \left[\frac{{}_{YL}d_{ik}}{{}_{YL}d_{ik'}}\right]^{\frac{2}{m-1}}}, \quad \mathbf{c}_k = \frac{\sum_{i=1}^{I} u_{ik}^m \mathbf{c}_i}{\sum_{i=1}^{I} u_{ik}^m}, \quad \mathbf{A}_k = \frac{\sum_{i=1}^{I} u_{ik}^m \mathbf{A}_i}{\sum_{i=1}^{I} u_{ik}^m}, \tag{8.21}$$

where $\mathbf{c}_k$, and $\mathbf{A}_k$ are, respectively, the apex (center) and the *panderance* matrix for the k-th centroid. An iterative algorithm for solving the equations (8.21) is constructed (see Yang and Liu, 1999).

Notice that the clustering model introduced by Hathaway, Bezdek, and Pedrycz (1996) utilizes the same clustering objective function as the *FKMCCFV* model when the so-called *panderance* matrix is a diagonal matrix (e.g., when, as is usually the case, there are non-interactive fuzzy numbers). Therefore, if we consider non-interactive fuzzy data, the models proposed by Hathaway, Bezdek, and Pedrycz (1996) (called the *parametric double fuzzy K-means clustering* model (*PDFKMC* model)) and Yang and Liu (1999) have the same performance.

8.4.3 Fuzzy K-means Clustering Model for Mixed Data (FKMCMD Model)

Yang, Hwang, and Chen (2004) propose a clustering model, called *fuzzy K-means clustering* model for *mixed data* (*FKMCMD* model) to classify *mixed* data, i.e., *symbolic* data and *LR-II type* fuzzy data. In particular, the suggested model is obtained by utilizing a new (squared) distance measure for mixed data in the traditional fuzzy clustering structure defined by Bezdek. Then, the objective function of the model is composed of two parts: the *symbolic component* in which a distance for symbolic data is considered and the *fuzzy component* characterized by a particular distance for fuzzy data.

Since the aim of our work is to analyze fuzzy clustering for only fuzzy data, we formalize the version of the *FKMCMD* model in which the empirical information is completely fuzzy (all the data are fuzzy). Then, we focus our attention on the (completely) fuzzy part of the model only.

To this purpose, Yang, Hwang, and Chen (2004) considered, first, a new parameterization of *LR-II type* fuzzy data. In particular, the authors generalized the parametrical representation of *symmetric* fuzzy data suggested by Hathaway, Bezdek, and Pedrycz (1996), by considering the same parameterization for trapezoidal fuzzy data. In this way, the (univariate) trapezoidal fuzzy data are characterized as follows:

$$x_i = (c_i, g_i, l_i, r_i), \quad i = 1,I, \tag{8.22}$$

where c_i is the *center*, g_i is the *inner diameter*, and l_i, r_i are, respectively, the *left* and *right outer radius* (see Figure 8.8).

Then, the authors proposed a *distance* measure between trapezoidal fuzzy data, say $x_i = (c_i, g_i, l_i, r_i)$ and $x_{i'} = (c_{i'}, g_{i'}, l_{i'}, r_{i'})$ $(i, i' = 1,I)$ by considering the Yang–Ko dissimilarity. By assuming that L and R are linear ($\lambda = \rho = 1/2$), the suggested distance is:

$$\begin{aligned} {}_{YHC}d_{ii'}^2 = {} & \left(\frac{2c_i - g_i}{2} - \frac{2c_{i'} - g_{i'}}{2}\right)^2 + \left(\frac{2c_i + g_i}{2} - \frac{2c_{i'} + g_{i'}}{2}\right)^2 + \left(\frac{2c_i - g_i}{2} - \frac{1}{2}l_i\right) - \left(\frac{2c_{i'} - g_{i'}}{2} - \frac{1}{2}l_{i'}\right)^2 \\ & + \left(\frac{2c_i + g_i}{2} + \frac{1}{2}r_i\right) - \left(\frac{2c_{i'} + g_{i'}}{2} + \frac{1}{2}r_{i'}\right)^2. \end{aligned} \tag{8.23}$$

By extending the previous (squared) distance (8.23) to a multi-dimensional case, in which the fuzzy data matrix is characterized in the following way $\mathbf{X} \equiv \{x_{ij} = (c_{ij}, g_{ij}, l_{ij}, r_{ij})_{LR-II} : i = 1,I; j = 1,J\}$, and after some algebraic manipulations, we can write the (squares) distance as follows:

$$\begin{aligned} {}_{YHC}d_{ii'}^2 = {} & \frac{1}{4}\left[\| 2(\mathbf{c}_i - \mathbf{c}_{i'}) - (\mathbf{g}_i - \mathbf{g}_{i'}) \|^2 + \| 2(\mathbf{c}_i - \mathbf{c}_{i'}) + (\mathbf{g}_i - \mathbf{g}_{i'}) \|^2 \right. \\ & \left. + \| 2(\mathbf{c}_i - \mathbf{c}_{i'}) - (\mathbf{g}_i - \mathbf{g}_{i'}) - (\mathbf{l}_i - \mathbf{l}_{i'}) \|^2 + \| 2(\mathbf{c}_i - \mathbf{c}_{i'}) + (\mathbf{g}_i - \mathbf{g}_{i'}) + (\mathbf{r}_i - \mathbf{r}_{i'}) \|^2\right], \end{aligned} \tag{8.24}$$

where $\mathbf{c}_i$, $\mathbf{g}_i$, $\mathbf{l}_i$, and $\mathbf{r}_i$ are, respectively, the vectors of the *center*, *inner diameter*, *left* and *right outer radius* of the ith unit, and $\mathbf{c}_{i'}$, $\mathbf{g}_{i'}$, $\mathbf{l}_{i'}$, and $\mathbf{r}_{i'}$ are the same vectors of the i'th unit.

Then, the clustering model proposed by Yang, Hwang, and Chen (2004) can be formalized in the following way:

$$\min: \sum_{i=1}^{I}\sum_{k=1}^{K} u_{ik}^{m}\,{}_{YHC}d_{ik}^{2} = \sum_{i=1}^{I}\sum_{k=1}^{K} u_{ik}^{m}\left\{\frac{1}{4}\left[\|2(\mathbf{c}_i-\mathbf{c}_k)-(\mathbf{g}_i-\mathbf{g}_k)\|^2+\|2(\mathbf{c}_i-\mathbf{c}_k)+(\mathbf{g}_i-\mathbf{g}_k)\|^2+\right.\right.$$

$$\left.\left.\|2(\mathbf{c}_i-\mathbf{c}_k)-(\mathbf{g}_i-\mathbf{g}_k)-(\mathbf{l}_i-\mathbf{l}_k)\|^2+\|2(\mathbf{c}_i-\mathbf{c}_k)+(\mathbf{g}_i-\mathbf{g}_k)+(\mathbf{r}_i-\mathbf{r}_k)\|^2\right]\right\}$$

$$\sum_{k=1}^{K} u_{ik} = 1, u_{ik} \geq 0, \tag{8.25}$$

where $\mathbf{c}_k$, $\mathbf{g}_k$, $\mathbf{l}_k$, and $\mathbf{r}_k$ are, respectively, the vectors of the *center*, *inner diameter*, *left* and *right outer radius* of the kth centroid.

By utilizing the Lagrangian multiplier method, we have for (8.25) the following *iterative solutions* of the model:

$$u_{ik} = \frac{1}{\sum_{k'=1}^{K}\left[\frac{{}_{YHC}d_{ik}}{{}_{YHC}d_{ik'}}\right]^{\frac{2}{m-1}}},\ \mathbf{c}_k = \frac{\sum_{i=1}^{I} u_{ik}^{m}(8\mathbf{c}_i-\mathbf{l}_i+\mathbf{r}_i+\mathbf{l}_k-\mathbf{r}_k)}{8\sum_{i=1}^{I} u_{ik}^{m}},\ \mathbf{g}_k = \frac{\sum_{i=1}^{I} u_{ik}^{m}(4\mathbf{g}_i+\mathbf{l}_i+\mathbf{r}_i-\mathbf{l}_k-\mathbf{r}_k)}{4\sum_{i=1}^{I} u_{ik}^{m}}$$

$$\mathbf{l}_k = \frac{\sum_{i=1}^{I} u_{ik}^{m}(-2\mathbf{c}_i+\mathbf{g}_i+\mathbf{l}_i+2\mathbf{c}_k-\mathbf{g}_k)}{\sum_{i=1}^{I} u_{ik}^{m}},\ \mathbf{r}_k = \frac{\sum_{i=1}^{I} u_{ik}^{m}(2\mathbf{c}_i+\mathbf{g}_i+\mathbf{r}_i-2\mathbf{c}_k-\mathbf{g}_k)}{\sum_{i=1}^{I} u_{ik}^{m}}. \tag{8.26}$$

On the basis of these solutions the authors constructed an iterative algorithm (cf. Yang, Hwang, and Chen, 2004).

8.4.4 Alternative Double Fuzzy K-numbers Clustering Model (ADFKNC Model)

Recently, Hung and Yang (2005) proposed a clustering model, called *alternative double fuzzy K-numbers clustering* model (*ADFKNC* model), to classify units. In particular, in the same way as the *DFKNC* model, the authors formalized the model for univariate fuzzy data, i.e., $x_i = (c_i, l_i, r_i)_{LR}, i = 1,I$, and assumed the membership function of the single fuzzy variable belonging to the *LR* family (8.3).

The authors proposed an exponential-type distance for LR fuzzy numbers based on the idea of Wu and Yang (2002) and discussed the robustness of this distance. The authors showed that the suggested distance is more robust than the distance proposed by Yang and Ko (cf. (8.14)). Then, they modified the *DFKNC* model proposed by Yang and Ko, integrating the new robust distance in the clustering model.

In particular, the proposed *distance* between each pair of fuzzy numbers, say x_i and $x_{i'}$, is:

$${}_{HY}d_{ii'}^{2}(\lambda,\rho) = 1-\exp(-b\,{}_{YC}\tilde{d}_{ii'}^{2}(\lambda,\rho)), \tag{8.27}$$

where ${}_{YC}\tilde{d}_{ii'}^{2}(\lambda,\rho) = 1/3\,{}_{YC}d_{ii'}^{2}(\lambda,\rho)$, b is a suitable positive constant (cf. Hung and Yang, 2005). Notice that the (squared) distance measure ${}_{HY}d_{ii'}^{2}(\lambda,\rho)$ is a monotone increasing function of ${}_{YC}d_{ii'}^{2}(\lambda,\rho)$ similar to that of Wu and Yang (2002).

By considering the (squared) distance (8.27), the authors formalized the *ADFKNC* model in the following way:

$$\min: \sum_{i=1}^{I}\sum_{k=1}^{K} u_{ik}^{m}\left[1-\exp(-b\,{}_{YC}\tilde{d}_{ik}^{2}(\lambda,\rho))\right]$$
$$\sum_{k=1}^{K} u_{ik}=1, u_{ik}\geq 0. \tag{8.28}$$

The iterative solutions are:

$$u_{ik}=\frac{1}{\sum_{k'=1}^{K}\left[\frac{1-\exp(-b\,{}_{YC}\tilde{d}_{ik}^{2}(\lambda,\rho))}{1-\exp(-b\,{}_{YC}\tilde{d}_{ik'}^{2}(\lambda,\rho))}\right]^{\frac{1}{m-1}}},\quad c_k=\frac{\sum_{i=1}^{I}u_{ik}^{m}[3c_i-\lambda(l_i-l_k)+\rho(r_i-r_k)]\exp(-b\,{}_{YC}\tilde{d}_{ik}^{2}(\lambda,\rho))}{3\sum_{i=1}^{I}u_{ik}^{m}\exp(-b\,{}_{YC}\tilde{d}_{ik}^{2}(\lambda,\rho))},$$

$$l_k=\frac{\sum_{i=1}^{I}u_{ik}^{m}(c_k+\lambda l_i-c_i)\exp(-b\,{}_{YC}\tilde{d}_{ik}^{2}(\lambda,\rho))}{\lambda\sum_{i=1}^{I}u_{ik}^{m}\exp(-b\,{}_{YC}\tilde{d}_{ik}^{2}(\lambda,\rho))},\quad r_k=\frac{\sum_{i=1}^{I}u_{ik}^{m}(c_i+\rho r_i-c_k)\exp(-b\,{}_{YC}\tilde{d}_{ik}^{2}(\lambda,\rho))}{\rho\sum_{i=1}^{I}u_{ik}^{m}\exp(-b\,{}_{YC}\tilde{d}_{ik}^{2}(\lambda,\rho))}. \tag{8.29}$$

In order to solve Equations (9.29) the authors suggested an iterative algorithm (cf. Hung and Yang, 2005).

8.4.5 Double Fuzzy Weighted K-means Clustering Model (DFWKMC Model)

In the *double fuzzy weighted K-means clustering* model (*DFWKMC* model) proposed by D'Urso and Giordani (2006a), a vector valued fuzzy variable was considered, with a symmetric LR membership function (see Section 8.4). The data are represented as follows:

$$\mathbf{x}_i=(\mathbf{c}_i,\mathbf{l}_i)_L,\quad i=1,I, \tag{8.30}$$

where $\mathbf{c}_i$ and $\mathbf{l}_i$ are, respectively, the center vector and the spread vector. The (symmetric) fuzzy data matrix is shown in formula (8.4).

For classifying (symmetric) units with fuzzy information the authors introduced a weighted (squared) distance measure between symmetric fuzzy data. This dissimilarity compared each pair of (symmetric) fuzzy data vector by considering, separately, the distances for the centers and the spreads and a suitable weighting system for such distance components. The proposed dissimilarity is formalized as follows:

$${}_{DG}d_{ii'}^{2}({}_{c}w,{}_{l}w)=({}_{c}w_{c}d_{ii'})^{2}+({}_{l}w_{l}d_{ii'})^{2}, \tag{8.31}$$

where ${}_{c}w, {}_{l}w$ are suitable weights for the center and spreads,

$${}_{c}d_{ii'}=\|\mathbf{c}_i-\mathbf{c}_{i'}\|,\qquad \textit{(Euclidean) center distance} \tag{8.32}$$
$${}_{l}d_{ii'}=\|\mathbf{l}_i-\mathbf{l}_{i'}\|,\qquad \textit{(Euclidean) spread distance.} \tag{8.33}$$

The weights can be fixed *subjectively* a priori by considering external or subjective conditions. However, the authors estimated the weights *objectively* within an appropriate clustering model. In particular, the authors suggested weighing the center and the spread distances differently by means of the following weights. As the membership function value of the centers is maximum, they suggested assuming that the center distance weight is higher than the spread distance one. Then, we have:

$${}_{c}w=v+z,\qquad \textit{center distance weight} \tag{8.34}$$
$${}_{l}w=v,\qquad \textit{spread distance weight} \tag{8.35}$$

such that ${}_{c}w+{}_{l}w=1$; $0\leq{}_{l}w\leq{}_{c}w$. It follows that $2v+z=1$; $v,z\geq 0$; $v\leq 0.5$ (${}_{c}w={}_{l}w\Leftrightarrow z=0$).

For computing the weights, the authors adopt an objective criterion; suitable weights are obtained by minimizing the loss function with regard to the optimal values of ${}_cw$ and ${}_lw$. In particular, the loss function is formalized as follows:

$$\sum_{i=1}^{I}\sum_{k=1}^{K} u_{ik}^{m} {}_{DG}d_{ik}^2 = \sum_{i=1}^{I}\sum_{k=1}^{K} u_{ik}^{m}\left[({}_cw\,{}_cd_{ik})^2+({}_lw\,{}_ld_{ik})^2\right] = \sum_{i=1}^{I}\sum_{k=1}^{K} u_{ik}^{m}\left[((v+z)\,{}_cd_{ik})^2+(v\,{}_ld_{ik})^2\right]$$
$$= \sum_{i=1}^{I}\sum_{k=1}^{K} u_{ik}^{m}\left[(1-v)^2{}_cd_{ik}^2+v^2{}_ld_{ik}^2\right], \quad (\text{since } 2v+z=1) \tag{8.36}$$

where ${}_cd_{ik}$ and ${}_ld_{ik}$ compare, respectively, the centers and spreads of the ith object and the k-th centroid.

The loss function (8.36) is minimized with regard to the membership degrees, the (fuzzy) centroids and the center and spread weights. Then, the *DFWKMC* model is characterized as follows:

$$\begin{aligned} &\min: \sum_{i=1}^{I}\sum_{k=1}^{K} u_{ik}^{m}\left[(1-v)^2\,{}_cd_{ik}^2+v^2\,{}_ld_{ik}^2\right] \\ &\sum_{k=1}^{K} u_{ik}=1, u_{ik}\geq 0 \\ &0\leq v\leq 0.5. \end{aligned} \tag{8.37}$$

The iterative solutions of the minimization problem in (8.37) are obtained by considering the Lagrangian multiplier method. In particular, the iterative solutions are:

$$u_{ik}=\frac{1}{\sum_{k'=1}^{K}\left[\frac{\left[(1-v)^2\,{}_cd_{ik}^2+v^2\,{}_ld_{ik}^2\right]}{\left[(1-v)^2\,{}_cd_{ik'}^2+v^2\,{}_ld_{ik'}^2\right]}\right]^{\frac{1}{m-1}}}, \quad v=\frac{\sum_{i=1}^{I}\sum_{k=1}^{K}u_{ik}^{m}\,{}_cd_{ik}^2}{\sum_{i=1}^{I}\sum_{k=1}^{K}u_{ik}^{m}({}_cd_{ik}^2+{}_ld_{ik}^2)}, \quad \mathbf{c}_k=\frac{\sum_{i=1}^{I}u_{ik}^{m}\mathbf{c}_i}{\sum_{i=1}^{I}u_{ik}^{m}},$$
$$\mathbf{l}_k=\frac{\sum_{i=1}^{I}u_{ik}^{m}\mathbf{l}_i}{\sum_{i=1}^{I}u_{ik}^{m}}. \tag{8.38}$$

To take into account the constraint $0 \leq v \leq 0.5$, the authors checked whether the optimal v is higher than 0.5. If so, we set $v = 0.5$.

When the optimal v is lower than 0.5, we can conclude that, taking into account (8.34) and (8.35), the weight of the center distance is higher than that of the spreads. It occurs when the differences among the spreads are more relevant than those of the centers. Under the assumption that the centers are more important than the spreads, the model automatically finds a suitable system of weights such that the role of the center distance in the minimization procedure is emphasized. In fact, when $v < 0.5$, ${}_cw >_l w$ whereas, when $v = 0.5$, ${}_cw =_l w$.

Notice that the *DFWKMC* model can be applied to *all* symmetric fuzzy data. In fact, it is assumed that the *shape* of the membership functions of the observed fuzzy data is *inherited*. Thus, the shape is supposed before starting the clustering procedure. Obviously, the shape of the membership functions of the symmetric fuzzy centroids, computed in the fuzzy clustering procedure, is inherited from the observed fuzzy data.

An iterative algorithm was also proposed by the authors for solving Equations (8.38).

8.4.6 A Comparative Assessment

In this section, in a comparative assessment we summarize the features (concerning the empirical and theoretical aspects) (cf. Table 8.1) and the performances of the previous clustering models. In particular, we compare the models by considering the following aspects:

Table 8.1 Features of some fuzzy clustering models for fuzzy data.

	Empirical information		Theoretical information
Model	Data	Specification approach	Distance, optimization problem and iterative solutions
DFKNC (Yang and Ko, 1996)	Univariate LR fuzzy data	—	External shape information of membership function; external weighting systems; constrained minimization problem; the algorithm does not guarantee the global optimum.
FKMCCFV (Yang and Liu, 1999)	Multivariate conic fuzzy data	Conjunctive	External shape information of membership function; distance also based on the comparison of *panderance* matrix; constrained minimization problem; the algorithm does not guarantee the global optimum.
FKMCMD (Yang, Hwang and Chen, 2004)	Multivariate LR-II type fuzzy data	Disjunctive	Riparameterization-based distance; constrained minimization problem; the algorithm does not guarantee the global optimum.
ADFKNC (Hung and Yang, 2005)	Univariate LR fuzzy data	—	External shape information of membership function; external weighting systems; 'exponential' distance, robustness; constrained minimization problem; the algorithm does not guarantee the global optimum.
DFWKMC (D'Urso and Giordani, 2006a)	Multivariate symmetric fuzzy data	Disjunctive	External shape information of membership function; internal weighting systems; constrained minimization problem; the algorithm does not guarantee that the global optimum.

(1) *typology of fuzzy data and specification approach*;

(2) *distance measures, optimization problem, and iterative solutions*;

(3) *performance.*

(1) The empirical features, i.e., the *typology of fuzzy data* (univariate or multivariate fuzzy data) and the connected *specification approach* (considered only in the multi-dimensional case), conjunctive or disjunctive, considered in the various fuzzy clustering models are shown in Table 8.1.

(2) The analyzed models are characterized by the utilization of different distances. In the distance measure utilized in the *DFKNC* model (Yang and Ko, 1996) particular parameters (λ and ρ), which summarize the shape of the left and right tails of the membership function, are considered. In this way, the information concerning the shape of the membership function is introduced by means of the distance in the clustering procedure. Furthermore, since this information is introduced by means of weights associated to the spreads, in this way, the center and spreads of the fuzzy data are suitably weighed. In fact, the parameters λ and ρ represent multiplicative coefficients that reduce the observed length of the spreads, accounting for the different degrees of membership associated to the various points lying within the upper (lower) bound and the center. Notice that this process of informational acquisition (concerning the shape of the membership function and the weighting systems regarding the components (center and left and right spreads) of the fuzzy data) is exogenous to the clustering model (*external weighting computation*). In fact, the shape of the membership function and then the parameters λ and ρ are assumed prior to starting the clustering process (for the selection of membership function see Section 8.3.2).

The distance utilized in the *DFKNC* model is incorporated in an "exponential" distance by Hung and Yang (2005) for defying a new robust distance. In this way, the *ADFKNC* model suggested by these authors and based on the new distance is robust, i.e., it neutralizes the disruptive effects of possible outliers and noises (in the fuzzy dataset) in the clustering process. In addition, the distance and the model suggested by Hung and Yang (2005) inherit all the features of the distance and model proposed by Yang and Ko (1996). Furthermore, notice that, by putting ${}_{YC}D_{ik} = \exp(-b\,{}_{YC}\tilde{d}^2_{ik}(\lambda,\rho))$, the prototypes of the *ADFKNC* model (shown in (8.29)) have extra weights ${}_{YC}D_{ik}$ (the other weights are u^m_{ik}) that decrease monotonically in ${}_{YC}\tilde{d}^2_{ik}(\lambda,\rho)$ with ${}_{YC}D_{ik} = 1$ as ${}_{YC}\tilde{d}^2_{ik}(\lambda,\rho) = 0$ and ${}_{YC}D_{ik} = 0$ as ${}_{YC}\tilde{d}^2_{ik}(\lambda,\rho) \to \infty$. Then, the *ADFKNC* prototypes assign reasonably different weights $u^m_{ik}\,{}_{YC}D_{ik}$ to outlier fuzzy data. For this reason, the *ADFKNC* model can be considered robust. Conversely, the *DFKNC* prototypes (i.e., centroids) (shown in (8.16)) use only the weights u^m_{ik} and then the *DFKNC* model is sensitive to outliers. In conclusion, we observe that membership degrees obtained by means of *ADFKNC* and *DFKNC* models coincide as $b \to 0$.

The distance utilized in the *FKMCCFV* (Yang and Liu, 1999) takes into account the particular type of fuzzy data analyzed in the clustering process (multivariate conic fuzzy data) by integrating the information concerning the center differences with the difference regarding the panderance matrix of each pair of fuzzy vectors. In this way, the authors introduce the information regarding the selected specification approach for the fuzzy conic vectors (conjunctive approach), by means of the panderance matrix. Also in this case the features of the membership function of the fuzzy data are assumed before starting the clustering procedure.

In the distance measure and in the complete fuzzy version of the *FKMCMD* model proposed by Yang, Hwang, and Chen (2004), the peculiar aspect is the consideration of a non-usual parameterization of the fuzzy data.

With regard to the *DFWKMC* model proposed by D'Urso and Giordani (2006a) the distance measure has the following features (D'Urso and Giordani, 2006a):

- It is a sum of two squared weighted Euclidean distances: the center distance and the spread distance.
- The weights are intrinsically associated with the component-distance (center and spread distance).
- By means of the weights, we can appropriately tune the influence of the two components of the fuzzy entity (center and spread) when computing the dissimilarity between each pair of fuzzy data.
- The weights can be defined by considering objective criteria. Then, they constitute specific parameters to be estimated within a clustering procedure.
- The weights are suitably tuned; in fact, by considering the dissimilarity within an appropriate clustering procedure and then the weights can be computationally estimated by taking into account, as one would expect, the condition that the center component of the fuzzy data has more or equal importance than the spread component. In this way, we leave out the "anomalous" situation in which the spread component, which represents the uncertainty around the value (mode or center of the fuzzy number), has more importance than the center value (center component), that represents the core information of each fuzzy data. In this way, we take into account the intuitive assumption of the fuzzy theory: the membership function value of the centers is maximum.
- The proposed weighted dissimilarity measure is used for making comparisons within a set of data rather than looking at a single pair of data. More specifically, this means that, for a given data-set, the weighting system is optimal only for the data-set involved.

Concerning the features of the model, we can point out the following aspects (D'Urso and Giordani, 2006a):

- The model can be applied to *all* symmetric fuzzy data. In fact, it assumes that the *shape* of the membership functions of the observed symmetric fuzzy data is *inherited*. Thus, the shape is supposed before starting the clustering procedure. Notice that, obviously, the shape of the membership functions of the symmetric fuzzy centroids, estimated in the fuzzy clustering procedure, is inherited from the observed fuzzy data.
- In the model, the weights constitute specific parameters to be objectively estimated *within* the clustering model (*internal weighting estimation*).

- By means of the model, it can appropriately tune the influence of the two components of the fuzzy data in the partitioning process. Furthermore, the model inherits all the other peculiarities of the "weighted" dissimilarity measure introduced in the objective function of the clustering procedure.

In conclusion, to solve the constrained optimization (minimization) problems pertaining to all analyzed fuzzy clustering models, the Lagrangian multiplier method is utilized. Furthermore, the iterative algorithms of the same models do not guarantee that the global optimum is obtained. For this reason, it is useful to start each algorithm by considering several starting points in order to check the stability of the solutions. For the different clustering models, the theoretical features are summarized in Table 1.

(3) In order to compare the performance of the previous models for multi-dimensional fuzzy data, regarding to the ability of finding homogeneous groups of units, a simulation study has been done by D'Urso and Giordani (2006a). In particular, they compare the following models: *FKMCCFV* (Yang and Liu, 1999) (we remember that, as we consider non-interactive fuzzy data, the *PDFKMC* model (Hathaway, Bezdek, and Pedrycz, 1996) and the *FKMCCFV* model have the same performance (see Section 8.4.3)), *FKMCMD* (Yang, Hwang, and Chen, 2004) and *DFWKMC*. In brief, the *DFWKMC* model performed, on average, better than the other ones, during the entire simulation. For the details on the simulation study see D'Urso and Giordani (2006a).

8.5 AN EXTENSION: FUZZY CLUSTERING MODELS FOR FUZZY DATA TIME ARRAYS

In this section we analyze an interesting extension of the fuzzy data matrix: the *three-way arrays* (same units $\times$ same (fuzzy) variables $\times$ occasions) with particular reference to the situation in which the occasions are times (Coppi and D'Urso, 2000). In this case, special attention is devoted to cluster in a fuzzy manner the so-called *fuzzy multivariate time trajectories*. For this reason, in the following subsections, we define the *fuzzy data time array* (three-way arrays with time occasions) and then the fuzzy multivariate time trajectories. Successively, we explain some *dissimilarity measures* and the *fuzzy clustering* for these type of fuzzy objects.

8.5.1 Fuzzy Data Time Array: Mathematical Representation and Distance Measures

8.5.1.1 Algebraic and Geometric Representation

An *LR fuzzy data time array* (same units $\times$ same (fuzzy) variables $\times$ times) is defined as follows:

$$\mathbf{X} \equiv \{x_{ijt} = (c_{ijt}, l_{ijt}, r_{ijt})_{LR} : i = 1,I; j = 1,J; t = 1,T\}, \tag{8.39}$$

where i, j and t denote the units, variables and times, respectively; $x_{ijt} = (c_{ijt}, l_{ijt}, r_{ijt})_{LR}$ represents the LR fuzzy variable j observed on the ith unit at time t, where c_{ijt} denotes the center and l_{ijt} and r_{ijt} the left and right spread, respectively, with the following membership function:

$$\mu(\tilde{u}_{ijt}) = \begin{cases} L\left(\dfrac{c_{ijt} - \tilde{u}_{ijt}}{l_{ijt}}\right) \tilde{u}_{ijt} \leq c_{ijt} (l_{ijt} > 0) \\ \\ R\left(\dfrac{\tilde{u}_{ijt} - c_{ijt}}{r_{ijt}}\right) \tilde{u}_{ijt} \geq c_{ijt} (r_{ijt} > 0), \end{cases} \tag{8.40}$$

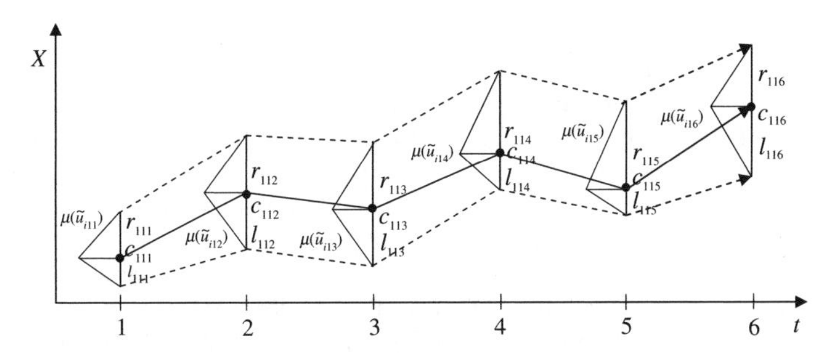

Figure 8.9 Example of triangular fuzzy (univariate) time trajectory in R^2 (for $t = 6$).

where L (and R) is a decreasing 'shape' function from $\Re^+$ to [0,1] with $L(0) = 1$; $L(z_{ijt}) < 1$ for all $z_{ijt} > 0, \forall i, j, t$; $L(z_{ijt}) > 0$ for all $z_{ijt} < 1 \forall i, j, t$; $L(1) = 0$ (or $L(z_{ijt}) > 0$ for all z_{ijt} and $L(+\infty) = 0$).

A particular case of LR fuzzy data time array is the triangular one (with triangular membership function). By combining the indices I, J and T, we can obtain from $\mathbf{X}$ the following stacked fuzzy matrices: $\mathbf{X} \equiv \{\mathbf{X}_i : i = 1, I\}$, $\mathbf{X} \equiv \{\mathbf{X}_t : t = 1, T\}$, $\mathbf{X} \equiv \{\mathbf{X}_j : j = 1, J\}$ with $\mathbf{X}_i \equiv \{x_{ijt} : j = 1, J; t = 1, T\}$, $\mathbf{X}_t \equiv \{x_{ijt} : i = 1, I; j = 1, J\}$, $\mathbf{X}_j \equiv \{x_{ijt} : i = 1, I; t = 1, T\}$. Let R^{J+1} be the vectorial space (space of units), where the axes are referred to the J variables and time. In this space, we represent each unit i by means of the following vectors, for each t: ${}_c\mathbf{y}_{it} = (c_{i1t}, \ldots, c_{ijt}, \ldots, c_{iJt}, t)'$, ${}_l\mathbf{y}_{it} = (l_{i1t}, \ldots, l_{ijt}, \ldots, l_{iJt}, t)'$, ${}_r\mathbf{y}_{it} = (r_{i1t}, \ldots, r_{ijt}, \ldots, r_{iJt}, t)'$.

(1) By fixing t, the scatters ${}_fN_I(t) \equiv \{({}_c\mathbf{y}_{it}, {}_l\mathbf{y}_{it}, {}_r\mathbf{y}_{it})\}_{i=1,I}$ represent the matrix $\mathbf{X}_t$. Letting t vary within its range, the scatters ${}_fN_I(t)$ are placed on T hyperplanes parallel to the sub-space R^J.
(2) By fixing i, the scatters ${}_fN_T(i) \equiv \{({}_c\mathbf{y}_{it}, {}_l\mathbf{y}_{it}, {}_r\mathbf{y}_{it})\}_{t=1,T}$ represent the matrix $\mathbf{X}_i$. Each scatter describes the LR fuzzy multivariate time trajectories of unit i across the time and $\{{}_fN_T(i) \equiv \{({}_c\mathbf{y}_{it}, {}_l\mathbf{y}_{it}, {}_r\mathbf{y}_{it})\}_{t=1,T}\}_{i=1,I}$ represent the set of the LR fuzzy multivariate time trajectories. Each LR fuzzy time trajectory ${}_fN_T(i)$ crosses the T hyperplanes parallel to R^J.

An example of geometrical representation of the (symmetrical) triangular version of the previous situations is shown in Figure 8.9 (Coppi and D'Urso, 2002).

Notice that problems of heterogeneity also involve the fuzzy data time array. To this purpose, we can suitably extend the data preprocessing procedures shown in Section 8.3.3. In addition, in a time framework, for stabilizing the variability of the time data, for instance, logarithmic transformations for the original data can also be taken into account.

8.5.1.2 Distance Measures Fuzzy Multivariate Time Trajectories

With reference to the fuzzy time array $\mathbf{X}$, we consider, for each type of $\mathbf{X}$, the following (squared) distances (Coppi and D'Urso, 2003):

$$ {}_1d^2_{ii't}(\lambda, \rho) = \| \mathbf{c}_{it} - \mathbf{c}_{i't} \|^2 + \| (\mathbf{c}_{it} - \lambda \mathbf{l}_{it}) - (\mathbf{c}_{i't} - \lambda \mathbf{l}_{i't}) \|^2 + \| (\mathbf{c}_{it} + \rho \mathbf{r}_{it}) - (\mathbf{c}_{i't} + \rho \mathbf{r}_{i't}) \|^2, \tag{8.41} $$

$$ \begin{aligned} {}_2d^2_{ii't}(\lambda, \rho) = & \| (\mathbf{c}_{it} - \mathbf{c}_{it-1}) - (\mathbf{c}_{i't} - \mathbf{c}_{i't-1}) \|^2 + \| [(\mathbf{c}_{it} - \lambda \mathbf{l}_{it}) - (\mathbf{c}_{it-1} - \lambda \mathbf{l}_{it-1})] - [(\mathbf{c}_{i't} - \lambda \mathbf{l}_{i't}) - \\ & (\mathbf{c}_{i't-1} - \lambda \mathbf{l}_{i't-1})] \|^2 + \| [(\mathbf{c}_{it} + \rho \mathbf{r}_{it}) - (\mathbf{c}_{it-1} + \rho \mathbf{r}_{it-1})] - [(\mathbf{c}_{i't} + \rho \mathbf{r}_{i't}) - (\mathbf{c}_{i't-1} + \rho \mathbf{r}_{i't-1})] \|^2 \\ = & \| {}_c\mathbf{v}_{it} - {}_c\mathbf{v}_{i't} \|^2 + \| ({}_c\mathbf{v}_{it} - \lambda_l \mathbf{v}_{it}) - ({}_c\mathbf{v}_{i't} - \lambda_l \mathbf{v}_{i't}) \|^2 + \| ({}_c\mathbf{v}_{it} + \rho_r \mathbf{v}_{it}) - ({}_c\mathbf{v}_{i't} + \rho_r \mathbf{v}_{i't}) \|^2, \end{aligned} \tag{8.42} $$

where $\lambda = \int_0^1 L^{-1}(\omega)\mathrm{d}\omega$, $\rho = \int_0^1 R^{-1}(\omega)\mathrm{d}\omega$, $\mathbf{c}_{it} = (c_{i1t}, \ldots, c_{ijt}, \ldots, c_{iJt})'$, $\mathbf{c}_{i't} = (c_{i'1t}, \ldots, c_{i'jt}, \ldots, c_{i'Jt})'$, $\mathbf{l}_{it} = (l_{i1t}, \ldots, l_{ijt}, \ldots, l_{iJt})'$, $\mathbf{l}_{i't} = (l_{i'1t}, \ldots, l_{i'jt}, \ldots, l_{i'Jt})'$, $\mathbf{r}_{it} = (r_{i1t}, \ldots, r_{ijt}, \ldots, r_{iJt})'$ and $\mathbf{r}_{i't} = (r_{i'1t}, \ldots, r_{i'jt}, \ldots, r_{i'Jt})'$; ${}_c\mathbf{v}_{it} = (\mathbf{c}_{it} - \mathbf{c}_{it-1})$, ${}_c\mathbf{v}_{i't} = (\mathbf{c}_{i't} - \mathbf{c}_{i't-1})$, ${}_l\mathbf{v}_{it} = (\mathbf{l}_{it} - \mathbf{l}_{it-1})$, ${}_l\mathbf{v}_{i't} = (\mathbf{l}_{i't} - \mathbf{l}_{i't-1})$, ${}_r\mathbf{v}_{it} = (\mathbf{r}_{it} - \mathbf{r}_{it-1})$ and ${}_r\mathbf{v}_{i't} = (\mathbf{r}_{i't} - \mathbf{r}_{i't-1})$ are, respectively, the vectors of the so-called *velocities* of the centers and left and right spreads pertaining to the fuzzy time trajectory of the ith and i'th units.

Notice that the concept of *velocity* can be defined in the following way. By considering the ith time trajectory of the centers, the velocity, in the time interval $[t-1, t]$, is ${}_c\mathbf{v}_{it} = (\mathbf{c}_{it} - \mathbf{c}_{it-1})/(t-(t-1)) = (\mathbf{c}_{it} - \mathbf{c}_{it-1})$. Then, for each variable j, ${}_cv_{ijt}$ can be greater (less) than zero according to whether the ith unit presents an increasing (decreasing) rate of change of its position in the time interval $[t-1, t]$; ${}_cv_{ijt} = 0$ if the unit does not change position from $t-1$ to t. For any "component" time trajectory ("center" time trajectory, "lower bound" time trajectory, "upper bound" time trajectory) the velocity pertaining to each pair of successive time points represents the slope of the straight line passing through them: if the velocity is negative (positive) the slope will be negative (positive) and the angle made by each segment of the trajectory with the positive direction of the t-axis will be obtuse (acute) (Coppi and D'Urso, 2003).

The squared Euclidean distance (8.41) compares the positions at time t of the centers and of the lower and upper bounds (center – left spread and center + right spread) between each pair of fuzzy time trajectories.

The squared Euclidean distance (8.42) compares the slopes (velocities) in each time interval $[t-1, t]$ of the segments of each "component" time trajectory concerning the ith unit with the corresponding slopes of the i'th unit. Notice that, the previous (squared) distances summarize the fuzziness embodied in each elementary observation of the fuzzy time array $\mathbf{X}$, through three parameters (center, left spread and right spread) and the shape of the corresponding membership functions (involving suitable values for the shape-parameters λ and ρ) (Coppi and D'Urso, 2003). On the basis of the above distances Coppi and D'Urso (2003) defined appropriate dissimilarity measures between fuzzy multivariate time trajectories, in the following way:

$$\sum_{t=1}^{T} ({}_1w_{t1}d_{ii't}(\lambda, \rho))^2 \quad \text{(instantaneous dissimilarity measure)}, \tag{8.43}$$

$$\sum_{t=2}^{T} ({}_2w_{t2}d_{ii't}(\lambda, \rho))^2 \quad \text{(velocity dissimilarity measure)}, \tag{8.44}$$

$$\sum_{s=1}^{2}\sum_{t} ({}_sw_{ts}d_{ii't}(\lambda, \rho))^2 \quad \text{(simultaneous dissimilarity measure)}, \tag{8.45}$$

where ${}_1w_t$, ${}_2w_t$, ${}_sw_t$ are suitable weights to be computed in each case (see Section 8.5.2).

In particular, the dissimilarity (8.43) takes into account the (squared) instantaneous distances (8.41), by considering the whole set of the T time occasions. Each occasion is weighted by means of ${}_1w_t$. This weight can be suitably determined in an objective way. The dissimilarity (8.44) considers, for all time intervals $[t-1, t]$, $t = 2,T$, the (squared) velocity distances (8.42). To each interval a weight ${}_2w_t$ is associated, whose value is computed in an objective manner. Finally, the dissimilarity measure (8.45) represents, in the observed time domain, a compromise between the (squared) instantaneous and velocity distances. The corresponding weighting system, ${}_sw_t$, which is determined within the appropriate clustering procedure, takes simultaneously into account the effects of the single time occasions and time intervals and of the two types of distances (instantaneous and velocity distances) (Coppi and D'Urso, 2003).

8.5.2 Dynamic Double Fuzzy K-means Clustering Models (D-DFKMC Models)

In order to classify a set of LR fuzzy time trajectories Coppi and D'Urso (2003) adopted a fuzzy approach. The adoption of a *fuzzy* clustering model for multivariate time trajectories is justified on the grounds of at least two considerations. First of all the "complexity" of the [fuzzy] trajectories (various observational times, several variables) suggests thinking in terms of "degrees" of membership to given clusters rather than in terms of total membership vs. non-membership. In fact, a crisp definition of clusters contrasts for example with the ambiguities presented in the following instances. (1) "Switching time trajectories" may occur, namely, trajectories showing a pattern typical of a given cluster during a certain time period and a completely different pattern (characteristic of another cluster) in another time period. (2) The time evolution of the variables defining an "object" (statistical unit) may follow a given pattern (belonging to a specific cluster) for a subgroup of variables, and a remarkably different pattern (another cluster) for a different subgroup of variables. Moreover, the following considerations support the adoption of the fuzzy approach. (1) Greater *adaptivity* in defining the "prototypes" (i.e., the "typical" multivariate trajectories). This can be better appreciated when the observed time patterns do not differ too much from each other. In this case, the fuzzy definition of the clusters allows us to single out underlying prototypes, if these are likely to exist in the given array of data. (2) Greater *sensitivity*, in capturing the details characterizing the time pattern of the individual units. In fact, the dynamics are often drifting or switching and the standard clustering approaches are likely to miss this underlying structure. On the contrary, the switches from one time state to another, which are usually vague and not focused on any particular time point, can be naturally treated by means of fuzzy clustering (Coppi, 2004).

By means of the so-called *cross-sectional double fuzzy K-means clustering* model (*CS-DFKMC* model), we classify LR fuzzy multivariate time trajectories taking into account their instantaneous (positional) features. In this case, the fuzzy clustering model can be formalized in the following way:

$$
\begin{aligned}
&\min: \sum_{i=1}^{I}\sum_{k=1}^{K} {}_1u_{ik}^m \sum_{t=1}^{T} \left({}_1w_t\, {}_1d_{ikt}(\lambda,\rho)\right)^2 \\
&= \sum_{i=1}^{I}\sum_{k=1}^{K} {}_1u_{ik}^m \sum_{t=1}^{T} \left({}_1w_t^2 (\| \mathbf{c}_{it} - \mathbf{c}_{kt} \|^2 + \| (\mathbf{c}_{it} - \lambda \mathbf{l}_{it}) - (\mathbf{c}_{kt} - \lambda \mathbf{l}_{kt}) \|^2 + \| (\mathbf{c}_{it} + \rho \mathbf{r}_{it}) - (\mathbf{c}_{kt} + \rho \mathbf{r}_{kt}) \|^2) \right) \\
&\sum_{k=1}^{K} {}_1u_{ik} = 1,\ {}_1u_{ik} \geq 0;\ \sum_{t=1}^{T} {}_1w_t = 1,\ {}_1w_t \geq 0 \quad \left(\lambda = \int_0^1 L^{-1}(\omega)\mathrm{d}\omega,\ \rho = \int_0^1 R^{-1}(\omega)\mathrm{d}\omega \right),
\end{aligned}
\tag{8.46}
$$

where ${}_1u_{ik}$ denotes the membership degree of the ith LR fuzzy multivariate time trajectory with regard to the kth cluster; ${}_1w_t$ is an instantaneous weight; $m > 1$ is a weighting exponent that controls the fuzziness of the obtained fuzzy partition (see Section 9.6); $\mathbf{c}_{kt}$, $\mathbf{l}_{kt}$ and $\mathbf{r}_{kt}$ denote, respectively, the vectors of the centers, left and right spreads of the LR fuzzy time trajectory of the kth centroid at time t.

By solving the previous constrained optimization problem, we obtain the following iterative solutions:

$$
\begin{aligned}
&{}_1u_{ik} = \frac{1}{\displaystyle\sum_{k'=1}^{K} \left[\frac{\sum_{t=1}^{T} ({}_1w_t\, {}_1d_{ikt}(\lambda,\rho))^2}{\sum_{t=1}^{T} ({}_1w_t\, {}_1d_{ik't}(\lambda,\rho))^2} \right]^{\frac{1}{m-1}}}, \quad {}_1w_t = \frac{1}{\displaystyle\sum_{t'=1}^{T} \left[\frac{\sum_{i=1}^{I}\sum_{k=1}^{K} {}_1u_{ik}^m\, {}_1d_{ikt}^2(\lambda,\rho)}{\sum_{i=1}^{I}\sum_{k=1}^{K} {}_1u_{ik}^m\, {}_1d_{ikt'}^2(\lambda,\rho)} \right]}, \\
&\mathbf{c}_{kt} = \frac{\sum_{i=1}^{I} {}_1u_{ik}^m [3\mathbf{c}_{it} - \lambda(\mathbf{l}_{it} - \mathbf{l}_{kt}) + \rho(\mathbf{r}_{it} - \mathbf{r}_{kt})]}{3\sum_{i=1}^{I} {}_1u_{ik}^m}, \quad \mathbf{l}_{kt} = \frac{\sum_{i=1}^{I} {}_1u_{ik}^m (\mathbf{c}_{kt} + \lambda \mathbf{l}_{it} - \mathbf{c}_{it})}{\lambda \sum_{i=1}^{I} {}_1u_{ik}^m}, \\
&\mathbf{r}_{kt} = \frac{\sum_{i=1}^{I} {}_1u_{ik}^m (\mathbf{c}_{it} + \rho \mathbf{r}_{it} - \mathbf{c}_{kt})}{\rho \sum_{i=1}^{I} {}_1u_{ik}^m}.
\end{aligned}
\tag{8.47}
$$

In a similar manner, by considering the velocity concept and the dissimilarity measure (8.44), Coppi and D'Urso (2003) formalize the so-called *longitudinal double fuzzy K-means clustering* model (*L-DFKMC* model) for classifying LR fuzzy multivariate time trajectories by taking into account their longitudinal (velocity) features.

The authors also defined the *mixed double fuzzy K-means clustering* model (*M-DFKMC* model) by considering, simultaneously, the instantaneous and longitudinal features of the LR fuzzy time trajectories, i.e., by utilizing the dissimilarity (8.45) in the clustering process (Coppi and D'Urso, 2003).

Notice that Coppi and D'Urso (2003) also formalize the *D-DFKMC* models for trapezoidal fuzzy data.

8.5.3 Other Dynamic Double Fuzzy Clustering Models

In the literature, other dynamic fuzzy clustering models for fuzzy time trajectories have been proposed by Coppi, D'Urso, and Giordani (2004, 2006a), i.e., the *dynamic double fuzzy clustering with entropy regularization* models (*D-DFCER* models) and the *dynamic double fuzzy K-medoids clustering* models (*D-DFKMDC* models).

In particular, in the *D-DFCER* models the authors minimize objective functions, which are the sum of two terms. The first term is a dynamic generalization of intra-cluster distance, in a fuzzy framework, that takes into account the instantaneous and/or longitudinal aspects of the LR fuzzy time-varying observations (the LR fuzzy multivariate time trajectories); in this way, they minimize the within-cluster dispersion (maximize the internal cohesion). The second term represents the Shannon entropy measure as applied to a fuzzy partition (entropy regularization); then, they maximize a given measure of entropy or, equivalently, minimize the converse of the entropy. Overall, they optimize the total functional depending on both the previous aspects (Coppi, 2004). The *D-DFKMDC* models classify LR fuzzy time trajectories and select, in the set of the observed LR fuzzy time trajectories, typical LR fuzzy time trajectories that synthetically represent the structural characteristics of the identified clusters (*medoid LR fuzzy time trajectories*). Then, contrary to the *D-DFKMC* models in which the typical LR fuzzy time trajectories are unobserved (*centroid LR fuzzy time trajectories*), in the *D-DFKMDC* models the typical LR fuzzy time trajectories belong to the set of the observed fuzzy trajectories (Coppi, D'Urso, and Giordani, 2006a).

8.6 APPLICATIVE EXAMPLES

As to the practical utilization of the fuzzy approach in the cluster analysis with imprecise data, several potential examples might be mentioned, ranging from engineering to social and psychological problems. In this section, in order to evaluate the empirical capabilities and the different performances of the fuzzy clustering models analyzed in Sections 8.4 and 8.5, several applicative examples are illustrated.

8.6.1 Univariate Case

For showing the applicative performances of the fuzzy clustering models for univariate fuzzy data (*ADFKNC* and *DFKNC* models), we utilize two data-sets.

8.6.1.1 Tea Data-set

We consider a data-set drawn by Hung and Yang (2005), regarding the evaluation of 70 kinds of Taiwanese tea. In particular, 10 experts evaluated each kind of tea by assigning and four criteria (attributes) – appearance, tincture, liquid color and aroma – five different quality levels: perfect, good, medium, poor, bad. These quality terms represent the imprecision and ambiguity inherent in human perception. Since the fuzzy sets can be suitably utilized for describing the ambiguity and imprecision in natural language, the authors defined these quality terms using triangular fuzzy numbers,

Table 8.2 Taiwanese tea data-set (Hung and Yang, 2005).

i	$\tilde{Y}=(c,l,r)$	i	$\tilde{Y}=(c,l,r)$
1	(0.8750, 0.2500, 0.1250)	36	(0.3750, 0.1875, 0.2500)
2	(0.5625, 0.2500, 0.1875)	37	(0.3750, 0.1875, 0.2500)
3	(0.5000, 0.2500, 0.2500)	38	(0.3750, 0.1875, 0.2500)
4	(0.5000, 0.2500, 0.2500)	39	(0.3750, 0.1875, 0.2500)
5	(0.5000, 0.2500, 0.2500)	40	(0.3125, 0.1250, 0.2500)
6	(0.4375, 0.1250, 0.1875)	41	(0.3125, 0.1250, 0.2500)
7	(0.4375, 0.1875, 0.2500)	42	(0.3125, 0.1250, 0.2500)
8	(0.4375, 0.1875, 0.2500)	43	(0.3125, 0.1250, 0.2500)
9	(0.4375, 0.1875, 0.2500)	44	(0.3125, 0.1250, 0.2500)
10	(0.4375, 0.1875, 0.2500)	45	(0.3125, 0.1250, 0.2500)
11	(0.4375, 0.1875, 0.2500)	46	(0.3125, 0.1250, 0.2500)
12	(0.4375, 0.1875, 0.2500)	47	(0.3125, 0.1875, 0.2500)
13	(0.4375, 0.1875, 0.2500)	48	(0.3125, 0.1250, 0.2500)
14	(0.4375, 0.1875, 0.2500)	49	(0.3125, 0.1250, 0.2500)
15	(0.4375, 0.1875, 0.2500)	50	(0.3125, 0.1250, 0.2500)
16	(0.4375, 0.1875, 0.2500)	51	(0.3125, 0.1250, 0.2500)
17	(0.4375, 0.1875, 0.2500)	52	(0.3125, 0.1250, 0.2500)
18	(0.4375, 0.1875, 0.2500)	53	(0.3125, 0.1250, 0.2500)
19	(0.4375, 0.1875, 0.2500)	54	(0.3125, 0.1250, 0.2500)
20	(0.4375, 0.1875, 0.2500)	55	(0.3125, 0.1250, 0.2500)
21	(0.3750, 0.1250, 0.2500)	56	(0.2500, 0.1250, 0.2500)
22	(0.3750, 0.1250, 0.2500)	57	(0.2500, 0.1250, 0.2500)
23	(0.3750, 0.1250, 0.1875)	58	(0.2500, 0.1250, 0.2500)
24	(0.3750, 0.1250, 0.2500)	59	(0.2500, 0.1250, 0.2500)
25	(0.3750, 0.1250, 0.2500)	60	(0.2500, 0.1250, 0.2500)
26	(0.3750, 0.1250, 0.2500)	61	(0.1875, 0.0625, 0.2500)
27	(0.3750, 0.1250, 0.2500)	62	(0.1875, 0.0625, 0.2500)
28	(0.3750, 0.1250, 0.2500)	63	(0.1250, 0.0625, 0.2500)
29	(0.3750, 0.1250, 0.2500)	64	(0.1250, 0.0625, 0.2500)
30	(0.3750, 0.1250, 0.2500)	65	(0.1250, 0.0625, 0.2500)
31	(0.3750, 0.1875, 0.2500)	66	(0.1250, 0.0625, 0.2500)
32	(0.3750, 0.1250, 0.2500)	67	(0.1250, 0.0625, 0.2500)
33	(0.3750, 0.1875, 0.2500)	68	(0.1250, 0.0625, 0.2500)
34	(0.3750, 0.1250, 0.2500)	69	(0.1250, 0.0625, 0.2500)
35	(0.3750, 0.1250, 0.2500)	70	(0.1250, 0.0625, 0.2500)

i.e.,: $\tilde{Y}=(1,0.25,0)$ (perfect), $\tilde{Y}=(0.75,0.25,0.25)$ (good), $\tilde{Y}=(0.5,0.25,0.25)$ (medium), $\tilde{Y}=(0.25,0.25,0.25)$(poor), $\tilde{Y}=(0,0,0.25)$ (bad) (Hung and Yang, 2005). Then, the authors obtained the univariate fuzzy data-set (shown in Table 8.2) by averaging, in a fuzzy manner, the fuzzy scores on the four attributes. Notice that the first tea (called White-tip Oolong) in Table 8.2 is the best and the most famous Taiwanese tea. In fact the expert (fuzzy) average evaluation is highest. For this reason, this type of tea can be considered a special tea and then an outlier in the tea data-set.

By applying the *ADFKNC* and *DFKNC* models (with m=2, K=5, λ=ρ=0.5, b=77.0130 (only for the *ADFKNC* model; it is obtained by using the formula suggested by Hung and Yang (2005)) we obtain the membership degrees shown in Table 8.3. In particular, the membership degrees obtained by utilizing the *DFKNC* model show that tea no.1 (special tea) belongs to cluster 1, the types of tea no. 2–20 belong more to cluster 2, the types of tea no. 21–39 belong more to cluster 3, the types of tea no. 40–60 belong more to cluster 4 and the types of tea no. 61–70 belong more to cluster 5. By applying the *ADFKNC* model, we have that tea no.1 (outlier) does not exactly belong to any cluster (in fact the membership degree is perfectly fuzzy), teas no. 2–20 belong more to cluster 1, teas no. 21–39 belong more to cluster 2, teas no. 40–55 belong more to cluster 3, teas no. 56–62 belong more to cluster 4 and teas no.

Table 8.3 Membership degrees.

i	*ADFKNC* model					*DFKNC* model				
1	0.2000	0.2000	0.2000	0.2000	0.2000	1.0000	0.0000	0.0000	0.0000	0.0000
2	0.3008	0.1936	0.1711	0.1674	0.1671	0.0609	0.5837	0.2129	0.1059	0.0365
3	0.5294	0.1620	0.1104	0.0998	0.0983	0.0134	0.7812	0.1374	0.0534	0.0147
4	0.5294	0.1620	0.1104	0.0998	0.0983	0.0134	0.7812	0.1374	0.0534	0.0147
5	0.5294	0.1620	0.1104	0.0998	0.0983	0.0134	0.7812	0.1374	0.0534	0.0147
6	0.7181	0.1552	0.0540	0.0379	0.0348	0.0035	0.7797	0.1720	0.0376	0.0072
7	0.9956	0.0025	0.0008	0.0006	0.0005	0.0004	0.9753	0.0195	0.0041	0.0008
8	0.9956	0.0025	0.0008	0.0006	0.0005	0.0004	0.9753	0.0195	0.0041	0.0008
9	0.9956	0.0025	0.0008	0.0006	0.0005	0.0004	0.9753	0.0195	0.0041	0.0008
10	0.9956	0.0025	0.0008	0.0006	0.0005	0.0004	0.9753	0.0195	0.0041	0.0008
11	0.9956	0.0025	0.0008	0.0006	0.0005	0.0004	0.9753	0.0195	0.0041	0.0008
12	0.9956	0.0025	0.0008	0.0006	0.0005	0.0004	0.9753	0.0195	0.0041	0.0008
13	0.9956	0.0025	0.0008	0.0006	0.0005	0.0004	0.9753	0.0195	0.0041	0.0008
14	0.9956	0.0025	0.0008	0.0006	0.0005	0.0004	0.9753	0.0195	0.0041	0.0008
15	0.9956	0.0025	0.0008	0.0006	0.0005	0.0004	0.9753	0.0195	0.0041	0.0008
16	0.9956	0.0025	0.0008	0.0006	0.0005	0.0004	0.9753	0.0195	0.0041	0.0008
17	0.9956	0.0025	0.0008	0.0006	0.0005	0.0004	0.9753	0.0195	0.0041	0.0008
18	0.9956	0.0025	0.0008	0.0006	0.0005	0.0004	0.9753	0.0195	0.0041	0.0008
19	0.9956	0.0025	0.0008	0.0006	0.0005	0.0004	0.9753	0.0195	0.0041	0.0008
20	0.9956	0.0025	0.0008	0.0006	0.0005	0.0004	0.9753	0.0195	0.0041	0.0008
21	0.0078	0.9816	0.0065	0.0024	0.0017	0.0001	0.0072	0.9867	0.0055	0.0005
22	0.0078	0.9816	0.0065	0.0024	0.0017	0.0001	0.0072	0.9867	0.0055	0.0005
23	0.0634	0.7895	0.0968	0.0301	0.0202	0.0013	0.0508	0.8679	0.0741	0.0059
24	0.0078	0.9816	0.0065	0.0024	0.0017	0.0001	0.0072	0.9867	0.0055	0.0005
25	0.0078	0.9816	0.0065	0.0024	0.0017	0.0001	0.0072	0.9867	0.0055	0.0005
26	0.0078	0.9816	0.0065	0.0024	0.0017	0.0001	0.0072	0.9867	0.0055	0.0005
27	0.0078	0.9816	0.0065	0.0024	0.0017	0.0001	0.0072	0.9867	0.0055	0.0005
28	0.0078	0.9816	0.0065	0.0024	0.0017	0.0001	0.0072	0.9867	0.0055	0.0005
29	0.0078	0.9816	0.0065	0.0024	0.0017	0.0001	0.0072	0.9867	0.0055	0.0005
30	0.0078	0.9816	0.0065	0.0024	0.0017	0.0001	0.0072	0.9867	0.0055	0.0005
31	0.0424	0.8695	0.0581	0.0179	0.0120	0.0007	0.0307	0.9257	0.0398	0.0031
32	0.0078	0.9816	0.0065	0.0024	0.0017	0.0001	0.0072	0.9867	0.0055	0.0005
33	0.0424	0.8695	0.0581	0.0179	0.0120	0.0007	0.0307	0.9257	0.0398	0.0031
34	0.0078	0.9816	0.0065	0.0024	0.0017	0.0001	0.0072	0.9867	0.0055	0.0005
35	0.0078	0.9816	0.0065	0.0024	0.0017	0.0001	0.0072	0.9867	0.0055	0.0005
36	0.0424	0.8695	0.0581	0.0179	0.0120	0.0007	0.0307	0.9257	0.0398	0.0031
37	0.0424	0.8695	0.0581	0.0179	0.0120	0.0007	0.0307	0.9257	0.0398	0.0031
38	0.0424	0.8695	0.0581	0.0179	0.0120	0.0007	0.0307	0.9257	0.0398	0.0031
39	0.0424	0.8695	0.0581	0.0179	0.0120	0.0007	0.0307	0.9257	0.0398	0.0031
40	0.0001	0.0002	0.9996	0.0001	0.0000	0.0002	0.0043	0.0186	0.9746	0.0023
41	0.0001	0.0002	0.9996	0.0001	0.0000	0.0002	0.0043	0.0186	0.9746	0.0023
42	0.0001	0.0002	0.9996	0.0001	0.0000	0.0002	0.0043	0.0186	0.9746	0.0023
43	0.0001	0.0002	0.9996	0.0001	0.0000	0.0002	0.0043	0.0186	0.9746	0.0023
44	0.0001	0.0002	0.9996	0.0001	0.0000	0.0002	0.0043	0.0186	0.9746	0.0023
45	0.0001	0.0002	0.9996	0.0001	0.0000	0.0002	0.0043	0.0186	0.9746	0.0023
46	0.0001	0.0002	0.9996	0.0001	0.0000	0.0002	0.0043	0.0186	0.9746	0.0023
47	0.0260	0.0578	0.8113	0.0840	0.0209	0.0007	0.0107	0.0382	0.9430	0.0074
48	0.0001	0.0002	0.9996	0.0001	0.0000	0.0002	0.0043	0.0186	0.9746	0.0023
49	0.0001	0.0002	0.9996	0.0001	0.0000	0.0002	0.0043	0.0186	0.9746	0.0023
50	0.0001	0.0002	0.9996	0.0001	0.0000	0.0002	0.0043	0.0186	0.9746	0.0023
51	0.0001	0.0002	0.9996	0.0001	0.0000	0.0002	0.0043	0.0186	0.9746	0.0023
52	0.0001	0.0002	0.9996	0.0001	0.0000	0.0002	0.0043	0.0186	0.9746	0.0023
53	0.0001	0.0002	0.9996	0.0001	0.0000	0.0002	0.0043	0.0186	0.9746	0.0023

Table 8.3 *(continued)*

i	*ADFKNC* model					*DFKNC* model				
54	0.0001	0.0002	0.9996	0.0001	0.0000	0.0002	0.0043	0.0186	0.9746	0.0023
55	0.0001	0.0002	0.9996	0.0001	0.0000	0.0002	0.0043	0.0186	0.9746	0.0023
56	0.0007	0.0009	0.0024	0.9951	0.0010	0.0056	0.0552	0.1283	0.6439	0.1670
57	0.0007	0.0009	0.0024	0.9951	0.0010	0.0056	0.0552	0.1283	0.6439	0.1670
58	0.0007	0.0009	0.0024	0.9951	0.0010	0.0056	0.0552	0.1283	0.6439	0.1670
59	0.0007	0.0009	0.0024	0.9951	0.0010	0.0056	0.0552	0.1283	0.6439	0.1670
60	0.0007	0.0009	0.0024	0.9951	0.0010	0.0056	0.0552	0.1283	0.6439	0.1670
61	0.0775	0.0843	0.1193	0.4162	0.3027	0.0051	0.0360	0.0672	0.1776	0.7141
62	0.0775	0.0843	0.1193	0.4162	0.3027	0.0051	0.0360	0.0672	0.1776	0.7141
63	0.0001	0.0001	0.0001	0.0001	0.9996	0.0002	0.0009	0.0014	0.0027	0.9949
64	0.0001	0.0001	0.0001	0.0001	0.9996	0.0002	0.0009	0.0014	0.0027	0.9949
65	0.0001	0.0001	0.0001	0.0001	0.9996	0.0002	0.0009	0.0014	0.0027	0.9949
66	0.0001	0.0001	0.0001	0.0001	0.9996	0.0002	0.0009	0.0014	0.0027	0.9949
67	0.0001	0.0001	0.0001	0.0001	0.9996	0.0002	0.0009	0.0014	0.0027	0.9949
68	0.0001	0.0001	0.0001	0.0001	0.9996	0.0002	0.0009	0.0014	0.0027	0.9949
69	0.0001	0.0001	0.0001	0.0001	0.9996	0.0002	0.0009	0.0014	0.0027	0.9949
70	0.0001	0.0001	0.0001	0.0001	0.9996	0.0002	0.0009	0.0014	0.0027	0.9949

63–70 belong more to cluster 5. Notice that these results reflect substantially the same results obtained by applying both *ADFKNC* and *DFKNC* models to the data-set without tea no.1. This shows that the *ADFKNC* model, conversely to the *DFKNC* model, is robust and able to tolerate the special tea (outlier). In fact, the *ADFKNC* model neutralizes and smoothes the disruptive effect of the outlier, preserving as almost invariant the natural clustering structure of the data-set.

8.6.1.2 Wine Data-set 1

We apply the *ADFKNC* and *DFKNC* models (with m=2, K=2, λ=ρ=0.5, b=0.4736 (only for the *ADFKNC* model)) to the "wine data-set" drawn by Coppi and D'Urso (2002). In particular, we take into account (triangular) fuzzy qualitative judgments on 15 wines (vintage 1965) (notice that in the original data-set we have different vintages 1964–1969; for the qualitative judgements see Hartigan (1975); for the fuzzy scores see Coppi and D'Urso, 2002). The analyzed data-set is shown in Table 8.4 and

Table 8.4 Wine data-set (Hartigan, 1975; Coppi and D'Urso, 2002).

Wines (vintage 1965)	Qualitative judgements	Fuzzy scores
1. Red Bordeaux Medoc and Graves	Worst	(3,3,1)
2. Red Bordeaux Saint Emilion and Pomerol	Poor	(4,1.5,1.5)
3. White Bordeaux Sauternes	Worst	(3,3,1)
4. White Bordeaux Graves	Worst	(3,3,1)
5. Red Burgundy	Poor	(4,1.5,1.5)
6. White Burgundy Cote de Beaune	Fair	(6,1,0.5)
7. White Burgundy Chablis	Poor	(4,1.5,1.5)
8. White Burgundy Beaujolais	Worst	(3,3,1)
9. Red Rhone North	Fair	(6,1,0.5)
10. Red Rhone South	Fair	(6,1,0.5)
11. White Loire	Poor	(4,1.5,1.5)
12. Alsace	Poor	(4,1.5,1.5)
13. Rhine	Poor	(4,1.5,1.5)
14 Moselle	Poor	(4,1.5,1.5)
15. California	Good	(8,1.75,0.25)

Table 8.5 Membership degrees and centroids.

i	ADFKNC model Membership degrees		DFKNC model Membership degrees	
1	0.3346	0.6654	0.0551	0.9449
2	0.0140	0.9860	0.0418	0.9582
3	0.3346	0.6654	0.0551	0.9449
4	0.3346	0.6654	0.0551	0.9449
5	0.0140	0.9860	0.0418	0.9582
6	0.9992	0.0008	0.9755	0.0245
7	0.0140	0.9860	0.0418	0.9582
8	0.3346	0.6654	0.0551	0.9449
9	0.9992	0.0008	0.9755	0.0245
10	0.9992	0.0008	0.9755	0.0245
11	0.0140	0.9860	0.0418	0.9582
12	0.0140	0.9860	0.0418	0.9582
13	0.0140	0.9860	0.0418	0.9582
14	0.0140	0.9860	0.0418	0.9582
15	0.5600	0.4400	0.8940	0.1060
	Centroids		Centroids	
	$c_1 = 6.0401, l_1 = 1.0145, r_1 = 0.4985$		$c_1 = 6.4180, l_1 = 1.1712, r_1 = 0.4507$	
	$c_2 = 3.8831, l_2 = 1.6731, r_2 = 1.4432$		$c_2 = 3.6481, l_2 = 2.0353, r_2 = 1.3200$	

Figure 8.10 (in which we also represent the centroids obtained with both models). Notice that the centroids computed by means of the *DFKNC* model are further away from each other than the centroids estimated by utilizing the *ADFKNC* model (see Table 8.5 and Figure 8.10).

8.6.2 Multivariate Case

In order to show the performances of the fuzzy clustering models for multivariate fuzzy data (i.e., the *DFWKMC* model, *FKMCCFV* model, and *FKMCMD* model) we consider two applications.

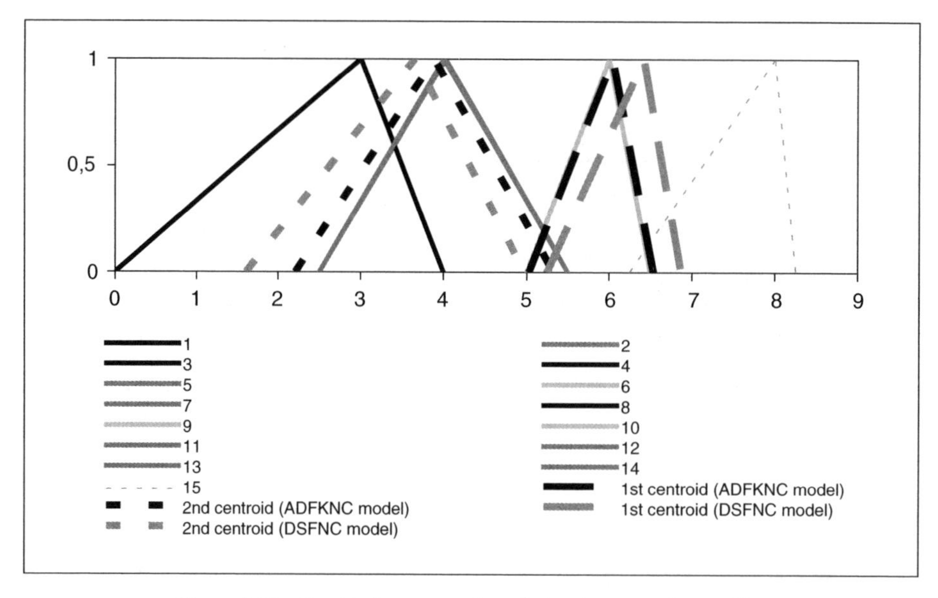

Figure 8.10 Graphical representation of wine data-set and centroids.

Table 8.6 Membership degrees (D'Urso and Giordani, 2006a).

Pat.	Memb.D.		Pat.	Memb.D.		Pat.	Memb.D.		Pat.	Memb.D.		Pat.	Memb.D.		Pat.	Memb.D.	
1	0.97	0.03	*19*	0.90	0.10	*37*	0.85	0.15	*55*	0.50	0.50	*73*	0.13	0.87	*91*	0.08	0.92
2	0.97	0.03	*20*	0.90	0.10	*38*	0.85	0.15	*56*	0.50	0.50	*74*	0.13	0.87	*92*	0.07	0.93
3	0.96	0.04	*21*	0.89	0.11	*39*	0.85	0.15	*57*	0.43	0.57	*75*	0.13	0.87	*93*	0.07	0.93
4	0.94	0.06	*22*	0.89	0.11	*40*	0.82	0.18	*58*	0.41	0.59	*76*	0.12	0.88	*94*	0.07	0.93
5	0.94	0.06	*23*	0.89	0.11	*41*	0.82	0.18	*59*	0.40	0.60	*77*	0.12	0.88	*95*	0.07	0.93
6	0.93	0.07	*24*	0.88	0.12	*42*	0.81	0.19	*60*	0.38	0.62	*78*	0.12	0.88	*96*	0.07	0.94
7	0.93	0.07	*25*	0.88	0.12	*43*	0.80	0.20	*61*	0.38	0.62	*79*	0.12	0.88	*97*	0.07	0.94
8	0.93	0.07	*26*	0.88	0.12	*44*	0.77	0.23	*62*	0.35	0.65	*80*	0.12	0.88	*98*	0.07	0.94
9	0.93	0.07	*27*	0.87	0.13	*45*	0.76	0.24	*63*	0.34	0.66	*81*	0.11	0.89	*99*	0.06	0.94
10	0.92	0.08	*28*	0.87	0.13	*46*	0.76	0.24	*64*	0.33	0.67	*82*	0.11	0.89	*100*	0.06	0.94
11	0.92	0.08	*29*	0.87	0.13	*47*	0.75	0.25	*65*	0.32	0.68	*83*	0.10	0.90	*101*	0.06	0.95
12	0.92	0.08	*30*	0.87	0.13	*48*	0.75	0.25	*66*	0.28	0.72	*84*	0.10	0.90	*102*	0.06	0.95
13	0.91	0.09	*31*	0.86	0.14	*49*	0.73	0.27	*67*	0.26	0.74	*85*	0.09	0.91	*103*	0.06	0.95
14	0.91	0.09	*32*	0.86	0.14	*50*	0.73	0.27	*68*	0.24	0.76	*86*	0.09	0.91	*104*	0.05	0.97
15	0.91	0.09	*33*	0.86	0.14	*51*	0.70	0.30	*69*	0.20	0.80	*87*	0.09	0.91	*105*	0.05	0.95
16	0.91	0.09	*34*	0.86	0.14	*52*	0.69	0.31	*70*	0.16	0.84	*88*	0.08	0.92	*106*	0.05	0.95
17	0.91	0.09	*35*	0.86	0.14	*53*	0.67	0.33	*71*	0.13	0.87	*89*	0.08	0.92	*107*	0.04	0.96
18	0.90	0.10	*36*	0.86	0.14	*54*	0.61	0.39	*72*	0.13	0.87	*90*	0.08	0.92	*108*	0.04	0.97

8.6.2.1 Blood Pressure Data

In this section, the *DFWKMC*, *FKMCCFV*, and *FKMCMD* models are applied to a "blood pressure data-set" drawn by D'Urso and Giordani (2006a) describing 108 patients in whom the daily systolic and diastolic blood pressures were observed (J=2). Notice that instead of considering numerical values (e.g., the average or median value of each variable during each day), the authors preferred to consider the minimum and maximum values daily registered so that the information to be managed is much more detailed (for the complete data-set, see D'Urso and Giordani, 2006a). The selected fuzziness coefficient and number of clusters are, respectively, $m = 2$ and $K = 2$. By considering the *DFWKMC* model, the optimal weights are $(1 - v) = 0.62$ and $v = 0.38$. Thus, the optimal weight of the center distance is considerably higher than the one of the spreads. Therefore, the model automatically determined a system of weights such that the role of the center distance was emphasized in order to make it more important than the spread one in clustering the fuzzy data. Since v is lower than 0.5, it follows that the differences in the spread scores are sensibly higher than those in the center scores. Nevertheless, the model chooses an objective system of weights such that the center scores, in which the membership function takes the highest value, play a more relevant role in assigning the patients to the clusters. The spread information plays the role of optimally tuning the memberships of the objects to the resulting clusters (D'Urso and Giordani, 2006a). The fuzzy partition obtained is explained in Table 8.6.

By applying the *FKMCCFV* and *FKMCMD* models (assuming a triangular representation of the fuzzy data) we obtain, respectively, a partition more and less fuzzy than the partition shown in Table 8.6.

8.6.2.2 Wine Data-set 2

We apply the *DFWKMC*, *FKMCCFV*, and *FKMCMD* models (fixing m=2, K=2) on a real data-set drawn by D'Urso and Giordani (2003) concerning 12 Greek white wines and a set of two (triangular)

Table 8.7 Input and membership degrees.

Greek white wine	Gallic Acid	Catechin	ADFKNC model Membership degrees		Fuzzy K-means (crisp data) Membership degrees	
1	(1.28, 0.15)	(3.02, 0.16)	0.1117	0.8883	0.1108	0.8892
2	(2.15, 0.30)	(0.96, 0.03)	0.0841	0.9159	0.0846	0.9154
3	(1.25, 0.19)	(-0.49, 0)	0.6215	0.3785	0.6244	0.3756
4	(0.00, 0.04)	(-0.52, 0)	0.9687	0.0313	0.9688	0.0312
5	(0.52, 0.10)	(0.05, 0.02)	0.7683	0.2317	0.7692	0.2308
6	(-0.77, 0)	(-0.52, 0)	0.9913	0.0087	0.9913	0.0087
7	(-0.71, 0.03)	(-0.52, 0)	0.9940	0.0060	0.9940	0.0060
8	(-0.77, 0)	(-0.17, 0.06)	0.9856	0.0144	0.9858	0.0142
9	(-0.77, 0)	(-0.34, 0.02)	0.9916	0.0084	0.9916	0.0084
10	(-0.73, 0.01)	(-0.41, 0.01)	0.9942	0.0058	0.9941	0.0059
11	(-0.77, 0)	(-0.52, 0)	0.9913	0.0087	0.9913	0.0087
12	(-0.67, 0.01)	(-0.52, 0)	0.9954	0.0046	0.9954	0.0046

fuzzy variables regarding the phenol concentrations, i.e., gallic acid and catechin (the preprocessed data are shown in Table 8.7). Notice that the observed spreads are very small (very close to zero), i.e., the data are approximately crisp. Then, the aim of this example is to show how the three fuzzy models work with (approximately) crisp data. In particular, in our application, these models provide, as one would expect, the same results as those of the traditional fuzzy K-means clustering model (for crisp data) (Bezdek, 1981). This shows the unbiased sensitivity of the models to the examined observational situations. In fact, all models capture the real structure of the data properly. The fuzzy partitions obtained by applying the *DFWKMC* model (*FKMCCFV* and *FKMCMD* models give the same results) and the traditional fuzzy K-means clustering model are pointed out in Table 8.7. For the *DFWKMC* model, the centroids are $\mathbf{c}_1 = (-0.4857\ -0.4034)$, $\mathbf{l}_1 = (0.0253\ 0.0115)$, $\mathbf{c}_2 = (1.6536\ 1.7083)$, $\mathbf{l}_2 = (0.2205\ 0.0835)$ (as one would expect, *FKMCCFV* and *FKMCMD* models give the same results and the traditional fuzzy K-means clustering model provides the same values of $\mathbf{c}_1$ and $\mathbf{c}_2$) and the weight is $v = 0.5$.

8.6.3 Three-way Case

This section is devoted to illustrating an applicative example of a fuzzy clustering model for fuzzy three-way data, i.e., fuzzy data time array.

8.6.3.1 Internet Banner Data

We show the results of an application of the *CS-DFKMC* model drawn by Coppi and D'Urso (2003). The available data (*Internet banner data*) refer to the subjective judgements of a sample of 20 Internet navigators concerning the advertising realized by means of different kinds of banners during the time. In fact, advertising on the Internet is usually done by means of "static" banners (which synthesize, in a single image, text and graphic), "dynamic" banners (characterized by a dynamic gif image, i.e., by a set of images visualized in sequence) and "interactive" banners (which induce the Internet-navigators to participate in polls, interactive games, and so on). The application is done by considering, for each time and each type of banner of each Web site, the median of the judgements expressed by the sample of navigators (Coppi and D'Urso, 2003). Thus, the fuzzy data time array has order $I = 18$ (Web sites) $\times$ $J = 3$ (types of banners) $\times$ $T = 6$ (consecutive periods, every fortnight). To take into account the subjective or linguistic vagueness expressed by the human perception a fuzzy coding has been considered. In particular

Table 8.8 Output of the CS-DFKMC model (cf. Coppi and D'Urso, 2003).

Fuzzy partition			
1. IOL.IT	0.02	0.94	0.04
2. KATAWEB.IT	0.04	0.06	0.90
3. TISCALINET.IT	0.08	0.16	0.76
4. TIN.IT	0.42	0.50	0.08
5. MSN.IT	0.17	0.73	0.10
6. VIRGILIO.IT	0.30	0.45	0.25
7. YAHOO.IT	0.87	0.09	0.04
8. ALTAVISTA.IT	0.36	0.55	0.09
9. EXCITE.IT	0.81	0.14	0.05
10. KATAMAIL.COM	0.45	0.16	0.39
11. ALTAVISTA.COM	0.07	0.07	0.86
12. INWIND.IT	0.04	0.07	0.89
13. SMSCASH.IT	0.06	0.16	0.78
14. IBAZAR.IT	0.89	0.04	0.07
15. REPUBBLICA.IT	0.89	0.05	0.06
16. MEDIASETONLINE.IT	0.05	0.77	0.18
17. YAHOO.COM	0.05	0.77	0.18
18. JUMPY.IT	0.15	0.54	0.31

Time weighting system
(0.158, 0.146, 0.179, 0.135, 0.173, 0.209).
Triangular case: $\lambda=\rho=1/2$
Cluster number: K=3
Fuzziness coefficient: m=1.35

the linguistic terms and their corresponding triangular fuzzy numbers are: *worst* (3, 3, 1); *poor* (4, 1.5, 1.5); *fair* (6, 1, 0.5); *good* (8, 1.75, 0.25); *best* (10, 2, 0) (Coppi and D'Urso, 2003). The output of the *CS-DFKMC* model is shown in Table 8.8, in which the time weighting system and the fuzzy partitions are reported. Notice that the authors utilize for the fuzzy data, a triangular membership function (then $\lambda = \rho = 0.5$); furthermore, in order to determine the number of clusters and the fuzziness coefficient they have suitably extended a cluster-validity criterion for a crisp data-set (i.e., the criterion suggested by Xie and Beni (1991) (see Coppi and D'Urso, 2003).

8.7 CONCLUDING REMARKS AND FUTURE PERSPECTIVES

In this chapter we focused our attention on fuzzy clustering models for fuzzy data. The main advantage of these models is the explicit consideration in the partitioning process of the imprecision (fuzziness, vagueness) of data. In particular, in this chapter, we defined the fuzzy data conceptually and mathematically and pointed at some preprocessing procedures and metrics for fuzzy data. Following this, starting with the so-called informational approach, we provided a survey of fuzzy clustering models for fuzzy univariate and multivariate data and for fuzzy three-way data (i.e., fuzzy data time array).

Summing up, the main features of the considered clustering approach are: the explicit recognition of the common informational nature of the ingredients of the data analytical procedure and of the uncertainty associated with them, here cast in the fuzzy perspective (the data and the clustering model); the adoption of a suitable class of membership functions representing the fuzziness of the observed data; the construction of appropriate metrics between fuzzy data, taking into account the vagueness; the extensive use of a generalized Bezdek criterion as the basis for the clustering process; the possibility of applying the clustering models in various observational settings, including the case where qualitative data are collected

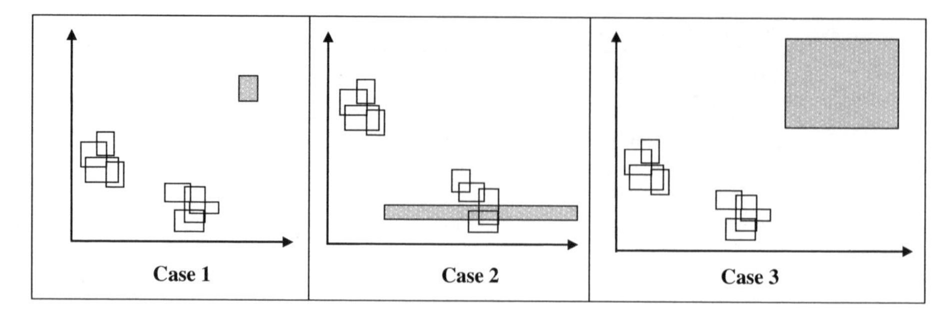

Figure 8.11 Examples of different cases of outlier fuzzy data.

– such as subjective judgments, ordinal categories and mixed data (obviously, this requires an adequate fuzzification of the qualitative data).

Concerning the fuzzy clustering models for fuzzy univariate, multivariate and time three-way data theoretical and mathematical aspects have been shown. Furthermore, in order to evaluate the empirical capabilities of the fuzzy clustering models for fuzzy univariate, multivariate, and three-way data, several applicative examples have been illustrated. The theoretical features, the comparative assessment and the applicative performances of the models showed the high flexibility of the fuzzy approach in the analysis of data with complex structure and then in the cognitive process with imprecise information.

In conclusion, we point out some future perspectives of research on fuzzy clustering for fuzzy data. For determining the "optimal" number of clusters (K) and the fuzziness coefficient (m), it would be useful to extend suitably the cluster-validity criteria for fuzzy clustering models of crisp data to a complete fuzzy framework; to this purpose, notice that the indices (utilized for fuzzy clustering for crisp data) belong to a class of *fuzziness-based cluster validity indices* (e.g., *partition coefficient* and *partition entropy* (Bezdek (1974a,b)) and can easily be utilized for the fuzzy clustering of fuzzy data.

It would also be interesting to study in depth the effects of outlier fuzzy data in the fuzzy clustering process. To this purpose, it is fundamental to define suitably, in a multivariate framework, the concept of anomalous (i.e., outlier) fuzzy data. In fact, the complex structure that characterizes the fuzzy data implies three possible cases (cf. Figure 8.11): outlier center–inlier spreads (case 1); inlier centers–outlier spreads (case 2); outlier center–outlier spreads (case 3).

Thus, in future research it would be interesting to propose robust fuzzy clustering models for multi-dimensional fuzzy data in which the three previous situations are suitably taken into account in order to neutralize and smooth the disruptive effects of possible fuzzy data with anomalous centers and/or spreads present in the fuzzy data matrix. In connection with this, following a typical interval approach (without fuzzy assumptions on the data), we remark that D'Urso and Giordani (2006b) suggested a robust fuzzy clustering model for interval valued data. This model represents an extension of Davè's model (1991) which uses a criterion similar to Ohashi's (1984) for interval valued data. Following the ideas of Ohashi and Davè, by means of the suggested robust fuzzy K-means clustering model for interval valued data, they introduce a special cluster, the noise cluster, whose role is to localize the noise and place it in a single auxiliary class. By assigning patterns to the noise class, they declare them to be outliers in the interval valued data-set (D'Urso and Giordani, 2006b).

An interesting research topic in fuzzy clustering for fuzzy data is also the fuzzy clustering with fuzzy linear (or nonlinear) prototypes. Some contributions are already present in literature (see, for example, D'Urso and Santoro, 2006). In particular, D'Urso and Santoro propose a fuzzy clusterwise linear regression model with fuzzy output, in which fuzzy regression and fuzzy clustering are integrated in a unique framework. In this way, the regression parameters (characterizing the linear prototypes of the clusters) and membership degrees are estimated simultaneously by optimizing one single objective function.

Other interesting future perspectives are: to study in depth the fuzzy clustering for interactive fuzzy data (Fullér and Majlender, 2004); to suggest fuzzy clustering models for fuzzy empirical information for classifying spatial units, by taking into account explicitly the autocorrelation among the spatial units; and to propose possibilistic clustering models for fuzzy data.

REFERENCES

Abonyi, J., Roubos, J.A., Szeifert, F. (2003) 'Data-driven generation of compact, accurate, and linguistically sound fuzzy classifiers based on a decision-tree initialization', *International Journal of Approximate Reasoning*, **32**, 1–21.

Alvo, M., Théberge, F. (2005) 'The problem of classification when the data are non-precise', *Austrian Journal of Statistics*, **34** (4), 373–388.

Auephanwiriyakul, S., Keller, J.M. (2002) 'Analysis and efficient implementation of a linguistic fuzzy c-means', *IEEE Transactions on Fuzzy Systems*, **10** (5), 563–582.

Bandemer, H. (2006) *Mathematics of Uncertainty. Ideas, Methods, Application Problems*, Springer-Verlag, Berlin.

Bellman, R., Kalaba, R., Zadeh, L.A. (1966) 'Abstraction and pattern recognition', *Journal of Mathematical Analysis and Application*, **2**, 581–586.

Bertoluzza, C., Corral, N., Salas, A. (1995)' On a new class of distance between fuzzy numbers', *Mathware and Soft Computing*, **2**, 71–84.

Bezdek, J.C. (1974a) 'Numerical taxonomy with fuzzy sets', *Journal of Mathematical Biology*, **1**, 57–71.

Bezdek, J.C. (1974b) 'Cluster validity with fuzzy sets', *Journal of Cybernetics*, **9**, 58–72.

Bezdek, J.C. (1981) *Pattern Recognition with Fuzzy Objective Function Algorithms*, Plenum Press, New York.

Bezdek, J.C. (1993) 'Fuzzy models – What are they, and why?', *IEEE Transactions on Fuzzy Systems*, **1**, 1–6.

Bloch, I. (1999) 'On fuzzy distances and their use in image processing under imprecision', *Pattern Recognition*, **32**, 1873–1895.

Celmins, A. (1987a) 'Least squares model fitting to fuzzy vector data', *Fuzzy Sets and Systems*, **22**, 245–269.

Celmins, A. (1987b) 'A. Multidimensional least-squares fitting of fuzzy models', *Mathematical Modelling*, **9**, 669–690.

Celmins, A. (1991) 'A practical approach to nonlinear fuzzy regression', *SIAM Journal of Scientific and Statistical Computing*, **12** (3), 521–546.

Chang, P.T., Lee, E.S., Konz, S.A. (1996) 'Applying fuzzy linear regression to VDT legibility', *Fuzzy Sets and Systems*, **80**, 197–204.

Chang, P.T., Huang, L.C., Lin, H.J. (2000) 'The fuzzy Delphi method via fuzzy statistics and membership function fitting and an application to the human resources', *Fuzzy Sets and Systems*, **112**, 511–520.

Chen, S.M. (1997) 'A new method for tool steel selection under fuzzy environment', *Fuzzy Sets and Systems*, **92**, 265–274.

Chen, S.M., Yeh, M.S., Hsio, P.Y. (1995) 'A comparison of similarity measures of fuzzy values', *Fuzzy Sets and Systems*, **72**, 79–89.

Cheng, C.H., Lin, Y. (2002) 'Evaluating the best main battle tank using fuzzy decision theory with linguistic criteria evaluation', *European Journal of Operational Research*, **142**, 174–186.

Chung, L.L.H., Chang, K.C.C. (2003) 'Evolutionary discovery of fuzzy concepts in data', *Brain and Mind*, **4**, 253–268.

Colubi, A., Gonzales Rodriguez, G., Montenegro, M., D'Urso, P. (2006) 'Two-samples test-based clustering for fuzzy random variables', *International Conference on Information Processing and Management of Uncertainty in Knowledge-Based Systems (IPMU 2006)*, 965–969, July 2–7, 2006, Paris.

Coppi, R. (2003) 'The fuzzy approach to multivariate statistical analysis', *Technical report, Dipartimento di Statistica, Probabilità e Statistiche Applicate, 11.*

Coppi, R., D'Urso, P. (2000) 'Fuzzy time arrays and dissimilarity measures for fuzzy time trajectories', in *Data Analysis, Classification and Related Methods* (eds. H.A.L. Kiers, J.P. Rasson, P.J.F. Groenen, M. Schader), Springer-Verlag, Heidelberg, pp. 273–278.

Coppi, R., D'Urso, P. (2002)' Fuzzy K-means clustering models for triangular fuzzy time trajectories', *Statistical Methods and Applications*, **11** (1), 21–40.

Coppi, R., D'Urso, P. (2003) 'Three-way fuzzy clustering models for LR fuzzy time trajectories', *Computational Statistics & Data Analysis'*, **43**, 149–177.

Coppi, R., D'Urso, P., Giordani, P. (2004) 'Informational Paradigm and Entropy-Based Dynamic Clustering in a Complete Fuzzy Framework', in *Soft Methodology in Random Information Systems* (eds. Angeles Gil, M., Lopez-Diaz, M.C., Grzegorzewski, P.) 463–470, Springer-Verlag Heidelberg.

Coppi, R., D'Urso, P., Giordani, P. (2006a) 'Fuzzy K-medoids clustering models for fuzzy multivariate time trajectories', COMPSTAT 2006, 28 August - 1 September, Rome, *Proceedings in Computational Statistics* (eds. A. Rizzi, M. Vichi), Physica-Verlag, 689–696.

Coppi, R., Gil, M.A., Kiers, H.A.L. (eds.) (2006b) 'The fuzzy approach to statistical analysis', *Computational Statistics & Data Analysis*, special issue, **51** (5).

Coppi, R., D'Urso, P., Giordani, P. (2007) 'Component models for fuzzy data', *Psychometrika*, in press.

de Carvalho, F.A.T. (2007) 'Fuzzy c-means clustering methods for symbolic interval data', *Pattern Recognition Letters*, **28** (4), 423–437.

Davè, R., (1991) 'Characterization and detection of noise in clustering', *Pattern Recognition Letters*, **12**, 657–664.

De Luca, A., Termini, S. (1972) 'A definition of non-probabilistic entropy in the setting of fuzzy set theory', *Information and Control*, **20**, 301–312.

Diamond, P., Kloeden, P. (1994) *Metrics Spaces of Fuzzy Sets. Theory and Applications*, World Scientific.

Di Lascio, L. et al. (2002) 'A fuzzy-based methodology for the analysis of diabetic neuropathy', *Fuzzy Sets and Systems*, **129**, 203–228.

Dubois, D., Prade, H. (1983) 'On distance between fuzzy points and their use for plausible reasoning', *International Conference on Systems, Man and Cybernetics*, 300–303.

Dunn, J.C. (1974) 'A fuzzy relative of the ISODATA process and its use in detecting compact well-separated clusters', *Journal of Cybernetics*, **3**, 32–57.

D'Urso, P., Gastaldi, T. (2002) 'An 'orderwise' polynomial regression procedure for fuzzy data', *Fuzzy Sets and Systems*, **130** (1), 1–19.

D'Urso, P., Giordani, P. (2003) 'Fitting of fuzzy linear regression models with multivariate response', *International Mathematical Journal*, **3** (6), 655–664.

D'Urso, P., Giordani, P. (2006a) 'A weighted fuzzy c-means clustering model for fuzzy data', *Computational Statistics & Data Analysis*, **50** (6), 1496–1523.

D'Urso, P., Giordani, P. (2006b) 'A robust fuzzy k-means clustering model for interval valued data', *Computational Statistics*, **21**, 251–269.

D'Urso, P., Santoro, A. (2006) 'Fuzzy clusterwise linear regression analysis with symmetrical fuzzy output variable', *Computational Statistics & Data Analysis*, **51** (5), 287–313.

El-Sonbaty, Y., Ismail, M.A. (1998) 'Fuzzy Clustering for Symbolic Data', *IEEE Transactions on Fuzzy Systems*, **6** (2), 195–204.

Finol J., Guo, Y.K., Jing, X.D. (2001) 'A rule based fuzzy model for the prediction of petrophysical rock parameters', *Journal of Petroleum Science & Engineering*, **29**, 97–113.

Fullér, R., Majlender, P. (2004) 'On interactive fuzzy numbers', *Fuzzy Sets and Systems*, **143**, 355–369.

Goetshel, R., Voxman, W. (1983) 'Topological properties of fuzzy numbers', *Fuzzy Sets and Systems*, **10**, 87–99.

Grzegorzewski, P. (2004) 'Distance between intuitionistic fuzzy sets and/or interval-valued fuzzy sets based on the Hausdorff metric', *Fuzzy Sets and Systems*, **148**, 319–328.

Hamdi, M., Lachiver, G., Michaud, F. (1999) 'A new predictive thermal sensation index of human response', *Energy and Buildings*, **29**, 167–178.

Hartigan, J.A. (1975) *Clustering Algorithms*, John Wiley & Sons, Inc., New York, USA.

Hathaway, R.J., Bezdek, J.C., Pedrycz, W. (1996) 'A parametric model for fusing heterogeneous fuzzy data', *IEEE Transactions on Fuzzy Systems*, **4** (3), 1277–1282.

Hougaard, J.L. (1999) 'Fuzzy scores of technical efficiency', *European Journal of Operational Research*, **115**, 529–541.

Huang, D., Chen, T., Wang, M.J.J. (2001) 'A fuzzy set approach for event tree analysis', *Fuzzy Sets and Systems*, **118**, 153–165.

Hung, W.-L., Yang, M-S. (2004) 'Similarity measures of intuitionistic fuzzy sets based on the Hausdorff metric', *Pattern Recognition Letters*, 1603–1611.

Hung, W.-L., Yang, M-S. (2005) 'Fuzzy clustering on LR-type fuzzy numbers with an application in Taiwanese tea evaluation', *Fuzzy Sets and Systems*, **150**, 561–577.

Kallithraka, S., Arvanitoyannis, I.S., Kefalas, P., El-Zajouli, A., Soufleros, E., Psarra, E. (2001) 'Instrumental and sensory analysis of Greek wines; implementation of principal component analysis (PCA) for classification according to geographical origin', *Food Chemistry*, **73**, 501–514.

Kaufman, A. (1973) *Introduction to the Theory of Fuzzy Subsets: Fundamental Theoretical Elements*, Academic Press, New York.

Kim, D.S., Kim, Y.K. (2004) 'Some properties of a new metric on the space of fuzzy numbers', *Fuzzy Sets and Systems*, **145**, 395–410.

Laviolette, M., Seaman, J.W., Barrett, J.D., Woodall, W.H. (1995) 'A probabilistic and statistical view of fuzzy methods (with discussion)', *Technometrics*, **37**, 249–292.

Lee, H.M. (1996) 'Applying fuzzy set theory to evaluate the rate of aggregative risk in software development', *Fuzzy Sets and Systems*, **79**, 323–336.

Li, Y., Olson, D.L., Qin, Z. (2007) 'Similarity measures between intuitionistic fuzzy (vague) sets: a comparative analysis', *Pattern Recognition Letters*, **28** (2), 278–285.

Lowen, R. Peeters, W. (1998) 'Distances between fuzzy sets representing grey level images', *Fuzzy Sets and Systems*, **99**, 135–150.

MacQueen, J.B. (1967) 'Some methods for classification and analysis of multivariate observations', *Proceedings of the Fifth Berkeley Symposium on Mathematical Statistics and Probability*, **2**, 281–297.

Moon, J.H., Kang, C.S. (1999) 'Use of fuzzy set theory in the aggregation of expert judgments', *Annals of Nuclear Energy*, **26**, 461–469.

Näther, W. (2000) 'On random fuzzy variables of second order and their application to linear statistical inference with fuzzy data', *Metrika*, **51**, 201–221.

Ohashi, Y. (1984) 'Fuzzy clustering and robust estimation', in *9th Meeting SAS Users Group Int.*, Hollywood Beach.

Pappis, C.P., Karacapilidis, N.I. (1993) 'A comparative assessment of measures of similarity of fuzzy values', *Fuzzy Sets and Systems*, **56**, 171–174.

Pedrycz, W. (2005) *Knowledge-based Clustering*, John Wiley & Sons, Inc. Hoboken, NJ, USA.

Pedrycz, W., Bezdek, J.C., Hathaway, R.J., Rogers, G.W. (1998) 'Two nonparametric models for fusing heterogeneous fuzzy data', *IEEE Transactions on Fuzzy Systems*, **6** (3), 411–425.

Raj, P. A., Kumar, D. N. (1999) 'Ranking alternatives with fuzzy weights using maximizing set and minimizing set', *Fuzzy Sets and Systems*, **105**, 365–375.

Ramot, D., Milo, R., Friedman, M., Kandel, A. (2001) 'On fuzzy correlations', *IEEE Transactions on Systems, Man, and Cybernetics part B*, **31** (3), 381–390.

Richei, A., Hauptmanns, U., Unger, H. (2001) 'The human error rate assessment and optimizing system HEROS-a new procedure for evaluating and optimizing the man-machine interface in PSA', *Reliability Engineering and System Safety*, **72**, 153–164.

Rosenfeld, A. (1979) 'Fuzzy digital topology', *Information and Control*, **40**, 76–87.

Ruspini, E. (1969) 'A new approach to clustering', *Information and Control*, **16**, 22–32.

Ruspini, E. (1970) 'Numerical methods of fuzzy clustering', *Information Sciences*, **2**, 319–350.

Ruspini, E. (1973) 'New experimental results in fuzzy clustering', *Information Sciences*, **6**, 273–284.

Ruspini, E.H., Bonissone, P., Pedrycz, W. (eds.) (1998) *Handbook of Fuzzy Computation*, Institute of Physics Publishing, Bristol and Philadelphia.

Sato, M., Sato, Y. (1995) 'Fuzzy clustering model for fuzzy data', *Proceedings of IEEE*, 2123–2128.

Shen, X.X., Tan, K.C., Xie, M. (2001) 'The implementation of quality function deployment based on linguistic data', *Journal of Intelligent Manufacturing*, **12**, 65–75.

Sii, H.S., Ruxton, T., Wang, J. (2001) 'A fuzzy-logic-based approach to qualitative safety modelling for marine systems', *Reliability Engineering and System Safety*, **73**, 19–34.

Szmidt, E., Kacprzyk, J. (2000) 'Distances between intuitionistic fuzzy sets', *Fuzzy Sets and Systems*, **114**, 505–518.

Takata, O., Miyamoto, S., Umayahara, K. (2001) 'Fuzzy clustering of data with uncertainties using minimum and maximum distances based on L1 metric', *Proceedings of Joint 9th IFSA World Congress and 20th NAFIPS International Conference*, July 25–28, 2001, Vancouver, British Columbia, Canada, 2511–2516.

Tran, L.T., Duckstein, L. (2002) 'Comparison of fuzzy numbers using a fuzzy distance measure', *Fuzzy Sets and Systems* **130** (3), 331–341.

Wang, X. (1997) 'New similarity measures on fuzzy sets and on elements', *Fuzzy Sets and Systems*, **85**, 305–309.

Wang, X., De Baets, D., Kerre, E. (1995) 'A comparative study of similarity measures', *Fuzzy Sets and Systems*, **73**, 259–268.

Wu, K.L., Yang, M.S. (2002) 'Alternative c-means clustering algorithms', *Pattern Recognition*, **35**, 2267–2278.

Xie, X.L., Beni, G. (1991) 'A validity measure for fuzzy clustering', *IEEE Transactions on Pattern Analysis Machine Intelligence*, **13**, 841–847.

Yang, M.S. (1993) 'A survey of fuzzy clustering', *Mathematical and Computer Modelling*, **18** (11), 1–16.

Yang, M.S., Ko, C.H. (1996) 'On a class of fuzzy c-numbers clustering procedures for fuzzy data', *Fuzzy Sets and Systems*, **84**, 49–60.

Yang, M.S., Ko, C.H. (1997) 'On a cluster-wise fuzzy regression analysis', *IEEE Transactions on Systems Man, and Cybernetics*, **27** (1), 1–13.

Yang, M.S., Liu, H.H. (1999) 'Fuzzy clustering procedures for conical fuzzy vector data', *Fuzzy Sets and Systems*, **106**, 189–200.

Yang, M.S., Hwang, P.Y., Chen, D.H. (2004) 'Fuzzy clustering algorithms for mixed feature variables', *Fuzzy Sets and Systems*, **141**, 301–317.

Yang, M.S., Hwang, P.Y., Chen, D.H. (2005) 'On a similarity measure between LR-type fuzzy numbers and its application to database acquisition', *International Journal of Intelligent Systems*, **20**, 1001–1016.

Yong, D., Wenkang, S., Feng, D., Qi, L. (2004) 'A new similarity measure of generalized fuzzy numbers and its application to pattern recognition', *Pattern Recognition Letters*, **25**, 875–883.

Zadeh, L.A. (1965) 'Fuzzy sets', *Information and Control*, **8**, 338–353.

Zadeh, L.A. (1973) 'Outline of a new approach to the analysis of complex system and decision process', *IEEE Transactions on Systems, Man and Cybernetics*, **3**, 28–44.

Zhang, C., Fu, H. (2006) 'Similarity measures on three kinds of fuzzy sets', *Pattern Recognition Letters*, **27** (12), 1307–1317.

Zimmermann, H. J. (2001) *Fuzzy Set Theory and its Applications*, Kluwer Academic Press, Dordrecht.

Zwich, R., Carlstein, E., Budescu, D.V. (1987) 'Measures of similarity among concepts: a comparative analysis', *International Journal of Approximate Reasoning*, **1**, 221–242.

9

Inclusion-based Fuzzy Clustering

Samia Nefti-Meziani[1] and Mourad Oussalah[2]

[1]*Department of Engineering and Technology, University of Manchester, Manchester, UK*
[2]*Electronics, Electrical and Computing Engineering, The University of Birmingham, Birmingham, UK*

9.1 INTRODUCTION

Clustering plays an important role in many engineering fields, especially data mining, pattern recognition, and image processing among others, where the aspect of looking for hidden patterns and classes is an issue (see, for instance, [18, 27] and references therein for an extensive survey). Typically, clustering methods divide a set of N observations, usually represented as vectors of real numbers, $\mathbf{x}_1, \mathbf{x}_2, \ldots, \mathbf{x}_N$ into c groups denoted by $K_1, K_2, \ldots, K_c$ in such a way that the members of the same group are more similar to one another than to the members of other groups. The number of clusters may be predefined as in the case of k-means type clustering or it may be set by the method. Various types of clustering have been developed in the literature. These include hierarchical clustering, graph theoretic based clustering, and those based on minimizing a criterion function [8, 18]. Typically, in standard clustering approaches, each datum belongs to one and only one cluster. Although this assumption is well justified in the case of compact and separated data, it is less justified in the case of overlapping data where a pattern can belong to more than one cluster. For instance, according to some features the datum belongs to class K_1 while according to another set of features it belongs to K_3. This provides solid motivational grounds for methods developed in fuzzy set theory where the degree of membership to a class is naturally graded within the unit interval. In this respect, a datum or instance of a data-set can belong to several classes concurrently, indicating the presence of nonzero overlapping with each of these classes.

On the other hand, various proposals have been put forward to extend the hierarchical clustering, graph theoretic clustering, and criterion-based minimization clustering to include fuzzy parameters, usually a matrix $\mathbf{U}$ indicating the degree of membership of each datum to a given class. The fuzziness can also be extended to data where each datum is described in terms of a fuzzy set. This can be justified either in situations in which data, even being single numeric values, are pervaded by uncertainty/imprecision. In such cases the fuzzy set is used as a tool to quantify such uncertainty/imprecision. In cases where the inputs are linguistic labels in which the exact numerical model is ill-known, the fuzziness arises naturally

Advances in Fuzzy Clustering and its Applications Edited by J. Valente de Oliveira and W. Pedrycz

from the intrinsic meaning of the words. This is particularly highlighted in rule-based systems in which the inputs of each rule are described by some linguistic quantifiers. In such cases the representation of those fuzzy sets is always an open issue.

For this purpose at least two streams of research can be distinguished. The first advocates the need to determine an exact model of the fuzzy set using some statistical or logical-based approaches. For instance, the probability–possibility transformations [12, 26] fail in this category. The use of optimization techniques can also be employed for such purposes [27, 28]. The second stream supports a more flexible view and acknowledges the subjective nature of the fuzzy set, which can be defined by the expert according to his prior knowledge [1, 27]. In this context, the use of parametric fuzzy sets is quite useful, especially from a computation perspective, and only a limited number of parameters are required to define the whole fuzzy set on its universe of discourse [13]. For instance, a triangular membership function requires only three parameters: modal value, left spread value, and right spread value. Gaussian type fuzzy sets require only two parameters: modal (or mean) value and standard deviation. Gaussian fuzzy sets are useful because of their differentiability, especially when the fuzzy systems are obtained by automated methods; thus any supervised learning algorithm can be used, such as various gradient descent learners and neuro-fuzzy systems [7].

Among clustering techniques developed in the fuzzy set literature, fuzzy C-means, with its various extensions, is among the most popular models. Its essence is based on finding clusters such that the overall distance from a cluster prototype to each datum is minimized. Extensions of fuzzy clustering tools to accommodate fuzzy data have been proposed by Pedrycz, Bezdek, Hathaway, and Rogers [30] where two parametric models have been put forward. Also, Pedrycz [29] developed appropriate clustering tools to deal with granular-like information, which constitutes a more complex form of fuzzy data.

However, when the inclusion concept is desired as a part of user requirements and constraints in the clustering problem, then the standard fuzzy clustering approach based on FCM shows its limitations. Strictly speaking, interest in the inclusion concept may arise from several standpoints. From a fuzzy set theoretical approach the inclusion corresponds to an entailment. Arguing that a fuzzy set A is included in fuzzy set B is therefore translated to a formal implication $B \rightarrow A$. In this respect, the knowledge induced by A is rather redundant to B since it can be inferred from the knowledge attached to B. It makes sense, therefore, to identify all those circumstances in which the inclusion relation holds. This is especially relevant in applications involving, for example, rule-base systems when simplification of (fuzzy) rules is crucial for implementation and flexibility purposes [19–20, 31–35]. Indeed, in fuzzy rule-based systems, the above entailment is modeled as an "if... then..." rule, therefore an inclusion-based algorithm would provide a powerful tool to reduce the number of rules in the database, which is very relevant to the user since the complexity increases with the number of rules. Such simplification is also motivated from a semantic viewpoint since a semantically unclear model is not easily verified by an expert in cases where the verification is desired as a part of the model validation [7, 14, 16]. From this perspective such an algorithm would allow us to improve the interpretability of fuzzy models by reducing their complexity in terms of the number of fuzzy sets, and thereby the number of fuzzy rules that are acquired by automated techniques.

Furthermore, in information retrieval systems, it may be desirable to discover all cases in which a given query document is either *exactly*, or *almost*, part of another document. Such a capability would be valuable in matters such as keeping track of ownership and intellectual property as well as related legacy issues.

The use of an inclusion index in any clustering-based approach encounters at least two main challenges. First, from a metric viewpoint, it is clear that any distance is symmetric while an inclusion is by definition an asymmetric relation. Therefore, its incorporation into a distance-based clustering is problematic and further assumptions need to be made. Secondly, as far as the optimization problem is concerned, without further assumptions, maximizing and/or minimizing the inclusion index would lead to an uninformative result corresponding either to the smallest pattern that may exist in the data-set – with respect to a given metric – or to the greatest. In either case this may coincide with the typical noise prototype and therefore a choice of appropriate constraints to avoid such vacuum situations is required.

The aim of this chapter is to investigate this issue and provide a possible construction for an inclusion index, which will be incorporated in the general scheme of the fuzzy C-means algorithm. Simulation results will be provided and compared with those obtained with the standard fuzzy C-means algorithm. The following section provides a background to fuzzy clustering and the fuzzy C-means algorithm. In Section 9.3 we elaborate on the construction of an inclusion index, while the derivation of the inclusion based fuzzy clustering is established in Section 9.4. In Section 9.5 two types of application are presented. In the first application, some simulation results using synthetic data are provided and compared with those supplied by the standard fuzzy C-means algorithm. The second application deals with a railway data-set where the proposed algorithm is used to provide qualitative assessment of the risk factors using track irregularity information.

9.2 BACKGROUND: FUZZY CLUSTERING

Fuzzy clustering partitions a data-set into a number of overlapping groups. One of the first and most commonly used fuzzy clustering algorithms is the fuzzy C-means (FCM) algorithm [2]. Since the introduction of FCM, many derivative algorithms have been proposed and different applications have been investigated. Typically, FCM allows the determination of the membership value u_{ij} of each datum $\mathbf{X}_i$ $(i = 1, \ldots, n)$ to cluster j based on the distance from datum $\mathbf{X}_i$ to cluster prototype $\mathbf{V}_j$ $(j = 1, \ldots, c)$. The number c of classes is expected to be known beforehand by the user. The obtained partition is optimal in the sense of minimizing the objective function

$$J = \sum_{i=1}^{n} \sum_{j=1}^{c} u_{ij}^{\alpha} d_{ij}^{2}, \tag{9.1}$$

subject to the constraint

$$\sum_{j=1}^{c} u_{ij} = 1. \tag{9.2}$$

The symbol d_{ij} indicates the distance from the cluster prototype $\mathbf{V}_j$ to the datum $\mathbf{X}_i$. The parameter $\alpha > 1$ is the fuzziness parameter, which controls the spread of the fuzzy sets induced by u_{ij}. The partition matrix $\mathbf{U}$, whose elements consist of the computed membership values u_{ij}, indicates the partitioning of the data-set into different clusters j or, equivalently, the membership grade of each datum $\mathbf{X}_i$ to the class j whose prototype is $\mathbf{V}_j$.

For any semi-definite and positive matrix $\mathbf{A}$, the distance d_{ij} can be written as

$$d_{ij}^{2} = (\mathbf{X}_i - \mathbf{V}_j)^{t}\, \mathbf{A}\, (\mathbf{X}_i - \mathbf{V}_j). \tag{9.3}$$

Thus, any choice of a matrix $\mathbf{A}$ induces a specific kind of distance interpretation and consequently generates its own meaning of cluster shape. For instance, if $\mathbf{A}$ is the identity matrix, d_{ij} corresponds to a Euclidean distance, which broadly induces spherical clusters. Gustafson and Kessel [17] have focused on the case where the matrix $\mathbf{A}$ is different for each cluster j. $\mathbf{A}_j$ is obtained from the covariance of data belonging to cluster j, while the determinant of each $\mathbf{A}_j$, which stands for the volume of the cluster, is kept constant. This enables the detection of ellipsoidal clusters. Bezdek, Coray, Gunderson, and Watson [3–4] have investigated the case where one of the eigenvectors of the matrix, which corresponds to the largest eigenvalue, is maximized. This allows the detection of linear clusters like lines or hyperplanes. Davé proposed a special formulation of the objective function that yields a better description of circular shape [10]. Later, he also proposed a method to deal with random noise in data-sets [11] by allocating an extra class corresponding to noise whose prototype vector is equally situated from each datum. Krishnapuram and Keller [21] put forward another formulation of J where the membership values are not normalized according to (9.2). Instead, the algorithm is implicitly constrained by the formulation of the objective function J.

The preceding paragraph indicates clearly that neither the formulation of the matrix **A**, nor that of the objective function J is completely fixed, and some flexibility is allowed. Moreover, the distance structure is often of limited capability in discriminating between various patterns when the feature space increases. To see this, suppose we use Euclidean distance, which is used in the standard FCM algorithm, to discriminate between the following two Gaussian membership functions: $G_1(0.2, 0.1)$ and $G_2(0.6, 0.1)$, and consider on the other hand two other Gaussian membership functions $G_3(0.2, 0.9)$ and $G_4(0.6, 0.9)$. Clearly, the use of Euclidean distance leads to

$$d_{G_1G_2} = \sqrt{(0.2 - 0.6)^2 + (0.1 - 0.1)^2} = 0.4$$

Similarly,

$$d_{G_3G_4} = \sqrt{(0.2 - 0.6)^2 + (0.9 - 0.9)^2} = 0.4.$$

In other words, despite a huge increase (from 0.1 to 0.9) of the spread of the distribution G_2, the distance remains unchanged, while intuitively as the spread increases, the overlapping area becomes more important. Thereby, we expect the two distributions to become closer to each other; however, this fact would not occur if the inclusion between the distributions is explicitly accounted for.

9.3 CONSTRUCTION OF AN INCLUSION INDEX

The development in this section is restricted to Gaussian fuzzy sets, but the general approach can also be applied to any other LR types of fuzzy sets since only the mean or modal value and the spread value matter. Besides, these values can be obtained from any LR fuzzy set [13], either directly or indirectly through appropriate transformation. Each fuzzy set μ_i is defined by three parameters, i.e., the mean m_i, the standard deviation σ_i and the height h_i. We assume that all fuzzy sets are normalized so that their heights are to equal 1. In this case each fuzzy set is characterized only by the two parameters m_i and σ_i. Because similar fuzzy sets would have similar parameters, a fuzzy clustering algorithm applied to data in this space could be used to detect groups of similar fuzzy sets.

We now turn our attention to fuzzy clustering. The goal here is to model the amount of mutual inclusion between two fuzzy sets. Let m and σ be the parameters corresponding to the mean and standard deviation, respectively, of a Gaussian fuzzy set G. Let us denote by G_i the Gaussian fuzzy set characterized by (m_i, σ_i). Let $Id(G_1, G_2)$ stand for the degree of inclusion of the Gaussian G_1 in the Gaussian G_2. Globally, we require that the more the former fuzzy set is a subset of another, the larger the value of $Id(G_1, G_2)$.

For any Gaussian fuzzy set G, it is known that almost 98 % of the fuzzy set is concentrated within the interval $[m - 3\sigma, m + 3\sigma]$. Each fuzzy set can therefore be represented as an interval centered at m_i and with length $6\,\sigma_i$, as shown in Figure 9.1. For the sake of clarity, we shall denote σ' (resp. σ'_i) the values of 3σ (resp. $3\sigma_i$).

Inspired from the fuzzy set theoretical inclusion, the inclusion grade of G_1 into G_2 can be assimilated to the ratio of $G_1 \cap G_2 / G_1$. From this perspective, the inclusion grade equals zero whenever fuzzy sets G_1 and G_2 are disjoint ($G_1 \cap G_2 = \emptyset$) and reaches the maximum value, which is one, whenever G_1 is (physically) included in G_2 ($G_1 \subset G_2$).

Now the use of the aforementioned interval interpretations of the Gaussians allows us to construct a model of previous ratio. In this respect, using the length of the interval as essence, an estimation of $Id(G_1, G_2)$ can be obtained as

$$Id(G_1, G_2) = \frac{L([m_1 - \sigma'_1, m_1 + \sigma'_1] \cap [m_2 - \sigma'_2, m_2 + \sigma'_2])}{2\sigma'_1}, \tag{9.4}$$

where L: $\Im \rightarrow \Re$ is a mapping from a set of intervals of real numbers $\Im$ to a set of real numbers, which assigns for each interval $[a, b]$ its length $(b - a)$.

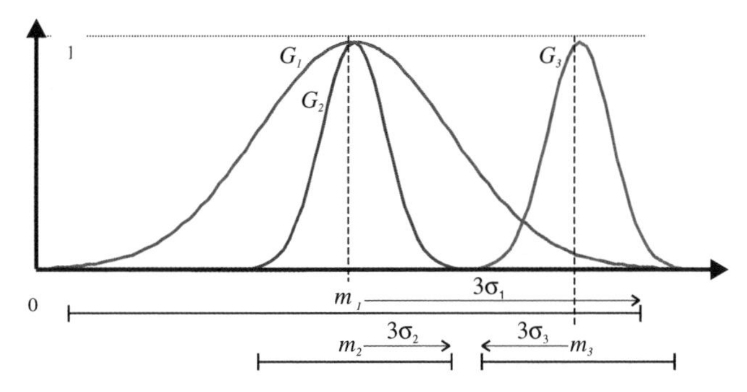

Figure 9.1 An interval representation for Gaussian fuzzy sets.

More specifically, Equation (9.4) can be rewritten as

$$Id_1(G_1, G_2) = \begin{cases} \dfrac{\sigma'_1 + \sigma'_2 - |m_1 - m_2|}{2\sigma'_1} & \text{if } 0 \leq \sigma'_1 + \sigma'_2 - |m_1 - m_2| \leq 2\sigma'_1 \\ 0 & \text{if } \sigma'_1 + \sigma'_2 - |\mathrm{m}_1 - \mathrm{m}_2| < 0 \\ 1 & \text{if } \sigma'_1 + \sigma'_2 - |m_1 - m_2| > 2\sigma'_1. \end{cases} \tag{9.5}$$

Alternatively, to avoid the use of absolute value in expression (9.5), we may use squared values of parameters instead. Therefore, a counterpart of (9.5) will be

$$Id_2(G_1, G_2) = \begin{cases} \dfrac{(\sigma'_1 + \sigma'_2)^2 - (m_1 - m_2)^2}{4(\sigma'_1)^2} & \text{if } 0 \leq (\sigma'_1 + \sigma'_2)^2 - (m_1 - m_2)^2 \leq 4(\sigma'_1)^2 \\ 0 & \text{if } (\sigma'_1 + \sigma'_2)^2 - (m_1 - m_2)^2 < 0 \\ 1 & \text{if } (\sigma'_1 + \sigma'_2)^2 - (m_1 - m_2)^2 > 4(\sigma'_1)^2. \end{cases} \tag{9.6}$$

The latter can be rewritten using step functions H as

$$\begin{aligned} Id_2(G_1, G_2) = {} & \frac{(\sigma'_1 + \sigma'_2)^2 - (m_1 - m_2)^2}{4(\sigma'_1)^2} H[(\sigma'_1 + \sigma'_2)^2 - (m_1 - m_2)^2] \\ & + \frac{4(\sigma'_1)^2 - (\sigma'_1 + \sigma'_2)^2 + (m_1 - m_2)^2}{4(\sigma'_1)^2} H[(\sigma'_1 + \sigma'_2)^2 - (m_1 - m_2)^2 - 4(\sigma'_1)^2], \end{aligned} \tag{9.7}$$

which holds almost everywhere.

Standard step functions H are defined as

$$H(x - a) = \begin{cases} 0 & \text{if } x < \mathrm{a} \\ \frac{1}{2} & \text{if } x = a \\ 1 & \text{if } x > \mathrm{a}. \end{cases} \tag{9.8}$$

From this perspective $Id_2(G_1, G_2)$ as defined in Equation (9.6), or equivalently Equation (9.7), corresponds to a straightforward interpretation of previous fuzzy set theoretical inclusion grade, and therefore preserves all the intuitive features and properties of the fuzzy set theoretical inclusion grade. Indeed, as soon as $|m_i - m_j| > 3(\sigma_i + \sigma_j)$, which means that the two fuzzy sets are far away from each other, $Id_2(G_i, G_j)$ vanishes. If the distributions have the same mean, i.e., $|m_i - m_j| = 0$ and $\sigma_i \leq \sigma_j$, then $Id_2(G_i, G_j)$ equals one, which corresponds to a physical inclusion of distribution G_i in G_j. Otherwise, the inclusion grade captures the overlap between the two distributions with respect to that of G_i and takes values in the unit interval, provided the parameters of the distributions also lie within the unit interval.

The index $Id_2(G_1, G_2)$ attains its maximum value when the first distribution is fully included in the second. For cluster merging purposes, however, it may be useful to assess not only whether one

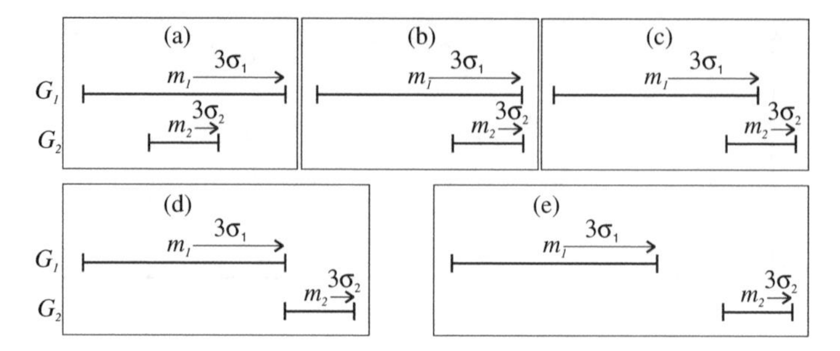

Figure 9.2 Illustration of inclusion for several pairs of fuzzy sets.

distribution is included in the other, but also the extent to which the including distribution is larger than the one included. In this case one may consider omitting the upper bound of the index. Therefore, a counterpart of $Id_2(G_1, G_2)$ would be

$$Id_3(G_1, G_2) = \begin{cases} \dfrac{(\sigma_1' + \sigma_2')^2 - (m_1 - m_2)^2}{4(\sigma_1')^2} & \text{if } (\sigma_1' + \sigma_2')^2 - (m_1 - m_2)^2 \geq 0 \\ 0 & \text{if } (\sigma_1' + \sigma_2')^2 - (m_1 - m_2)^2 < 0 \end{cases} \tag{9.9}$$

or equivalently, using step functions,

$$Id_3(G_1, G_2) = \frac{(\sigma_1' + \sigma_2')^2 - (m_1 - m_2)^2}{4(\sigma_1')^2} H[(\sigma_1' + \sigma_2')^2 - (m_1 - m_2)^2] \tag{9.10}$$

Loosely speaking, the latter captures not only the inclusion of G_1 in G_2 but also the extent to which the spread of G_1 is smaller than that of G_2. It holds that as soon as a physical inclusion $G_1 \subset G_2$ occurs, then $Id_3(G_1, G_2) \geq 1$. Otherwise, the index still captures the degree of overlap between the two distributions with respect to the distribution G_1.

Consequently, Id_3 offers a better performance when the relative size of distributions is desired as part of the evaluation index, which in turn endows the inclusion index with discrimination power.

Especially, as exhibited in Figure 9.2, the inclusion index Id_3 allows us to discriminate between the different physical scenarios of inclusion induced by the different pairs of fuzzy sets shown. In this case the formulation in Equation (9.10) leads to the following ordering (subscript 3 is omitted from Id_3 for the purpose of clarity, i.e., $Id_{(a)}$ stands for evaluation of index Id_3 in the case of a pair of distributions pertaining to Figure 9.2(a)):

$$Id_{(a)}(G_2, G_1) \geq Id_{(b)}(G_2, G_1) \geq Id_{(c)}(G_2, G_1) \geq Id_{(d)}(G_2, G_1) = Id_{(e)}(G_2, G_1) = 0, \tag{9.11}$$

which sounds in full agreement with the intuition regarding the given examples.

9.4 INCLUSION-BASED FUZZY CLUSTERING

We now introduce a fuzzy clustering algorithm that incorporates the inclusion index. Strictly speaking, the introduction of the inclusion constraint in the clustering scheme needs to be accomplished only in the global sense, since otherwise trivial vacuum cases occur. Therefore, we must additionally constrain the fuzzy clustering algorithm in order to detect the inclusion of fuzzy sets in the cluster prototype. Moreover, the inclusion concept is somehow hidden in the distance structure. This accounts for inclusion in only a global sense, which is sound for a given definition of optimality [23–24].

The aim is to obtain class prototypes such that all elements of that class are roughly included in the Gaussian class prototype in the sense of the evaluation given by index Id_3. Ideally, for each class, we would look for a Gaussian prototype that maximizes the overall inclusion degrees over the set of Gaussians belonging to that class, or equivalently maximizes the total sum of inclusion degrees of each Gaussian into that prototype, while each inclusion index is weighted by the corresponding value of the membership value u_{ij}^{α}. Consequently, the problem boils down to maximizing the quantity $\sum_{j=1}^{c}\sum_{i=1}^{n} Id_3(G_i, Gv_j).u_{ij}^{\alpha}$, or equivalently minimizing the objective function J:

$$J = -\sum_{j=1}^{c}\sum_{i=1}^{n} Id_3(G_i, Gv_j).u_{ij}^{\alpha}, \tag{9.12}$$

subject to

$$\sum_{j=1}^{c} u_{ij} = 1,$$

where G_i corresponds to the ith Gaussian of initial datum, with mean m_i and standard deviation σ_i, and Gv_j corresponds to the Gaussian prototype of the jth class whose mean and standard deviation are mv_j and σv_j, respectively.

Unfortunately, the optimization problem (9.12) and (9.2) without further constraints would lead to the extreme solution where the spread (standard deviation) of all the prototypes tend to their maximal values.

To circumvent this effect, the maximization (9.13) can be balanced by minimizing the distance from each prototype to all elements (Gaussians) of the same class as the above prototype. A possible formulation of such reasoning consists of a linear combination of objective functions (9.1) and (9.12). This leads to the following counterpart of (10.12) and (10.2)

$$\text{minimize } J = -\sum_{j=1}^{c}\sum_{i=1}^{n} Id_3(G_i, Gv_j).u_{ij}^{\alpha} + w\sum_{j=1}^{c}\sum_{i=1}^{n} (x_i - v_j)^T A(x_i - v_j)u_{ij}^{\alpha}, \quad w > 0 \tag{9.13}$$

subject to (9.2).

In (9.13), each Gaussian G_i is interpreted as a vector x_i with coordinates $(m_i, 3\sigma_i)$. Similarly, the jth prototype can be represented as a vector with coordinates $(mv_j, 3\sigma v_j)$.

Equation (9.13) indicates a balance between maximizing inclusion indices of each element in the prototype Gv_j, and minimizing the distance from the prototype to these elements. The weight w is used to normalize the distance and inclusion evaluations as neither the distances d_{ij} nor Id_3 are normalized entities, and to quantify the relative importance of both factors with respect to each other. It should be noted that the value of w that ensures a rational behavior for the above optimization is not unique as will be pointed out later. On the one hand, choosing w relatively too large makes the distance minimization requirement a predominant part in the objective function (9.13), which in turn makes the above optimization closer to the standard fuzzy C-means algorithm. On the other hand, taking w relatively too small makes the inclusion evaluation a predominant part in the objective function (9.13), so that (9.13) tends toward the optimization (9.12), which in turn induces prototypes with maximum spread as already mentioned.

Using matrix formulation, let $B_1 = \begin{pmatrix} 1 & 0 \\ 0 & 0 \end{pmatrix}$ and $B_2 = \begin{pmatrix} 0 & 0 \\ 0 & 1 \end{pmatrix}$, then the Id_3 expression can be rewritten as

$$Id_3(G_i, Gv_j) = \frac{1}{4}(x_i^T B_2 x_i)^{-1} S_{i,j} H(S_{i,j}) \tag{9.14}$$

with

$$S_{i,j} = (x_i + v_j)^T B_2(x_i + v_j) - (x_i - v_j)^T B_1(x_i - v_j). \tag{9.15}$$

We can now calculate the updated equations that lead to the optimal solution. By combining Equations (9.13)–(9.15) and Equation (9.2), and by using the Lagrange multipliers β_i, we obtain

$$J(U,V,\beta) = -\frac{1}{4}\sum_{j=1}^{c}\sum_{i=1}^{n}(x_i^T B_2 x_i)^{-1} S_{i,j} H(S_{i,j}).u_{ij}^{\alpha} + w\sum_{j=1}^{c}\sum_{i=1}^{n}(x_i - v_j)^T A(x_i - v_j)u_{ij}^{\alpha} + \sum_{i=1}^{n}\beta_i\left(\sum_{j=1}^{c} u_{ij} - 1\right). \tag{9.16}$$

Setting the derivative of J with respect to U, V, and β_j (see Appendix 9A.1) to zero, leads to the following optimal solutions

$$u_{i,j} = \frac{1}{\sum_{k=}^{c} \left[\frac{-\frac{1}{4}(x_i^T B_2 x_i)^{-1} S_{i,k} H(S_{i,k}) + w(x_i - v_k)^T A(x_i - v_k)}{-\frac{1}{4}(x_i^T B_2 x_i)^{-1} S_{i,j} H(S_{i,j}) + w(x_i - v_j)^T A(x_i - v_j)} \right]^{1/\alpha - 1}} \tag{9.17}$$

$$E.v_j = F, \tag{9.18}$$

where

$$E = \sum_{i=1}^{n} \frac{1}{2}(x_i^T B_2 x_i)^{-1}(B_1 - B_2) H(S_{i,j}) u_{ij}^{\alpha} + 2wA.u_{ij}^{\alpha} \tag{9.19}$$

$$F = \sum_{i=1}^{n} \left[\frac{1}{2}(x_i^T B_2 x_i)^{-1}(B_2 + B_1) x_i.H(S_{i,j}) u_{ij}^{\alpha} + 2wA.x_i.u_{ij}^{\alpha} \right]. \tag{9.20}$$

The matrix A in the above equations coincides with the identity matrix, which makes the distance metric equivalent to Euclidean distance.

Note that we need to ensure that solution v_j does not make $S_{i,j}$ vanish since the underlying step function is not differentiable for $S_{i,j} = 0$. This requirement is usually satisfied, since numeric optimization provides an approximate result. In order to simplify the resolution of Equation (9.18), where quantity $H(S_{ij})$ is also a function of v_j to be used in computing v_j, we evaluate $S_{i,j}$ thereby $H(S_{ij})$, using the previous estimate of v_j (from the previous step of the iterative algorithm). The matrix solution is then $v_j = E^{-1}F$. This approximation is well justified by at least three arguments. First, $H(S_{ij})$ is a binary variable and so any small error in v_j, if relatively close to the true v_j^*, has no influence at all on the estimation of $H(S_{ij})$. Secondly, as the number of iterations increases the estimations of v_j tend to be very close to each other, which offers appealing justification to the first argument. Finally, this reasoning is very similar in spirit to iterative approaches applied for solving linear/nonlinear systems in optimization problems [15].

A possible initialization of v_j consists of using the fuzzy C-means algorithm. The proposed inclusion-based clustering algorithm can then be summarized as follows.

Inclusion-based fuzzy clustering algorithm

Step 1 Fix the number of clusters c, the parameter of fuzziness α, the value of w and initialize the matrix **U** by using the fuzzy C-means algorithm.
Step 2 Use (9.18) to determine v_j.
Step 3 Determine the new matrix **U** using the previous evaluation of $S_{i,j}$ and Equation (9.17).
Step 5 Test if matrix **U** is stable: if stable, stop; else return to step 2.

Alternatively, Equation (9.18) can be computed using an approximated model, where $H(S_{ij})$ are evaluated at the previous step. The previous algorithm is then modified as follows.

Approximated inclusion-based fuzzy clustering algorithm

Step 1 Fix the number of clusters c, the parameter of fuzziness α, the value of w and initialize the matrix **U** and the prototypes v_j by using the fuzzy C-means algorithm.
Step 2 Evaluate quantities $S_{i,j}$ and $H(S_{i,j})$ by using the estimates v_j.
Step 3 Determine the new matrix **U** using Equation (9.17).
Step 4 Determine prototypes v_j using Equation (9.18).
Step 5 Test if matrix **U** is stable: if stable, stop; else return to step 2.

A possible evaluation of parameter w can be determined if we normalize all data to be within the unit interval. The values of components of v_j should then also lie within the unit interval. Consequently, with some manipulations, expanding Equation (9.18) and sorting out an inequality independent of **U** leads to

$$w \geq \frac{1}{\max_i \sigma_i^2} \text{ and } w \geq \frac{\max_i \sigma_i^2}{\max_i \sigma_i^2 (1 - \min_i \sigma_i)(1 + \min_i \sigma_i)},$$

so

$$w \geq \max\left(\frac{\max_i \sigma_i^2}{\max_i \sigma_i^2 (1 - \min_i \sigma_i)(1 + \min_i \sigma_i)}, \frac{1}{\max_i \sigma_i^2}\right). \quad (9.21)$$

Thus a potential candidate of w is the lower bound in Equation (9.21), which corresponds to the right-hand side of the inequality (9.21).

9.5 NUMERICAL EXAMPLES AND ILLUSTRATIONS

The fuzzy inclusion clustering developed in the previous section might be very appealing in several applications involving fuzzy control or approximate reasoning (see, for instance, [1, 6, 9, 22, 35]), as far as the redundancy of fuzzy sets is concerned, especially where fuzzy rules are concerned. To demonstrate the feasibility and the performance of the developed algorithm, two types of application are presented. The first involves synthetic data of randomly generated Gaussian membership functions. The second deals with an industrial application in railways involving the prediction of safety factors, where the IFC algorithm was used to provide a qualitative assessment of risk.

9.5.1 Synthetic Data

To illustrate the performance of the algorithm, we first considered a synthetic data-set corresponding to Gaussian membership functions that need to be clustered into one, two, and three classes, respectively. The data-set consists of a fixed number of Gaussian membership functions obtained through randomly generated mean and spread values, from a uniform distribution in [0,1].

The aim is to obtain prototypes that best capture the inclusion concept among the elements of the same class. Furthermore, in order to evaluate the performance of the algorithm a comparison is carried out with the standard fuzzy C-means algorithm in the same plot. In all these examples, the application of the fuzzy C-means is performed assuming that data x_i ($i = 1$ to N) consist of the initial Gaussian membership functions represented as vectors of their mean and standard deviation. So, the obtained class prototypes v_j ($j = 1$ to c) are also represented through their mean and the standard deviation.

In the first experiment we deliberately set the mean value of the Gaussian membership function to a constant value while the standard deviation term (3σ) varies randomly within the interval [0 5], so the x-axis is restricted by $[m - 3\sigma \; m + 3\sigma]$, which corresponds here to the interval [0 10]. We generated six Gaussian membership functions using this methodology. The aim is then to look at a partition provided by both algorithms when we restrict the number of partitions to one. Figure 9.3 shows both the initial membership functions and the centroid or the prototype vector $v_j (j = 1)$ pertaining to the underlying class, represented as a thick line. When the inclusion context is desired as part of our requirement, the fuzzy inclusion clustering (FIC) algorithm supplies more highly intuitive and appropriate results than the standard fuzzy C-means (FCM) algorithm in the sense that the inclusion relation between the prototype and each of the initial inputs (Gaussian membership functions) is better highlighted in the case of FIC. Indeed, broadly speaking, the FCM in such cases tends toward the averaging operation. Note that the value of the weight factor "w" used in this experiment is equal to its lowest upper bound as suggested by Equation (9.21). In other words, the underlying objective function would be more balanced by the inclusion index rather than the distance index. Note that in Figure 9.3(b), the prototype, or equivalently

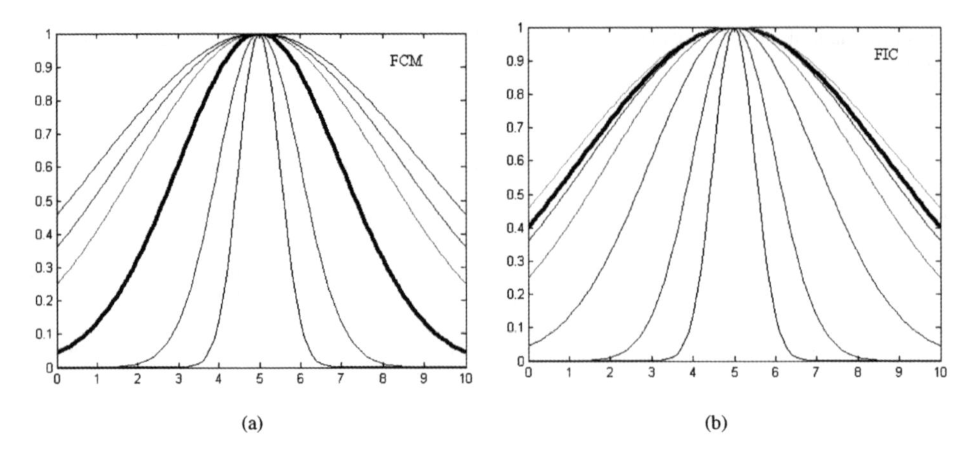

Figure 9.3 Example of (b) fuzzy inclusion clustering (FIC) compared with (a) fuzzy C-means algorithm in the case of one-class clustering. Inputs are randomly generated Gaussian with fixed mean while variance takes the random value in [0 5].

the including distribution, does not coincide completely with the distribution with the highest spread. This is due essentially to the effect of the distance part in the objective function (9.13).

Notice that when using only one single-class classification problem the information provided by the matrix **U** is not relevant as it takes a value one for one datum while it vanishes elsewhere, indicating there is only one single input that belongs to the class, which is obviously not true. This is mainly due to the constraint (9.2) in the optimization problem. This observation is still valid in the case of FIC as the constraint (9.2) is also accounted for in the optimization problem pertaining to it.

In Figure 9.4, we randomly generated six distributions by letting the standard deviation act as a random variable of some uniform distribution, while the mean values are again kept fixed and constant for each class. The results in terms of centroid characteristic or the two prototype vector $v_j (j = 1, 2)$ are exhibited.

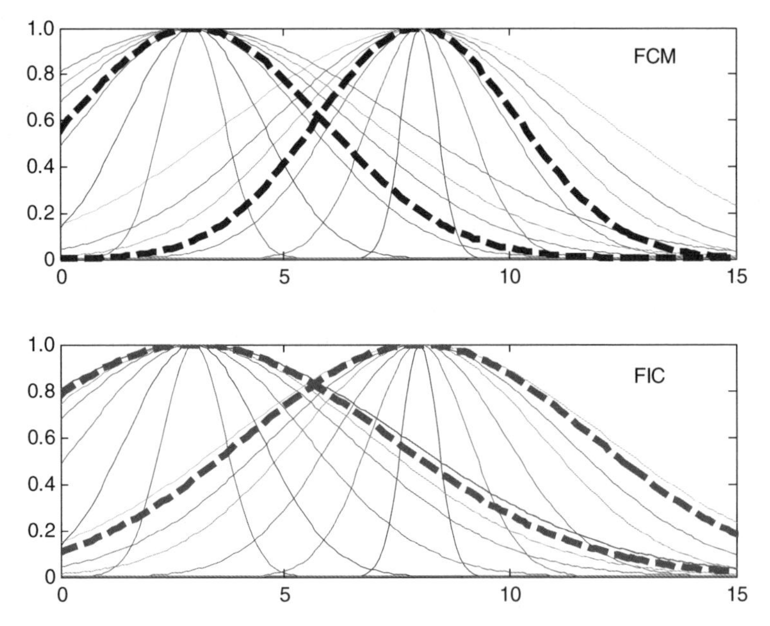

Figure 9.4 Example of fuzzy inclusion clustering (FIC) compared with fuzzy C-means algorithm in the case of two-class clustering.

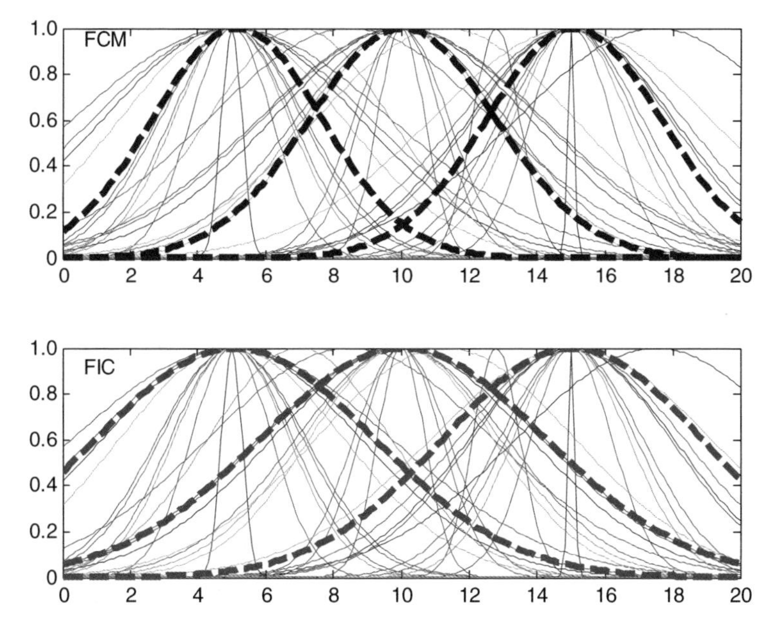

Figure 9.5 Example of fuzzy inclusion clustering (FIC) compared with fuzzy C-means algorithm in the case of three-class clustering.

Again it is clear that the FIC offers better results in terms of agreement with the inclusion concept. On the other hand, notice that from the partition viewpoint both algorithms provide equal results. Indeed, when looking at the matrix **U**, all elements of the same class (distributions having the same mean value) are always assigned the highest value of membership grade. However, when the characteristic of class prototype matters, the tendency of FCM to average the elements of the same class is further highlighted in this example.

In Figure 9.5, a three-class classification problem is considered. In this example, the initial data-set is obtained by considering randomly generated spreads through uniform probability function within the range of [0 5]. We considered 10 distributions with fixed mean value for each class; that is, there are 30 distributions in total. Furthermore, five noisy data whose means are randomly generated within the range of the x-axis ([0 20]) have been added to the above data-set.

Notice that the resulting membership functions do not necessarily coincide with the greatest membership function in terms of variance because the inclusion concept is taken into account only in a global sense. These results are roughly in agreement with the intuition ground.

Figure 9.6 illustrates the result shown in the example pointed out in Figure 9.5 when using the mean–variance scale representation instead. This highlights the tendency of the FCM algorithm to average the inputs, including the variance values, while the FIC algorithm tends to capture the maximum value.

In order to compare the performances of both algorithms, we compute the inclusion index for each class as well as the validity partition index given by the classification entropy index [2,5]. Namely, for each class, we compute the quantities

$$Id_j = \sum_{i=1}^{n} Id(x_i, v_j).u_{ij} \qquad (j = 1 \text{ to } c) \tag{9.22}$$

and

$$H(U,\ c) = \frac{\sum_{k=1}^{n}\sum_{i=1}^{c} u_{ik}\log_a(u_{ik})}{c-n}. \tag{9.23}$$

The smaller the value of $H(U, c)$, the better the performance of the classification, while higher values of Id_j indicate enhanced consideration of the inclusion aspect in the classification in the sense that

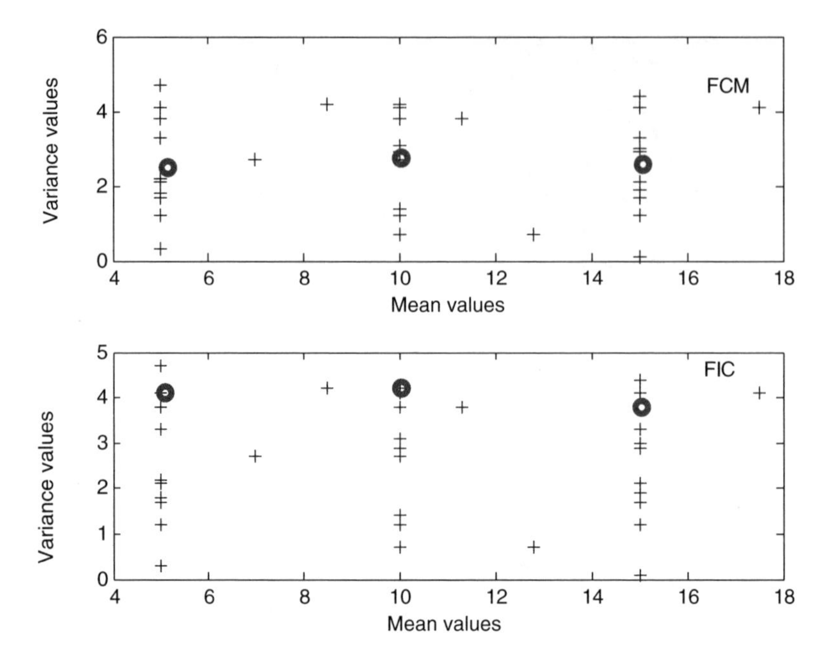

Figure 9.6 Mean–variance representation of the results in the case of three-class clustering.

physical inclusion elements pertaining to a given class in the underlying class prototype are better highlighted.

Table 9.1 summarizes the results of both algorithms (FCM and FIC), and illustrates that the FIC algorithm always outperforms the FCM algorithm when the inclusion evaluation matters.

Notice that in the one-class clustering, the value of classification entropy is labeled NA for undefined. This is due to FCM instability in cases where some elements of **U**, whose logarithmic value is undefined, are zero valued. As can be seen from Table 9.1, when the performance is quantified in terms of the inclusion index and the classification entropy, then the FIC algorithm always outperforms FCM.

Regarding the parameters employed in both algorithms concerning the maximum number of iterations and the tolerance indicating the minimum level of improvement, one should notice that the same values of these parameters were used in both algorithms: 100 iterations at most and a tolerance of 10^{-6}, which is very common in the literature. The influence of this choice is not very significant if the values assigned to these two parameters were relatively close to the above values. However, low values of maximum number of iterations would obviously lead to a divergence of the algorithm(s) unless the tolerance level is increased, which in turn may leave the reliability of the result open to question. On the other hand, setting the maximum number of iterations very high and/or the tolerance level very low would significantly increase the computational complexity of the algorithms, which may make the algorithm(s) unfeasible

Table 9.1 Comparison of FCM and FIC algorithms.

	One-class clustering		Two-class clustering		Three-class clustering	
	FCM	FIC	FCM	FIC	FCM	FIC
Inclusion index *Id*	2.4	4.1	1.1	1.9	3.1	6.6
			3.2	3.9	3.2	6.3
					2.8	5.9
Classification entropy index $H(U,c)$	NA	0	0.41	0.18	0.34	0.06

under some circumstances. The weighting factor "w" involved in the fuzzy inclusion clustering algorithm, which plays the role of a penalty term, is crucial. In cases where the value assigned to this parameter is far away from that supplied by the inequality (9.21) then the superiority of the FIC algorithm can be questioned. Consequently, a prudent attitude should be taken when deciding to change the value of this weighting factor.

9.5.2 Application to Railtrack Safety Prediction

In a railway system, the problems of predicting system malfunctions or safety modes are of paramount importance. Traditional ways of predicting railway safety based on reliability models are shown to be very expensive in terms of computational complexity and data requirement, which make them inefficient under certain circumstances. On the other hand, unsupervised classification of the inputs can provide a sound indication of safety trends. For this purpose, in this application, the inputs consist of track irregularities in terms of vertical displacement, horizontal displacement, and curve and variation of the gauge, while the output consists of safety factor or a 'cyclic top', which is known to contribute significantly to derailment risk. The latter takes its values within the unit interval where zero value stands for very low risk and one for very high risk. Besides, as the number of inputs is concerned, a wavelet transform was used to reduce the dimension as well as the size of the data-set. A neural network architecture was used to learn the outputs using historical records of data-set [25]. The block diagram shown in Figure 9.7 illustrates the overall methodology.

In order to provide a labeling evaluation of the risk in terms of "low risk," "medium risk," and "high risk," a classification of the outcomes supplied by the neural network module is necessary. In each risk scenario the operator takes appropriate action(s). For this purpose, our inclusion fuzzy clustering algorithm was used. From the outcomes of the neural network architecture, we first generate a histogram with 10 bins. This finite number of bins provides a first partitioning of the outcomes, whose ranges lie within the unit interval, which will be translated into the three-class risk partitioning using the FIC algorithm. For the latter, the number of outputs falling in each bin provides an indication of the spread of the distribution centered around the location of the bin on the x-axis. Figure 9.8 illustrates an instance of results. In this example, 100 outputs are considered, whose distribution in terms of the associated histogram is plotted in Figure 9.8(a). The classification problem boils down to partitioning the 10 classes issued from the histogram into the appropriate "low risk," "medium risk," and "high risk" classes using the inclusion fuzzy clustering. Figure 9.8(b) illustrates the class prototypes as well as the associated inputs represented in terms of Gaussian membership functions whose modal values correspond to the location of bins, while the spreads are proportional to the height of the associated bin. It should be pointed out that the underlying methodology allows us to achieve a prudent attitude in terms of risk allocation, which is in agreement with common-sense reasoning in safety-based installations. Indeed, from this perspective, the use of the inclusion-based fuzzy clustering provides an augmented possibility of accounting for less fair events due to obvious observation that widespread prototype distribution contains a larger number of elements.

The above methodology has shown to be very effective both in terms of computational perspective due to the use of the neural network based approach, and also its usability from the operator's perspective as it

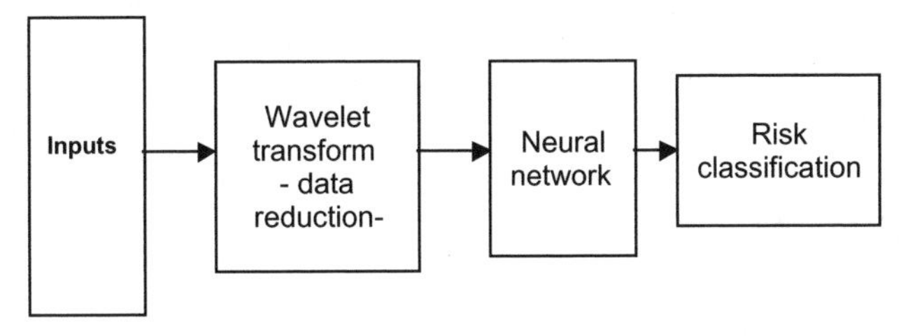

Figure 9.7 Block diagram for risk assessment.

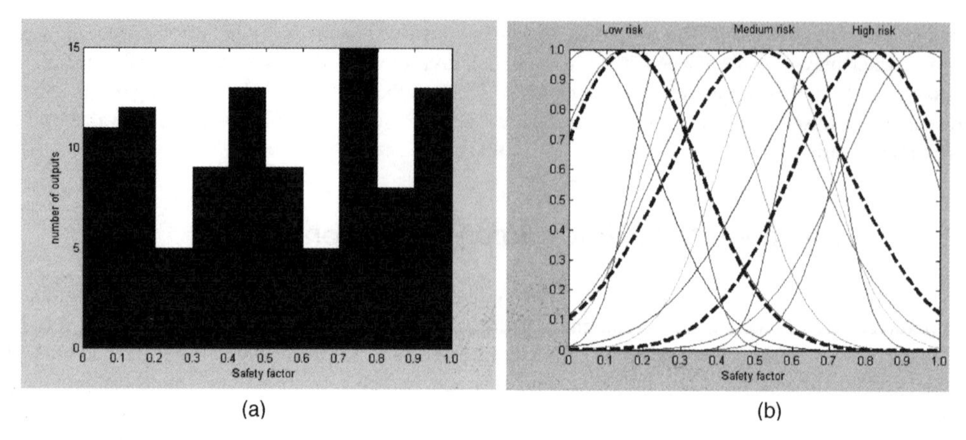

Figure 9.8 Example of risk partitioning; (a) histogram of outcomes from neural network architecture; (b) Result of three-class clustering for risk labelling identification.

provides immediate qualitative assessment of the risk at a given instance, which triggers appropriate actions by the operator. For instance, in the case where low risk is obtained, there is no specific action to be taken by the operator, so the underlying scheduled tasks will move ahead. On the other hand, if "medium risk" occurs, the operator takes the following actions. First, the section of the track where the medium safety indicator occurred will be sent to all train operators with a recommendation to speed down the trains" speeds. Second, a reinvestigation of the results will be carried out to reinforce the decision. Third, a decision to send a team of employees to investigate and possibly accomplish maintenance operation on the track will be issued. Finally in the case of a high-risk situation, the main action to be taken is to stop the circulation of the trains nearby the region of track where high-risk evaluation was obtained, and issue immediate notice to investigate the track and perform related maintenance tasks.

9.6 CONCLUSIONS

In this chapter, we have investigated a fuzzy clustering algorithm based on the inclusion concept that allows the determination of the class prototype that best covers all the patterns of that class. An inclusion index has been proposed, which captures the inclusion of fuzzy set data, mainly Gaussian membership functions. The suggested inclusion index has been incorporated into a fuzzy clustering optimization problem, where the relationship between fuzzy data and class prototype is described in terms of inclusion. This optimization problem is based on the idea of looking for a distance structure hidden in the inclusion concept. More specifically, a matrix A, supporting the distance structure, is determined for each class, such that the total distance from each class prototype to all patterns coincides with the amount of the degrees of inclusion of all patterns in that prototype. The algorithm was illustrated by some numerical examples and comparison with standard fuzzy C-means has been presented. The algorithm provides an appealing framework when the reduction of redundancy of fuzzy sets or simplification of rule base is desired. Finally, an industrial application in railways involving the prediction of safety factors has also been examined, where the inclusion fuzzy clustering algorithm was used to partition the outcomes issued from neural network architecture into classes of "low risk," "medium risk," and "high risk."

ACKNOWLEDGEMENTS

The authors would like to thank Railtrack and Serco Railtest for providing the measured track data. Also the authors are grateful to Dr Kaymak Uzay from Rotterdam University for his contributions and discussion of an early version of this chapter. Finally, the authors thank the editors for their suggestions to enhance the quality of this chapter.

APPENDIX 9A.1

By Starting from

$$\text{minimize } J = -\sum_{j=1}^{c}\sum_{i=1}^{n} Id_3(G_i, Gv_j).u_{ij}^{\alpha} + w\sum_{j=1}^{c}\sum_{i=1}^{n}(x_i - v_j)^T A(x_i - v_j)u_{ij}^{\alpha},$$

Subject to

$$\sum_{j=1}^{c} u_{ij} = 1$$

we have arrived at to the following augmented Lagrangian

$$J(U, V, \beta) = -\sum_{j=1}^{c}\sum_{i=1}^{n} Id_3(G_i, Gv_j).u_{ij}^{\alpha} + w\sum_{j=1}^{c}\sum_{i=1}^{n}(x_i - v_j)^T A(x_i - v_j)u_{ij}^{\alpha} + \sum_{i=1}^{n}\beta_i\left(\sum_{j=1}^{c}u_{ij} - 1\right) \tag{9A.1}$$

Substituting the Id_3 expression in (A10.1) leads to

$$J(U,V,\beta) = -\frac{1}{4}\sum_{j=1}^{c}\sum_{i=1}^{n}(x_i^T B_2 x_i)^{-1} S_{i,j} H(S_{i,j}).u_{ij}^{\alpha} + w\sum_{j=1}^{c}\sum_{i=1}^{n}(x_i - v_j)^T A(x_i - v_j)u_{ij}^{\alpha} + \sum_{i=1}^{n}\beta_i\left(\sum_{j=1}^{c}u_{ij} - 1\right) \tag{9A.2}$$

with

$$S_{i,j} = (x_i + v_j)^T B_2(x_i + v_j) - (x_i - v_j)^T B_1(x_i - v_j). \tag{9A.3}$$

The necessary conditions for optimality are found by setting the derivatives of J with respect to its parameters to zero. Therefore, provided $S_{i,j} \neq 0$, which ensures the derivability of the step function $H(S_{ij})$

$$\frac{\partial J}{\partial u_{ij}} = -\frac{\alpha}{4}(x_i^T B_2 x_i)^{-1} S_{i,j} H(S_{i,j})u_{ij}^{\alpha-1} + w\alpha(x_i - v_j)^T A(x_i - v_j)u_{ij}^{\alpha-1} + \beta_i = 0 \tag{9A.4}$$

$$\frac{\partial J}{\partial \beta_i} = \left(\sum_{j=1}^{c} u_{i,j} - 1\right) = 0 \tag{9A.5}$$

$$\frac{\partial J}{\partial v_j} = -\frac{1}{2}\sum_{i=1}^{n} x_i^T B_2 x_i)^{-1}[B_2(x_i + v_j) + B_1(x_i - v_j)]H(S_{i,j})u_{ij}^{\alpha} - 2w\sum_{i=1}^{n} A(x_i - v_j)u_{ij}^{\alpha} = 0. \tag{9A.6}$$

Equation (9A.6) was obtained by noticing that the derivative of step function $H(x)$ with respect to variable x is a Dirac function $\delta(x)$, which is zero valued for all non-zero values of x. Consequently using the standard function derivative rules the result (9A.6) becomes is straightforward.

Equation (9A.4) entails

$$u_{i,j} = \left[\frac{\beta_i}{\frac{\alpha}{4}(x_i^T B_2 x_i)^{-1} S_{i,j} H(S_{i,j}) - w\alpha(x_i - v_j)^T A(x_i - v_j)}\right]^{1/\alpha-1}. \tag{9A.7}$$

Now using Equation (9A.5), (9A.7) is equivalent to

$$u_{i,j} = \frac{1}{\sum_{k=1}^{c}\left[\frac{-\frac{1}{4}(x_i^T B_2 x_i)^{-1} S_{i,k} H(S_{i,k}) + w(x_i - v_k)^T A(x_i - v_k)}{-\frac{1}{4}(x_i^T B_2 x_i)^{-1} S_{i,j} H(S_{i,j}) + w(x_i - v_j)^T A(x_i - v_j)}\right]^{1/\alpha-1}}. \quad (9A.8)$$

In order to determine the prototype vector v_j, Equation (9A.6) can be rewritten as

$$\left[\sum_{i=1}^{n}\frac{1}{2}(x_i^T B_2 x_i)^{-1}(B_1 - B_2)H(S_{i,j})u_{ij}^{\alpha} + 2wA.u_{ij}^{\alpha}\right] v_j = \sum_{i=1}^{n}\left[\frac{1}{2}(x_i^T B_2 x_i)^{-1}(B_2 + B_1)x_i.H(S_{i,j})u_{ij}^{\alpha} + 2wA.x_i u_{ij}^{\alpha}\right].$$

So

$$E.v_j = F \quad (9A.9)$$

where

$$E = \sum_{i=1}^{n}\frac{1}{2}(x_i^T B_2 x_i)^{-1}(B_1 - B_2)H(S_{i,j})u_{ij}^{\alpha} + 2wA.u_{ij}^{\alpha}$$

$$F = \sum_{i=1}^{n}\left[\frac{1}{2}(x_i^T B_2 x_i)^{-1}(B_2 + B_1)x_i.H(S_{i,j})u_{ij}^{\alpha} + 2wA.x_i.u_{ij}^{\alpha}\right].$$

REFERENCES

[1] R. Babuška (1998) *Fuzzy Modeling for Control*. Boston: Kluwer Academic Publishers.

[2] J. C. Bezdek (1981) *Pattern Recognition with Fuzzy Objective Function*. New York: Plenum Press.

[3] J. C. Bezdek, C. Coray, R. Gunderson and J. Watson (1981a) 'Detection and characterization of cluster substructure - Part I. Linear structure: fuzzy *c*-lines'. *SIAM Journal of Applied Mathematics*, **40**(2), 339–357.

[4] J. C. Bezdek, C. Coray, R. Gunderson and J. Watson (1981b) 'Detection and characterization of cluster substructure - Part II. Fuzzy *c*- varieties and convex combination thereof'. *SIAM Journal of Applied Mathematics*, **40**(2), 358–372.

[5] J.C. Bezdek and N. R. Pal (1998) 'Some new indices of cluster validity'. *IEEE Transactions on Systems, Man and Cybernetics – Part B*, **28**(3), 301–315.

[6] J. C. Bezdek, T. R. Reichherzer, G. S. Lim and Y. Attikiouzel (1998) 'Multiple prototype classifier design'. *IEEE Transactions on Systems, Man and Cybernetics – Part C*, **28**(1), 67–79.

[7] D. Chakraborty and N. R. Pal (2001) 'Integrated feature analysis and fuzzy rule-based system identification in a neuro-fuzzy paradigm'. *IEEE Transactions on Systems, Man and Cybernetics – Part B*, **31**(3), 391–400.

[8] C. L. Chang (1974) 'Finding prototypes for nearest neighbor classifiers'. *IEEE Transaction on Computers*, **23**(11), 1179–1184.

[9] S. L. Chiu (1994) 'Fuzzy model identification based on cluster estimation'. *Journal of Intelligent Fuzzy Systems*, **2**(3), 267–278.

[10] R. N. Davé (1990) 'Fuzzy shell-clustering and applications to circle detection in digital images'. *International Journal of General Systems*, **16**, 343–355.

[11] R. N. Davé (1991) 'Characterization and detection of noise in clustering'. *Pattern Recognition Letters*, **12**, 657–664.

[12] D. Dubois and H. Prade (1989) 'Fuzzy sets, probability and measurement'. *European Journal of Operational Research*, **40** 135–154.

[13] D. Dubois, E. Kerre, R. Mesiar and H. Prade (2000) 'Fuzzy interval analysis'. In (D. Dubois and H. Prade, eds) *Fundamentals of Fuzzy Sets, The Handbooks of Fuzzy Sets Series, Volume 7*. Kluwer Academic Publishers, pp. 483–81.

[14] A. E. Gaweda and J. M. Zurada (2003) 'Data-driven linguistic modeling using relational fuzzy rules', *IEEE Transactions on Fuzzy Systems*, **11**(1), 121–134.

[15] P. E. Gill, W. Murray and M. Wright (1981) *Practical Optimization*. New York: Academic Press.

[16] S. Guillaume (2001) 'Designing fuzzy inference systems from data: an interpretability-oriented review'. *IEEE Transactions on Fuzzy Systems*, **9**(3), 426–443.

[17] D. E. Gustafson and W. C. Kessel (1979) 'Fuzzy clustering with fuzzy covariance matrix'. In *Proceedings of the IEEE Conference on Decision and Control*, San Diego, CA, pp. 761–766.

[18] A. K. Jain, M. N. Murty and P. J. Flynn (1999) 'Data clustering: a review'. *ACM Computing Surveys*, **31**, 265–322.

[19] U. Kaymak and M. Setnes (2002) 'Fuzzy clustering with volume prototypes and adaptive cluster merging'. *IEEE Transactions on Fuzzy Systems*, **10**(6), 705–712.

[20] U. Kaymak and R. Babuška (1995) 'Compatible cluster merging for fuzzy modeling'. In *Proceedings of the Fourth IEEE International Conference on Fuzzy Systems* (*FUZZ-IEEE/IFES'95*), Yokohama, Japan, vol. 2, 897–904.

[21] R. Krishnapuram and J. Keller (1993) 'A possibilistic approach to clustering'. *IEEE Transactions on Fuzzy Systems*, **1**, 98–109.

[22] A. M. Luciano and M. Savastano (1997) 'Fuzzy identification of systems with unsupervised learning', *IEEE Transactions on Systems, Man and Cybernetics – Part B*, **27**(1), 138–141.

[23] S. Nefti, M. Oussalah, K. Djouani and J. Pontnau (1998) 'Clustering and adaptive robot navigation in an unknown environment'. *In Proccedings of the IEEE International Conference on Intelligent Systems (INES'98)*, Vienna, Austria, pp. ???–???

[24] Samia Nefti-Meziani and Mourad Oussalah, K. Djouani and J. Pontnau (2001) 'Intelligent adaptive mobile robot navigation'. *Journal of Intelligent and Robotic Systems*, **30**(4), 311–329.

[25] Samia Nefti-Meziani and Mourad Oussalah (2004) 'A neural network approach for railway safety prediction'. In *Proceedings of IEEE Systems, Man and Cybernetics, Netherlands*, **4**, 3915–3920.

[26] M. Oussalah (2000) 'On the probability/possibility transformations: a comparative analysis'. *Int. Journal of General Systems*, **29**, 671–718.

[27] W. Pedrycz (1994) 'Why triangular membership functions'? *Fuzzy Sets and Systems*, **64**, 21–30.

[28] W. Pedrycz (1995) *Fuzzy Sets Engineering*. CRC Press, Boca Raton, Florida, USA.

[29] W. Pedrycz (2005) *Knowledge-Based Clustering: From Data to Information Granules. From Data to Information Granules*. John Wiley & Sons, Canada.

[30] W. Pedrycz, J. C. Bezdek, R. J. Hathaway and G. W. Rogers (1998) 'Two nonparametric models for fusing heterogeneous fuzzy data'. *IEEE Transactions on Fuzzy Systems*, **6**(3), 411–425.

[31] G. S. V. Raju and J. Zhou (1993) 'Adaptive hierarchical fuzzy controllers'. *IEEE Transactions on Systems, Man and Cybernetics*, **23**(4), 973–980.

[32] R. Rovatti and R. Guerrieri (1996) 'Fuzzy sets of rules for system identification'. *IEEE Transactions on Fuzzy Systems*, **4**(2), 89–102.

[33] M. Setnes, R. Babuška, U. Kaymak and H. R. van Nauta Lemke (1998) 'Similarity measures in fuzzy rule base simplification'. *IEEE Transactions on Systems, Man and Cybernetics – Part B*, **28**(3), 376–386.

[34] M. Setnes, R. Babuška and H. B. Verbruggen (1998) 'Rule-based modeling: precision and transparency'. *IEEE Transactions on Systems, Man and Cybernetics – Part C*, **28**(1), 165–169.

[35] C.-T. Sun (1994) 'Rule-base structure identification in an adaptive-network-based fuzzy inference system'. *IEEE Transactions on Fuzzy Systems*, **2**(1), 64–73.

10
Mining Diagnostic Rules Using Fuzzy Clustering

Giovanna Castellano, Anna M. Fanelli and Corrado Mencar

Department of Computer Science, University of Bari, Italy

10.1 INTRODUCTION

Medical diagnosis is a field where fuzzy set theory can be applied with success, due to the high prominence of sources of uncertainty that should be taken into account when the diagnosis of a disease has to be formulated (Hughes, 1997; Mordeson and Malik, 2000). Furthermore, a proper application of fuzzy set theory can bridge the gap between the numerical world, in which symptoms are often observed and measured, and the symbolic world, in which knowledge should be expressed so as to be easy read and understood by human users (Zadeh, 1996).

The medical diagnosis problem is inherently generally a classification problem, where for each vector of symptom measurements one or a set of possible diagnoses are associated, eventually with different degrees of evidence (typicality). Such a diagnostic problem can be conveniently solved tackled by means of fuzzy set theory, leading to the so-called Fuzzy Diagnosis systems (Kuncheva and Steimann, 1999). In this case, the classifier is based on fuzzy rules, which can provide useful knowledge to physicians for the diagnosis of diseases when a set of symptoms is observed. Therefore, an important issue is the definition of such fuzzy rules, which can be set by domain experts or automatically mined from available data.

Many different techniques have been proposed for deriving fuzzy rules from data. In particular fuzzy clustering algorithms are powerful tools since they are able to discover automatically multi-dimensional relationships from data in the form of fuzzy relations. Such fuzzy relations are typically used to define fuzzy classification rules. Unfortunately, most fuzzy clustering techniques are designed to acquire accurate knowledge from data, paying little or no attention to the interpretability (i.e., readability by human users) of the extracted knowledge. Indeed, interpretability is a fundamental factor for the acceptability and the usability of a medical diagnosis system (Nauck and Kruse, 1999).

To emphasize the importance of interpretability in fuzzy medical diagnosis, in this chapter we describe a framework, which enables the extraction of transparent diagnostic rules through fuzzy clustering. The methodology underlying the framework relies on two clustering steps. In the first step, a clustering algorithm is applied to the multi-dimensional data, in order to discover the hidden relationships among data. The second clustering step operates at the level of each input dimension

Advances in Fuzzy Clustering and its Applications Edited by J. Valente de Oliveira and W. Pedrycz

to enable the definition of interpretable fuzzy sets to be used in the definition of diagnostic fuzzy rules. Differently from standard clustering schemes, this approach is able to extract fuzzy rules that satisfy interpretability constraints, so that a linguistic interpretation of diagnostic rules is immediate. Moreover, the framework is quite general and does not depend on specific clustering algorithms, which could be chosen according to different needs.

The chapter is organized as follows. In Section 10.2 we briefly overview the flourishing research area of fuzzy medical diagnosis. Section 10.3 motivates the need for interpretability in fuzzy rule-based systems when used in specific applicative areas such as medical diagnosis. In Section 10.4 we describe a fuzzy clustering framework for deriving interpretable fuzzy rules for medical diagnosis. In Section 10.5 the framework is applied for classifying different types of aphasia disease. Concluding remarks are given in Section 10.6.

10.2 FUZZY MEDICAL DIAGNOSIS

The diagnosis as a medical activity is aimed at stating if a patient suffers from a specific disease and, if the answer is positive, the specialist will provide a specific treatment. Based on a collection of observed symptoms, i.e., any information about the patient's state of health, a physician has to find a list of diagnostic possibilities for the patient.

The medical diagnosis process is inherently characterized by vagueness and uncertainty – both symptoms information and the resulting diagnoses are pervaded by imprecision. Therefore the so-called *medical knowledge*[1], expressing the relationships that exist between symptoms and diagnoses, is typically made of imprecise formulations (Kovalerchuk, Vityaev, and Ruiz, 2000). This imprecision is not a consequence of human inability, but it is an intrinsic part of expert knowledge acquired through protracted experience. As a consequence, any formalism disallowing uncertainty is not suitable to capture medical knowledge.

Pioneering research on computer aided medical diagnosis (Lusted, 1965) demonstrated the fundamental inadequacy of conventional mathematical methods for coping with the analysis of biological systems, and led "to accept as unavoidable a substantial degree of fuzziness in the description of the behavior of biological systems as well as in their characterization" (Zadeh, 1969). Of course, diseases can be considered fuzzy in that it is possible to have a disease to some degree: an expert attempts to classify a patient into some disease category using limited vague knowledge consisting primarily of elicited linguistic information[2].

Fuzzy set theory, introduced as a formal framework that allows us to capture the meaning of vague concepts, represents a natural way to fulfill two main requirements of medical diagnosis: on one side, preserving as much as possible the natural uncertainty embedded in medical knowledge; on the other side, providing a reliable diagnostic tool to the physician. In addition, according to the "Computing with Words" paradigm advocated by Zadeh (Zadeh, 1996), fuzzy set theory can bridge the gap between the numerical world, in which often symptoms are observed and measured, and the symbolic world, in which knowledge should be expressed so as to be easy to read and understand by human users.

During three decades of research done in computer-aided medicine, fuzzy set theory, used successfully with complex industrial control problems, has been widely applied on a variety of medical fields, leading to a proliferation of fuzzy medical systems (Phuong and Kreinovich, 2001; Szczepaniak, Lisboa, and Kacprzyk, 2000; Steimann, 1997). In particular a great deal of work has been done in anesthesia monitoring (Asbury and Tzabar, 1995; Linkens, 1996; Mason, Linkeno, and Edwards, 1997) and cardiology (Grauel, Ludwig and Klene, 1998; Sigura, Sigura, Kazui, and Harada, 1998; Kundu, Nasipuri, and Basu, 1998).

[1]The term *medical knowledge* was introduced for the first time in (Perez-Ojeda, 1976).
[2]Actually, this fuzziness arises in many classification domains. The medical domain is a typical extreme case, whereby most or even all information available is of a linguistic nature.

Diagnosis of diseases, however, remains the most prominent medical application domain in which fuzzy set theory has been effective, since it was found to be appropriate to model uncertainties found in medical diagnostic information (Adlassnig, 1986). Therefore a new research area was raised in the medical domain, known as Fuzzy Medical Diagnosis (Steimann and Adlassing, 2000; Kuncheva and Steimann, 1999). Good surveys of fuzzy approaches to medical diagnosis can be found in (Lisboa and Kacprzyk, Szczepaniak, 2000). Among these, early work by Adlassnig (Adlassnig, 1998) has been particularly influential in this domain, together with other fuzzy approaches to medical diagnosis (e.g., Belacel, Vincke, Schieff, and Bowassel, 2001; Kopecky, Hayde, Prusa, and Adlassnig, 2001; Kilic, Uncu, and Turksen, 2004; Nakashima *et al.*, 2005; Santos, 1993).

Fuzzy medical diagnosis concerns solving the problem of disease diagnosis (classification) through a system that relies on a set of fuzzy rules that use the particular representation of medical knowledge through fuzzy sets and hence can provide useful knowledge to physicians for the diagnosis of diseases when a set of symptoms is observed. For example, a fuzzy diagnostic rule can be of the form "*If clump thickness is HIGH and ⋯ and mitoses is MED-LOW then cancer is BENIGN (99%) or MALIGN* (1%)," where linguistic terms appearing in the antecedent and consequent parts are represented by fuzzy sets.

Fuzzy diagnostic rules can be manually defined by a human expert or automatically extracted from data. A lot of techniques have been proposed to derive fuzzy rules from available data. Among these, fuzzy clustering algorithms are commonly used (as in Abe, Thawonmas, and Kayama, 1999; Gomez-Skarmeta, Delgado, and Vila, 1999; Pedrycz, 1998; Tsekourasa, Sarimveisb, Kavaklia, and Befasb, 2005; Setnes, 2000), since they are able to discover multi-dimensional relationships from numerical observations, in the form of fuzzy relations that are subsequently employed as building blocks to define fuzzy rules.

Among the vast number of fuzzy systems based on rules discovered through fuzzy clustering, only a few of them (e.g., Nauck and Kruse, 1999; Delgado *et al.* 1999; John and Innocent, 2005; Castellano, Fanelli, and Mencar, 2003a; Roubos and Setnes, 2001) retain the original essence of fuzzy set theory, that lies in capturing complex relationships with transparent representation. Conversely, most of the proposed approaches to discover fuzzy rules from data aim to achieve high accuracy, disregarding the linguistic interpretability of fuzzy rules. Nevertheless, interpretability is highly desirable when fuzzy systems are applied to medical diagnosis, since fuzzy rules should provide a clear and understandable description of the medical knowledge to the physician.

The issue of interpretability, which is of fundamental importance in medical diagnosis, is addressed in depth in the following section.

10.3 INTERPRETABILITY IN FUZZY MEDICAL DIAGNOSIS

Intelligent systems, often used as decision support tools for aiding physicians in the diagnostic process, should be able to explain the relationships between the given set of inputs (i.e., symptoms, measurements, etc.) and the diagnosis provided in output. Put in other words, the knowledge base underlying the decision support system should be interpretable, i.e., accessible to the final user. In some applicative fields, such as the control field, interpretability has a lower priority w.r.t. accuracy, unless the produced knowledge base has to be used for explaining the behavior of a controlled plant. In many other applications involving decision support, interpretability of the knowledge base underlying the system is of prominent importance, as the user must be "convinced" on a decision suggested by the system. An interpretable knowledge base can be easily verified and related to the user domain knowledge so as to assess its reliability. As a consequence, debugging of the diagnostic system is facilitated and both the knowledge base and the related learning algorithm can be easily improved. In other words, when the decisional system relies on an interpretable knowledge base, it can be easily validated for its maintenance and for its evolution in view of changes in the external world (van de Merckt and Decaestecker, 1995).

Interpretability of intelligent systems represents an open-ended study in artificial intelligence known as the "Comprehensibility Problem" (Giboin, 1995). In this context, Michalski formulated the so-called

"Comprehensibility Postulate" that helps to give the right direction in the study of interpretability in intelligent systems (Michalski, 1983):

> The results of computer induction (i.e., empirical learning) should be symbolic descriptions of given entities, semantically and structurally similar to those a human expert might produce observing the same entities. Components of these descriptions should be comprehensible as single "chunks" of information, directly interpretable in natural language, and should relate quantitative and qualitative concepts in an integrated fashion.

Some considerations can be drawn from this postulate. First of all, the system supporting the user in the decision process should use symbols to describe the inference process. Black box models such as neural networks are very useful in many application areas but might not be fully justfiable in contexts like medical diagnosis, because of the inaccessibility of the acquired knowledge. Secondly, symbols used by the decisional system should be structurally and semantically similar to those produced by a human expert. As far as the structure of symbols is concerned, a simple yet powerful way to describe relationships between entities and, hence, knowledge, is represented by rules (Herrmann, 1997). Hence, rule-based systems are particularly suited to represent interpretable knowledge. Besides the structural facet, the semantics of symbols is of critical importance to achieve interpretability. Symbols represent "chunks" of information and should be directly mapped into natural language terms, often referring to vaguely defined concepts. Natural language terms are actually used in medical diagnosis to describe the medical knowledge from which a diagnosis is formulated. Fuzzy logic offers the possibility of formally representing terms with vague semantics, thus it plays an outstanding role in describing interpretable knowledge.

Actually, the association of linguistic terms to fuzzy relations is a very delicate task. Indeed, natural language terms implicitly bear a semantics that is shared among all speakers of that language. Let us call this implicit semantics "metaphor." As an example, the metaphor of the linguistic term "TALL" (referred to the height of a human body) is commonly shared by all English-speaking people. The communication of such term from person A to person B immediately highlights in B's mind a (fuzzy) set of body heights that highly matches the (fuzzy) set of body heights in A's mind. To achieve interpretability, fuzzy systems should adopt fuzzy sets with a semantics that is also shared by their users. This necessary condition does not heavily restrict the flexibility of such models: their learning ability should be able to adjust the semantics of fuzzy sets to adapt better to data (i.e., to improve their accuracy). This is common in human beings: the semantics of the term "TALL" in person A is not required to be perfectly the same of that in B, but only highly matching, since both A and B might have matured the concept of tallness on the basis of a different experience (Mencar, Castellano, and Fanelli, 2005).

Based on these considerations, we can state that fuzzy logic systems have a high potential for satisfying the Comprehensibility Postulate, i.e., for being interpretable. However, fuzzy systems acquired from data without necessary constraints could provide fuzzy yet inaccessible knowledge. Interpretability of fuzzy systems is a flourishing research direction (see Casillas, Cordon, Herera, and Magdalena, (2003) for related work), which "re-discovers" Zadeh's original intention of modeling linguistic terms, recently evolved in the definition of the so-called "Precisiated Natural Language," and the "Computing With Words" machinery for natural language-based inference (Zadeh, 2004). Interpretability issues typically arise when fuzzy rules are automatically acquired from data. In such cases, a common approach for ensuring interpretability is to impose a number of constraints to drive the learning process so that the acquired fuzzy sets can be easily associated to linguistic terms. Several interpretability constraints have been described in literature, which apply at different levels of the knowledge structure (Mencar, 2005):

- constraints on individual fuzzy sets;
- constraints on families of fuzzy sets belonging to the same universe of discourse;
- constraints on fuzzy relations, i.e., combinations (often the Cartesian product) of single fuzzy sets;
- constraints on individual fuzzy rules;
- constraints on the entire knowledge base.

Some interpretability constraints are application-oriented, while others are more of general-purpose. Some of them are even in conflict, thus witnessing the blurriness of the interpretability notion as there is no general agreeement on the choice of interpretability constraints. In order to achieve a high level of interpretability in fuzzy diagnostic rules, some general-purpose interpretability constraints should be applied on individual fuzzy sets, such as those described in Valente de Oliveira (1999a) and summarized below.

Normality. Each fuzzy set has at least one element with full membership. Normality is required when fuzzy sets represent possibility distributions (Dubois and Prade, 1997), and sub-normal fuzzy sets have nonzero degree of inclusion to the empty set, i.e., they are partially inconsistent (Pedrycz and Gomide, 1998). Normal fuzzy sets are related to interpretability since they guarantee the existence of at least one element of the universe which fully satisfy the semantics of the associated linguist term.

Convexity. Each fuzzy set is characterized by a prototype and the membership value of an element monotonically decreases as its distance to the prototype increases. Convex fuzzy sets are desirable in interpretable fuzzy models because they represent elementary concepts, whose degree of evidence is directly related to the similarity of elements to some prototype. Convex fuzzy sets are also useful in representing fuzzy numbers and fuzzy intervals, thus enabling a homogeneous representation of both qualitative and quantitative concepts.

Coverage. Fuzzy sets should cover the entire universe of discourse so that each element is well represented by some linguistic term. Usually, a strict coverage is preferable, which requires that for each element of the domain there exists at least one fuzzy set whose membership degree is greater than a predefined threshold;

Distinguishability. Two fuzzy sets of the same universe of discourse should not overlap too much, so that they can be associated to linguistic terms with well-separated metaphors. Such a property can be formalized by requiring the possibility function to be smaller than a predefined threshold for each couple of fuzzy sets (Mencar, Castellano, Bargiela, and Fanelli, 2004).

Extreme fuzzy sets. The extreme values of a universe of discourse should be prototypes for some fuzzy sets, which are called "leftmost" and "rightmost" fuzzy sets respectively. This constraint ensures that extreme points of the universe of discourse are prototypes of some limit concepts, such as "very low/very high" and "initial/final." Limit concepts are not required when the only quantitative concepts are used, while they are very common in human qualitative reasoning where extreme values are indicated by some linguistic terms.

Justifiable number of fuzzy sets. The number of fuzzy sets on each dimension, as well as the number of fuzzy rules, should be limited to 7 ± 2 in order to preserve interpretability. This limit derives from the psychological limit concerning the number of entities (independent of their informational complexity) that can be simultaneously stored in human short-term memory (Miller, 1956).

Interpretability constraints force the process of extraction of fuzzy rule to (totally or partially) satisfy a set of properties that are necessary in order to label fuzzy sets with linguistic terms. A preliminary survey of methods to keep interpretability in fuzzy rule extraction is given in Casillas, Cordon, Herrera, and Magdalena, (2003) and Guillame (2001). Constrained extraction of fuzzy rules can be achieved through a learning process, which may belong to one of the following categories:

(1) *Regularized learning algorithms.* These learning algorithms are aimed to extract fuzzy rules so as to optimize an objective function that promotes the derivation of an accurate knowledge base but penalizes those solutions that violate interpretability constraints, as in Valente de Oliveira (1999a). The objective function is properly encoded so as to be employed in classical constrained optimization techniques (e.g., Lagrangian multipliers method).

(2) *Genetic algorithms*. Fuzzy rules are properly encoded into a population of individuals that evolve according to an evolutionary cycle, which involves a selection process that fosters the survival of accurate and interpretable rules, as in Jin, Von Seelen and Sendhoff (1999). Genetic algorithms are especially useful when the interpretability constraints cannot be formalized as simple mathematical functions that can be optimized with the use of classical optimization techniques (e.g., gradient descent, least square methods, etc.). Moreover, multi-objective genetic algorithms are capable of dealing with several objective functions simultaneously (e.g., one objective function that evaluates accuracy and another to assess interpretability). The drawback of genetic algorithms is their inherent inefficiency that restricts their applicability. However, the adoption of genetic algorithms may rise efficiency issues in some application contexts.

(3) *Ad hoc learning algorithms*. Interpretable fuzzy rules are extracted by means of a learning algorithm that encodes in its schema the interpretability constraints, as in Chow, Altung, and Trussel (1999). Such types of algorithms can exhibit good performance at the expense of a more complex design.

In the next section, an algorithmic framework belonging to the last category is presented in detail.

10.4 A FRAMEWORK FOR MINING INTERPRETABLE DIAGNOSTIC RULES

The problem of medical diagnosis can be formalized as a classification problem, where a set of M diagnoses are defined for a certain medical problem and formalized as class labels:

$$\mathbf{D} = \{d_1, d_2, \ldots, d_M\}.$$

In order to assign a diagnosis to a patient, a set of symptoms (or any other useful information, such as clinical measurements, test results, etc.) are measured and formalized as an n-dimensional real vector $\mathbf{x} = (x_1, x_2, \ldots, x_n)$. To perform a diagnosis, a classifier is required to perform a mapping:

$$D : \mathbf{X} \subseteq \mathbb{R}^n \to \mathbf{D}. \tag{10.1}$$

The domain $\mathbf{X}$ defines the range of possible values for each symptom. Without loss of generality, the domain can be defined as a hyper-interval:

$$\mathbf{X} = [m_1, M_1] \times [m_2, M_2] \times \cdots \times [m_n, M_n].$$

In fuzzy diagnosis, the classifier relies on fuzzy set theory. Fuzzy sets can be used at different stages of the classifier design, e.g., as fuzzy inputs, fuzzy classes, or fuzzy rules. The latter are of prominent importance in medical diagnosis since they can represent vague knowledge in a nicely readable form.

In the presented framework, we focus on mining diagnostic fuzzy rules of the following schema:

$$\text{IF } \mathbf{x} \text{ is } \mathbf{G}_r \text{ THEN } \tilde{D}(\mathbf{x}) \text{ is } d_1(v_{r1}), \ldots, d_M(v_{rM}) \tag{10.2}$$

where $\mathbf{G}_r$ is a n-dimensional fuzzy relation. The degree of membership of the measurement vector $\mathbf{x}$ in $\mathbf{G}_r$ defines the strength of the rth rule and can be interpreted as a degree of similarity of the measurement vector $\mathbf{x}$ with respect to the prototype vector[3] defined by $\mathbf{G}_r$. Such fuzzy rules are used to perform a fuzzy classification task by realizing a mapping of the form:

$$\tilde{D} : \mathbf{x} \in \mathbf{X} \to (v_1, v_2, \ldots, v_M) \in [0, 1]^M$$

where each v_j denotes the degree to which a diagnosis d_j is assigned to a patient whose symptoms measurements are represented by $\mathbf{x}$. Such degree of membership can have different semantics, depending on the way it is calculated (Dubois and Prade, 1997). In our work, the membership value v_j is interpreted as a degree of typicality of case $\mathbf{x}$ with respect to diagnosis d_j.

[3] A prototype of a fuzzy set is an element with maximum membership. For normal fuzzy sets (like those used throughout this chapter) the membership degree of a prototype is equal to 1.0

Each rule of the schema (10.2) defines a local model for the function $\tilde{D}$. In particular, the rth rule entails the fuzzy classifier to have:

$$\tilde{D}(\mathbf{x}) = (v_{r1}, v_{r2}, \ldots, v_{rM})$$

when $\mathbf{x}$ is a prototype of $\mathbf{G}_r$. When a set of R rules is given, the membership grades of each diagnosis are defined by the following inference formula:

$$v_j = \frac{\sum_{r=1}^{R} \mathbf{G}_r(\mathbf{x}) v_{rj}}{\sum_{r=1}^{R} \mathbf{G}_r(\mathbf{x})}$$

where $\mathbf{G}_r(\mathbf{x})$ denotes the degree of membership of vector $\mathbf{x}$ to the fuzzy relation $\mathbf{G}_r$. If required, the fuzzy decision $\tilde{D}$ can be made crisp as in (10.1) by defining:

$$D(\mathbf{x}) = d_j \Leftrightarrow v_j = \max\ (v_1 \ldots v_M)$$

In the case of tiers, i.e., points assigned to different classes with the same membership degree, just one is chosen according to a selected criterion (e.g., randomly).

Fuzzy relations $\mathbf{G}_r$ are defined as the Cartesian product of one-dimensional fuzzy sets, that is:

$$\mathbf{G}_r = G_{r1} \times G_{r2} \times \cdots \times G_{rn}$$

where:

$$G_{ri} : [m_i, M_i] \rightarrow [0, 1]$$

is a one-dimensional fuzzy set that must models a qualitative property on its respective domain.

Here we present an algorithmic framework, called DC*f* (double clustering framework) (Castellano, Fanelli, and Mencar, 2005) for deriving fuzzy relations through clustering, with the additional feature of guaranteeing the satisfaction of the interpretability constraints described in the previous section. With such constraints verified, fuzzy relations can be easily attached with linguistic terms, so that the derived diagnostic rules express meaningful relationships between symptoms and diagnoses in natural language. To achieve both accuracy and interpretability, DC*f* is designed to combine the advantages of both multi-dimensional and one-dimensional clustering. Indeed, multi-dimensional clustering captures the multi-dimensional relationships existing among data, but the fuzzification of the resulting clusters may result in fuzzy sets that cannot be associated with qualitative linguistic labels. Conversely, one-dimensional clustering provides interpretable fuzzy sets but may loose information about the multi-dimensional relations underlying the data. The integration of one-dimensional and multi-dimensional clustering enables derivation of fuzzy relations that lead to interpretable fuzzy rules.

Specifically, DC*f* performs three main steps:

(1) *Data clustering*. Clustering is performed in the multi-dimensional space of numerical data to embrace similar data into clusters, providing a number of multi-dimensional prototypes (Figure 10.1);
(2) *Prototype clustering*. Prototypes obtained from the first clustering step are further clustered along each dimension of the input space, so as to obtain a number of one-dimensional prototypes for each feature (Figure 10.2);
(3) *Cluster fuzzification*. Multi-dimensional and one-dimensional prototypes provide useful information to derive fuzzy relations that can be conveniently represented by fuzzy sets (Figure 10.3 and 10.4). Moreover, such fuzzy sets are built in accordance with the interpretability constraints that allow the derivation of transparent diagnostic rules.

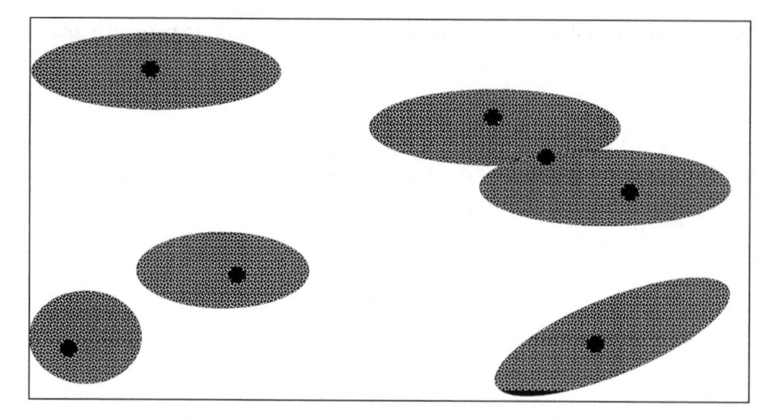

Figure 10.1 The first step of DC*f*. The available data are clustered providing multi-dimensional prototypes.

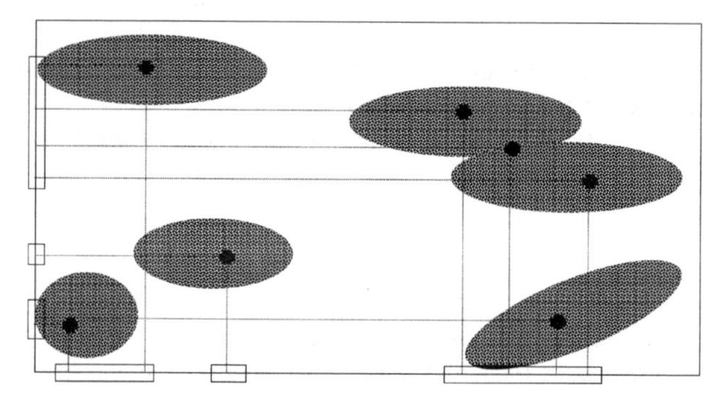

Figure 10.2 The second step of DC*f*. The multi-dimensional prototypes are projected onto each dimension and then clustered.

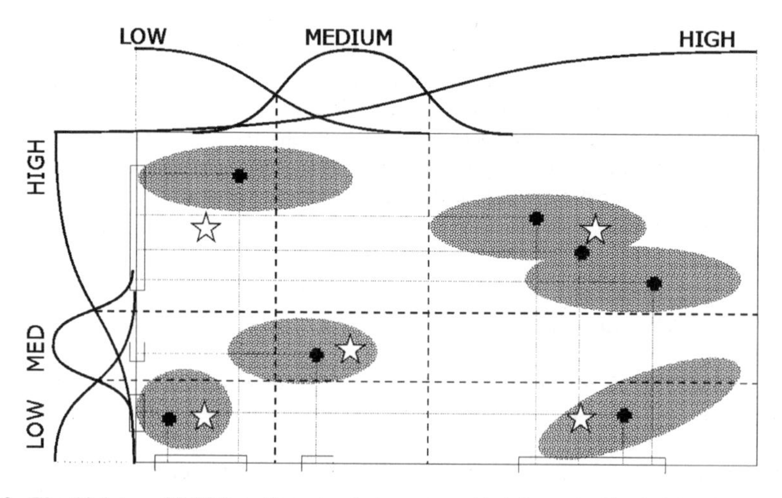

Figure 10.3 The third step of DC*f*. One-dimensional clusters provide information for defining fuzzy sets for each feature, which meet the interpretability constraints. One-dimensional fuzzy sets are combined, but only representative combinations (marked with stars) are selected.

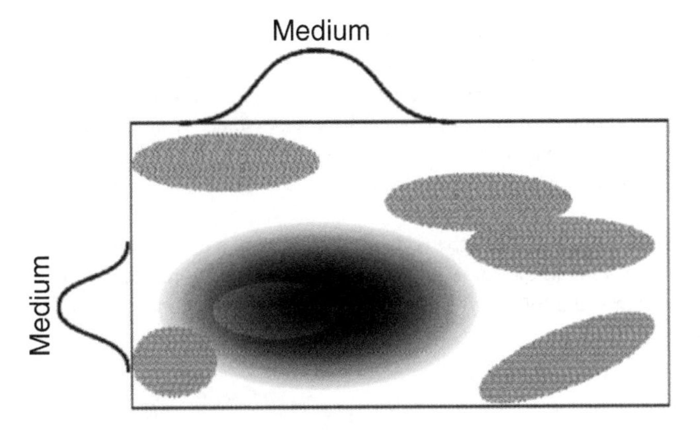

Figure 10.4 The membership function of a multi-dimensional cluster (shaded area). The cluster covers an aggregate of data and is represented by linguistically meaningful terms.

Formally, the Double Clustering framework can be described as follows. Let

$$T = \{x_l \in \mathbf{X} \ : \ l = 1, 2, \ldots, N\}$$

be a data-set of available examples. The first step of DC*f* performs a multi-dimensional clustering on the data-set T, providing a collection of multi-dimensional prototypes:

$$\mathbf{c}_1, \mathbf{c}_2, \ldots, \mathbf{c}_p \in \mathbf{X}$$

being $c_{\mathbf{h}} \ldots = (c_h^{(1)}, c_h^{(2)}, \ldots, c_h^{(n)}), \ h = 1, 2, \ldots, p.$

The multi-dimensional prototypes are then projected onto each dimension, resulting in n sets:

$$C^{(i)} = \left\{c_h^{(i)} \in [m_i, M_i] \ : \ h = 1, 2, \ldots, p\right\}$$

for $i = 1, 2, \ldots, n.$

In the second step of DC*f*, the points of each $C^{(i)}$ are subject to one-dimensional clustering, yielding to one-dimensional prototypes:

$$P^{(i)} = \left\{p_1^{(i)}, p_2^{(i)}, \ldots, p_{K_i}^{(i)}\right\}$$

K_i being the number of clusters in the ith dimension. To choose the number of clusters for each dimension, a trade off between accuracy and interpretability has to be considered. Since DC*f* is mainly designed to provide interpretable rules, it is advisable to choose K_i so as to satisfy the *justifiable number of fuzzy sets* interpretability constraint. Of course, if higher accuracy is needed, a higher value of K_i should be chosen.

The last step of DC*f* involves the derivation of fuzzy relations. This is achieved by first fuzzifying the one-dimensional clusters defined by the prototypes in each $P^{(i)}$ and then by aggregating one-dimensional fuzzy sets to form multi-dimensional fuzzy relations. Precisely, for each dimension $i = 1, 2, \ldots, n$, the K_i extracted prototypes are transformed into as many interpretable fuzzy sets. Different types of membership functions can be used to characterize fuzzy sets. Here, Gaussian fuzzy sets are considered, with the following membership functions:

$$A_k^{(i)}(x) = \exp\left[-\frac{(x - \omega_k^{(i)})^2}{2(\sigma_k^{(i)})^2}\right]$$

for $k = 1, 2, \ldots, K_i$.

The choice of Gaussian membership functions within DC*f* is motivated by their symmetry w.r.t. their respective prototypes, which can lead to a more natural assignment of linguistic terms. Moreover, the membership function of a Gaussian fuzzy set is strictly decreasing as the distance between any element and the prototype increases. As a consequence, it is always possible to compare any two elements of the

Universe of Discourse in terms of the property represented by the fuzzy set. As an example, given a Gaussian fuzzy set representing the property TALL, and any two subjects A and B, it is always possible to verify whether A is taller than B or vice versa. On the other hand, if other types of membership functions are used, which do not provide for strict monotonicity w.r.t. distance from the prototypes (e.g., triangular fuzzy sets), subjects A and B may have the same degree of tallness even if their effective height is highly different. For such reasons, Gaussian fuzzy sets may be more powerful for representing the semantics of linguistic terms.

The definition of the centers $\omega_k^{(i)}$ and the widths $\sigma_k^{(i)}$ should take into account the information provided by the clustering stages and, at the same time, should meet the required interpretability constraints. To satisfy both requirements, the following cut points are defined (note that there are $K_i + 1$ cut points for each dimension):

$$t_k^{(i)} = \begin{cases} 2m_i - t_1^{(i)} & \text{for} \quad k = 0 \\ \left(p_k^{(i)} + p_{k+1}^{(i)}\right)/2 & \text{for} \quad 0 < k < K_i \\ 2M_i - t_{K_i-1}^{(i)} & \text{for} \quad k = K_i. \end{cases}$$

Cut points are used to define centers and widths of the Gaussian membership functions according to the following relations:

$$\omega_k^{(i)} = \frac{t_{k-1}^{(i)} + t_k^{(i)}}{2}$$

and

$$\sigma_k^{(i)} = \frac{t_k^{(i)} - t_{k-1}^{(i)}}{2\sqrt{-2\ln\varepsilon}}$$

where ε is the maximum allowed overlap between two adjacent fuzzy sets.

It is easy to show that for each dimension, these fuzzy sets meet all the mentioned interpretability constraints. Multi-dimensional fuzzy relations can be then formed by combining one-dimensional fuzzy sets, one for each feature. However, this would lead to an exponentially high number of fuzzy relations, that is:

$$\prod_{i=1}^{n} K_i \geq 2^n$$

which violates the compactness interpretability constraint. Furthermore, many of such fuzzy relations would be useless as they would not represent available data. To avoid such a combinatorial explosion, only fuzzy relations that better represent the multi-dimensional prototypes $\mathbf{c}_h$ are considered. The selection of such clusters is accomplished on each dimension by considering, for each $h = 1, 2, \ldots, p$, the fuzzy set in the ith dimension with highest membership value on the ith projection of the hth prototype (see also Figure 10.3). The index of such a fuzzy set is defined as:

$$\overline{k_h^{(i)}} = \underset{k=1,2,\ldots,K_i}{\arg\max}\, A_k^{(i)}\left(c_h^{(i)}\right).$$

Once representative fuzzy sets are chosen, multi-dimensional fuzzy relations can be defined as usual. Specifically, the semantics of each relation is defined by the Cartesian product of the selected fuzzy sets, that is: $G_{hi} = A^{(i)}_{\overline{k_h^{(i)}}}$ while its linguistic notation is defined as:

$$\mathbf{G}_h \equiv v^{(1)} \text{ is } \ell A^{(1)}_{\overline{k_h^{(1)}}} \text{AND} \ldots \text{AND} v^{(n)} \text{is } \ell A^{(n)}_{\overline{k_h^{(n)}}}$$

for $h = 1, 2, \ldots, p$, where $\ell A^{(1)}_{\overline{k_h^{(1)}}}$ is the linguistic term (usually an adjective such as "high," "small," etc.) of the $\overline{k_h^{(1)}}$th fuzzy set defined for the jth feature. It is possible that two or more clusters coincide, hence the total number R of derived fuzzy relations is upper bounded by the number of multi-dimensional prototypes p.

The last stage of DC*f* aims to define diagnostic rules from the discovered fuzzy relations. More specifically, each distinct fuzzy relation constitutes the antecedent of a fuzzy rule whose consequent part is defined as:

$$v_{rj} = \frac{\sum_{l=1}^{N} \mathbf{G}_r(\mathbf{x}_l) \cdot \chi(d(\mathbf{x}_l), d_j)}{\sum_{k=1}^{N} \mathbf{G}_r(\mathbf{x}_l)}, \quad \begin{array}{l} r = 1, 2, \ldots, R \\ j = 1, 2, \ldots, M \end{array}$$

where $d(\mathbf{x}_l)$ is the diagnosis (class) associated to the lth example of the data set by an external supervisor, and:

$$\chi(d(\mathbf{x}_l), d_j) = 1 \quad \text{if } d(\mathbf{x}_l) = d_j, \quad \text{otherwise } 0.$$

Due to the method employed to calculate membership grades v_{rj}, their semantics can be interpreted as a degree of typicality of diagnosis d_j when the measures of symptoms are in the neighborhood of the prototype of $\mathbf{G}_r$.

DC*f* can be customized by choosing appropriate clustering algorithms, either for the first or the second step. To perform the first step of DC*f*, any clustering algorithm that provides a set of prototypes can be used, such as the fuzzy C-means (Bezdek, 1981) and all its variants (e.g., the Gustafson–Kessel algorithm, Gustafson and Kessel (1979), the gath-geva algorithm (Abonyi, Babuška, and Szeifert, 2002), the conditional fuzzy C-means (Pedrycz, 1996), etc.), and the possibilistic C-means (Krishnapuram and Keller, 1993). The second step of DC*f* does not require a fuzzy clustering scheme, since fuzzification in each dimension is performed only in the third step of DC*f*. Moreover, the number of points to be clustered on each dimension in the second step is very low (i.e., it coincides with the number of multi-dimensional clusters extracted in the first step), hence a simple crisp clustering algorithm can be adopted to perform prototype clustering.

The choice of specific clustering algorithms defines a particular implementation of DC*f*. Here, two possible implementations of DC*f* are briefly described:

Fuzzy double clustering. A first implementation of DC*f* integrates the fuzzy C-means algorithm for the multi-dimensional clustering (first step) and a hierarchical clustering scheme for the prototype clustering (second step). The hierarchical clustering is simple and quite efficient for one-dimensional numerical data. This implementation of DC*f*, that we call *F*DC (*Fuzzy* double clustering) is particularly suited to enhance existing fuzzy clustering algorithms in order to mine interpretable fuzzy rules from data (Castellano, Fanelli and Mencar, 2002).

Crisp double clustering. The implementation of DC*f* based on fuzzy double clustering is very straightforward and turns out to be useful when an existing fuzzy clustering application is to be wrapped so as to accommodate interpretability constraints. However, the use of a fuzzy clustering scheme requires computation and storing of the partition matrix, which is actually unused in the second step. To reduce such computational effort, it is more convenient to use a vector quantization technique in place of the fuzzy clustering algorithm in the multi-dimensional data clustering stage of DC*f*. This leads to another implementation of DC*f*, called *C*DC (*crisp* double clustering), in which a vector quantization algorithm that follows the Linde–Buzo–Gray (LBG) formulation (Linde and Gray, 1980) is used to accomplish the first clustering step and, like in *F*DC, a hierarchical clustering algorithm is used for the second step. Details about the *C*DC can be found in (Castellano, Fanelli, and Mencar, 2003b).

10.5 AN ILLUSTRATIVE EXAMPLE

To show the effectiveness of the double clustering framework in mining diagnostic rules from data that are both interpretable and accurate, a real-world diagnostic problem was considered. In particular, data from

Table 10.1 The AAT scores used in the simulation.

AAT score	Description
P1	Articulation and prosody (melody of speech)
P5	Syntactic structure (structure of sentences, grammar)
N0	Repetition
C1	Written language – reading aloud

the Aachen Aphasia Test (AAT), publicly available[4], were used to derive aphasia diagnostic rules. The original AAT data-set consists of 265 cases with several attributes, including AAT scores, several nominal attributes (including the diagnosis) and images of lesion profiles (full details can be found in Axer *et al.* (2002a)).

The data-set was preprocessed by selecting only 146 cases corresponding to the four most common aphasia diagnoses: Broca (motor or expressive aphasia), Wernicke (sensory or receptive aphasia), Anomic (difficulties in retrieval of words), and Global (total aphasia). In addition, only a selection of the AAT scores was taken into account, according to feature selection suggested Axer *et al.* (2000b). The selected AAT scores are illustrated in Table 10.1.

To perform simulations, we run the **CDC** version of DC*f* by varying the number of multi-dimensional prototypes and the number of fuzzy sets per dimension. Each simulation was repeated according to the 20-fold stratified cross-validation strategy. The average number of discovered rules is reported in Table 10.2, while classification results are summarized in Tables 10.3 and 10.4 reporting, respectively, the mean classification error on the test set and on the training set. It can be noted that the classification error decreases both on the test set and training set as the number of multi-dimensional prototypes – and consequently the number of rules – increases. On the other hand, when the number of fuzzy sets per input increases (i.e., it is greater than four), we observe an increase of classification error. Such a trend can be justified by the presence of noise that is captured by finely grained fuzzy partitioning. Summarizing, a good trade off between accuracy and interpretability can be achieved by selecting three to four fuzzy sets per input so as to avoid overfitting, and a number of prototypes from 10 to 20 so as to limit the number

Table 10.2 Average number of discovered rules.

	Fuzzy sets per input					
N	2	3	4	5	6	7
2	2.0	—	—	—	—	—
4	3.8	4.0	4.0	—	—	—
6	5.4	6.0	6.0	6.0	6.0	—
8	6.8	8.0	8.0	8.0	8.0	8.0
10	5.7	9.7	10.0	10.0	10.0	10.0
12	6.2	10.4	12.0	12.0	12.0	12.0
14	6.7	11.9	13.9	14.0	14.0	14.0
16	7.5	12.8	15.8	16.0	16.0	16.0
18	7.3	14.5	17.3	18.0	18.0	18.0
20	7.1	15.3	18.8	19.8	19.8	19.9
22	7.4	15.3	20.0	21.4	21.7	21.8
24	7.7	15.5	21.6	23.4	23.7	24.0

N = Number of multi-dimensional prototypes.

[4]http://fuzzy.iau.dtu.dk/aphasia.nsf/htmlmedia/database.html.

Table 10.3 Mean classification error on the test set.

	Fuzzy sets per input					
N	2	3	4	5	6	7
2	54.9%	—	—	—	—	—
4	50.3%	33.5%	40.9%	—	—	—
6	35.9%	24.1%	28.0%	25.1%	30.5%	—
8	34.7%	26.7%	24.3%	30.7%	29.2%	35.5%
10	30.9%	18.3%	22.1%	21.1%	23.4%	25.0%
12	31.1%	20.4%	20.7%	24.9%	25.7%	20.8%
14	27.2%	16.6%	19.6%	20.1%	23.0%	18.4%
16	27.6%	16.9%	19.3%	17.9%	20.3%	25.7%
18	32.4%	17.0%	16.0%	19.6%	19.3%	19.1%
20	30.2%	16.4%	18.7%	23.7%	20.4%	22.1%
22	29.2%	21.3%	21.7%	23.1%	22.1%	25.0%
24	27.1%	21.8%	16.6%	16.4%	19.0%	17.8%

N = Number of multi-dimensional prototypes.

of rules. The achieved classification results are quite stable, as shown by low values of standard deviations on the training sets (reported in Table 10.6). Conversely, standard deviations on the test sets (see Table 10.5) are higher but less significant since the mean cardinality of the test sets is very small (about 20 examples).

To appreciate the interpretability of the diagnostic rules discovered by DC*f*, in Figure 10.5 we report the seven rules generated by **CDC** with 10 multi-dimensional prototypes and three fuzzy sets per input. The membership functions of each fuzzy set are shown in Figure 10.6. It can be seen that all interpretability constraints are met. In particular, the number of rules, as well as the number of fuzzy sets per input, are sufficiently small, so that the resulting knowledge base is easy to read and understand. The high interpretability of this rule base is balanced by a good classification accuracy of the resulting classifier that provides a mean classification error of 19.2 % on the test set.

For comparison, we solved the same diagnostic problem by means of NEFCLASS[5](Nauck and Kruse, 1997). NEFCLASS was applied on the same data-set (with 20-fold cross validation), with asymmetric

Table 10.4 Mean classification error on the training set.

	Fuzzy sets per input					
N	2	3	4	5	6	7
2	52.5%	—	—	—	—	—
4	44.9%	34.2%	37.5%	—	—	—
6	33.3%	23.2%	25.2%	25.3%	30.7%	—
8	31.7%	22.0%	22.3%	25.5%	28.1%	32.7%
10	28.5%	18.2%	18.7%	20.2%	22.3%	22.3%
12	30.7%	18.7%	19.2%	20.5%	21.8%	23.5%
14	29.0%	16.9%	16.5%	16.8%	17.2%	18.7%
16	29.1%	15.4%	16.1%	17.0%	17.4%	18.5%
18	30.3%	14.3%	12.1%	15.0%	15.6%	16.9%
20	28.9%	15.4%	13.8%	15.4%	16.7%	17.2%
22	28.9%	17.2%	14.0%	15.7%	15.7%	16.5%
24	27.6%	16.0%	13.7%	14.2%	14.6%	15.0%

N = Number of multi-dimensional prototypes.

[5]We used the NEFCLASS-J package, available at http://fuzzy.cs.uni-magdeburg.de/nefclass/nefclass-j/.

Table 10.5 Standard deviation of the classification error on the test set.

	Fuzzy sets per input					
N	2	3	4	5	6	7
2	10.8%	—	—	—	—	—
4	18.7%	19.2%	14.9%	—	—	—
6	19.7%	23.2%	16.1%	16.4%	—	
8	16.5%	25.4%	19.9%	22.7%	17.7%	13.2%
10	17.2%	18.8%	19.5%	15.3%	14.5%	17.4%
12	14.6%	18.8%	20.0%	19.0%	17.0%	12.9%
14	16.7%	18.2%	20.4%	19.4%	20.5%	17.0%
16	19.4%	16.0%	23.3%	19.7%	15.3%	23.1%
18	19.7%	16.6%	16.4%	17.0%	15.1%	12.0%
20	12.7%	14.6%	18.7%	18.7%	17.3%	18.4%
22	11.8%	17.2%	19.4%	18.8%	17.8%	20.6%
24	20.5%	19.9%	18.8%	22.8%	19.9%	13.5%

N = Number of multi-dimensional prototypes.

Table 10.6 Standard deviation of the classification error on the training set.

	Fuzzy sets per input					
N	2	3	4	5	6	7
2	1.7%	—	—	—	—	—
4	7.7%	3.6%	5.8%	—	—	—
6	5.1%	3.2%	5.0%	5.0%	9.9%	—
8	5.3%	2.6%	3.6%	3.6%	3.2%	5.2%
10	4.8%	2.8%	3.1%	2.6%	3.6%	4.3%
12	4.5%	1.5%	3.2%	3.0%	4.1%	4.1%
14	3.3%	3.2%	3.1%	2.6%	1.7%	2.3%
16	3.4%	2.0%	2.4%	2.7%	2.3%	2.8%
18	3.4%	3.2%	2.4%	2.5%	2.2%	2.4%
20	4.6%	2.6%	3.3%	2.4%	2.3%	1.7%
22	2.9%	2.3%	3.3%	2.4%	2.2%	2.7%
24	3.9%	2.2%	3.4%	1.7%	2.1%	1.5%

N = Number of multi-dimensional prototypes.

1. If P1 is LOW AND P5 is LOW AND N0 is LOW AND C1 is LOW Then APHASIA is ANOMIC (0), BROCA (0.019), GLOBAL (0.98), WERNICKE (0.003)
2. If P1 is MEDIUM AND P5 is LOW AND N0 is LOW AND C1 is LOW Then APHASIA is ANOMIC (0), BROCA (0.1), GLOBAL (0.85), WERNICKE (0.053)
3. If P1 is MEDIUM AND P5 is MEDIUM AND N0 is MEDIUM AND C1 is HIGH Then APHASIA is ANOMIC (0.001), BROCA (0.96), GLOBAL (0.008), WERNICKE (0.035)
4. If P1 is HIGH AND P5 is LOW AND N0 is LOW AND C1 is LOW Then APHASIA is ANOMIC (0), BROCA (0.019), GLOBAL (0.91), WERNICKE (0.069).
5. If P1 is HIGH AND P5 is HIGH AND N0 is LOW AND C1 is MEDIUM Then APHASIA is ANOMIC (0.004), BROCA (0.097), GLOBAL (0.053), WERNICKE (0.85).
6. If P1 is HIGH AND P5 is HIGH AND N0 is MEDIUM AND C1 is HIGH Then APHASIA is ANOMIC (0.13), BROCA (0.14), GLOBAL (0.001), WERNICKE (0.72).
7. If P1 is HIGH AND P5 is HIGH AND N0 is HIGH AND C1 is HIGH Then ANOMIC (0.61), BROCA (0.053), GLOBAL (0), WERNICKE (0.34).

Figure 10.5 A set of rules discovered by CDC.

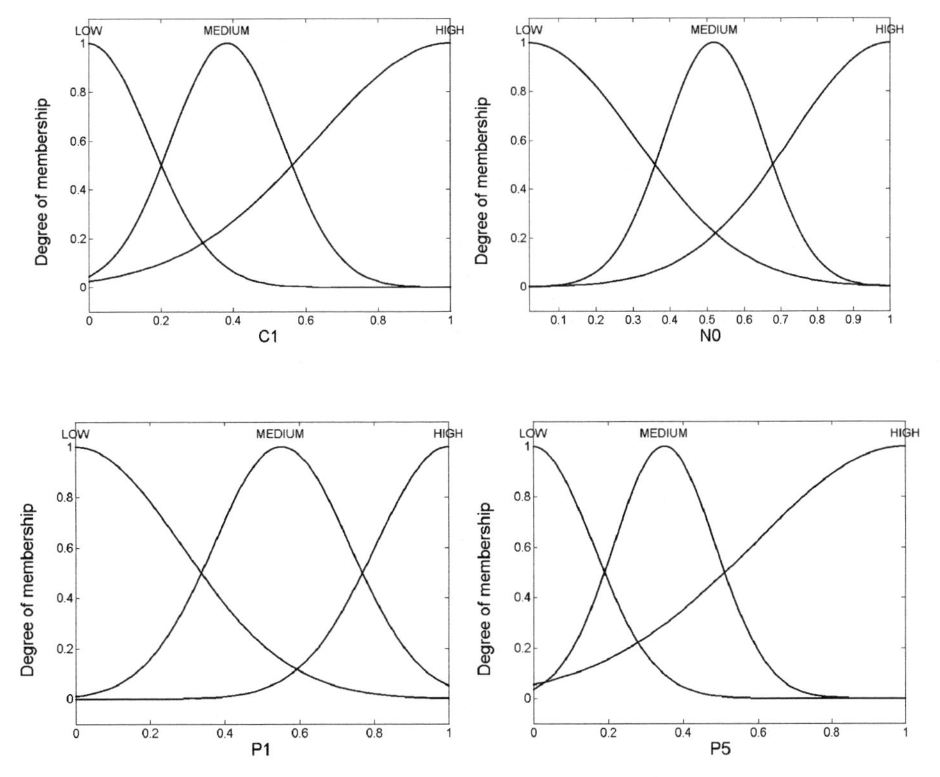

Figure 10.6 The fuzzy sets derived for each feature.

triangular membership functions constrained so as they hold the same order during training, and they always overlap and intersect at membership value 0.5. The maximum number of training epochs was fixed at 5000. Two rule selection strategies were used: "best rules" and "best per class." No pruning strategy was adopted, nor interactive management of the knowledge base. The results of NEFCLASS training are reported in Table 10.7.

Experimental results show that **CDC**, with a proper choice of multi-dimensional prototypes and fuzzy sets per dimensionis, leads to significantly higher accuracy than NEFCLASS. In addition, the number of rules generated by **CDC** is considerably smaller than the number of rules generated by NEFCLASS, which strongly requires an additional process for pruning the rule base to preserve legibility. Finally, it should be observed that **CDC** uses Gaussian membership functions, which are completely specified by

Table 10.7 Simulation results obtained by NEFCLASS.

Fuzzy sets per input	2	3	4	5	6	7
		Best rules				
Err.	25.4%	20.0%	16.7%	23.7%	29.2%	28.7%
Rules	11.0	11.7	29.5	34.9	48.4	55.4
		Best per class				
Err.	31.5%	21.1%	17.7%	24.2%	30.7%	32.0%
Rules	7.0	11.7	29.3	34.5	46.1	55.5

two parameters (center + width), whilst NEFCLASS uses asymmetrical triangular membership functions, which need three parameters to be specified (center + left width + right width).

10.6 CONCLUDING REMARKS

The success of the adoption of novel technologies and paradigms in critical application areas heavily depends on their ability to provide human-centric systems that support users in accomplishing complex tasks. Fuzzy logic (in the broad sense) has a great potential, because of its ability of representing and processing that kind of knowledge that humans routinely use to decide and act in complex situations. However, fuzzy logic, as any other mathematical tool, must be used with the necessary care to deliver useful tools to support users' decisions. This is especially important in situations, such as medical diagnosis, where trusting exclusively on machines is ethically unacceptable. Research on interpretability goes in the direction of giving to fuzzy tools the necessary conditions for being useful in such situations.

In this chapter we studied in detail the issue of interpretability in fuzzy medical diagnosis and emphasize how the adoption of formal constraints in the process of mining diagnostic rules can greatly increase the interpretability of a fuzzy diagnostic system. From this perspective, we have presented a clustering-based framework for mining interpretable fuzzy rules from data, which can be effectively employed in medical diagnosis. An illustrative example of disease diagnosis was presented to highlight how the mining of interpretable diagnostic rules can lead to the prediction of diagnoses with a high degree of typicality thus providing useful information for patient-tailored treatments.

The research on interpretability is still open-ended, and involves several disciplines other than computer science, such as cognitive science, psychology and philosophy. On the computational level, a promising research direction concerns improving existing systems (or devising new ones) to extract useful knowledge with a higher degree of automation.

REFERENCES

Abe, S., Thawonmas, R. and Kayama, M.A. (1999) 'Fuzzy classifier with ellipsoidal regions for diagnosis problems'. *IEEE Transactions on Systems, Man, and Cybernetics- Part C: Applications and Reviews*, **29**(1), 140–149.

Abonyi, J., Babuška, R. and Szeifert, F. (2002) 'Modified Gath–Geva fuzzy clustering for identification of Takagi–Sugeno fuzzy models'. *IEEE Transactions on Systems, Man and Cybernetic-Part B*, **32**(5), 612–621.

Adlassnig, K.P. (1986) 'Fuzzy set theory in medical diagnosis' *IEEE Trans. on Systems, Man and Cybernetics* **16**, 260–265.

Adlassnig, K.P. (1998) 'A fuzzy logical model of computer assisted medical diagnosis' *Meth. Inf. Med* **19**, 141–148.

Asbury, A.J., and Tzabar, Y. (1995) 'Fuzzy-logic – new ways of thinking anesthesia'. *British Journal of Anaesthesia*, **75**(1), 1–2.

Axer, H. *et al.* (2000a) 'The aphasia database on the web: description of a model for problems of classification in medicine'. In *Proc. of ESIT 2000*, pp. 104–111.

Axer, H. *et al.* (2000b) 'Aphasia classification using neural networks'. In *Proc. of ESIT 2000*, pp. 111–115Belacel, N., Vincke, N., Scheiff, J.M. and Boulassel, M.R. (2001) 'Acute leukemia diagnosis aid using multicriteria fuzzy assignment methodology'. *Comput. Meth. Prog. Med.* **64**(2), 145–151.

Bezdek, J. (1981) *Pattern Recognition with Fuzzy Objective Function Algorithms*. Plenum.

Casillas, J., Cordon, O., Herrera, F. and Magdalena, L. (eds.) (2003) *Interpretability Issues in Fuzzy Modeling*. Springer.

Castellano, G., Fanelli, A.M. and Mencar, C. (2002) 'A double-clustering approach for interpretable granulation of data'. In *Proc. 2nd IEEE Int. Conf. Systems, Man and Cybernetics (IEEE SMC 2002)*, Vol. 2, pp. 483–487.

Castellano, G., Fanelli, A.M. and Mencar, C. (2003a) 'A fuzzy clustering approach for mining diagnostic rules'. In *Proc. 2nd IEEE Int. Conf. Systems, Man and Cybernetics (IEEE SMC 2003)*, pp. 2007–2012.

Castellano, G., Fanelli, A.M. and Mencar, C. (2003b) 'Fuzzy granulation of multidimensional data by a crisp double clustering algorithm'. In *Proc. 7th World Multi-Conference on Systemics, Cybernetics and Informatics*, pp. 372–377.

Castellano, G., Fanelli, A.M. and Mencar, C. (2005) 'DCf : a double clustering framework for fuzzy information granulation'. In *Proc. IEEE Int. Conf. Granular Computing (IEEE GrC 2005)*, vol. 2, pp. 397–400.

Chow, M.Y., Altug, S. and Trussel, H.J. (1999) 'Heuristic constraints enforcement for training a fuzzy/neural architecture-part I: Foundation'. *IEEE Transactions on Fuzzy Systems*, **7**(2), 143–150.

Delgado, M., Sanchez, D., Martin-Bautista, M.J. and Vila, M.A. (1999) 'Mining association rules with improved semantics in medical databases'. *Artificial Intelligence in Medicine*, **21**, 241–245.

Dubois, D. and Prade, H. (1997) 'The three semantics of fuzzy sets'. *Fuzzy Sets and Systems*, **90**, 141–150.

Giboin, A. (1995) 'ML Comprehensibility and KBS explanation: stating the problem collaboratively'. In *Proc. of IJCAI'95 Workshop on Machine Learning and Comprehensibility*, pp. 1–11.

Gomez-Skarmeta, A.F., Delgado, M. and Vila, M.A. (1999) 'About the use of fuzzy clustering techniques for fuzzy model identification'. *Fuzzy Sets and Systems* **106**, 179–188.

Grauel, A., Ludwig, L.A. and Klene, G. (1998) 'ECG diagnostics by fuzzy decision making'. *Int. Journal of Uncertainty, fuzziness and Knowledge-based systems* **6**(2).

Guillaume, S. (2001) 'Designing fuzzy inference systems from data: an interpretability-oriented review'. *IEEE Transactions on Fuzzy Systems* **9**(3), 426–443.

Gustafson, E. and Kessel, W. (1979) 'Fuzzy clustering with a covariance matrix'. In *Proc. IEEE Conference on Decision Control*, pp. 761–766.

Herrmann, C.S. (1997) 'Symbolic reasoning about numerical data: a hybrid approach'. *Applied Intelligence*, **7**, 339–354.

Hughes, C. (1997) 'The representation of uncertainty in medical expert systems'. *Medical Informatics* **14**(4), 339–354.

Jin, Y., Von Seelen, W. and Sendhoff, B. (1999) 'On generating FC3 fuzzy rule systems from data using evolution strategies'. *IEEE Transactions on Systems, Man and Cybernetics-Part B*, **29**(6), 829–845.

John, R.I. and Innocent, P.R. (2005) 'Modeling uncertainty in clinical diagnosis using fuzzy logic'. *IEEE Transactions on Systems, Man and Cybernetics-Part B: Cybernetics*, **35**(6), 1340–1350.

Kilic, K., Uncu, O. and Turksen, I.B. (2004) 'Different approaches of fuzzy structure identification in mining medical diagnosis rules'. In *Proc. of 2004 IEEE Conference on Cybernetics and Intelligent Systems*, vol. 1, pp. 563–568.

Kopecky, D., Hayde, M., Prusa, A.R. and Adlassnig, K.P. (2001) 'Knowledge-based interpretation of toxoplasmosis serology results including fuzzy temporal concepts-The ToxoNet system'. *Medinfo* **10**, (1), 484–488.

Kovalerchuk, B., Vityaev, E. and Ruiz, J.F. (2000) 'Consistent knowledge discovery in medical diagnosis'. *IEEE Engineering in Medicine and Biology Magazine*, **19**(4), 26–37.

Krishnapuram, R. and Keller, J. (1993) 'A possibilistic approach to clustering'. *IEEE Transactions on Fuzzy Systems*, **1**(2), 98–110.

Kuncheva, L.I. and Steimann, F. (1999) 'Fuzzy Diagnosis (Editorial)'. *Artificial Intelligence in Medicine*, **16**(2), 121–128.

Kundu, M., Nasipuri, M. and Basu, D.K. (1998) 'A knowledge-based approach to ECG interpretation using fuzzy logic'. *IEEE Transactions on Systems, Man and Cybernetics-Part B*, **28**, 237–243.

Linde, A.B.Y. and Gray, R.M. (1980) 'An Algorithm for Vector Quantization Design'. *IEEE Trans. on Comm.*, **28**, 84–95.

Linkens, D.A., Shieh, J.S. and Peacock, J.E. (1996) 'Hierarchical fuzzy modeling for monitoring depth of anesthesia'. *Fuzzy Sets and Systems*, **79**(1), 43–57.

Lusted, L.B. (1965) 'Computer techniques in medical diagnosis'. In *Computers in Biomedical Research* (eds. Stacy, R.W. and Waxman, B.D.), pp. 319–338. Academic Press.

Mason, D.G., Linkens, D.A. and Edwards, N.D. (1997) 'Self-learning fuzzy logic control in medicine'. *Lecture Notes in Artificial Intelligence*, **1211**, 300–303.

Mencar, C. (2005) *Theory of Fuzzy Information Granulation: Contributions to Interpretability Issues*. PhD thesis, University of Bari, Italy.

Mencar, C., Castellano, G., Bargiela, A. and Fanelli, A.M. (2004) 'Similarity vs. possibility in measuring fuzzy sets distinguishability'. In *Proc. 5th International Conference on Recent Advances in Soft Computing (RASC 2004)*, pp. 354–359.

Mencar, C., Castellano, G. and Fanelli, A.M. (2005) 'Some fundamental interpretability issues in fuzzy modeling'. In *Proc. of Fourth Conference of the European Society for Fuzzy Logic and Technology and Rencontres Francophones sur la Logique Floue et ses applications (Joint, E.U.SFLAT-LFA 2005)*, pp. 100–105.

Michalski, R.S. (1983) 'A theory and methodology of inductive learning'. *Artificial Intelligence*, **20**, 111–161.

Miller, G.A. (1956) 'The magical number seven, plus or minus two: some limits on our capacity for processing information'. *The Psychological Review*, **63**, 81–97.

Mordeson, J.N. and Malik, D.S. (2000) *Fuzzy Mathematics in Medicine*. Physica-Verlag.

Nauck, D. and Kruse, R. (1999) 'Obtaining interpretable fuzzy classification rules from data'. *Artificial Intelligence in Medicine*, **16**(2), 129–147.

Nauck, D. and Kruse, R. (1997) 'A Neuro-Fuzzy Method to learn Fuzzy Classification Rules from Data'. *Fuzzy Sets and Systems* **89**, 277–288.

Nakashima, T., Schaefer, G., Yokota, Y., Shao Ying, Zhu and Ishibuchi, H. (2005) 'Weighted fuzzy classification with integrated learning method for medical diagnosis'. In *Proc. of 27th Annual International Conference of the Engineering in Medicine and Biology Society (IEEE-EMBS 2005)*, pp. 5623–5626.

Pedrycz, W. (1996) 'Conditional fuzzy c-means'. *Pattern Recognition Letters* **17**, 625–631.

Pedrycz, W. (1998) 'Fuzzy set technology in knowledge discovery'. *Fuzzy Sets and Systems*, **98**, 279–290.

Pedrycz, W. and Gomide, F. (1998) *An Introduction to Fuzzy Sets. Analysis and Design*. MIT Press.

Perez-Ojeda, A. (1976) *Medical knowledge network. A database for computer aided diagnosis*. Medical knowledge network. A database for computer aided diagnosis Master Thesis. Department of Industrial Engineering. University of Toronto, Canada.

Phuong, N.H. and Kreinovich, V. (2001) 'Fuzzy logic and its applications in medicine'. *International Journal of Medical Informatics*, **62**, 165–173.

Roubos, H. and Setnes, M. (2001) 'Compact and transparent fuzzy models and classifiers through iterative complexity reduction'. *IEEE Transactions on Fuzzy Systems*, **9**(4), 516–524.

Santos, E. Jr. (1993) 'On modeling time and uncertainty for diagnosis through linear constraint satisfaction'. In *Proc. Int. Congr. Computer Systems Applied Mathematics Workshop Constraint Processing*, pp. 93–106.

Setnes, M. (2000) 'Supervised fuzzy clustering for rule extraction'. *IEEE Transactions on Fuzzy Systems*, **8**(4), 416–424.

Sigura, T., Sigura, N., Kazui, T. and Harada, Y. (1998) 'A self-tuning effect of membership functions in a fuzzy logic-based cardiac pacing system'. *Journal of Medical Engineering and Technology*, **22**(3), 137–143.

Steimann, F. (1997) 'Fuzzy set theory in medicine (Editorial)'. *Artificial Intelligence in Medicine*, **11**, 1–7.

Steimann, F. and Adlassnig, K.P. (2000) 'Fuzzy Medical Diagnosis'. [Online].Available at http://www.citeseer.nj.-nec.com/160037.html in Rupini, E., Bonissone, P., Pedrycz, W. (1998) *Handbook of Fuzzy Computation*, Oxford University Press and Institute of Physics Publishing.

Szczepaniak, P.S., Lisboa, P.J.G. and Janusz Kacprzyk, J. (2000) *Fuzzy Systems in Medicine*. Phisica Verlag.

Tsekourasa, G., Sarimveisb, H., Kavaklia, E. and Bafasb, G. (2005) 'A hierarchical fuzzy-clustering approach to fuzzy modelling'. *Fuzzy Sets and Systems*, **150**, 245–266.

Valente de Oliveira, J. (1999a) 'Semantic constraints for membership function optimization'. *IEEE Transactions on Systems, Man and Cybernetics-Part A*, **29**(1), 128–138.

Valente de Oliveira, J. (1999) 'Towards neuro-linguistic modeling: constraints for optimization of membership functions'. *Fuzzy Sets and Systems*, **106**, 357–380.

van de Merckt, T.C. and Decaestecker, C. (1995) 'Multiple-knowledge representations in concept learning'. *Lecture Notes in Artificial Intelligence* **914**.

Zadeh, L. (1969) 'Biological application of the theory of fuzzy sets and systems'. In *Proc. Int. Symp. Biocybernetics of the Central Nervous System*, Brown & Co.

Zadeh, L. (1996) 'Fuzzy Logic = Computing with Words'. *IEEE Trans. on Fuzzy Systems*, **2**, 103–111.

Zadeh, L. (2004) 'Precisiated natural language (PNL)'. *AI Magazine*, **25**(3), 74–91.

11

Fuzzy Regression Clustering

Mika Sato-Ilic

Faculty of Systems and Information Engineering, University of Tsukuba, Japan

11.1 INTRODUCTION

Fuzzy regression clustering is a consortium of hybrid techniques combining regression analysis with fuzzy clustering. Regression analysis is a well-known method of data analysis the analytic capability of which is constantly challenged by problems of complexity and uncertainty of real data. Such data have emerged with inevitable local features of data that are nonignorable features. Many researches relating to regression analysis of these data have been greatly encouraged from the following two points: the first is the local statistical methods in which the purpose is to obtain a better fitness of the regression. The conventional statistical weighted regression analyses and related regression analyses using ideas of kernel, penalty, and smoothing are typical examples (Draper and Smith, 1966; Hastie and Tibshirani 1990; McCullagh and Nelder, 1989; Diggle, Heagerty, Liang, and Zeger, 2002). The second is spatial analysis whose target data is spatially distributed with local features. The main scope of this analysis is the implementation of regression analysis assessing the significance of the local relationship of observations in an attribute (or variable) space or a geographical space. This analysis is not designed for the pursuit of the better fitness.

The fuzzy regression clustering described in this chapter is based on the second point and the spatial features are captured by a classification structure of data. In clustering, the classification structure is usually obtained as exclusive clusters (groups) of objects in the observation space with respect to attributes (or variables). Exclusive clusters mean that objects that belong to a cluster are completely different from objects that belong to other clusters. Therefore, the features of each obtained cluster are assumed to be independent of each other. However, such a classification structure is inadequate to explain the spatial features since the spatial features are relatively related in space with respect to attributes (or variables). Fuzzy clustering can solve this problem and obtain the spatial features as degrees of belongingness of objects to fuzzy clusters. Fuzzy clustering can obtain not only the belonging status of objects but also how much the objects belong to the clusters. That is, in a fuzzy clustering result, there exist objects that belong to several fuzzy clusters simultaneously with certain degrees. In other words, the features of each obtained fuzzy cluster are relative to each other. Therefore, the result of fuzzy clustering is validated for use as the spatial feature of data. Moreover, we exploit some advantages of fuzzy clustering. Compared with conventional clustering, fuzzy clustering is well known as a robust and efficient way to reduce computation cost to obtain the result.

Advances in Fuzzy Clustering and its Applications Edited by J. Valente de Oliveira and W. Pedrycz

The models of fuzzy regression clustering discussed in this chapter are hybrid models that combine models of regression analysis and fuzzy clustering. By incorporating the aspects of nonlinearity of data obtained as fuzzy clustering into the regression that are capable of obtaining the linear structure, the mainstream of the fuzzy regression clustering is capturing the linear structure under the intrinsically distributed classification structure of data.

This chapter is organized as follows. Section 11.2 covers the basic issues of a statistical weighted regression model. Associated by the statistical weighted regression model, we describe a geographically weighted regression model (Brunsdon, Fotheringham, and Charlton, 1998) that is a typical spatial regression analysis. Section 11.3 develops models of fuzzy regression clustering. First, we briefly review typical fuzzy clustering methods and discuss several models of fuzzy regression clustering. In order to show a performance of the models of fuzzy regression clustering, Section 11.4 is concerned with results of analyses of residuals for the models. Examples are given in Section 11.5 to demonstrate the validity of the models of fuzzy regression clustering. We describe some conclusions in Section 11.6.

11.2 STATISTICAL WEIGHTED REGRESSION MODELS

An observed data that is composed of n objects and p independent variables is denoted as $\tilde{X} = (x_{ia}),\ i = 1, \cdots, n,\ a = 1, \cdots, p.\boldsymbol{y} = (y_1, \cdots, y_n)^t$ consists of n objects for a dependent variable.

The model of multiple regression analysis is defined as follows:

$$\boldsymbol{y} = X\boldsymbol{\beta} + \boldsymbol{\varepsilon}, \tag{11.1}$$

where

$$\boldsymbol{y} = \begin{pmatrix} y_1 \\ \vdots \\ y_n \end{pmatrix},\quad X = \begin{pmatrix} 1 & x_{11} & \cdots & x_{1p} \\ \vdots & \vdots & \vdots & \vdots \\ 1 & x_{n1} & \cdots & x_{np} \end{pmatrix},\quad \boldsymbol{\beta} = \begin{pmatrix} \beta_0 \\ \beta_1 \\ \vdots \\ \beta_p \end{pmatrix},\quad \boldsymbol{\varepsilon} = \begin{pmatrix} \varepsilon_1 \\ \vdots \\ \varepsilon_n \end{pmatrix}.$$

$\boldsymbol{\varepsilon}$ shows residual and we assume

$$\varepsilon_i \sim N(0, \sigma^2), \tag{11.2}$$

that is, ε_i is a normally distributed random variable, with mean zero and variance σ^2. We also assume ε_i and $\varepsilon_j,\ (i \neq j)$ are independent.

There sometimes occurs a case in which some of the n observations are "less reliable" compared to the other observations. For example, some of the observations have a large variance compared to the other observations. In this case, model (11.1) is not adaptable for the assumption (11.2).

In order to avoid the heteroscedastic residuals, a weighted regression analysis using the weighted least squares have been proposed (Draper and Smith, 1966; Dobson, 1990; Mandel, 1964). In the weighted regression analysis, we assume $\boldsymbol{\varepsilon}$ in Equation (11.1) as follows:

$$E\{\boldsymbol{\varepsilon}\} = \boldsymbol{0},\ V\{\boldsymbol{\varepsilon}\} = A\sigma^2,\ \boldsymbol{\varepsilon} \sim N(0, A\sigma^2). \tag{11.3}$$

where A is represented using a unique nonsingular symmetric matrix W that satisfies the following:

$$A = W^2,\ E\{\tilde{\boldsymbol{\varepsilon}}\} = \boldsymbol{0},\ \tilde{\boldsymbol{\varepsilon}} \equiv W^{-1}\boldsymbol{\varepsilon}. \tag{11.4}$$

$E\{\cdot\}$ shows a vector of expected values for $\cdot$ and $V\{\cdot\}$ is a variance-covariance matrix of $\cdot$. Equation (11.3) shows that we assume the heteroscedastic residuals in the regression model (11.1). Then the weighted regression model is defined as follows:

$$W^{-1}\boldsymbol{y} = W^{-1}X\boldsymbol{\beta} + W^{-1}\boldsymbol{\varepsilon} = W^{-1}X\boldsymbol{\beta} + \tilde{\boldsymbol{\varepsilon}}. \tag{11.5}$$

From Equation (11.4),

$$\begin{aligned} V\{\tilde{\varepsilon}\} &= E\{W^{-1}\varepsilon\varepsilon^t W^{-1}\} \\ &= W^{-1}E\{\varepsilon\varepsilon^t\}W^{-1} \\ &= W^{-1}A\sigma^2 W^{-1} \\ &= W^{-1}WWW^{-1}\sigma^2 \\ &= I\sigma^2, \end{aligned} \tag{11.6}$$

where I is an identity matrix. From Equation (11.6), it can be seen that we obtain the estimate of the weighted least squares of $\boldsymbol{\beta}$ under the homoscedastic variance for residuals. From Equation (11.5), the estimate of the weighted least squares of $\boldsymbol{\beta}$ can be obtained as follows:

$$\tilde{\boldsymbol{\beta}} = (X^t A^{-1} X)^{-1} X^t A^{-1}\mathbf{y}. \tag{11.7}$$

Using the idea of spatial analysis for geographically obtained data, a geographically weighted regression model has been proposed (Brunsdon, Fotheringham, and Charlton, 1998). The estimate of regression coefficient to an area k is obtained as:

$$\hat{\boldsymbol{\beta}}_k = (X^t W_k X)^{-1} X^t W_k \mathbf{y}, \tag{11.8}$$

where

$$W_k = \begin{pmatrix} \alpha_{1k} & \cdots & 0 \\ \vdots & \ddots & \vdots \\ 0 & \cdots & \alpha_{nk} \end{pmatrix}, \quad k = 1, \cdots, K.$$

K shows the number of areas and α_{ik} shows the weight of an object i to an area k estimated using kernel estimates. That is, the weight W_k is defined as

$$\alpha_{ik} = \phi(d_{ik}), \quad i = 1, \cdots, n, \; k = 1, \cdots, K,$$

using the kernel function ϕ that satisfies the following conditions:

$$\begin{aligned} &\phi(0) = 1, \\ &\lim_{d \to \infty} \{\phi(d)\} = 0, \\ &\phi \;\; : \text{monotone decreasing function}, \end{aligned}$$

where d_{ik} shows a distance between ith object and kth area. The following are examples of ϕ:

$$\alpha_{ik} = \exp\left(\frac{-d_{ik}}{r}\right),$$

$$\alpha_{ik} = \exp\left(\frac{-d_{ik}^2}{2r^2}\right),$$

where r shows the control parameter influenced by the range of each area.

Equation (11.8) is obtained by minimizing

$$\hat{S}_k \equiv (\mathbf{y} - X\boldsymbol{\beta}_k)^t W_k (\mathbf{y} - X\boldsymbol{\beta}_k) \equiv \hat{\boldsymbol{\delta}}_k^t \hat{\boldsymbol{\delta}}_k,$$

$$\hat{\boldsymbol{\delta}}_k \equiv W_k^{\frac{1}{2}}(\mathbf{y} - X\boldsymbol{\beta}_k) = (\hat{\delta}_{1k}, \cdots \hat{\delta}_{nk})^t,$$

where

$$W_k^{\frac{1}{2}} = \begin{pmatrix} \sqrt{\alpha_{1k}} & \cdots & 0 \\ \vdots & \ddots & \vdots \\ 0 & \cdots & \sqrt{\alpha_{nk}} \end{pmatrix}.$$

11.3 FUZZY REGRESSION CLUSTERING MODELS

In order to obtain a more reliable solution from the regression model, several methods using the idea of fuzzy logic (Zadeh, 1965; Zadeh, 2005) have been proposed (Tanaka and Watada, 1988; Hathaway and Bezdek, 1993; Sato-Ilic, 2003, 2004). Within these methods, there are several methods using the result of fuzzy clustering for the weights in the weighted regression model. In this section, we discuss the regression models using fuzzy clustering.

11.3.1 Fuzzy Clustering

First, we explain the fuzzy clustering algorithm used in this chapter. Fuzzy c-means (FCM) (Bezdek, 1987) is one of the methods of fuzzy clustering that minimizes the weighted within-class sum of squares:

$$J(U, \boldsymbol{v}_1, \cdots, \boldsymbol{v}_K) = \sum_{i=1}^{n} \sum_{k=1}^{K} u_{ik}^m d^2(\boldsymbol{x}_i, \boldsymbol{v}_k), \ i = 1, \cdots, n, \ k = 1, \cdots, K, \tag{11.9}$$

where u_{ik} shows the degree of belongingness of objects to clusters, n is the number of objects, and K is the number of clusters. In general, the state of fuzzy clustering is represented by a partition matrix $U = (u_{ik})$ whose elements satisfy the following conditions:

$$u_{ik} \in [0, 1], \ \sum_{k=1}^{K} u_{ik} = 1. \tag{11.10}$$

$\boldsymbol{v}_k = (v_{ka}), \ k = 1, \cdots, K, \ a = 1, \cdots, p$ denotes the values of the centroid of a cluster $k, \boldsymbol{x}_i = (x_{ia}), \ i = 1, \cdots, n, \ a = 1, \cdots, p$ shows ith object, and $d^2(\boldsymbol{x}_i, \boldsymbol{v}_k)$ is the square Euclidean distance between $\boldsymbol{x}_i$ and $\boldsymbol{v}_k$. p is the number of variables. The exponent m that determines the degree of fuzziness of the clustering is chosen from $(1, \infty)$ in advance.

The purpose of FCM is to obtain the solutions U and $\boldsymbol{v}_1, \cdots, \boldsymbol{v}_K$ that minimize Equation (11.9). If $\boldsymbol{v}_k = \sum_{i=1}^{n} u_{ik}^m \boldsymbol{x}_i / \sum_{i=1}^{n} u_{ik}^m$, then Equation (11.9) is shown as:

$$J(U) = \sum_{k=1}^{K} \left(\sum_{i=1}^{n} \sum_{j=1}^{n} u_{ik}^m u_{jk}^m d_{ij} / (2 \sum_{l=1}^{n} u_{lk}^m) \right). \tag{11.11}$$

When $m = 2$, the algorithm in which the objective function is Equation (11.11) is known as the FANNY algorithm (Kaufman and Rousseeuw, 1990).

11.3.2 Fuzzy c-regression Model

The fuzzy c-regression model (Hathaway and Bezdek, 1993) obtains the following estimate of weighted least squares:

$$\tilde{\tilde{\boldsymbol{\beta}}}_k = (\tilde{X}^t (U_k)^m \tilde{X})^{-1} \tilde{X}^t (U_k)^m \boldsymbol{y}, \tag{11.12}$$

that minimizes

$$S = \sum_{k=1}^{K} S_k, \ S_k \equiv (\boldsymbol{y} - \tilde{X}\boldsymbol{\beta}_k)^t (U_k)^m (\boldsymbol{y} - \tilde{X}\boldsymbol{\beta}_k) \equiv \boldsymbol{\delta}_k^t \boldsymbol{\delta}_k, \tag{11.13}$$

$$\boldsymbol{\delta}_k \equiv (U_k)^{\frac{m}{2}} (\boldsymbol{y} - \tilde{X}\boldsymbol{\beta}_k) = (\delta_{1k}, \cdots, \delta_{nk})^t, \ m > 1, \tag{11.14}$$

where

$$(U_k)^m = \begin{pmatrix} u_{1k}^m & \cdots & 0 \\ \vdots & \ddots & \vdots \\ 0 & \cdots & u_{nk}^m \end{pmatrix}.$$

In general, we assume the following conditions in order to avoid $u_{ik} = 0$ in Equation (11.13):

$$u_{ik} \in (0, 1), \quad \sum_{k=1}^{K} u_{ik} = 1. \tag{11.15}$$

11.3.3 Fuzzy Cluster Loading Model

A fuzzy clustering method is offered to contract clusters with uncertainty boundaries; this method allows one object to belong to some overlapping clusters with some grades. However, replaced by the representativeness of fuzzy clustering to real complex data, the interpretation of such a fuzzy clustering can cause us some confusion because we sometimes think that objects that have a similar degree of belongingness can together form one more cluster. In order to obtain the interpretation of the obtained fuzzy clusters, we have proposed a fuzzy cluster loading (Sato-Ilic, 2003) that can show the relationship between the clusters and the variables.

The model is defined as:

$$u_{ik} = \sum_{a=1}^{p} x_{ia} z_{ak} + \varepsilon_{ik}, \; i = 1, \cdots, n, \; k = 1, \cdots, K, \tag{11.16}$$

where u_{ik} is the obtained degree of belongingness of objects to clusters by using a fuzzy clustering method and is assumed to satisfy the conditions in Equation (11.15). z_{ak} shows the fuzzy degree that represents the amount of loading of a cluster k to a variable a and we call this fuzzy cluster loading. This parameter shows how each cluster can be explained by each variable. ε_{ik} is an error.

The model (11.16) is rewritten as

$$\boldsymbol{u}_k = \tilde{X} \boldsymbol{z}_k + \boldsymbol{\varepsilon}_k, \quad k = 1, \cdots, K, \tag{11.17}$$

using

$$\boldsymbol{u}_k = \begin{pmatrix} u_{1k} \\ \vdots \\ u_{nk} \end{pmatrix}, \quad \boldsymbol{z}_k = \begin{pmatrix} z_{1k} \\ \vdots \\ z_{pk} \end{pmatrix}, \quad \boldsymbol{\varepsilon}_k = \begin{pmatrix} \varepsilon_{1k} \\ \vdots \\ \varepsilon_{nk} \end{pmatrix},$$

where, we assume

$$\varepsilon_{ik} \sim N(0, \sigma^2). \tag{11.18}$$

The estimate of least squares of $\boldsymbol{z}_k$ for Equation (11.17) is obtained as follows:

$$\tilde{\boldsymbol{z}}_k = (\tilde{X}^t \tilde{X})^{-1} \tilde{X}^t \boldsymbol{u}_k, \tag{11.19}$$

by minimizing

$$\boldsymbol{\varepsilon}_k^t \boldsymbol{\varepsilon}_k = \varepsilon_{1k}^2 + \cdots + \varepsilon_{nk}^2. \tag{11.20}$$

Using Equation (11.17) and

$$U_k^{-1} = \begin{pmatrix} u_{1k}^{-1} & \cdots & 0 \\ \vdots & \ddots & \vdots \\ 0 & \cdots & u_{nk}^{-1} \end{pmatrix},$$

the model (11.16) can be rewritten again as:

$$\mathbf{1} = U_k^{-1}\tilde{X}z_k + \boldsymbol{e}_k, \quad \boldsymbol{e}_k \equiv U_k^{-1}\boldsymbol{\varepsilon}_k, \quad k = 1, \cdots, K, \tag{11.21}$$

where,

$$\mathbf{1} = \begin{pmatrix} 1 \\ \vdots \\ 1 \end{pmatrix}, \quad \boldsymbol{e}_k = \begin{pmatrix} e_{1k} \\ \vdots \\ e_{nk} \end{pmatrix}.$$

By minimizing

$$\boldsymbol{e}_k^t \boldsymbol{e}_k = \boldsymbol{\varepsilon}_k^t (U_k^{-1})^2 \boldsymbol{\varepsilon}_k, \tag{11.22}$$

we obtain the estimate of least squares of z_k for Equation (11.21) as follows:

$$\tilde{\tilde{z}}_k = (\tilde{X}^t (U_k^{-1})^2 \tilde{X})^{-1} \tilde{X}^t U_k^{-1} \mathbf{1}. \tag{11.23}$$

From Equation (11.22), $\tilde{\tilde{z}}_k$ is the estimate of weighted least squares of z_k in the weighted regression analysis. Equation (11.22) can be rewritten as follows:

$$\boldsymbol{e}_k^t \boldsymbol{e}_k = \boldsymbol{\varepsilon}_k^t (U_k^{-1})^2 \boldsymbol{\varepsilon}_k = (u_{1k}^{-1}\varepsilon_{1k})^2 + \cdots + (u_{nk}^{-1}\varepsilon_{nk})^2. \tag{11.24}$$

From Equation (11.24) and condition (11.15), we can see that if an object i belongs to the cluster k with a large degree, that is, a larger u_{ik} has a smaller $(u_{ik}^{-1}\varepsilon_{ik})^2$, since

$$a > b \Rightarrow (a^{-1}c)^2 < (b^{-1}c)^2, \quad a, b, c > 0.$$

If an object clearly belongs to a cluster then the weight works to the local regression over the cluster to fit better for the object. In other words, we avoid the objects that are vaguely situated on the clustering by treating them as noise of the data. It is shown where the degree of belongingness over the K clusters is close to $1/K$ for all of the objects. So, $\tilde{\tilde{z}}_k$ is obtained considering not only the fitness of the model but also considering the classification structure of the data $\tilde{X}$.

11.3.4 Kernel Fuzzy Cluster Loading Model

The kernel method has been discussed in the context of support vector machines (Cristianini and Shawe-Taylor, 2000), the efficient advantage of which is widely recognized in many areas. The essence of the kernel method is an arbitrary mapping from a lower dimension space to a higher dimension space. Note that the mapping is an arbitrary mapping, so we do not need to find the mapping, this is called the kernel trick.

Suppose an arbitrary mapping Φ:

$$\Phi: R^n \to F,$$

where F is a higher dimension space than R^n.

We assume

$$k(\boldsymbol{x}, \boldsymbol{y}) = \Phi(\boldsymbol{x})^t \Phi(\boldsymbol{y}),$$

where k is the kernel function that is defined in R^n and $\boldsymbol{x}, \ \boldsymbol{y} \in R^n$.

Typical examples of the kernel function are as follows:

$$\begin{aligned} k(\boldsymbol{x}, \boldsymbol{y}) &= \exp\left(-\frac{\| \boldsymbol{x} - \boldsymbol{y} \|}{2\sigma^2}\right) \text{ (Gaussian kernel)} \\ k(\boldsymbol{x}, \boldsymbol{y}) &= (\boldsymbol{x} \cdot \boldsymbol{y})^d \text{ (polynomial kernel of degree } d) \\ k(\boldsymbol{x}, \boldsymbol{y}) &= \tanh(\alpha(\boldsymbol{x} \cdot \boldsymbol{y}) + \beta) \text{ (Sigmoid kernel).} \end{aligned} \tag{11.25}$$

By the introduction of this kernel function, we can analyze the data in F without finding the mapping Φ explicitly.

From Equation (11.23), we can obtain the following:

$$\begin{aligned} z_k &= (\tilde{X}^t (U_k^{-1})^2 \tilde{X})^{-1} \tilde{X}^t U_k^{-1} \mathbf{1} \\ &= ((U_k^{-1}\tilde{X})^t (U_k^{-1}\tilde{X}))^{-1} (U_k^{-1}\tilde{X})^t \mathbf{1} \\ &\equiv (C_k^t C_k)^{-1} C_k^t \mathbf{1}, \end{aligned} \tag{11.26}$$

where $C_k = (c_{ia(k)}),\ c_{ia(k)} \equiv u_{ik}^{-1} x_{ia},\ i = 1, \cdots, n,\ a = 1, \cdots, p.$

Using $\boldsymbol{c}_{a(k)}^t = (c_{ia(k)}, \cdots, c_{na(k)})$, we can represent Equation (11.26) as follows:

$$z_k = (\boldsymbol{c}_{a(k)}^t \boldsymbol{c}_{b(k)})^{-1} (\boldsymbol{c}_{a(k)}^t \mathbf{1}),\ a, b = 1, \cdots, p, \tag{11.27}$$

where $C_k^t C_k = (\boldsymbol{c}_{a(k)}^t \boldsymbol{c}_{b(k)}),\ \ C_k^t \mathbf{1} = (\boldsymbol{c}_{a(k)}^t \mathbf{1}),\ \ a, b = 1, \cdots, p.$

Then we consider the following mapping Φ:

$$\Phi\text{: } R^n \to F,\ \boldsymbol{c}_{a(k)} \in R^n. \tag{11.28}$$

From Equations (11.27) and (11.28), the fuzzy cluster loading in F is as follows:

$$\hat{\tilde{z}}_k = (\Phi(\boldsymbol{c}_{a(k)})^t \Phi(\boldsymbol{c}_{b(k)}))^{-1} (\Phi(\boldsymbol{c}_{a(k)})^t \Phi(\mathbf{1})),\ a, b = 1, \cdots, p, \tag{11.29}$$

where $\hat{\tilde{z}}_k$ shows the fuzzy cluster loading in F.

Using the kernel representation $k(\boldsymbol{x}, \boldsymbol{y}) = \Phi(\boldsymbol{x})^t \Phi(\boldsymbol{y})$, Equation (11.29) is rewritten as follows:

$$\hat{\tilde{z}}_k = (k(\boldsymbol{c}_{a(k)}, \boldsymbol{c}_{b(k)}))^{-1} (k(\boldsymbol{c}_{a(k)}, \mathbf{1})),\ a, b = 1, \cdots, p. \tag{11.30}$$

From this, using the kernel method, we can estimate the fuzzy cluster loading in F (Sato-Ilic, 2003).

11.3.5 Weighted Fuzzy Regression Model

We have proposed another model for weighted regression using a fuzzy clustering result obtained as classification of the data consisting of independent variables (Sato-Ilic, 2004). The model is defined as:

$$\boldsymbol{y} = U_k \tilde{X} \boldsymbol{\beta}_k + \hat{\boldsymbol{e}}_k, \tag{11.31}$$

where

$$U_k = \begin{pmatrix} u_{1k} & \cdots & 0 \\ \vdots & \ddots & \vdots \\ 0 & \cdots & u_{nk} \end{pmatrix},\quad \hat{\boldsymbol{e}}_k = \begin{pmatrix} \hat{e}_{1k} \\ \vdots \\ \hat{e}_{nk} \end{pmatrix},$$

under an assumption of

$$\hat{e}_{ik} \sim N(0, \sigma^2). \tag{11.32}$$

The estimate of least squares of $\boldsymbol{\beta}_k$ in Equation (11.31) is obtained as

$$\hat{\tilde{\boldsymbol{\beta}}}_k = (\tilde{X}^t U_k^2 \tilde{X})^{-1} \tilde{X}^t U_k \boldsymbol{y}. \tag{11.33}$$

Notice that Equation (11.12) is different from Equation (11.33), even if we put $m = 2$ in Equation (11.12). The structure of Equation (11.12) is essentially the same as the estimate of conventional weighted regression analysis shown in Equation (11.7). That is, in Equation (11.13), we assume that weights of objects in the dependent variable are the same as the weights of objects in the independent variables. However, in model (11.31), we consider that the clustering result for the data of a dependent variable is not always the same as the clustering result for the data of independent variables. So, we multiplied the weights only to $\tilde{X}$ in Equation (11.31), since U_k is obtained as a classification result of $\tilde{X}$.

Figure 11.1 shows an example of the difference. In this figure, the abscissa shows values of the independent variable and the ordinate shows values of the dependent variable for five objects $\{x_1, x_2, x_3, x_4, x_5\}$. Solid circles in this figure show the clustering result with respect to the bivariate, dependent variable and independent variable. In this case, we assume that the classification structure of data of the dependent variable is the same as the classification structure of data of the independent

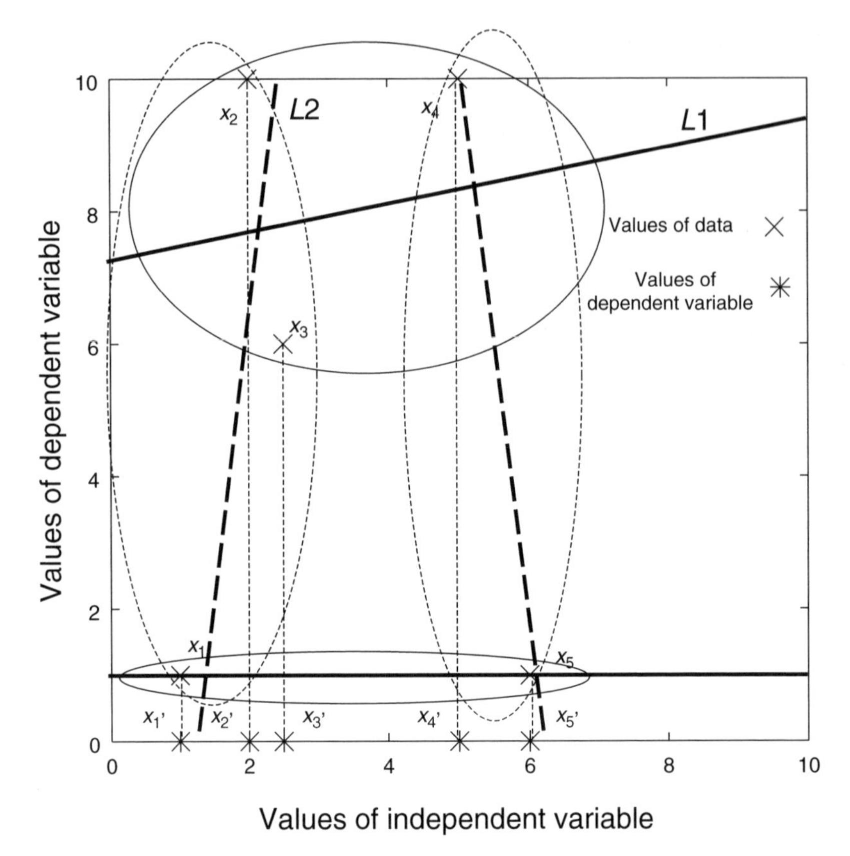

Figure 11.1 Comparison between regressions by change of classification region.

variable. On the other hand, the dotted circles show the clustering result of data with respect to independent variable. Two solid lines show the result of regression for clusters obtained by using both dependent and independent variables. Two dotted lines show the result of regression for clusters obtained by using only the independent variable. The regressions represented by the solid line between objects x_1 and x_5 and the dotted line between objects x_4 and x_5 are perfectly fitted. Therefore, it is sufficient to investigate the difference of fitness between the regressions $L1$ and $L2$.

Table 11.1 shows difference of coefficient of determination for regressions between $L1$ and $L2$. From this table, we can see that $L2$ has a better fitness than $L1$. In other words, the regression that uses a clustering result of data of an independent variable has a better fitness than the regression that assumes that the classification structure of a dependent variable is the same as the classification structure of an independent variable.

Table 11.1 Coefficient of determination.

Regression	Coefficient of determination
$L1$	0.13
$L2$	0.49

$L1$: Regression using clustering result of data for dependent and independent variables.
$L2$: Regression using clustering result of data for independent variable.

11.4 ANALYSES OF RESIDUALS ON FUZZY REGRESSION CLUSTERING MODELS

In this section, we show several results of analyses of residuals for the models of fuzzy regression clustering discussed in previous sections (Sato-Ilic, 2005). Basically, the analysis of residuals is based on an idea that if the fitted regression model is correct, then the obtained residuals should satisfy the assumptions that we have made. The assumptions are concerned with the normality and homoscedasticity of variance for residuals. We discuss a statistical test score on the normality of residuals and check the homoscedasticity of the variance for the residuals by using the analysis of residuals (Draper and Smith, 1966). That is, we investigate whether the estimates of the regression coefficients are satisfactory as the best linear unbiased estimators or not and whether they are maximum likelihood estimators or not, since the estimates of the regression coefficients in the target models are obtained using the weighted least squares method or the least squares method.

For examining residuals shown in Equations (11.14) and (11.32), 90 data are randomly generated for the normal distribution with mean equal to 0 and standard deviation equal to 1 as values for a dependent variable. We denote the 90 data with respect to a dependent variable as

$$\boldsymbol{y} = (y_1, \cdots, y_{90})^t, \tag{11.34}$$

where $y_i \sim N(0,1)$, $i = 1, \cdots, 90$. We also randomly generated the 90 data with respect to two independent variables as

$$\tilde{X} = (x_{ia}),\ i = 1, \cdots, 90,\ a = 1, 2, \tag{11.35}$$

where

$$x_{ia} \sim N(0,1),\ i = 1, \cdots, 30,\ a = 1, 2,$$
$$x_{ia} \sim N(5,1),\ i = 31, \cdots, 60,\ a = 1, 2,$$
$$x_{ia} \sim N(10,1),\ i = 61, \cdots, 90,\ a = 1, 2.$$

That is, we randomly generated 30 data from the first to 30th for the normal distribution with mean equal to 0 and standard deviation equal to 1. For the 31st to 60th, we use the normal distribution with mean equal to 5 and standard deviation equal to 1 in order to generate the second group of data. For the third group, 61st to 90th, the normal distribution with mean equal to 10 and standard deviation equal to 1 is used.

The principal ways for obtaining the residuals shown in Equation (11.32) are as follows:

(1) Using the data $\tilde{X}$ shown in Equation (11.35) to the FANNY method, obtain the degree of belongingness for the clusters shown in Equation (11.10). The number of clusters is assumed to be three. In the numerical experiments in this section, we use the "R" package for obtaining a result with the FANNY method.[1]

(2) Applying the data shown in Equations (11.34) and (11.35) and the fuzzy clustering result shown in Equation (11.10) to the model (11.31), we obtain the estimate of regression coefficients, $\hat{\hat{\boldsymbol{\beta}}}_k$, shown in Equation (11.33).

(3) Using the obtained $\hat{\hat{\boldsymbol{\beta}}}_k$, $k = 1, 2, 3$, we calculate the following estimates

$$\hat{\boldsymbol{y}}_k = U_k \tilde{X} \hat{\hat{\boldsymbol{\beta}}}_k.$$

(4) We obtain the residuals between $\boldsymbol{y}$ and $\hat{\boldsymbol{y}}_k$ as

$$\hat{\boldsymbol{e}}_k = \boldsymbol{y} - \hat{\boldsymbol{y}}_k,\ k = 1, 2, 3. \tag{11.36}$$

Since we use the least squares method, minimizing $\hat{\boldsymbol{e}}_k^t \hat{\boldsymbol{e}}_k$, $\forall k$ is the same as minimizing $\hat{\boldsymbol{e}}$, where

$$\hat{\boldsymbol{e}} = \sum_{k=1}^{3} \hat{\boldsymbol{e}}_k^t \hat{\boldsymbol{e}}_k.$$

Therefore, we investigate for $\hat{\boldsymbol{e}}_k$, $k = 1, 2, 3$.

[1] The R Project for Statistical Computing. [http://www.r-project.org/].

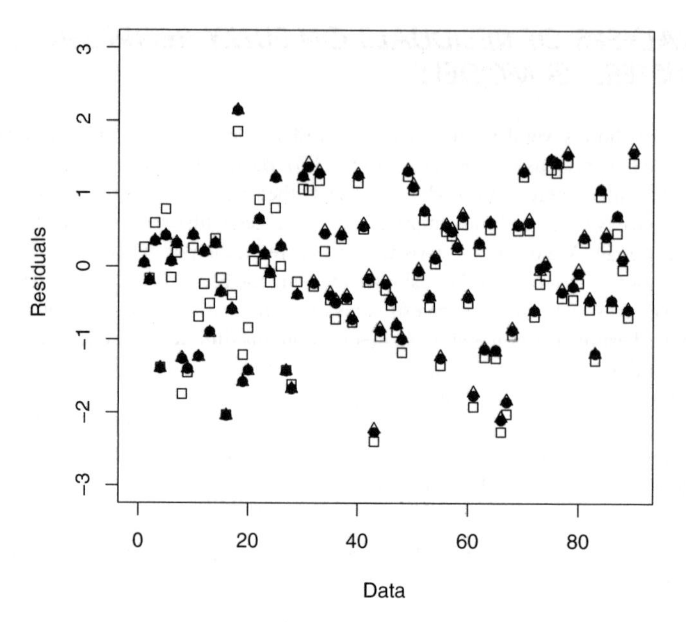

Figure 11.2 Residuals for weighted regression model based on fuzzy clustering.

In the same way, from Equation (11.14) the residuals $\boldsymbol{\delta}_k,\ k = 1, 2, 3$ can be obtained as

$$\boldsymbol{\delta}_k = U_k \boldsymbol{y} - U_k \tilde{X} \tilde{\tilde{\boldsymbol{\beta}}}_k, \tag{11.37}$$

where $\tilde{\tilde{\boldsymbol{\beta}}}_k$ is the obtained estimate of regression coefficients when $m = 2$ in Equation (11.12).

Figures 11.2 and 11.3 show the results on the analysis of residuals for homoscedasticity of variance for residuals. In these figures, the abscissa shows 90 data and the ordinate shows residuals for each cluster. Squares show residuals for cluster 1, black dots show residuals for cluster 2, and triangles are residuals for

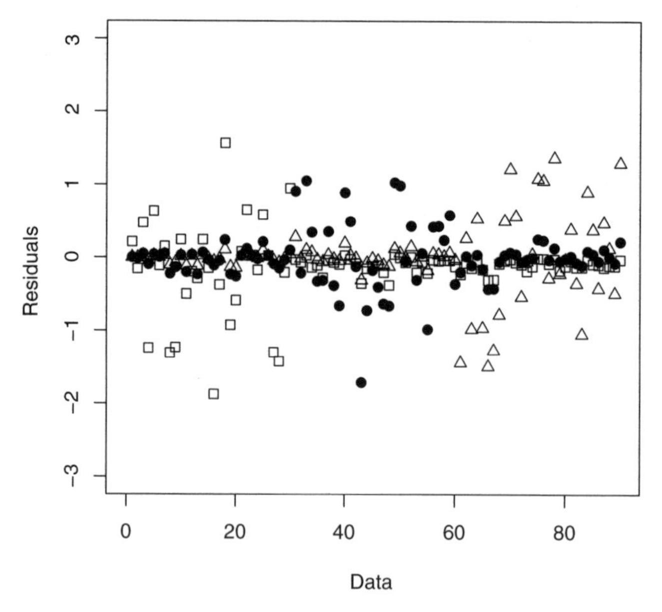

Figure 11.3 Residuals for fuzzy c-regression model.

cluster 3. Figure 11.2 shows the residuals shown in Equation (11.36) and Figure 11.3 shows the residuals shown in Equation (11.37). From Figure 11.2, we can clearly see the homoscedasticity of variance for residuals for all of the three clusters.

However, from Figure 11.3, we can see heteroscedastic variance for residuals for each cluster. That is, variance of residuals for each cluster has bias. For the first cluster, the variance of residuals is large for the first 30 objects. The middle range objects from the 31st to 60th have large variance of residuals for the second cluster. The third cluster gives large variance for the last 30 objects. The bias seems to be caused by the assumption of the use of the same classification structure over the data with respect to the dependent variable and the data with respect to independent variables, because it seems that 30 data for each cluster cause the difference in variance in Figure 11.3.

In order to investigate the normality of the residuals shown in Equations (11.36) and (11.37), we use a test of goodness-of-fit. Figure 11.4 shows the values of the chi-square test statistic. From the results shown in Figure 11.4, it can be seen that the test does not reject the normality hypothesis with a significance level of 0.05 for all residuals for the weighted regression model based on fuzzy clustering that are shown as $\hat{\boldsymbol{e}}_1, \hat{\boldsymbol{e}}_2$, and $\hat{\boldsymbol{e}}_3$. However, the test rejects the normality hypothesis with a significance level of 0.05 for the residuals of the fuzzy c-regression model for cluster 1 that is shown as $\boldsymbol{\delta}_1$. From the result shown in Figure 11.4, we cannot say that the estimate shown in Equation (11.33) is not a maximum likelihood estimator.

In order to investigate the features of residuals shown in Equation (11.18) of the model (11.17) and compare the features of residuals shown in Equation (11.21), we again discuss the statistical test score on the normality of the residuals and check the homoscedasticity of the variance for the residuals by using the analysis of residuals. The difference between the two models (11.17) and (11.21) is whether to use the weights or not, in order to estimate the fuzzy cluster loading. Model (11.21) uses the weights and model (11.17) does not use the weights. In model (11.21), we obtain the estimate of the regression coefficients

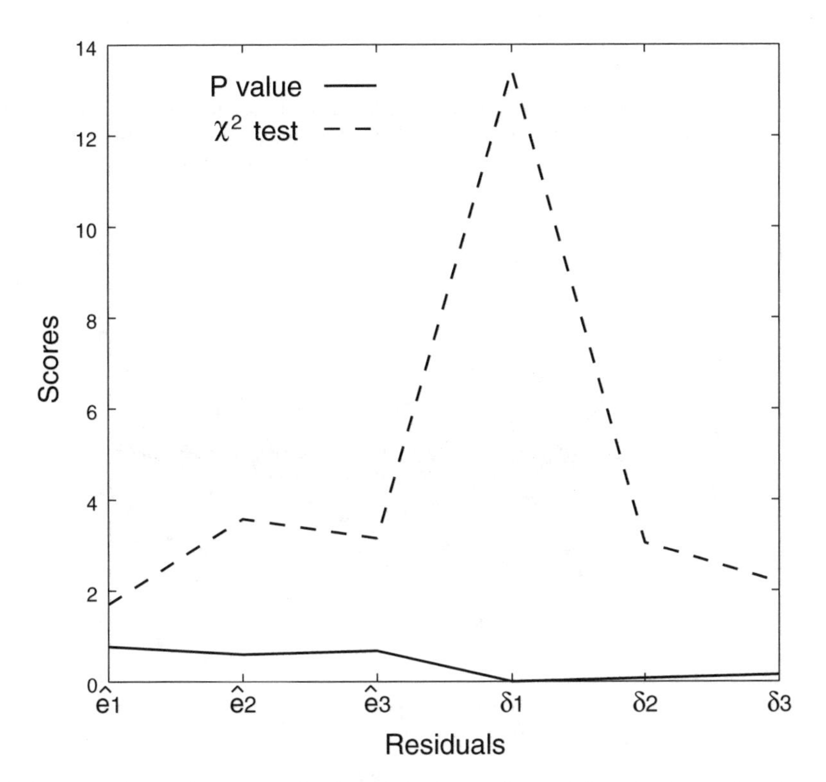

Figure 11.4 Results in test of goodness-of-fit for weighted regression model based on fuzzy clustering and fuzzy c-regression model.

that minimizes $(u_{1k}^{-1}\varepsilon_{1k})^2 + \cdots + (u_{nk}^{-1}\varepsilon_{nk})^2$ shown in Equation (11.24). In model (11.17), we obtain the estimate of the regression coefficients that minimizes $\varepsilon_{1k}^2 + \cdots + \varepsilon_{nk}^2$ shown in Equation (11.20).

We randomly generated 100 data with respect to two independent variables for the normal distribution with mean equal to 0 and standard deviation equal to 1. That is, if we denote the 100 data with respect to independent variables as

$$\tilde{X} = (x_{ia}),\ i = 1, \cdots, 100,\ a = 1, 2, \tag{11.38}$$

then

$$x_{ia} \sim N(0, 1),\ i = 1, \cdots, 100,\ a = 1, 2.$$

The principal ways for obtaining the residuals shown in Equation (11.18) are as follows:

(1) We apply the data $\tilde{X}$ shown in Equation (11.38) to the FANNY method and obtain the degree of belongingness for clusters shown in Equation (11.15). The number of clusters is assumed to be three.
(2) Applying the data shown in Equation (11.38) and the fuzzy clustering result shown in Equation (11.15) to model (11.17), we obtain the estimate of fuzzy cluster loadings (regression coefficients), $\tilde{z}_k$, shown in Equation (11.19).
(3) Using the obtained $\tilde{z}_k,\ k = 1, 2, 3$, we calculate the following estimates

$$\hat{\boldsymbol{u}}_k = \tilde{X}\tilde{z}_k.$$

(4) Then we obtain the residuals between $\boldsymbol{u}_k$ and $\hat{\boldsymbol{u}}_k$ as

$$\boldsymbol{\varepsilon}_k = \boldsymbol{u}_k - \hat{\boldsymbol{u}}_k,\ k = 1, 2, 3. \tag{11.39}$$

Next, we input the data shown in Equation (11.38) and the fuzzy clustering result shown in Equation (11.15) into Equation (11.21), and obtain the estimate of fuzzy cluster loadings (regression coefficients), $\tilde{\tilde{z}}_k$ shown in Equation (11.23). The residuals $\boldsymbol{e}_k,\ k = 1, 2, 3$ shown in Equation (11.21) can be obtained as

$$\boldsymbol{e}_k = \boldsymbol{1} - U_k^{-1}\tilde{X}\tilde{\tilde{z}}_k. \tag{11.40}$$

The results on the analysis of residuals for homoscedasticity of variance for residuals are shown in Figures 11.5 and 11.6. In these figures, the abscissa shows 100 data and the ordinate shows residuals for each cluster. Figure 11.5 shows the residuals shown in Equation (11.40) and Figure 11.6 shows the

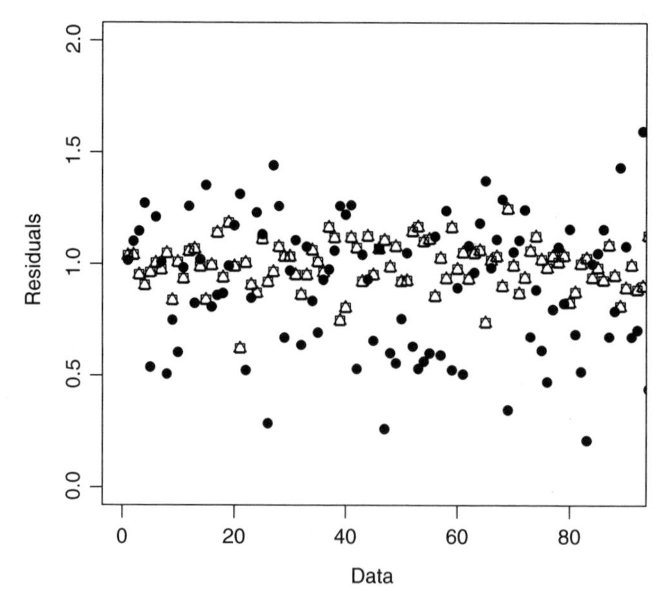

Figure 11.5 Residuals for weighted fuzzy cluster loading model.

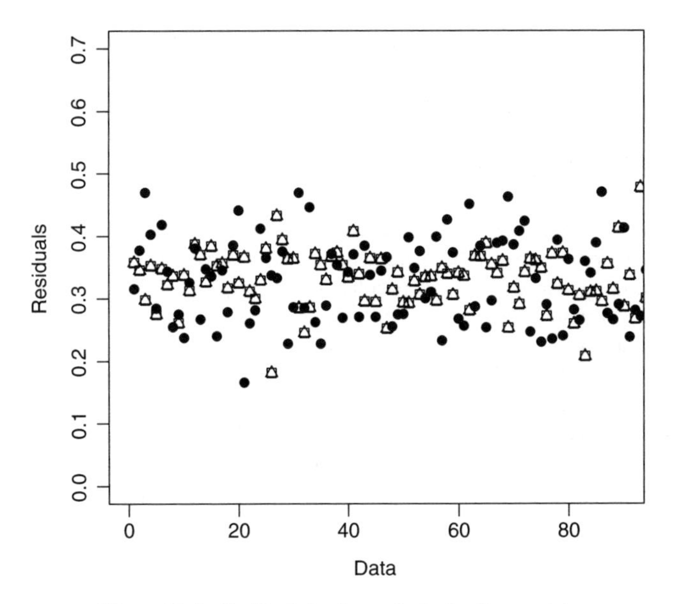

Figure 11.6 Residuals for fuzzy cluster loading model.

residuals shown in Equation (11.39). Squares show residuals for cluster 1, black dots show residuals for cluster 2, and triangles are residuals for cluster 3. From Figures 11.5 and 11.6, we can clearly see the homoscedasticity of variance for residuals for all of the three clusters.

Next, we investigate the normality of the residuals shown in Equations (11.39) and (11.40) using a test of goodness-of-fit. Figure 11.7 shows the values of the chi-square test statistic. From the results shown in Figure 11.7, we can see that the test does not reject the normality hypothesis with a significance level of 0.05 for all residuals for the weighted fuzzy cluster loading model which are shown as $\boldsymbol{e}_1$, $\boldsymbol{e}_2$, and $\boldsymbol{e}_3$.

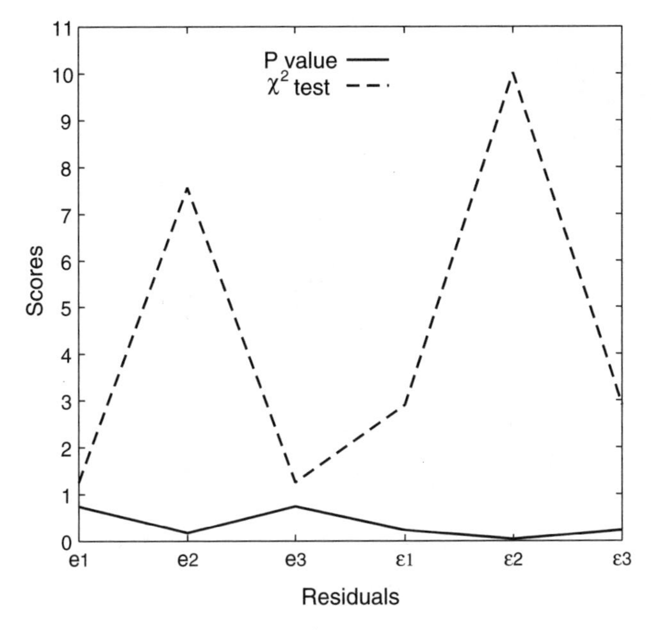

Figure 11.7 Results in test of goodness-of-fit for weighted fuzzy cluster loading model and fuzzy cluster loading model.

However, the test rejects the normality hypothesis with a significance level of 0.05 for the residuals of the fuzzy cluster loading model for cluster 2 that is shown as ε_2. From the result shown in Figure 11.7, we cannot say that the estimate shown in Equation (11.23) is not a maximum likelihood estimator.

11.5 NUMERICAL EXAMPLES

The data is the Fisher iris data (Fisher, 1936) that consists of 150 samples of iris flowers with respect to four variables, sepal length, sepal width, petal length, and petal width. The samples are observed from three kinds of iris flowers, iris setosa, iris versicolor, and iris virginica. We use the data of sepal length as values of a dependent variable, and the data with respect to sepal width, petal length, and petal width are treated as values of independent variables. First, we apply the data of the independent variables for fuzzy c-means method. We use the "R" package for obtaining a result of the fuzzy c-means method. The number of clusters is assumed to be three.

Figure 11.8 shows the result of the fuzzy c-means method. (a) shows a result of cluster 1, (b) is a result of cluster 2, and (c) is a result of cluster 3. In these figures, three axes show the three variables that are sepal width, petal length, and petal width. The locations of dots show the observational values of each iris flower and the cubes show the observation space. The gray tone shows the degree of belongingness for the clusters and the darker color means a larger degree of belongingness for the cluster. From this figure, it can be seen that there are three clusters that can show that the three kinds of iris flowers and the three clusters are vaguely situated in relation to each other.

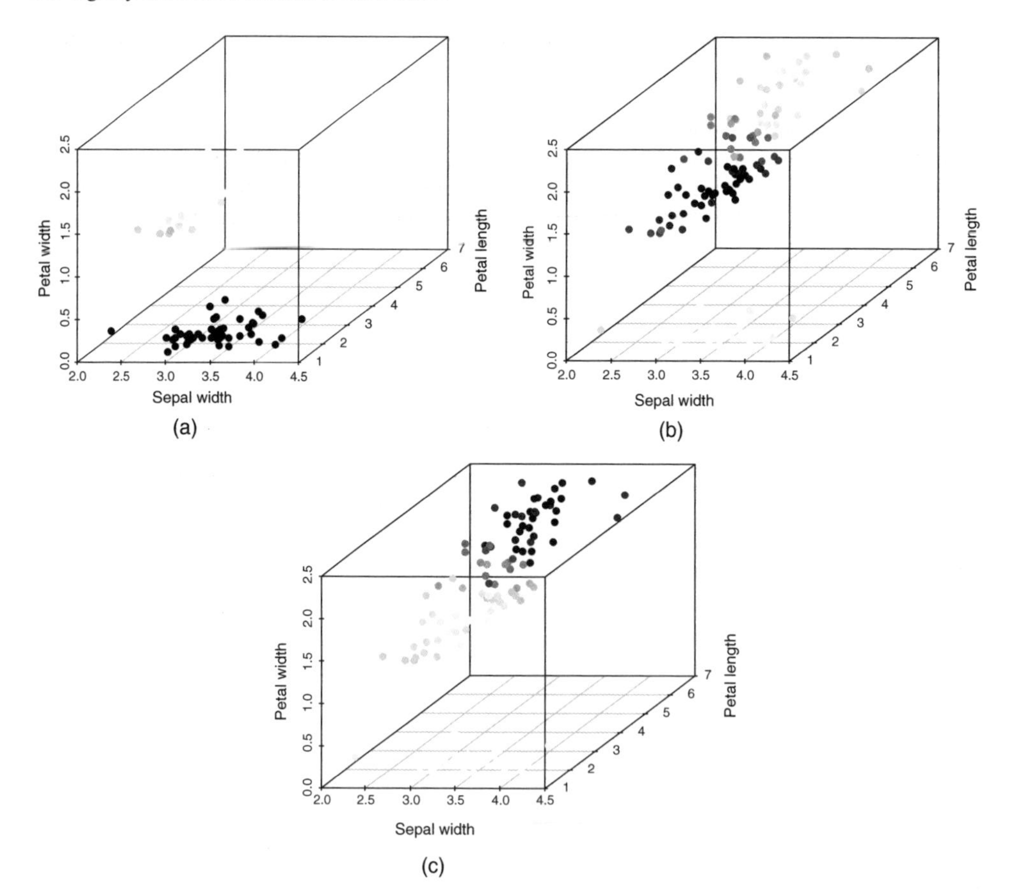

Figure 11.8 Results of the fuzzy c-means for iris data: (a) for cluster 1, (b) for cluster 2, and (c) for cluster 3.

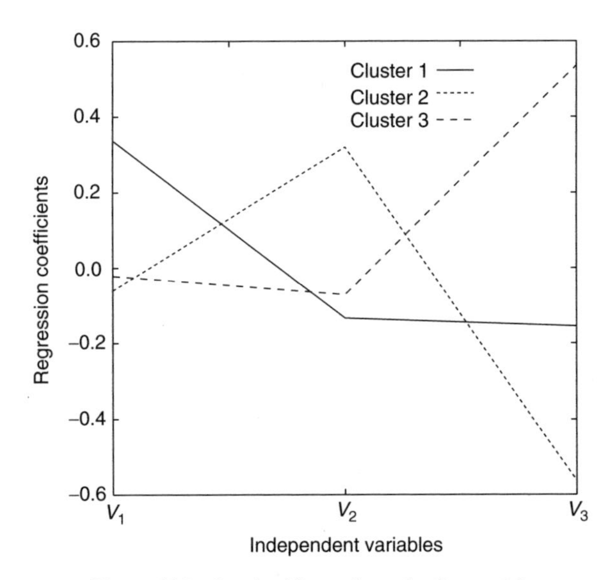

Figure 11.9 Result of fuzzy cluster loading model.

Using the result of the fuzzy c-means shown in Figure 11.8, Figures 11.9 and 11.10 show the results of regression coefficients of the fuzzy cluster loading model and the weighted fuzzy cluster loading model shown in Equations (11.19) and (11.23), respectively. In these figures, the abscissa shows three independent variables. v_1 shows sepal width, v_2 is petal length, and v_3 shows petal width. The ordinate shows values of the regression coefficients that are fuzzy cluster loadings. Each cluster is shown by a different line. From these figures, we can see the significant explanation of v_3 (petal width) for both of the results. This is also captured by using the fuzzy clustering shown in Figure 11.8. That is, from Figure 11.8, we can see a tendency for the classification result with respect to the three variables to be almost the same as the classification with respect to only v_3 (petal width). This feature is more clearly revealed by considering the weights obtained as the fuzzy clustering result shown in Figure 11.10.

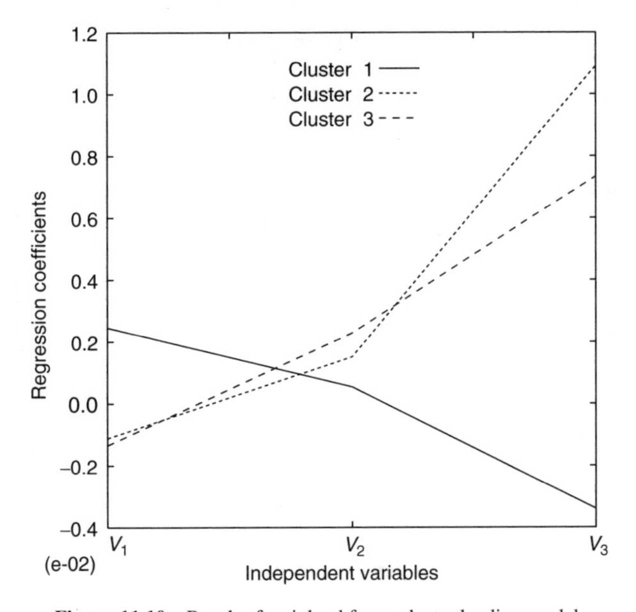

Figure 11.10 Result of weighted fuzzy cluster loading model.

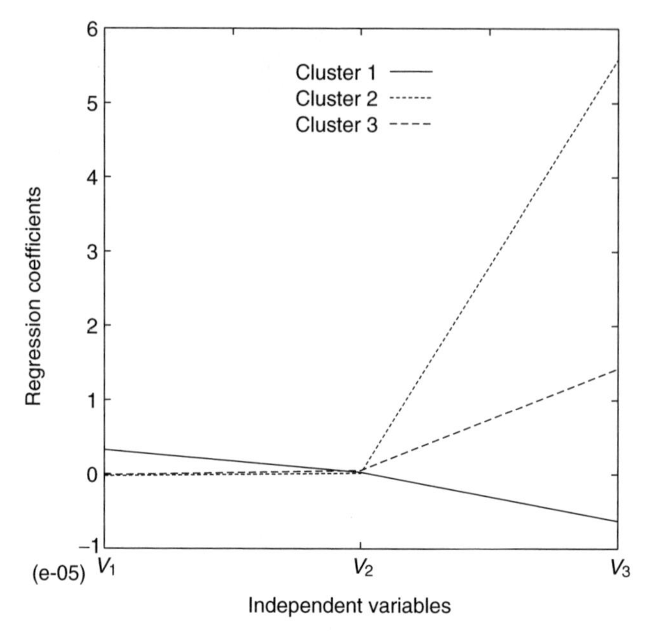

Figure 11.11 Result of kernel fuzzy cluster loading model.

Moreover, Figure 11.11 shows the result of the kernel fuzzy cluster loading model shown in Equation (11.30). We use the polynomial kernel of degree 2 shown in Equation (11.25) as the kernel function in Equation (11.30). This result shows a clearer significance for the v_3 (petal width) when compared to the result of Figure 11.10.

Figure 11.12 shows the result of weighted fuzzy regression model shown in Equation (11.33). We use the data for a dependent variable as sepal length in this case. As a comparison, Figure 11.13 shows the

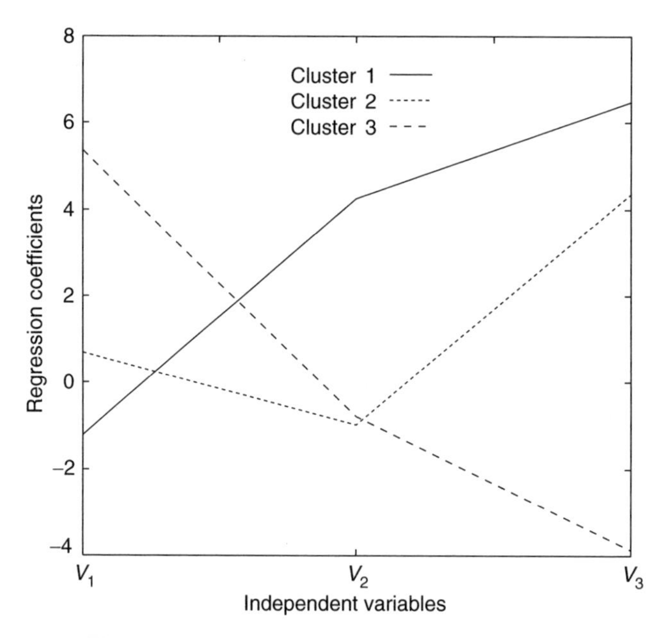

Figure 11.12 Result of weighted fuzzy regression model.

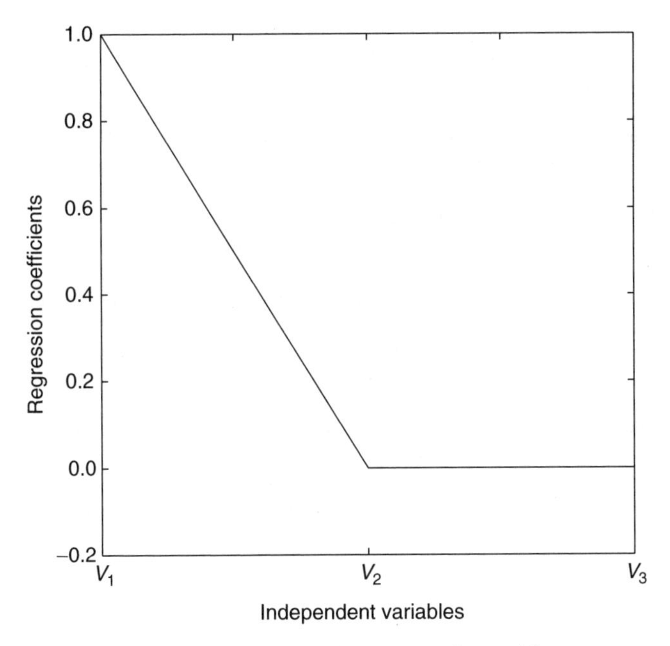

Figure 11.13 Result of linear regression model.

result of regression coefficients using a conventional linear regression model shown in Equation (11.1). From the result of the conventional regression model shown in Figure 11.13, we simply obtain that the sepal length (dependent variable) is highly correlated with v_1 (sepal width) that seems obvious. However, using the fuzzy classification structure, the result shown in Figure 11.12 can show that each kind of iris flower has a different tendency for the correlation of the three independent variables into the dependent variable.

11.6 CONCLUSION

This chapter discusses several regression models using the results of fuzzy clustering. Following conventionally defined terms related to these models, they are weighted regression models and also nonlinear regression models. Despite many weighted regression models and nonlinear regression models, the problem of how to estimate the weights and how to define the nonlinearity are still core issues. Fuzzy clustering might offer a solution in which the weights are estimated by the degree of belongingness of the fuzzy clustering result and the nonlinearity is assumed to be defined as the fuzzy classification structure. Considering the fact that conventional weighted regression models have mainly challenged the correction of local heteroscedastic variance of observation, the use of classification features in regression models is deemed to be reasonable.

REFERENCES

Bezdek, J.C. (1987) *Pattern Recognition with Fuzzy Objective Function Algorithms.* Plenum Press.

Brunsdon, C., Fotheringham, S. and Charlton, M. (1998) 'Geographically Weighted Regression Modeling Spatial Non-Stationarity'. *Journal of the Royal Statistical Society,* **47**, 431–443.

Cristianini, N. and Shawe-Taylor, J. (2000) *An Introduction to Support Vector Machines and Other Kernel-Based Learning Methods.* Cambridge University Press.

Diggle, P.J., Heagerty, P., Liang, K. and Zeger, S.L. (2002) *Analysis of Longitudinal Data.* Oxford University Press.
Dobson, A.J. (1990) *An Introduction to Generalized Linear Models.* Chapman & Hall.
Draper, N.R. and Smith, H. (1966) *Applied Regression Analysis.* John Wiley & Sons, Inc., New York.
Fisher, R.A. (1936) 'The use of multiple measurements in taxonomic problems'. *Annals of Eugenics* **7**, 179–188.
Hastie, T.J. and Tibshirani, R.J. (1990) *Generalized Additive Models.* Chapman & Hall.
Hathaway, R.J. and Bezdek, J.C. (1993) 'Switching regression models and fuzzy clustering'. *IEEE Trans. Fuzzy Syst.*, **1**, 3, 195–204.
Kaufman, L. and Rousseeuw, P.J. (1990) 'Finding Groups in Data'. John Wiley & Sons, Ltd, Chichester, UK.
Mandel, J. (1964) *The Statistical Analysis of Experimental Data.* Dover Publications.
McCullagh, P. and Nelder, J.A. (1989) *Generalized Linear Models.* Chapman & Hall.
Sato-Ilic, M. (2003) 'On kernel based fuzzy cluster loadings with the interpretation of the fuzzy clustering result'. *International Journal of Computational and Numerical Analysis and Applications*, **4**, 3, 265–278.
Sato-Ilic, M. (2004) 'On fuzzy clustering based regression models'. *NAFIPS 2004 International Conference*, 216–221.
Sato-Ilic, M. (2005) 'On features of fuzzy regression models based on analysis of residuals'. *Intelligent Engineering Systems through Artificial Neural Networks*, **15**, 341–350.
Tanaka, H. and Watada, J. (1988) 'Possibilistic linear systems and their application to the linear regression model'. *Fuzzy Sets and Systems*, **27**, 275–289.
Zadeh, L.A. (1965) 'Fuzzy sets'. *Information and Control*, **8**, 338–353.
Zadeh, L.A. (2005) 'Toward a generalized theory of uncertainty (GTU) – an outline'. *Information Sciences*, **172**, 1–40.

12 Implementing Hierarchical Fuzzy Clustering in Fuzzy Modeling Using the Weighted Fuzzy C-means

George E. Tsekouras

Department of Cultural Technology and Communication, University of the Aegean, Mytilene, Greece

12.1 INTRODUCTION

Fuzzy modeling has been viewed as an alternative to more traditional modeling paradigms in order to deal with complex, ill defined, and less "tractable" systems. As many scholars would agree (Delgado, Gomez-Skarmeta, and Vila, 1996; Sugeno and Yasukawa, 1993; Yoshinari, Pedrycz, and Hirota, 1993; Pedrycz, 2005), the most beneficial step throughout a fuzzy modeling procedure is to detect the underlying data structure and to translate it into a collection of fuzzy rules. Fuzzy clustering can be very helpful in learning fuzzy rules from data, since it is able to partition the available data-set into natural groupings. By using fuzzy clustering in fuzzy modeling we do not have to be concerned about the fuzzy partition of the feature space, since it is derived as a direct result of the clustering process (Sugeno and Yasukawa, 1993). Thus, the modeling procedure becomes structure-free, since the clustering process reveals the model's structure (Yoshinari, Pedrycz, and Hirota, 1993). Moreover, fuzzy clustering offers an unsupervised learning platform, which provides great flexibility throughout the model design (Delgado, Gomez-Skarmeta, and Vila, 1996). The above benefits can well be justified when we have a lot of data without any other information about the system. In this case, fuzzy clustering becomes a reliable tool to help construct accurate fuzzy models starting from scratch. In the literature, a plethora of methods have been proposed to incorporate different fuzzy clustering techniques into fuzzy modeling (Chiu, 1994; Wong and Chen, 1999; Yao, Dash, Tan, and Liu, 2000; Angelov, 2004; Panella and Gallo, 2005).

One of the most widely used clustering approaches in fuzzy modeling is the fuzzy C-means (FCM) algorithm (Bezdek, 1981). However, the implementation of the fuzzy C-means suffers from three major problems. The first one is related to the fact that the fuzzy C-means requires a priori knowledge of the

Advances in Fuzzy Clustering and its Applications Edited by J. Valente de Oliveira and W. Pedrycz

number of clusters. Since each cluster is usually assigned to a specific fuzzy rule, the number of rules must be known in advance. There are two general modeling frameworks that can be used to resolve this problem. According to the first framework, the modeling procedure constitutes an iterative process, where the number of rules gradually increase until the model's performance meets a predefined accuracy. Related algorithms have been developed by Kim, Park, Ji, and Park (1997), Kim, Park, Kim, and Park (1998), Chen, Xi, and Zhang, (1998), Kukolj and Levi (2004), and Tsekouras (2005). The second modeling framework is based on using optimal fuzzy clustering (Pal and Bezdek, 1995). Such approaches were proposed by Sugeno and Yasukawa (1993), Emani, Turksen, and Goldenberg (1998), Tsekouras, Sarimveis, Kavakli, and Bafas (2005), Chen and Linkens (2000, 2001), and Linkens and Chen (1999).

The second problem concerns the dependence of fuzzy C-means on the initialization of the iterative process. To deal with this issue, we need to obtain some information about possible structures that may exist in the training data sample. A feasible way to accomplish this is to develop a clustering unit to preprocess the available data. This unit detects a number of properly identified clusters that can be used as initial conditions by the fuzzy C-means. The preprocessing clustering unit is usually constructed by employing agglomerative hierarchical clustering (Emani, Turksen, and Goldenberg, 1998), self-organizing clustering networks (Linkens and Chen, 1999; Chen and Linkens, 2000, 2001; Tsekouras, 2005), or nearest neighbor clustering (Tsekouras, Sarimveis, Kavakli, and Bafas, 2005). The preprocessing clustering unit along with the fuzzy C-means constitutes a hierarchical clustering scheme.

Finally, the third problem concerns the interpretability of the fuzzy model. An accurate fuzzy model is not as functional as a fuzzy model that maintains its interpretability. On the other hand, a fully interpretable fuzzy model may not be not accurate enough to fully describe a real system. Thus, there is a trade off between accuracy and interpretability, which has to be taken into account during the modeling process. A lot of work has been done in this direction (Chao, Chen, and Teng, 1996; Setnes, Babuska, Kaymak, and van Nauta Lemke, 1998; Valente de Oliveira, 1999; Chen, and Linkens, 2004; Guillaume and Charnomordic, 2004; Cassilas, Gordon, deJesus, and Herrera, 2005; Zhou and Gan, 2006). When we use the fuzzy C-means to generate fuzzy rules, each rule corresponds to a specific cluster. Or, to be more precise, each rule corresponds to a specific fuzzy set in each dimension. That is, the fuzzy sets are not shared by the rules (Guillaume, 2001). However, in order to maintain interpretability, similar fuzzy sets have to be merged. This directly implies that the structure of the rule base could be disturbed. A feasible way to resolve this problem is to perform, at a later stage, some model simplification (Setnes, Babuska, Kaymak, and van Nauta Lemke, 1998).

This chapter presents a systematic approach to fuzzy modeling that takes into account all the aforementioned problems. To deal with the first problem the algorithm adopts the first modeling framework mentioned above. To cope with the second problem, we use a two-level hierarchical clustering scheme, where at the first level the self-organizing map (SOM) (Kohonen, 1988) is employed to preprocess the data. The basic design issue of the hierarchical clustering scheme is the weighted fuzzy C-means (Geva, 1999; Bezdek, 1981), in the second level of the hierarchy. As will be discussed later on in this chapter, the weighted C-means is able to resolve certain difficulties involved when the classical fuzzy C-means is applied. Moreover, our simulations showed that it provides very reliable initialization capabilities with respect to the model parameters. Finally, to solve the third problem we employ a standard model simplification technique.

This chapter is organized as follows: Section 12.2 briefly outlines the essence of the Takagi and Sugeno's fuzzy model to be used throughout this chapter. In Section 12.3 we elaborate on the modeling method. Section 12.4 discusses the simulation results and Section 12.5 offers concluding remarks.

12.2 TAKAGI AND SUGENO'S FUZZY MODEL

One of the most well-known fuzzy models is the Takagi and Sugeno (TS) model (Takagi and Sugeno, 1985). We briefly outline its main features. Let us consider that the input space is a p-dimensional vector space, denoted as: $\boldsymbol{X} = \boldsymbol{X}_1 \times \boldsymbol{X}_2 \times \cdots \times \boldsymbol{X}_p$. Therefore, a p-dimensional input data vector $\boldsymbol{x} \in \boldsymbol{X}$ can be represented as $\boldsymbol{x} = [x_1, x_2, \ldots, x_p]^T$ with $x_j \in \boldsymbol{X}_j$ $(1 \le j \le p)$. Assuming that there exist c fuzzy rules, each fuzzy rule is described by the relationship:

R^i: If x_1 is Ω_1^i and x_2 is Ω_2^i and ... and x_p is Ω_p^i

$$\text{then } y^i = b_0^i + b_1^i x_1 + \ldots + b_p^i x_p \quad (1 \le i \le c) \tag{12.1}$$

where b_j^i is a real number and Ω_j^i is a fuzzy set. Here, the fuzzy set membership functions are Gaussian functions of the form

$$\Omega_j^i(x_j) = \exp\left[-\left(\frac{x_j - v_j^i}{\sigma_j^i}\right)^2\right] \quad (1 \le i \le c,\ 1 \le j \le p). \tag{12.2}$$

Finally, the inferred output of the model is calculated as:

$$\tilde{y} = \frac{\sum_{i=1}^{c} \omega^i(\boldsymbol{x}) y^i}{\sum_{i=1}^{c} \omega^i(\boldsymbol{x})} \tag{12.3}$$

where

$$\omega^i(\boldsymbol{x}) = \min_j \{\Omega_j^i(x_j)\} \quad (1 \le j \le p,\ 1 \le i \le c). \tag{12.4}$$

12.3 HIERARCHICAL CLUSTERING-BASED FUZZY MODELING

In the context of fuzzy modeling two basic steps dominate the model's overall design procedure: (a) structure identification and (b) parameter estimation. Structure identification is strongly related to the determination of the number of rules. Moreover, it provides initial values for the model parameters. On the other hand, parameter estimation refers to the calculation of the appropriate model parameter values that relate to the detailed description of the real system.

Minimizing a certain error-based objective function carries out the parameter estimation. To do this, we usually employ classical optimization tools. Thus, the key point is to provide the above minimization procedure with good initial conditions, meaning that we mainly focus on the development of an efficient structure identification strategy.

The modeling algorithm presented here achieves the above task by using a hierarchical clustering structure. The algorithm uses N input–output data pairs of the form $(\boldsymbol{x}_k; y_k)$ with $\boldsymbol{x}_k \in \Re^p$ and $k = \{1, 2, \ldots, N\}$, and consists of the five design steps shown in Figure 12.1. In the first step we construct the data preprocessing clustering unit. In this chapter, the SOM (Kohonen, 1988) is applied to partition the input space into an initial number of clusters. Then, we assign weight factors to the resulting cluster centers. The cluster centers accompanied with their weights constitute a new data-set, which is further clustered by the weighted fuzzy C-means, in the second step of the algorithm. The above two steps define the hierarchical clustering scheme. The third step projects on each axis the clusters obtained by the weighted fuzzy C-means and provides an initial estimation of the premise model parameters. The respective consequent parameters are calculated by the least-squares method. The fourth step utilizes a gradient-descent based approach to optimize the model parameters. The last three steps are implemented through an iterative process, where the model's complexity is increasing during each iteration until the model's performance lies within acceptable levels. With the fuzzy model constructed, the algorithm proceeds to the fifth step, where the model simplification takes place. This procedure is accomplished by applying a fuzzy set merging technique and a rule elimination–combination process.

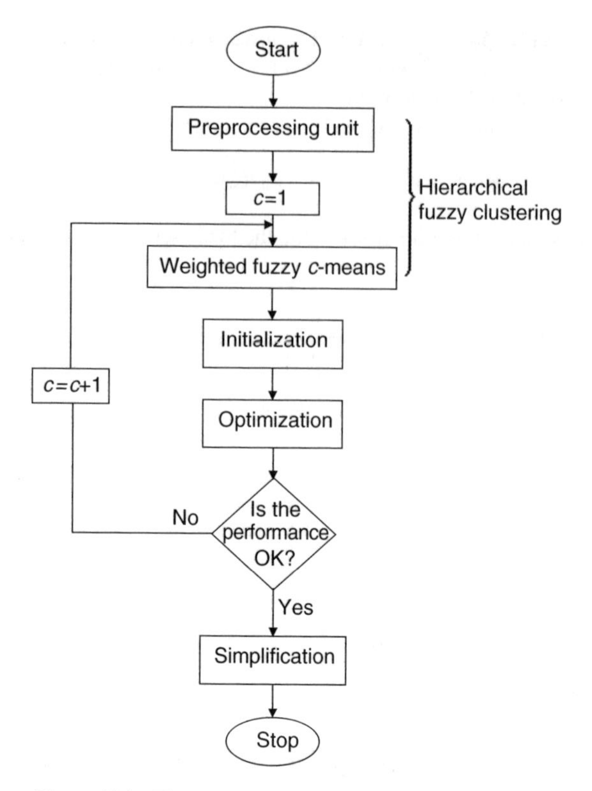

Figure 12.1 The overall structure of the modeling algorithm.

Before proceeding with the theoretical description of each design step, we must analyze in more detail the main criteria used in the above modeling scheme. The first criterion refers to the weighted fuzzy c-means. Similar approaches (Chen and Linkens, 2000; Linkens and Chen, 1999) utilized the classical fuzzy C-means to elaborate the cluster centers produced by the pre-processing unit. Apparently, the implementation of the fuzzy C-means assumes that all these centers are of equal significance. However, this may be not true because two different initial clusters may contain different number of training instances. Thus, by applying the fuzzy C-means we delete all the information that relates the original data-set to the above centers. Therefore, the real data structure may not be discovered. To resolve this problem we assign weights to the centers and we elaborate them using the weighted version of the fuzzy C-means. Each of the above weights is defined by the cardinality of the respective initial cluster. In this way, all the necessary information related to the original data-set is taken into account during the whole model's design process. The second criterion concerns the SOM. The SOM strongly depends on the parameter initialization. The main impact of this fact is that some clusters may contain a small number of data or, at worst, no data at all due to the competitive nature of the learning process (winner takes all). However, the use of the weighted fuzzy C-means cancels the contribution of these clusters to the final fuzzy partition, since the respective weights will be close to zero or equal to zero, respectively. Thus, we can benefit from the advantages provided by the SOM algorithm such as its simplicity and its speed. At this point, it has to be noticed that the implementation of the weighted fuzzy C-means is too fast, since the algorithm is applied to a substantially smaller data-set than the original one. Therefore, the computational time needed to run the hierarchical clustering scheme is approximately equal to the time needed by the SOM. Hence, whenever we have a system with a large number of inputs, the computational complexity of the hierarchical clustering method is approximately equal to the complexity of the SOM, i.e., it is a fast approach that is easy to implement.

12.3.1 Preprocessing Clustering Unit Design

Here, we apply the self-organizing map (SOM) (Kohonen, 1988) algorithm to group all the input training data vectors into a number of clusters. The SOM is a two-layer feed-forward neural network that can discover topological structures hidden in the data, by using unsupervised competitive learning. The implementation of the SOM is based on defining an appropriate distance norm function. This distance measures the dissimilarity between data vectors and thus it enables the SOM to classify efficiently the data into natural structures. The distance norm usually used is the Euclidean distance, which for the p-dimensional space is defined as:

$$||\boldsymbol{x}-\boldsymbol{y}|| = \left|\sum_{j=1}^{p} |x_j - y_j|^2\right|^{1/2} \tag{12.5}$$

where $\boldsymbol{x}$ and $\boldsymbol{y}$ are p-dimensional data vectors. The above distance function is a subclass of the general Minkowski q-norm distance that is given by the equation

$$||\boldsymbol{x}-\boldsymbol{y}||_q = \left|\sum_{j=1}^{p} |x_j - y_j|^q\right|^{1/q}.$$

The SOM algorithm is briefly described below.

The Self-Organizing Map Algorithm

Suppose that we are given a set of N input data vectors $\boldsymbol{X} = \{\boldsymbol{x}_1, \boldsymbol{x}_2, \ldots, \boldsymbol{x}_N\} \subset \Re^p$. Also suppose that the SOM network consists of n neurons, where the synaptic weight of the lth neuron is a point in the p-dimensional space: $\boldsymbol{v}_l = [v_{l1}, v_{l2}, \ldots, v_{lp}]^T$.

Step 1 Choose random values for the neuron synaptic weights $\boldsymbol{v}_l$ $(1 \le l \le n)$. Also, select a value for the maximum number of iterations $t_{\max}$, and set $t = 0$.

Step 2 For $k = 1$ to N

determine the best matching neuron (l_0) to the current data vector $(\boldsymbol{x}_k)$ according to the nearest neighbor rule:

$$||\boldsymbol{x}_k - \boldsymbol{v}_{l_0}(t)||^2 = \min_{1 \le l \le n}\{||\boldsymbol{x}_k - \boldsymbol{v}_l(t)||^2\}.$$

Update the synaptic weight of each neuron:

$$\boldsymbol{v}_l(t+1) = \boldsymbol{v}_l(t) + \eta(t) h_{l,l_0,k}(t)\,(\boldsymbol{x}_k - \boldsymbol{v}_l(t)). \tag{12.6}$$

End for

Step 3 If no noticeable changes in the synaptic weights are observed or $t > t_{\max}$ then stop. Else set $t = t + 1$ and go to step 2.

In the above algorithm, all the distances are calculated using (12.5). The function $h_{l,l_0,k}(t)$ defines the later interaction phenomenon between the neurons l and l_0. Here we use the following form of this function,

$$h_{l,l_0,k}(t) = \begin{cases} 1, \textit{if } l = l_0 \\ 0, \textit{if } l \ne l_0. \end{cases} \tag{12.7}$$

In (12.6), $\eta(t)$ is the learning rate parameter, which is required to decrease monotonically,

$$\eta(t) = \eta_0\left(1 - \frac{t}{t_{\max}}\right) \tag{12.8}$$

where η_0 is an initial value of η. The above algorithm classifies the input training data into n clusters, where the lth cluster is given as

$$A^l = \{x_k \in X : \| x_k - \upsilon_l \|^2 = \min_{1 \le j \le n} [\| x_k - \upsilon_j \|^2]\} \ (1 \le l \le n, \ 1 \le k \le N)$$

and its center element is the υ_l. Then, the weight of significance that is assigned to υ_l can simply be determined using the cluster cardinalities:

$$w_l = \frac{\aleph(A^l)}{\sum_{j=1}^{n} \aleph(A^j)} \quad (1 \le l \le n) \tag{12.9}$$

where $\aleph(A^l)$ is the cardinality of the lth cluster. Thus, each weight is the normalized cardinality of the corresponding cluster. Since the set $A = A^1 \cup A^2 \cup \ldots \cup A^n$ is a crisp partition on X, it follows that:

$$\sum_{j=1}^{n} \aleph(A^j) = N.$$

Finally, the set $\{\upsilon_l, w_l\}$ $(1 \le l \le n)$ is considered as a new training data-set, which will be further clustered in the next step of the algorithm.

12.3.2 Implementation of the Weighted Fuzzy c-means

The objective function for the weighted fuzzy C-means is

$$J_m(U, V) = \sum_{l=1}^{n} \sum_{i=1}^{c} w_l (u_{il})^m \|\upsilon_l - v_i\|^2 \tag{12.10}$$

where c is the number of the final clusters that coincides with the number of rules, $U = \{[u_{il}], 1 \le i \le c, \ 1 \le l \le n\}$ is the partition matrix, $V = \{[v_i], \ 1 \le i \le c\}$ the cluster center matrix, υ_l $(1 \le l \le n)$ are the data to be clustered, $m \in (1, \infty)$ is a factor to adjust the membership degree weighting effect, and w_l is given in Equation (12.9). The parameter m is the fuzzification coefficient and its values must be greater than unity. When m equals unity the partition becomes crisp. On the other hand, as m increases the overlapping degree between clusters also increases and the partition becomes fuzzy. A wide range of applications has shown that choosing a value in the interval [1.5, 4] provides very credible results (Pal and Bezdek, 1995). In this chapter we use the value $m = 2$, since it is the one being commonly used in the literature.

The scope is to minimize $J_m(U, V)$ under the following constraint

$$\sum_{i=1}^{c} u_{il} = 1 \quad \forall l.$$

The final prototypes and the membership functions that solve this constrained optimization problem are given by the equations:

$$v_i = \frac{\sum_{l=1}^{n} w_l (u_{il})^m \upsilon_l}{\sum_{l=1}^{n} w_l (u_{il})^m} \qquad 1 \le i \le c \tag{12.11}$$

$$u_{il} = \frac{1}{\sum_{j=1}^{c} \left(\frac{\|\upsilon_l - v_i\|}{\|\upsilon_l - v_j\|} \right)^{2/(m-1)}}, \qquad 1 \le i \le c, \ 1 \le l \le n. \tag{12.12}$$

Equations (12.11) and (12.12) constitute an iterative process, which is described next.

The Weighted Fuzzy c-Means Algorithm

Step 1 Select the number of clusters c, a value for the factor m, and initial values for the prototypes $\boldsymbol{v}_1, \boldsymbol{v}_2, \ldots, \boldsymbol{v}_c$.

Step 2 Employ Equation (12.12) to calculate the membership values u_{il} $(1 \leq i \leq c,\ 1 \leq l \leq n)$.

Step 3 Calculate the updated cluster center values $\boldsymbol{v}_1^{new}, \boldsymbol{v}_2^{new}, \ldots, \boldsymbol{v}_c^{new}$ using Equation (12.11).

Step 4 If $\max_i\{\|\boldsymbol{v}_i - \boldsymbol{v}_i^{new}\|_{err}\} \leq \varepsilon$ then stop. Else go to step 2.

In the case where all the weights are equal the weighted fuzzy C-means is equivalent to the classical fuzzy C-means algorithm.

12.3.3 Model Parameter Initialization

The centers v_j^i $(1 \leq j \leq p,\ 1 \leq i \leq c)$ of the fuzzy sets in Equation (12.2) are obtained by simply projecting the final cluster centers $\boldsymbol{v}_i (i = 1, 2, \ldots, c)$ in each axis. To calculate the respective standard deviations (σ_j^i) we utilize the fuzzy covariance matrix

$$\boldsymbol{F}_i = \frac{\sum_{l=1}^{n} w_l\,(u_{il})^m\,(\boldsymbol{v}_l - \boldsymbol{v}_i)\,(\boldsymbol{v}_l - \boldsymbol{v}_i)^T}{\sum_{l=1}^{n} w_l\,(u_{il})^m}\,(1 \leq i \leq c).$$

Then, the standard deviations are,

$$\sigma_j^i = [\text{Diag}(\boldsymbol{F}_i)]^{1/2} \quad 1 \leq j \leq p,\ 1 \leq i \leq c. \tag{12.13}$$

After the premise parameters have been initialized, we can expand the output of the model, given in Equation (12.3), into the following fuzzy basis functions (FBFs) form

$$\tilde{y}_k = \sum_{i=1}^{c} p^i(\boldsymbol{x}_k)\, y_k^i \tag{12.14}$$

with

$$p^i(\boldsymbol{x}_k) = \omega^i(\boldsymbol{x}_k) \Big/ \sum_{j=1}^{c} \omega^j(\boldsymbol{x}_k) \tag{12.15}$$

where y_k^i is the value of the consequent part of the ith fuzzy rule when $\boldsymbol{x}_k$ is the input vector. For the N input–output data pairs, the consequent parameters are obtained through the minimization of the function:

$$J_1 = \sum_{k=1}^{N} (y_k - \tilde{y}_k)^2. \tag{12.16}$$

To minimize J_1 we employ the well-known least squares method. To this end, the values of the premise and consequent parameters are used as an initial choice for the tuning process that follows.

12.3.4 Model Parameter Optimization

In this step the system parameters obtained in the previous section are fine tuned using the gradient descent algorithm. The objective function used for this purpose comes in the form

$$J_2 = \frac{1}{2N} \sum_{k=1}^{N} (y_k - \tilde{y}_k)^2. \tag{12.17}$$

By applying the gradient descent method to minimize J_2, the values of the premise parameters of the fuzzy model can be adjusted by the learning rules:

$$\Delta v_j^i = \frac{\eta_1}{N} \sum_{k=1}^{N} \left\{ (y_k - \tilde{y}_k)\ (y_k^i - \tilde{y}_k)\ \frac{p^i(\boldsymbol{x}_k)}{\omega^i(\boldsymbol{x}_k)} \frac{\partial[\omega^i(\boldsymbol{x}_k)]}{\partial v_j^i} \right\} \tag{12.18}$$

and

$$\Delta \sigma_j^i = \frac{\eta_2}{N} \sum_{k=1}^{N} \left\{ (y_k - \tilde{y}_k)\ (y_k^i - \tilde{y}_k)\ \frac{p^i(\boldsymbol{x}_k)}{\omega^i(\boldsymbol{x}_k)} \frac{\partial[\omega^i(\boldsymbol{x}_k)]}{\partial \sigma_j^i} \right\}. \tag{12.19}$$

Based on (12.2) and (12.4) the partial derivatives in (12.18) and (12.19) can be easily derived. Similarly, for the consequent parameters the learning formule are

$$\Delta b_0^i = \frac{\eta_3}{N} \sum_{k=1}^{N} [(y_k - \tilde{y}_k)\, p^i(\boldsymbol{x}_k)] \tag{12.20}$$

and

$$\Delta b_j^i = \frac{\eta_3}{N} \sum_{k=1}^{N} [(y_k - \tilde{y}_k)\, p^i(\boldsymbol{x}_k)\, x_{kj}]. \tag{12.21}$$

In the above equations the parameters $\eta_1, \eta_2,$ and η_3 are the gradient-descent learning rates.

12.3.5 Model Simplification

Interpretability can be easily maintained when the number of rules is small enough to be comprehensible, each rule is consistent, and the membership function forms provide linguistic meaning to the user. However, the model obtained by the above steps may exhibit redundant behavior in terms of highly overlapping fuzzy sets, which are not distinguishable. Therefore, it is difficult to assign any meaning to them. Three issues characterize this behavior (Setnes, Babuska, Kaymak, and van Nauta Lemke, 1998): (a) similarity between fuzzy sets, (b) similarity of a fuzzy set to the universal set, and (c) similarity of a fuzzy set to a singleton set. To take into account the above issues, we have to define a suitable and efficient similarity measure between fuzzy sets. There are two types of similarity measures: set-theoretic measures and geometric based measures. Set-theoretic measures are able to detect similarities between overlapping sets, while geometric measures can be efficiently used for distinct fuzzy sets. Then, the simplification procedure consists of two processes: (i) fuzzy set merging and (ii) fuzzy rule elimination and combination.

12.3.5.1 Merging of Fuzzy Sets

The most representative set-theoretic similarity measure is of the form

$$S(A,\, B) = \frac{|A \cap B|}{|A \cup B|} \tag{12.22}$$

where A, B are fuzzy sets and $|\cdot|$ stands for the fuzzy set cardinality. More specifically, the cardinality of a fuzzy set A is calculated as

$$|\mathrm{A}| = \sum_{x \in A} u_A(x)$$

where $u_A(x)$ is the membership degree of x in the set A.

We can merge A and B whenever $S(A,B) \geq \phi$, with $\phi \in [0,1]$. According to this merging process, the parameters of the new fuzzy set are the mean values of the respective parameters of A and B. When we use Gaussian membership functions, however, the calculation of the intersection between two fuzzy sets is very complex because of their nonlinear shape. However, we can approximate a Gaussian function using triangular (Chao, Chen, and Teng, 1996) or trapezoidal (Chen and Linkens, 2004) membership functions. Here, we use the approach developed by Chao, Chen, and Teng, (1996).

Consider two Gaussian fuzzy sets $A = \{v_A, \sigma_A\}$ and $B = \{v_B, \sigma_B\}$. We distinguish the following four cases:

(a) $v_A = v_B$ and $\sigma_A \geq \sigma_B$ then: $S(A,B) = \sigma_B/\sigma_A$. (12.23)

(b) $|\sigma_A - \sigma_B|\sqrt{\pi} \leq v_A - v_B \leq (\sigma_A + \sigma_B)\sqrt{\pi}$ and $v_A > v_B$ then

$$S(A,B) = \frac{(c_1 + c_2)h_1}{2(\sigma_A + \sigma_B)\sqrt{\pi} - (c_1 + c_2)h_1}, \tag{12.24}$$

where

$$c_1 = \frac{\sigma_A(v_B - v_A) + \sigma_A(\sigma_A + \sigma_B)\sqrt{\pi}}{\sigma_A + \sigma_B}, \quad c_2 = \frac{\sigma_B(v_B - v_A) + \sigma_B(\sigma_A + \sigma_B)\sqrt{\pi}}{\sigma_A + \sigma_B}$$

and $h_1 = \dfrac{(v_B - v_A) + (\sigma_A + \sigma_B)\sqrt{\pi}}{(\sigma_A + \sigma_B)\sqrt{\pi}}$.

(c) $v_A - v_B \leq |\sigma_B - \sigma_A|\sqrt{\pi}$, $v_A > v_B$ and $\sigma_A \leq \sigma_B$ then,

$$S(A,B) = \frac{c_1h_1 + c_2h_2 + c_3h_3}{2(\sigma_A + \sigma_B)\sqrt{\pi} - (c_1h_1 + c_2h_2 + c_3h_3)} \tag{12.25}$$

where

$$c_1 = \frac{\sigma_A(v_B - v_A) + \sigma_A(\sigma_A + \sigma_B)\sqrt{\pi}}{\sigma_A + \sigma_B}, \quad c_2 = \frac{\sigma_A(v_B - v_A) + \sigma_A(\sigma_B - \sigma_A)\sqrt{\pi}}{\sigma_B - \sigma_A},$$

$$c_3 = 2\sigma_A\sqrt{\pi} - (c_1 + c_2), h_1 = \frac{(v_B - v_A) + (\sigma_A + \sigma_B)\sqrt{\pi}}{(\sigma_A + \sigma_B)\sqrt{\pi}},$$

$$h_2 = \frac{(v_B - v_A) + (\sigma_B - \sigma_A)\sqrt{\pi}}{(\sigma_B - \sigma_A)\sqrt{\pi}} \text{ and } h_3 = h_1 + h_2.$$

(d) $v_A - v_B > (\sigma_A + \sigma_B)\sqrt{\pi}$ and $v_A > v_B$ then $S(A,B) = 0$.

For the interested reader, a more detailed analysis of the above approach is given by Chao, Chen, and Teng (1996).

12.3.5.2 Rule Elimination and Combination

After the fuzzy set merging process takes place, some rules may also exhibit redundancy. For example, when a fuzzy membership function is always close to zero, then the firing degree of the respective rule is always close to zero. Therefore, such kinds of rules have to be deleted. Moreover, we can use the above similarity measures to relate two fuzzy rules. This is accomplished by determining the similarities between the fuzzy sets that participate in these rules and belong to same universe of discourse. Then, if the minimum of these similarities becomes greater than a certain threshold, the two fuzzy rules are merged (Chen and Linkens, 2004). In this case, the parameter values for the new rule are obtained as the mean values of the respective parameters of the original rules.

12.4 SIMULATION STUDIES

In this section, we test the algorithm with respect to three issues: (a) the performance of the *initial model*, which is the model obtained before the gradient-descent is applied, (b) the performance of the *final model*, which is the resulting model after the gradient-descent implementation, and (c) the performance of the *simplified model*. If we want to assign qualitative meanings to the above three models we can make the following comments. The *initial model* is the result of the weighted fuzzy C-means application. Therefore, its performance is directly related to the initialization capabilities offered by the weighted fuzzy C-means, as far as the gradient-descent is concerned. Roughly speaking, this issue is one of the most important issues in the modeling method, since providing good initial conditions to the gradient-descent we expect to obtain an accurate and reliable fuzzy model. The *final model* reflects the impact of the gradient-descent on the iterative training process. The performance of this model is a quantitative measure of the overall prediction capabilities of the algorithm. Finally, the *simplified model* is the interpretable fuzzy model. Its performance is expected to be inferior when compared with the performance of the final model. However, this model is more functional than the previous one.

12.4.1 Nonlinear Static Function Approximation

In this example we study the function approximation capabilities of the algorithm. The function is described by:

$$y = (1 + x_1^{-2} + x_2^{-1.5})^2, \quad 1 \le x_1, x_2 \le 5.$$

To build the fuzzy model 50 input–output data pairs were used. To run the SOM we used $n = 15$ clusters. The model's performance index was calculated in terms of the mean-square error (MSE):

$$MSE = \sum_{k=1}^{N} (y_k - \tilde{y}_k)^2 \Big/ N.$$

The implementation of the iterative process produced a final model with $c = 4$ fuzzy rules. The MSE of this model was equal to 0.0019, while the respective performance of the initial fuzzy model was 0.0581. The premise parameters of the final model are depicted in Table 12.1. The rules are:

$$\begin{aligned}
R^1 &: \text{if } x_1 \text{ is } \Omega_1^1 \text{ and } x_2 \text{ is } \Omega_2^1 \text{ then } y = 14.38603 - 1.96250\, x_1 - 0.97448\, x_2 \\
R^2 &: \text{if } x_1 \text{ is } \Omega_1^2 \text{ and } x_2 \text{ is } \Omega_2^2 \text{ then } y = 12.52732 - 0.15062\, x_1 - 0.11260\, x_2 \\
R^3 &: \text{if } x_1 \text{ is } \Omega_1^3 \text{ and } x_2 \text{ is } \Omega_2^3 \text{ then } y = 7.13280 - 4.64877\, x_1 + 1.27211\, x_2 \\
R^4 &: \text{if } x_1 \text{ is } \Omega_1^4 \text{ and } x_2 \text{ is } \Omega_2^4 \text{ then } y = 6.05905 - 1.26086\, x_1 - 0.15812\, x_2.
\end{aligned} \quad (12.26)$$

To test the performance of the final model, we carried out a 10-fold cross-validation procedure, where we generated 10 fuzzy models. Each of this models consisted of four fuzzy rules. The mean and the standard

Table 12.1 Premise parameter values of the final model for the static function example.

	Ω_1^i		Ω_2^i	
i	v_1^i	σ_1^i	v_2^i	σ_2^i
1	2.20167	2.18865	−0.52216	1.68843
2	2.94248	1.60927	3.08207	2.41844
3	−0.36051	1.75746	2.66382	1.09466
4	2.75782	0.60507	1.64110	0.32758

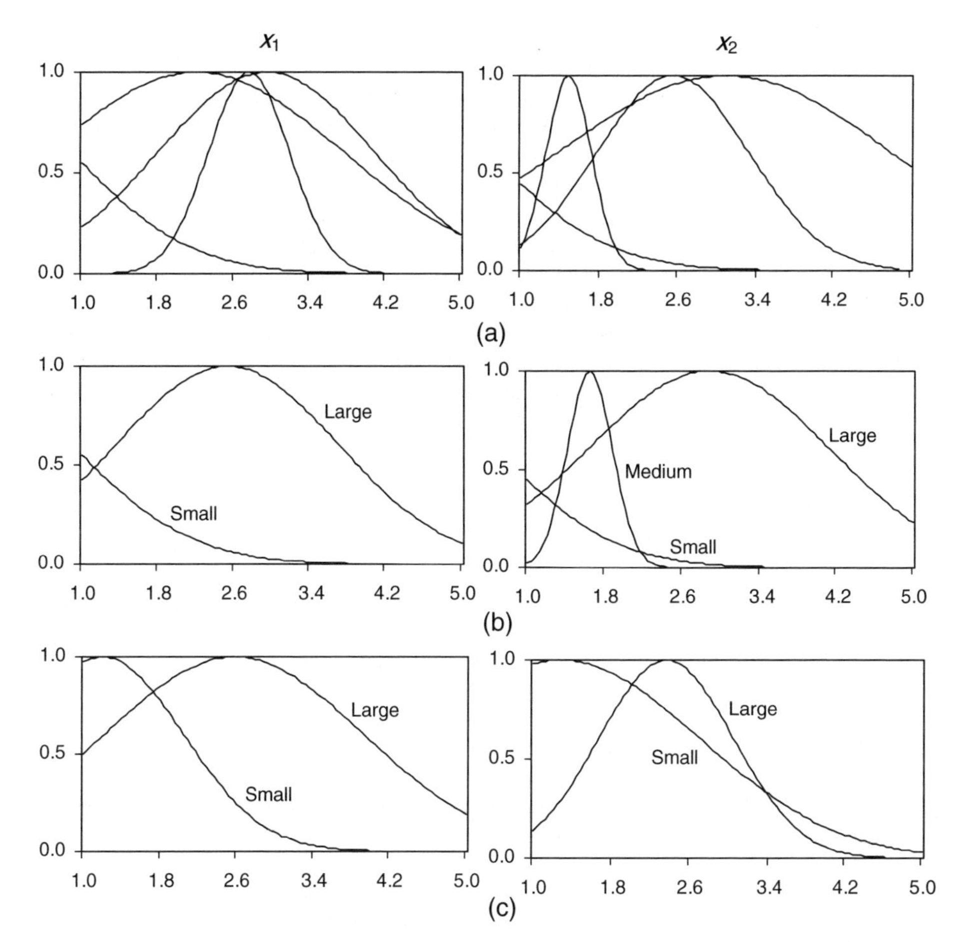

Figure 12.2 Fuzzy sets for the static function example of: (a) the final fuzzy model, (b) the four-rule simplified fuzzy model, and (c) the two-rule simplified fuzzy model.

deviation of the performance index for the training data were 0.0021 and 0.00101, while for the test data were 0.00949 and 0.00722, respectively. Thus, the performance of the final model in (12.26) lies within the limits provided by the cross-validation procedure.

The next experiment concerns the model simplification. Initially, we performed the model simplification using the fuzzy set merging process, only. Then, we applied the gradient descent to optimize the consequent parameters. This procedure gave a four-rule simplified fuzzy model. Figure 12.2(a) illustrates the input space fuzzy sets before the model simplification, and Figure 12.2(b) after the implementation of this process. The resulting rule base is:

$$\begin{aligned}
&R^1\text{: if } x_1 \text{ is } \textit{large} \text{ and } x_2 \text{ is } \textit{small} && \text{then } y = 9.80056 - 0.87872\,x_1 - 0.64935\,x_2\\
&R^2\text{ : if } x_1 \text{ is } \textit{large} \text{ and } x_2 \text{ is } \textit{large} && \text{then } y = 2.34883 - 0.22419\,x_1 + 0.03410\,x_2\\
&R^3\text{ : if } x_1 \text{ is } \textit{small} \text{ and } x_2 \text{ is } \textit{large} && \text{then } y = 13.5639 - 4.52583\,x_1 - 0.85026\,x_2\\
&R^4\text{ : if } x_1 \text{ is } \textit{large} \text{ and } x_2 \text{ is } \textit{medium} && \text{then } y = 6.17259 - 0.61819\,x_1 - 1.06178\,x_2.
\end{aligned}$$

The fuzzy set labels in the above rules are shown in Figure 12.2(b). The MSE obtained by the four-rule simplified fuzzy model was equal to 0.0132, which is inferior to the original model. Comparing Figures 12.2(a) and 12.2(b), we can easily notice that while the interpretability has been improved, some fuzzy sets are subnormal in the universe of discourse. The reason behind this is that we have determined the model parameters using an unconstrained optimization process. Thus, the resulting local minimum may not be able to guarantee that the fuzzy partition can provide meaningful semantic information for the input space membership functions. A systematic way to solve this problem is to perform constrained model parameter estimation (Valente de Oliveira, 1999), but in this case the computational cost will increase. Here we applied the fuzzy set merging and the rule combination–elimination process and we obtained a two-rule simplified fuzzy model. The MSE for this model was equal to 0.1239, which is substantially inferior when compared to the performance of the four-rule simplified model. The resulting input space fuzzy partition is depicted in Figure 12.2(c). From this figure it is obvious that the interpretability has been further improved and the subnormal sets have been absorbed. However, some fuzzy sets still have significant overlap. This fact directly implies that it is very difficult to obtain a fully interpretable model through an automated process, retaining at the same time its accurate performance. With this regard, the use of constrained optimization might be useful, but this approach is beyond the scope of this chapter. The rule base of the two-rule simplified fuzzy model is:

$$R^1 : \text{ if } x_1 \text{ is } \textit{large} \text{ and } x_2 \text{ is } \textit{small} \text{ then } y = 5.20606 - 0.42646\,x_1 - 0.49237\,x_2$$
$$R^2 : \text{ if } x_1 \text{ is } \textit{small} \text{ and } x_2 \text{ is } \textit{large} \text{ then } y = 14.55039 - 4.26635\,x_1 - 2.20197\,x_2.$$

Finally, Table 12.2 compares the performances of the models generated here with the performances of other methods that exist in the literature. This table depicts some very interesting results. First, the performances of the final model and the mean of the 10-fold cross-validation analysis are the best of all the other performances. Moreover, these performances were achieved by using a small number of rules and parameters. Secondly, the four-rule simplified model exhibited very accurate behavior, which is strongly comparable with methods that do not use any simplification procedure. Thirdly, the initial model's performance is also comparable with other performances that use an optimization approach to estimate the model parameter values. The last remark should be strongly emphasized, since it directly indicates that the initialization capabilities provided by the weighted fuzzy C-means were accurate.

Table 12.2 Performance comparison for the static function example.

Model	No. of rules	No. of Parameters	MSE
Chen and Linkens (2004) (final model)	4	28	0.0043
Chen and Linkens (2004) (simplified model)	4	22	0.0078
Emani, Turksen, and Goldenberg (1998)	8	91	0.0042
Kim, Park, Ji, and Park (1997)	3	21	0.0197
Kim, Park, Kim, and Park (1998)	3	21	0.0090
Lee and Ouyang (2003)	10	—	0.0148
Nozaki, Iishibuchi, and Tanaka (1997)	25	125	0.0085
Sugeno and Yasukawa (1993) (initial model)	6	65	0.3180
Sugeno and Yasukawa (1993) (final model)	6	65	0.0790
Tsekouras, Sarimveis, Kavakli, and Bafas (2005)	6	42	0.0110
Tsekouras (2005)	6	30	0.0051
This model (initial model)	4	28	0.0581
This model (final model)	4	28	0.0019
This model (10-fold training data)	4	28	0.0021
This model (10-fold test data)	4	28	0.0095
This model (four-rule simplified)	4	22	0.0132
This model (two-rule simplified)	2	14	0.1239

12.4.2 Mackey-Glass System

In this section we use the proposed modeling algorithm to predict the Mackey–Glass time series. The Mackey–Glass time series is generated by the following time-delay differential equation

$$\frac{\mathrm{d}x(t)}{\mathrm{d}t} = \frac{0.2\,x(t-\tau)}{1+x^{10}(t-\tau)} - 0.1\,x(t).$$

When the parameter τ is large, the system displays a chaotic behavior. In our simulations we set $\tau = 17$ and generated a sample of 1000 points, which are depicted in Figure 12.3. The first 500 points were used as training data, and the last 500 points as test data to validate the fuzzy model. The input variables of the model were: $x(k-18), x(k-12), x(k-6)$, and $x(k)$, while the output was the point $x(k+6)$. To measure the model's performance we employed the root mean-square error (RMSE):

$$RMSE = \sqrt{\sum_{k=1}^{N} (y_k - \tilde{y}_k)^2 \Big/ N}.$$

For implementing the SOM algorithm we used $n = 12$ initial clusters. The iterative process gave $c = 5$ rules. The performances of the initial and final model on the test data were equal to 0.0232 and 0.0052, respectively. The performance of the final model on the training data-set was 0.005. Figure 12.4 shows the predictions of the final model when compared with the original output values.

To further test the model's performance, we applied a 10-fold cross-validation. According to this analysis, the mean and the standard variation for the training data were equal to 0.00532 and 0.00061, and for the test data 0.00565 and 0.00098, respectively.

In the next experiment, we applied the fuzzy set merging and the rule elimination–combination processes. The resulting simplified fuzzy partition of the input space is depicted in Figure 12.5. The simplified model consisted of the following $c = 3$ fuzzy rules,

R^1: if x_1 is *large* and x_2 is *large* and x_3 is *large* and x_4 is *medium*
then $y = 1.96854 - 0.41024\,x_1 - 1.02690\,x_2 - 0.06285\,x_3 + 0.50012\,x_4$

R^2: if x_1 is *large* and x_2 is *medium* and x_3 is *small* and x_4 is *small*
then $y = 1.55146 - 0.04672\,x_1 - 1.40628\,x_2 + 0.65099\,x_3 + 0.18154\,x_4$

R^3: if x_1 is *small* and x_2 is *small* and x_3 is *medium* and x_4 is *large*
then $y = -0.16365 + 0.38428\,x_1 + 0.29107\,x_2 + 0.74229\,x_3 + 0.12890\,x_4$.

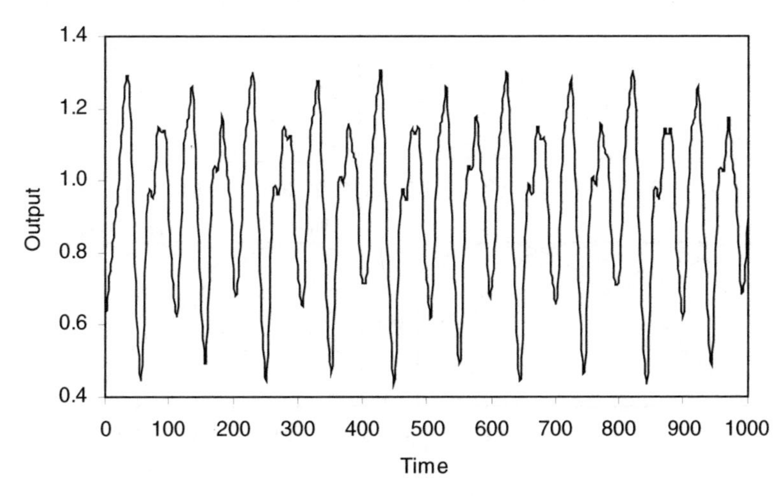

Figure 12.3 A sample of 1000 points for the Mackey–Glass example.

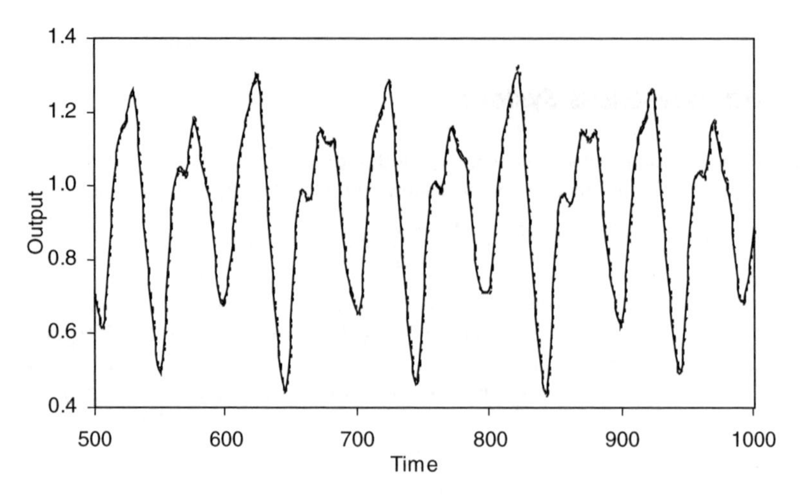

Figure 12.4 Original (solid line) and predicted (dashed line) values of the final model for the Mackey–Glass example.

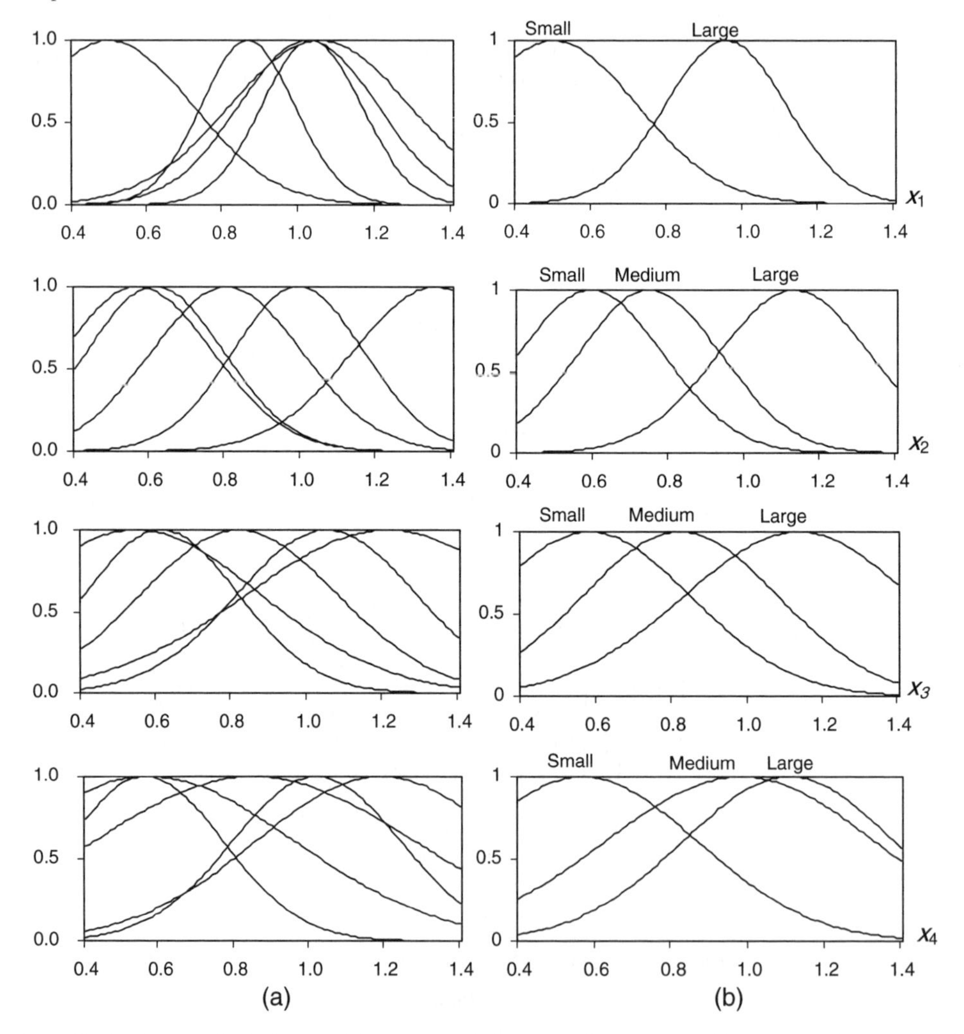

Figure 12.5 Fuzzy sets for the Mackey–Glass example of: (a) the final fuzzy model and (b) the simplified fuzzy model.

Table 12.3 Performance comparison for the Mackey–Glass example.

Model	No. of rules	No. of parameters	RMSE
ANFIS (Jang, 1993)	16	104	0.0016
DeSouza, Vellasco, and Pacheco (2002)	—	64	0.0065
Kim and Kim (1997)	9	81	0.0264
Kukolij (2002)	9	117	0.0061
Tsekouras, Sarimveis, Kavakli, and Bafas (2005)	6	78	0.0041
This model (initial model-test data)	5	65	0.0232
This model (final model-test data)	5	65	0.0052
This model (final model-training data)	5	65	0.0050
This model (10-fold training data)	5	65	0.0053
This model (10-fold testing data)	5	65	0.0056
This model (simplified model)	3	37	0.0251

To obtain the simplified rule base we optimized only the consequent model parameters using (12.20) and (12.21). The RMSE for the simplified model was 0.0251. In Figure 12.5(b) there still exist some fuzzy sets with a high overlapping degree. It is clear that we can further increase the model's interpretability, but as shown in the previous example the performance of the model will substantially decrease. Finding a balance between interpretability and accurate performance is a very challenging problem. So far, no systematic methodology exists to address this issue because the problem differs from application to application.

Table 12.3 compares the three models with other methods. While the simplified model utilizes the smallest number of rules and parameters, it eventually achieves an accurate performance. On the other hand, the initial model outperformed the simplified model, as well as the model developed by Kim and Kim (1997), indicating once again the effective impact of the weighted fuzzy C-means. Finally, the final model's performance is one of the best performances reported in this table.

12.5 CONCLUSIONS

In this chapter, we extended the fuzzy clustering-based fuzzy modeling, by incorporating the weighted fuzzy C-means into the model's designing procedure. This task was accomplished through the implementation of a hierarchical clustering scheme, which consisted of two levels. At the first level the SOM algorithm was employed to preprocess the available training data. At the second level, the weighted fuzzy C-means was applied. As was shown, the utilization of the weighted fuzzy C-means exhibits three appealing features. First, it is able to deal with several difficulties that appear when dealing with the classical fuzzy *c*-means. Secondly, it cancels the undesired effects produced by the application of the preprocessing unit. Thirdly, it provides reliable initialization capabilities for the model parameters. Finally, we presented several experimental studies illustrating the impact of the above features on the performances of the final as well as the simplified model.

REFERENCES

Angelov, P. (2004) 'An approach for fuzzy rule-base adaptation using on-line fuzzy clustering'. *International Journal of Approximate Reasoning*, **35**, 275–289.

Bezdek, J.C. (1981) *Pattern Recognition with Fuzzy Objective Function Algorithms*. Plenum Press, NY. Cassilas, J., Gordon, O., deJesus, M.J. and Herrera, F. (2005) 'Genetic tuning of fuzzy rule deep structures preserving interpretability and its interaction with fuzzy rule set reduction'. *IEEE Transactions on Fuzzy Systems*, **13**, 13–29.

Chao, C.T., Chen, Y.J. and Teng, C.C. (1996) 'Simplification of fuzzy-neural systems using similarity analysis'. *IEEE Transactions on SMC-B*, **26**, 344–354.

Chen, M.Y. and Linkens, D.A. (2000) 'A fuzzy modeling approach using hierarchical neural networks'. *Neural Computing and Applications*, **9**, 44–49.

Chen, M.Y. and Linkens, D.A. (2001) 'A systematic neuro-fuzzy modeling framework with application to material property prediction'. *IEEE Transactions on SMC-B*, **31**, 781–790.

Chen, M.Y. and Linkens, D.A. (2004) 'Rule-base self-generation and simplification for data-driven fuzzy models'. *Fuzzy Sets and Systems*, **142**, 243–265.

Chen, J.Q., Xi, Y.G. and Zhang, Z.J. (1998) 'A clustering algorithm for fuzzy model identification'. *Fuzzy Sets and Systems*, **98**, 319–329.

Chiu, S. (1994) 'Fuzzy model identification based on cluster estimation'. *Journal of Intelligent and Fuzzy Systems*, **2**, 267–278.

Delgado, M., Gomez-Skarmeta, A.F. and Vila, A. (1996) 'On the use of hierarchical clustering in fuzzy modeling'. *International Journal of Approximate Reasoning*, **14**, 237–257.

De Souza, F.J., Vellasco, M.M.R. and Pacheco, M.A.C. (2002) 'Hierarchical neuro-fuzzy quadtree models'. *Fuzzy Sets and Systems*, **130**, 189–205.

Emani, M.R., Turksen, I.B. and Goldenberg, A.A. (1998) 'Development of a systematic methodology of fuzzy logic modeling'. *IEEE Transactions on Fuzzy Systems*, **6**, 346–361.

Geva, A.B. (1999) 'Hierarchical-fuzzy clustering of temporal-patterns and its application for time series prediction'. *Pattern Recognition Letters*, **20**, 1519–1532.

Guillaume, S. (2001) 'Designing fuzzy inference systems from data: an interpretability-oriented overview'. *IEEE Transactions on Fuzzy Systems*, **9**, 426–443.

Guillaume, S. and Charnomordic, B. (2004) 'Generating an interpretable family of fuzzy partition from data'. *IEEE Transactions on Fuzzy Systems*, **12**, 324–335.

Jang, J.S.R. (1993) 'ANFIS: Adaptive-network-based fuzzy inference system'. *IEEE Transactions on SMC*, **23**, 665–685.

Kim, D. and Kim, C. (1997) 'Forecasting time series with genetic fuzzy predictor ensemble'. *IEEE Transactions on Fuzzy Systems*, **5**, 523–535.

Kim, E., Park, M., Ji, S. and Park, M. (1997) 'A new approach to fuzzy modeling'. *IEEE Transactions on Fuzzy Systems*, **5**, 328–337.

Kim, E., Park, M., Kim, S. and Park, M. (1998) 'A transformed input-domain approach to fuzzy modeling'. *IEEE Transactions on Fuzzy Systems*, **6**, 596–604.

Kohonen, T. (1988) *Self-Organization and Associative Memory*, 3rd edn, Springer, NY.

Kukolj, D. (2002) 'Design of adaptive Takagi-Sugeno-Kang fuzzy models'. *Applied Soft Computing*, **2**, 89–103.

Kukolj, D. and Levi, E. (2004) 'Identification of complex systems based on neural networks and Takagi-Sugeno fuzzy model'. *IEEE Transactions on SMC-Part B*, **34**, 272–282.

Lee, S.J. and Ouyang, C.S. (2003) 'A neuro-fuzzy system modeling with self-constructing rule generation and hybrid SVD-based learning'. *IEEE Transactions on Fuzzy Systems*, **11**, 341–353.

Linkens, D.A. and Chen, M.Y. (1999) 'Input selection and partition validation for fuzzy modeling using neural networks'. *Fuzzy Sets and Systems*, **107**, 299–308.

Nozaki, K., Iishibuchi, H. and Tanaka (1997) 'H A simple but powerful method for generating fuzzy rules from numerical data'. *Fuzzy Sets and Systems*, **86**, 251–270.

Pal, N.R. and Bezdek, J.C. (1995) 'On clustering validity for the fuzzy C-means model'. *IEEE Transasctions on Fuzzy Systems*, **3**, 370–379.

Panella, M. and Gallo, A.S. (2005) 'An input–output clustering approach to the synthesis of ANFIS networks'. *IEEE Transactions on Fuzzy Systems*, **13**, 69–81.

Pedrycz, W. (2005) *Knowledge-Based Clustering*, John Wiley & Sons, Inc., NJ.

Setnes, M., Babuska, R., Kaymak, U. and van Nauta Lemke, H.R. (1998) 'Similarity measures in fuzzy rule base simplification'. *IEEE Transactions on SMC-B*, **28**, 376–386.

Sugeno, M. and Yasukawa, T. (1993) 'A fuzzy-logic-based approach to qualitative modeling'. *IEEE Transactions Fuzzy Systems*, **1**, 7–31.

Takagi, T. and Sugeno, M. (1985) 'Fuzzy identification of systems and its application to modeling and control'. *IEEE Transactions on SMC*, **15**, 116–132.

Tsekouras, G.E. (2005) 'On the use of the weighted fuzzy C-means in fuzzy modeling'. *Advances in Engineering Software*, **36**, 287–300.

Tsekouras, G., Sarimveis, H., Kavakli, E. and Bafas, G. (2005) 'A hierarchical fuzzy-clustering approach to fuzzy modeling'. *Fuzzy Sets and Systems*, **150**, 245–266.

Valente de Oliveira, J. (1999) 'Semantic constraints for membership function optimization'. *IEEE Transactions on SMC-A*, **29**, 128–138.

Wong, C.C. and Chen, C.C. (1999) 'A hybrid clustering and gradient descent approach for fuzzy modeling'. *IEEE Transactions on SMC-B*, **29**, 686–693.

Yao, J., Dash, M., Tan, S.T. and Liu, H. (2000) 'Entropy-based fuzzy clustering and fuzzy modeling', *Fuzzy Sets and Systems*, **113**, 381–388.

Yoshinari, Y., Pedrycz, W. and Hirota, K. (1993) 'Construction of fuzzy models through clustering techniques'. *Fuzzy Sets and Systems*, **54**, 157–165.

Zhou, S.M. and Gan, J.Q. (2006) 'Constructing accurate and parsimonious fuzzy models with distinguishable fuzzy sets based on an entropy measure'. *Fuzzy Sets and Systems*, **157**, 1057–1074.

13 Fuzzy Clustering Based on Dissimilarity Relations Extracted from Data

Mario G.C.A. Cimino, Beatrice Lazzerini, and Francesco Marcelloni

Dipartimento di Ingegneria dell'Informazione: Elettronica, Informatica, Telecomunicazioni, University of Pisa, Pisa, Italy

13.1 INTRODUCTION

Clustering algorithms partition a collection of data into a certain number of clusters (groups, subsets, or categories). Though there is no universally agreed definition, most researchers describe a cluster by considering the internal homogeneity and the external separation (Xu and Wunsch, 2005), i.e., patterns in the same cluster should be similar to each other, while patterns in different clusters should not (Jain, Murty, and Flynn 1999; Su and Chou, 2001). Thus, the correct identification of clusters depends on the definition of similarity. Typically, similarity (more often dissimilarity) is expressed in terms of some distance function, such as the Euclidean distance or the Mahalanobis distance. The choice of the (dis)similarity measure induces the cluster shape and therefore determines the success of a clustering algorithm on the specific application domain. For instance, the Euclidean and Mahalanobis distances lead clustering algorithms to determine hyperspherical-shaped or hyperellipsoidal-shaped clusters, respectively. Typically, when we apply a clustering algorithm, we do not know a priori the most natural and effective cluster shapes for the specific data-set. Each data-set is characterized by its own data distribution and therefore requires cluster shapes different from other data-sets. Nevertheless, we have to choose the dissimilarity measure before starting the clustering process. For instance, when we apply the classical fuzzy C-means (FCM) (Bezdek, 1981), which is one of the best known partitional fuzzy clustering algorithms, we decide a priori to use the Euclidean distance and therefore to identify hyperspherical-shaped clusters. To overcome this problem, in the literature, several approaches have been proposed. For instance, density-based clustering algorithms determine on-line the shape of clusters. In density-based clustering, clusters are regarded as regions in the data space in which the objects are dense. These regions may have an arbitrary shape and the points inside a region may be arbitrarily distributed. To determine if a region is

Advances in Fuzzy Clustering and its Applications Edited by J. Valente de Oliveira and W. Pedrycz

dense, we need to define the concept of neighborhood, based on a priori defined proximity (see, for instance, DBSCAN, Ester, Kriegel, Sander, and Xu, 1996, or OPTICS, Ankerst, Breunig, Kriegel, and Sander, 1999). Though proximity can be defined in terms of any dissimilarity measure, applications of density-based clustering algorithms proposed in the literature adopt a distance function to determine spatial proximity. Thus, again, though the shapes of clusters may be different from each other, they still depend on the a priori choice of a distance.

When applying clustering to data with irregular distribution, as is often the case for image segmentation and pattern recognition (Jain and Flynn, 1993), distance functions cannot adequately model dissimilarity (Valentin, Abdi, O'Toole, and Cottrell, 1994; Kamgar-Parsi and Jain, 1999; Santini and Jain, 1999; Latecki and Lakamper, 2000). Consider, for example, the pixels of an image made up of distinguishable elements with irregular-shaped contours (for instance, bikes, cars, houses, trees). The dissimilarity between pixels should be small (large) when the pixels belong to the same image element (different image elements).

To solve this problem, some approaches have been proposed in the literature. For example, Jarvis and Patrick (1973) defined the dissimilarity between two points as a function of their context, i.e., the set of points in the neighborhood of each such point. Michalski, Stepp, and Diday (1983) used predefined concepts to define the "conceptual similarity" between points. Yang and Wu (2004) proposed adopting a total similarity related to the approximate density shape estimation as objective function of their clustering method. Jacobs, Weinshall, and Gdalyahu (2000) observed that classification systems, which can model human performance or use robust image matching methods, often exploit similarity judgement that is non-metric. Makrogiannis, Economou, and Fotopoutos (2005) introduced a region dissimilarity relation that combines feature-space and spatial information for color image segmentation.

Pedrycz (2005) suggested exploiting some auxiliary information (*knowledge-based hints*), which reflect some additional sources of domain knowledge, in order to guide the clustering process. He, first, proposed a general taxonomy of knowledge-based hints. Then, he discussed some clustering algorithms which partition the data-set guided by these hints. In particular, he considered a partially supervised version of the classical FCM algorithm (Pedrycz and Waletzky, 1997), which uses some labeled patterns as knowledge-based hints: these labeled patterns serve as reference elements in modeling the cluster shapes. Further, he discussed a proximity-based fuzzy clustering algorithm where knowledge-based hints are represented by proximity values between pairs of patterns (Pedrycz, Loia, and Senatore, 2004). Similarly, Lange, Law, Jain, and Buhmann (2005) or Law, Topchy, and Jain (2005) proposed exploiting a priori knowledge about a desired model via two types of pairwise constraints: must-link and must-not-link constraints. The two constraints correspond to the requirements that two objects should and should not be associated with the same label, respectively.

A different approach proposed extracting the dissimilarity relation directly from the data by guiding the extraction process itself with as little supervision as possible (Pedrycz *et al.*, 2001). Following this approach, Hertz, Bar-Hillel, and Weinshall (2004) suggested learning distance functions by using a subset of labeled data. In particular, they trained binary classifiers with margins, defined over the product space of pairs of images, to discriminate between pairs belonging to the same class and pairs belonging to different classes. The signed margin is used as a distance function. Both support vector machines and boosting algorithms are used as product space classifiers. Using some benchmark databases from the UCI repository, the authors showed that their approach significantly outperformed existing metric learning methods based on learning the Mahalanobis distance.

Recently, some methods have been proposed to exploit pairwise dissimilarity information for learning distance functions (Xing, Ng, Jordan, and Russell, 2003). Tsang, Cheung, and Kwok (2005), for instance, proposed learning distance metric from a subset of pairwise dissimilarity values by a kernelized version of the relevant component analysis method. Chang and Yeung (2005) formulated the metric learning problem as a kernel learning problem, which is efficiently solved by kernel matrix adaptation.

Similarly, in this chapter, we will discuss how the dissimilarity relation can be extracted directly from a few pairs of data with known dissimilarity values rather than from pairs of data with known labels. We will discuss the application of two different techniques based on, respectively, neural networks and fuzzy systems. More precisely, we use a multilayer perceptron (MLP) with supervised learning (Haykin, 1999) and a Takagi–Sugeno (TS) fuzzy system (Takagi and Sugeno, 1985). The rules of the TS are identified by using the method proposed by Setnes and Roubos (2000). Once the MLP has been trained and the TS has been identified, the two models can associate a dissimilarity value with each pair of patterns in the data-set. This relation, extracted from the data, can be exploited by a relational clustering algorithm to partition the data-set into a suitable number of clusters.

In real applications, clusters are generally overlapped and their boundaries are fuzzy rather than crisp. The identification of these clusters demands appropriate fuzzy relational clustering algorithms. To the best of our knowledge, fuzzy relational clustering algorithms proposed in the literature require dissimilarity relations which are symmetric and irreflexive (Bezdek, Keller, Krisnapuram, and Pal, 1999). On the other hand, the generalization performed by the MLP and by the TS may produce a relation that is neither symmetric nor irreflexive. For instance, a pattern not included in the training set may be judged slightly dissimilar to itself. Further, the dissimilarity value between pattern $\mathbf{x}_i$ and pattern $\mathbf{x}_j$ may be different from the dissimilarity value between pattern $\mathbf{x}_j$ and pattern $\mathbf{x}_i$. Thus, though fuzzy relational algorithms may work correctly on the relation produced by the two models, sometimes they converge to solutions which are not sound.

Actually, we observed that some of the best known fuzzy clustering algorithms, when applied to the dissimilarity relations extracted by the MLP and the TS, converge to a partition composed completely superimposed clusters, that is, each pattern in the data-set belongs to all clusters with the same membership grade. To overcome this unstable behavior, in a previous paper we proposed a new approach to fuzzy relational clustering: (Corsini, Lazzerini, and Marcelloni, 2005) starting from the definition of relational clustering algorithm, we transformed a relational clustering problem into an object clustering problem. This transformation allows us to apply any object clustering algorithm to partition sets of objects described by relational data. In the implementation based on the classical FCM algorithm and denoted ARCA (Corsini, Lazzerini, and Marcelloni, 2005), we verified that this approach produces partitions similar to the ones generated by the other fuzzy relational clustering algorithms, when these converge to a sound partition. On the other hand, as FCM has proved to be one of the most stable fuzzy clustering algorithms, ARCA is appreciably more stable than the other fuzzy relational clustering algorithms. In this chapter we show the effectiveness of the combinations MLP–ARCA and TS–ARCA using a synthetic data-set and the Iris data-set, respectively. We describe how these combinations achieve very good clustering performance using a limited number of training samples. Further, we show how the TS can provide an intuitive linguistic description of the dissimilarity relation. Finally, we discuss the performance obtained by the combination TS–ARCA on three real data-sets from the UCI repository.

We wish to point out that the combination of supervised and unsupervised learning discussed in this chapter is intended for use in all cases in which the dissimilarity relation can be learnt from a reasonably small portion of samples, which form the training set. The method works, in principle, with any kind of data-set. In fact, as it unfolds the entire data onto as many dimensions as the number of data points in order to transform the relational clustering to object-based clustering, it is more appropriate for moderate-size data-sets, typically containing up to a few hundreds of patterns.

13.2 DISSIMILARITY MODELING

Our approach to fuzzy clustering is based on extracting the dissimilarity measure that drives the clustering strategy from a small set of known similarities. Thus, we have to generate a model that, given a pair $(\mathbf{x}_i, \mathbf{x}_j)$ of input data, outputs the dissimilarity degree $d_{i,j}$ between $\mathbf{x}_i$ and $\mathbf{x}_j$. The generation of this model is a typical identification problem, which has been tackled by different techniques such as classical mathematical theory, support vector machines, neural networks, and fuzzy modeling. In this work, we discuss the application of two of these techniques: neural networks and fuzzy modeling. In particular, to model the dissimilarity relation, we used a multilayer perceptron (MLP) neural network with supervised learning and a Takagi–Sugeno (TS) fuzzy system. We assume that the patterns are described by numerical features (possibly, nonnumerical features are appropriately transformed into numerical ones) and the dissimilarity degrees[1] between a few pairs of patterns are known. Let $T = \{\mathbf{z}_1, \ldots, \mathbf{z}_N\}$ be the set of

[1]Actually, our method can deal with both similarity and dissimilarity relations.

known data, where $\mathbf{z}_n = [\mathbf{x}_i, \mathbf{x}_j, d_{i,j}] \in \Re^{2F+1}$. In the following two sections, we briefly describe the MLP and the TS used in our experiments.

13.2.1 The Multilayer Perceptron

We use a standard feedforward three-layer MLP neural network. Each neuron is equipped with a sigmoidal nonlinear function. The standard back-propagation algorithm with a dynamically decreasing learning rate is used as a learning scheme. Errors less than 0.001 are treated as zero. Initial weights are random values in the range $\lfloor -1/\sqrt{m}, 1/\sqrt{m} \rfloor$, with m being the number of inputs to a neuron. As described by Corsini, Lazzerini, and Marcelloni (2006), to determine the best structure of the neural network with respect to the generalization capability, we performed a number of experiments with two-layer and three-layer MLP and with a different number of neurons for each hidden layer. For the data-sets discussed in this chapter, we observed that the best generalization properties are obtained by using an architecture with 20 and eight neurons for the first and second hidden layers, respectively. Further, we experimentally verified that this result is quite independent of the size of the training set, at least for the sizes used in the experiments.

13.2.2 The Takagi-Sugeno System

The rules of the TS have the following form:

$$r_i\text{: If } X_{1,1} \text{ is } A_{i,1,1} \text{ and} \ldots X_{1,F} \text{ is } A_{i,1,F} \text{ and } X_{2,1} \text{ is } A_{i,2,1} \text{ and} \ldots X_{2,F} \text{ is } A_{i,2,F}$$
$$\text{then } d_i = \mathbf{a}_{i,1}^T \mathbf{X}_1 + \mathbf{a}_{i,2}^T \mathbf{X}_2 + b_i, \qquad i = 1..R$$

where R is the number of rules, $\mathbf{X}_e = [X_{e,1}, \ldots, X_{e,F}]$, with e=1, 2, are the two input variables of F components that represent the pair of patterns whose dissimilarity has to be evaluated, $A_{i,e,1}, \ldots, A_{i,e,F}$ are fuzzy sets defined on the domain of $X_{e,1}, \ldots, X_{e,F}$, respectively, $\mathbf{a}_{i,e}^T = [a_{i,e,1}, \ldots, a_{i,e,F}]$, with $a_{i,e,f} \in \Re$, and $b_i \in \Re$. The model output d, which represents the dissimilarity between two input patterns, is computed by aggregating the conclusions inferred from the individual rules as follows:

$$d = \frac{\sum_{i=1}^{R} \beta_i d_i}{\sum_{i=1}^{R} \beta_i} \tag{13.1}$$

where $\beta_i = \prod_{f=1}^{F} A_{i,1,f}(x_{j,f}) \prod_{f=1}^{F} A_{i,2,f}(x_{k,f})$ is the degree of activation of the ith rule, when the pair $(\mathbf{x}_j, \mathbf{x}_k)$ is fed as input to the rule.

A TS model is built through two steps, called the *structure identification* and the *parameter identification* (Babuška, 1996). The structure identification determines the number of rules and the variables involved in the rule antecedents. The parameter identification estimates the parameters that define, respectively, the membership functions of the fuzzy sets in the antecedents and the consequent functions. The number of rules is generally computed by exploiting a clustering algorithm (Angelov and Filev, 2004; Abonyi, Babuška, and Szeifert, 2002). More precisely, the number of rules coincides with the number of clusters of the input-output space partition, which results to be the best with respect to an appropriate validity index. The parameter identification is obtained by first computing the fuzzy sets in the antecedent of the rules, and then estimating the parameters of the mathematical functions in the consequent (Angelov and Filev, 2004). One of the most used clustering algorithms to identify the structure of a TS is the classical FCM with Euclidean distance. As the FCM algorithm finds the fuzzy partition starting from a fixed number of clusters, and the number of clusters determines the number of rules that compose the fuzzy model, a criterion has to be adopted to determine the optimal number of clusters. The most common approach is to identify an interval of possible values of the number R of clusters and execute the FCM for each value in the interval. Each execution is therefore assessed against a validity index. Several different validity indexes have been proposed in the literature (Bezdek, Keller, Krisnapuram, and Pal, 1999). The most used among these indexes are the Xie and Beni's index (XB, Xie, and Beni, 1991), the Fukuyama and Sugeno's index (Pal and Bezdek, 1995), the Gath and Geva's index (Gath and Geva, 1989), and the Rezaee, Lelieveldt, and Reiber's index (Rezaee, Lelieveldt, and Reiber, 1998). As is well known in the literature, there does not exist a validity index which is good for each data-set (Pal and Bezdek, 1995). In order to choose the most reliable index for the data-sets used in the experiments, we compared the aforementioned validity indexes against the TS accuracy obtained with the number of rules determined by the indexes. We observed that the XB index guarantees the best results. The Xie–Beni index is defined as

$$XB(U,V,T) = \frac{\sum_{i=1}^{R}\sum_{n=1}^{N} u_{i,n}^2 \parallel \mathbf{z}_n - \mathbf{v}_i \parallel^2}{N(\min_{j \neq k} \parallel \mathbf{v}_j - \mathbf{v}_k \parallel^2)},$$

where V is the vector of cluster prototypes $\mathbf{v}_i$ and U is the fuzzy partition matrix whose generic element $u_{i,n}^2$ represents the grade of membership of $\mathbf{z}_n$ to cluster i. The numerator of the fraction measures the compactness of the clusters while the denominator measures the degree of separation of the cluster prototypes. For compact and well-separated clusters we expect small values of XB. We execute the FCM algorithm with increasing values of the number R of clusters for values of the fuzzification constant m in $\{1.4, 1.6, 1.8, 2.0\}$ and plot the Xie–Beni index versus R. We choose, as the optimal number of clusters, the value of R corresponding to the first distinctive local minimum.

The combination of the FCM and the Xie–Beni index helps determining only the rules that describe important regions of the input/output space, thus leading to a moderate number of rules. Fuzzy sets $A_{i,e,f}$ are obtained by projecting the rows of the partition matrix U onto the fth component of the input variable $\mathbf{X}_e$ and approximating the projections by triangular membership functions $A_{i,e,f}(l_{i,e,f}, m_{i,e,f}, r_{i,e,f})$ with $l_{i,e,f} < m_{i,e,f} < r_{i,e,f}$ real numbers on the domain of definition of $X_{e,f}$. We computed the parameter $m_{i,e,f}$, which corresponds to the abscissa of the vertex of the triangle, as the weighted average of the $X_{e,f}$ components of the training patterns, the weights being the corresponding membership values. Parameters $l_{i,e,f}$ and $r_{i,e,f}$ were obtained as intersection of the $X_{e,f}$ axis with the lines obtained as linear regression of the membership values of the training patterns, respectively, on the left and the right sides of $m_{i,e,f}$. Obviously, if $l_{i,e,f}$ and $r_{i,e,f}$ are beyond the extremes of the definition domain of variable $X_{e,f}$, the sides of the triangles are truncated in correspondence to the extremes. The use of triangular functions allows easy interpretation of the fuzzy sets in linguistic terms. Once the antecedent membership functions have been fixed, the consequent parameters $[\mathbf{a}_{i,1}, \ \mathbf{a}_{i,2}, \ b_i]$, $i = 1..R$, of each individual rule i are obtained as a local least squares estimate.

The strategy used so far to build the TS is aimed at generating a rule base characterized by a number of interesting properties, such as a moderate number of rules, membership functions distinguishable from each other, and space coverage, rather than at minimizing the model error. We experimentally verified that this TS could show a poor performance, in particular for training sets composed of a high number of pairs. Thus, we apply a genetic algorithm (GA) to tune simultaneously the parameters in the antecedent and consequent parts of each rule in a global optimization. To preserve the good properties of the fuzzy model, we impose that no gap exists in the partition of each input variable. Further, to preserve distinguishability we allow the parameters that define the fuzzy sets to vary within a range around their initial values. Each chromosome represents the entire fuzzy system, rule by rule, with the antecedent and consequent parts (see Figure 13.1). Each rule antecedent consists of a sequence of $2 \cdot F$ triplets (l, m, r) of real numbers representing triangular membership functions, whereas each rule consequent contains $2 \cdot F + 1$ real numbers corresponding to the consequent parameters. The fitness value is the inverse of the mean square error (MSE) between the predicted output and the desired output over the training set.

We start with an initial population composed of 70 chromosomes generated as follows. The first chromosome codifies the system generated by the FCM, the others are obtained by perturbing the first chromosome randomly within the ranges fixed to maintain distinguishability. At each generation, the arithmetic crossover and the uniform mutation operators are applied with probabilities 0.8 and 0.6, respectively. Chromosomes to be mated are chosen by using the well-known roulette wheel selection method. At each generation, the offspring are checked against the aforementioned space coverage

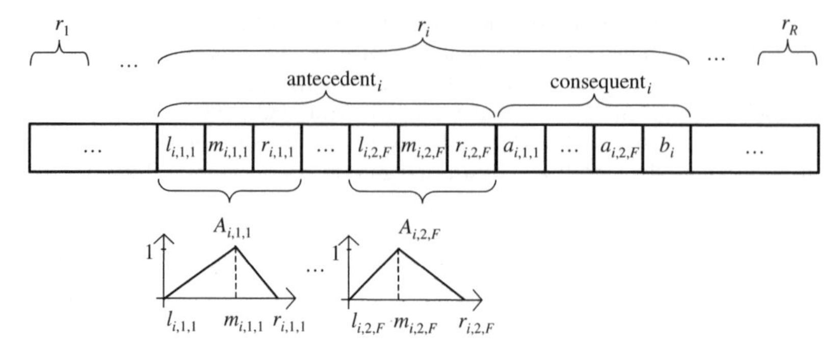

Figure 13.1 The chromosome structure.

criterion. To speed up the convergence of the algorithm without significantly increasing the risk of premature convergence to local minima, we adopt the following acceptance mechanism: 40 % of the new population is composed of offspring, whereas 60 % consists of the best chromosomes of the previous population. When the average of the fitness values of all the individuals in the population is greater than 99.9 % of the fitness value of the best individual or a prefixed number of iterations has been executed (6000 in the experiments), the GA is considered to have converged.

The fairly large size of the population and the mutation probability higher than usual have been chosen to counteract the effect of the strong exploitation of local linkages. Indeed, due to real coding (Wright, 1991) and to constraints imposed on the offspring so as to maintain distinguishability, exploitation could lead to a premature convergence to sub-optimal solutions. The values of the GA parameters used in the experiments reduce this risk. To strengthen this consideration, we observed in the experiments that, varying the data-set, the values of the GA parameters do not need to be changed.

13.2.3 MLP versus TS

To compare the two approaches, we used the synthetic data-set shown in Figure 13.2 and the Iris data-set (UCI, 2006). The first data-set was chosen because clustering algorithms, which measure the dissimilarity

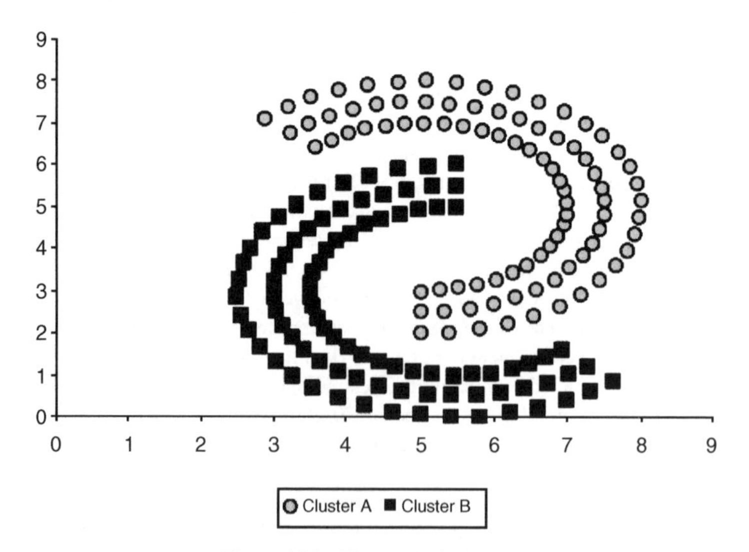

Figure 13.2 The synthetic data-set.

between two points as the distance between the two points, cannot partition it correctly. Indeed, both the Euclidean and the Mahalanobis distances that induce, respectively, spherical and ellipsoidal cluster shapes lead, for instance, the FCM algorithm and the GK algorithm (Gustafson and Kessel, 1979) to partition the data-set incorrectly (Corsini, Lazzerini, and Marcelloni, 2006).

For each data-set, we carried out five experiments. In these experiments, we aimed to assess how much the size of the training pool affected the performance of the MLP and the TS. For this purpose, we randomly extracted a pool of patterns (called the *training pool*) from the data-set. This pool was composed of 5 %, 10 %, 15 %, 20 %, and 25 % of the data-set, respectively, in the five experiments. Then, we built the training set by selecting a given number of pairs of patterns from the training pool. More precisely, assume that C is the number of clusters, which we expect to identify in the data-set. Then, for each pattern $\mathbf{x}_i$ in the training pool, we formed $q \cdot C$ pairs $(\mathbf{x}_i, \mathbf{x}_j)$, with $q \in [1..8]$, by randomly selecting $q \cdot C$ patterns $\mathbf{x}_j$ of the training pool as follows: q patterns were chosen among those with dissimilarity degree lower than 0.5 with $\mathbf{x}_i$, and the remaining $q \cdot (C\text{-}1)$ patterns were chosen among those with dissimilarity degree higher than 0.5.

This choice tries to provide the same number of training samples for pairs of points belonging to different clusters as for pairs of points belonging to the same cluster. Obviously, since we do not know a priori the membership of each point to a class, this choice is only an approximation. However, we experimentally verified that it provides reliable results. It is obvious that increasing values of q leads to better classification performance, but also to increasing execution times. Obviously, we assumed the dissimilarity degrees between all the pairs that can be built from patterns in the training pool were known. This assumption, which is not actually necessary, was made to test the effects of q on the performance of the MLP and the TS. For the two data-sets, we observed that $q = 5$ provides a good trade off between classification accuracy and execution time. Let $d_{i,j}$ be the degree of dissimilarity between $\mathbf{x}_i$ and $\mathbf{x}_j$. We inserted both $[\mathbf{x}_i, \mathbf{x}_j, d_{i,j}]$ and $[\mathbf{x}_j, \mathbf{x}_i, d_{i,j}]$ into the training set.

We carried out the five experiments described above and, for each experiment, we executed 10 trials. For the sake of simplicity, in the experiments, we used only 0 and 1 to express the dissimilarity degree of two input points belonging to the same class or to different classes, respectively. Please note that we use the knowledge about classes just to assign dissimilarity degrees to pairs of points in the training pool.

To assess the generalization properties, for each trial and each experiment we tested the two models on all possible pairs of points in the data-set and measured the percentage of the point pairs with dissimilarity degree lower than (higher than) 0.5 for pairs of points belonging (not belonging) to the same class.

Tables 13.1 and 13.2 show the percentages of correct dissimilarity values obtained by applying the MLP to the synthetic and the Iris data sets. In the tables, the columns show, respectively, the percentage of points composing the training pool and the percentage (in the form (mean $\pm$ standard deviation)) of pattern pairs with correct dissimilarity.

Tables 13.3 and 13.4 show the percentages of correct dissimilarity values obtained by applying the TS system to the synthetic and the Iris data-sets. In the tables, the columns indicate, respectively, the percentage of points composing the training pool, the number of rules of the TS model (in the form (mean $\pm$ standard deviation)) and the percentage of correct dissimilarity values before and after the GA optimization. It can be observed that the application of the GA sensibly improves the percentage of correct dissimilarity values generated by the TS model independently of the cardinality of the training pool.

Table 13.1 Percentage of point pairs with correct dissimilarity values (MLP system on the synthetic data-set).

Training pool	Correct dissimilarity values
5%	70.1% ± 5.2%
10%	73.8% ± 4.5%
15%	81.5% ± 4.3%
20%	85.1% ± 3.3%
25%	89.8% ± 2.2%

Table 13.2 Percentage of point pairs with correct dissimilarity values (MLP system on Iris data-set).

Training pool	Correct dissimilarity values
5%	81.2% ± 3.2%
10%	85.5% ± 3.8%
15%	88.1% ± 3.4%
20%	90.4% ± 3.5%
25%	90.7% ± 2.7%

As shown in the tables, the two approaches have similar performance. Both the MLP and the TS achieve about 90 % of correct dissimilarity values with 25 % of the points. We have to consider that the percentage of total pairs of points included in the training set is much lower than the percentage of total points in the training pool. For instance, for the synthetic data-set, a training pool composed of 25 % of the points corresponds to a training set composed of 2.78 % of the dissimilarity values. Taking this into account, the percentages achieved by the two approaches are undoubtedly remarkable.

As regards the computational overhead, to achieve the results shown in Tables 13.1 and 13.2, the identification of the best performing architecture of the MLP has required several experiments. We used architectures with both two layers and three layers and with a different number of neurons for each hidden layer. For the two-layer architecture, we used 10, 20, 30, 40, 50, 60, and 70 neurons in the hidden layer and for the three-layer architecture, we used 12, 16, 20, 24, and 28 neurons in the first hidden layer and 4, 6, 8, 10, and 12 in the second hidden layer (Corsini, Lazzerini, and Marcelloni, 2006). For each architecture, we trained the MLP and evaluated the percentage of point pairs with correct dissimilarity. Similarly, to determine the structure of the TS system, we executed the FCM algorithm with increasing values of the number R of clusters for different values of the fuzzification constant m and assessed the goodness of each resulting partition using the Xie–Beni index. Since the execution of the FCM is generally faster than the learning phase of an MLP, the determination of the TS structure is certainly quicker than the identification of the MLP architecture.

Once the structure has been identified, the TS requires the execution of the GA for tuning the membership functions and the consequent parameters so as to minimize the mean square error. As is well known in the literature, GAs are generally computationally heavy. We verified, however, that the GA used in this work performs a good optimization after a reasonable number of iterations. As an example, Figure 13.3 shows the percentage of correct dissimilarity values versus the number of generations in five trials with the Iris data-set and a training pool of 25 %. We can observe that a thousand generations allow the genetic algorithm to achieve a good approximation of the dissimilarity relation. If we consider that, as discussed in the next section, we can obtain good clustering results with 70–75 % of correct dissimilarity values, we can stop the genetic algorithm after a few hundreds of generations. This solution provides the further advantage of preventing overfitting problems, which may occur for small and unrepresentative training sets. The results shown in Tables 13.3 and 13.4 were obtained by stopping the GA after 2000 generations. Thus, we can conclude that the generation of the TS requires less effort than the generation of the MLP. Indeed, the determination of the best MLP network requires iteration through a number of MLP architectures with a different number of hidden layers and of nodes for each layer. The different networks

Table 13.3 Percentage of point pairs with correct dissimilarity values (TS system on the synthetic data-set).

Training pool	Number of rules	Correct dissimilarity values before GA	Correct dissimilarity values after GA
5%	10.5 ± 3.3	61.8% ± 6.7%	69.6% ± 7.6%
10%	10.1 ± 3.2	66.3% ± 3.8%	75.7% ± 5.2%
15%	11.7 ± 3.2	65.6% ± 6.8%	82.6% ± 4.2%
20%	12.6 ± 2.9	67.8% ± 3.0%	85.3% ± 3.9%
25%	14.2 ± 1.5	69.7% ± 2.8%	90.4% ± 3.5%

Table 13.4 Percentage of pattern pairs with correct dissimilarity values (TS system on Iris data-set).

Training pool	Number of Rules	Correct dissimilarity values before GA	Correct dissimilarity values after GA
5%	8.9 ± 2.4	$80.0\% \pm 4.1\%$	$80.5\% \pm 4.5\%$
10%	6.4 ± 1.6	$82.7\% \pm 4.8\%$	$87.7\% \pm 3.1\%$
15%	4.8 ± 0.6	$80.8\% \pm 3.0\%$	$90.2\% \pm 2.2\%$
20%	4.4 ± 0.8	$78.5\% \pm 7.0\%$	$91.6\% \pm 2.0\%$
25%	4.7 ± 0.5	$80.7\% \pm 4.7\%$	$91.6\% \pm 1.8\%$

are compared against accuracy. Each architecture has to be trained and this operation generally requires a considerable amount of time, depending on the number of layers and neurons for each layer. On the contrary, the determination of the TS structure requires iteration of the execution of FCM with different values of the number of clusters. The execution of FCM is certainly faster than the training of the MLP network and also the number of executions of FCM needed is generally smaller than the number of MLP networks to be trained. On the other hand, the generation of the TS systems requires the execution of the GA, which is quite time consuming. We have to consider, however, that the GA is executed just one time.

Finally, unlike the MLP, the TS allows describing the dissimilarity relation intuitively. Figure 13.4 shows the antecedent and the consequent of the rules that compose a TS model (after the optimization performed by GA) generated with the training pool composed of 15 % of the synthetic data-set. Here, we have associated a label with each fuzzy set based on the position of the fuzzy set in the universe of definition.

Since each rule defines its fuzzy sets, which may be different from the other rules, we used the following method to assign a meaningful linguistic label to each fuzzy set. First, we uniformly partition the universes of discourse into G triangular fuzzy sets (denoted as *reference terms* in the following) and associate a meaningful label with each fuzzy set. In the example, labels L, ML, M, MH, and H denote, respectively, *low*, *medium-low*, *medium*, *medium-high*, and *high* (see Figure 13.5). Then, we compute the similarity between each fuzzy set used in the rules and the reference terms using the formula

$$S_{i,e,f,l} = \frac{\left|A_{i,e,f} \cap P_{l,e,f}\right|}{\left|A_{i,e,f} \cup P_{l,e,f}\right|},$$

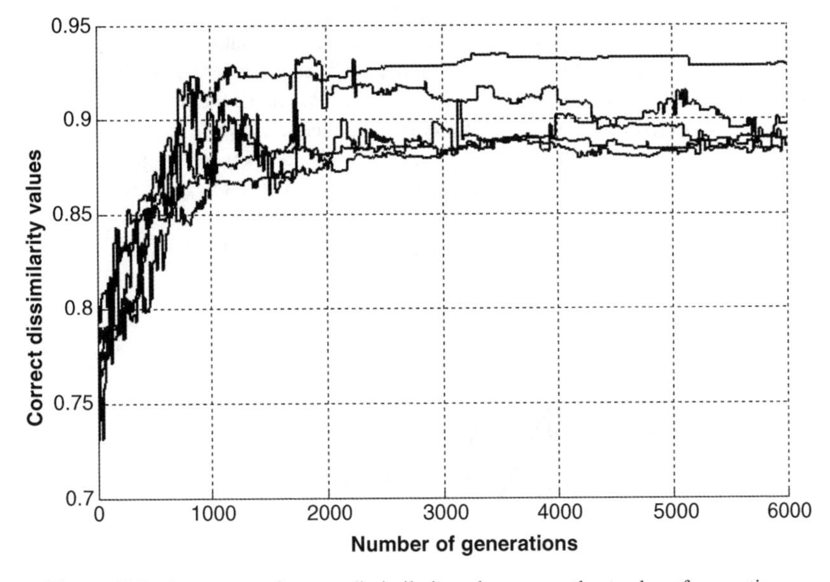

Figure 13.3 Percentage of correct dissimilarity values versus the number of generations.

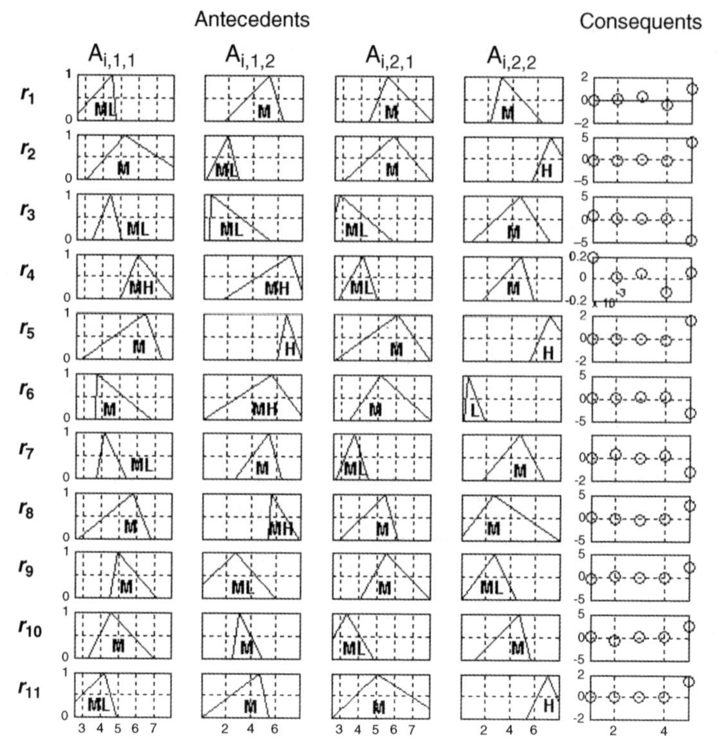

Figure 13.4 Rules after GA (synthetic data-set).

where $A_{i,e,f}$ and $P_{l,e,f}$ are, respectively, a fuzzy set and a reference term defined on the domain of input $X_{e,f}$ (Sugeno and Yasukawa, 1993). Finally, if there exists a value of $S_{i,e,f,l}$, with $l = 1..G$, larger than a fixed threshold τ, the reference term $P_{l,e,f}$ is associated with $A_{i,e,f}$ (if there exist more $P_{l,e,f}$ with $S_{i,e,f,l} > \tau$, then $A_{i,e,f}$ is associated with the $P_{l,e,f}$ corresponding to the highest $S_{i,e,f,l}$); otherwise, $A_{i,e,f}$ is added to the reference terms after associating a meaningful label with it. This association is carried out as follows. We first determine the reference term $P_{l,e,f}$ more similar to $A_{i,e,f}$. Then we generate four fuzzy sets. Two fuzzy sets are obtained by halving and doubling the support of $P_{l,e,f}$. We name the two fuzzy sets *very* $P_{l,e,f}$ and *more or less* $P_{l,e,f}$, respectively. The other two fuzzy sets are generated as $(P_{l,e,f} + P_{l,e,f-1})/2$, if $f \neq 0$, and $(P_{l,e,f} + P_{l,e,f+1})/2$, if $f \neq F$ (in the cases $f = 0$ and $f = F$, no fuzzy set is generated). The results of $(P_{l,e,f} + P_{l,e,f-1})/2$ and $(P_{l,e,f} + P_{l,e,f+1})/2$ are two triangular fuzzy sets defined as

$$\left(\frac{l_{l,e,f} + l_{l,e,f-1}}{2}, \frac{m_{l,e,f} + m_{l,e,f-1}}{2}, \frac{r_{l,e,f} + r_{l,e,f-1}}{2}\right)$$

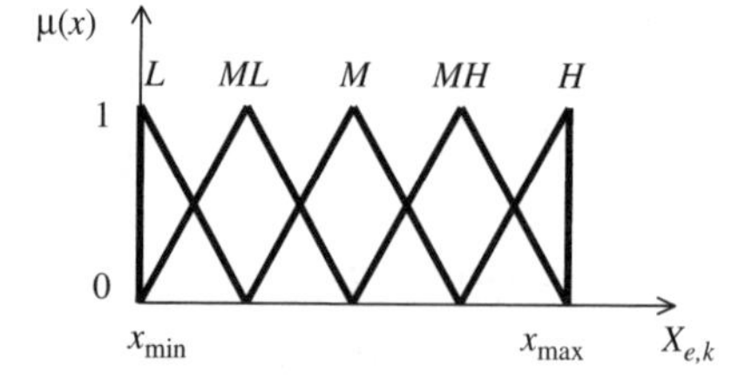

Figure 13.5 Reference terms for a generic input variable $X_{e,k}$.

Table 13.5 The qualitative model.

Rule	$X_{1,1}$	$X_{1,2}$	$X_{2,1}$	$X_{2,2}$	$\bar{d}_{i,j}$
r_1	*ML*	*M*	M	M	0.92
r_2	*M*	*ML*	*M*	H	0.73
r_3	*ML*	*ML*	*ML*	M	0.00
r_4	MH	MH	*ML*	M	1.00
r_5	*M*	*H*	*M*	H	0.00
r_6	*M*	MH	*M*	L	0.70
r_7	*ML*	*M*	*ML*	M	0.00
r_8	*M*	MH	*M*	M	0.73
r_9	*M*	*ML*	*M*	ML	0.32
r_{10}	*M*	M	*ML*	M	0.50
r_{11}	*ML*	*M*	M	H	0.38

and

$$\left(\frac{l_{l,e,f}+l_{l,e,f+1}}{2},\frac{m_{l,e,f}+m_{l,e,f+1}}{2},\frac{r_{l,e,f}+r_{l,e,f+1}}{2}\right),$$

respectively. We name these two fuzzy sets as $P_{l,e,f}-P_{l,e,f-1}$ and $P_{l,e,f}-P_{l,e,f+1}$, respectively. For instance, if $P_{l,e,f}=ML$, we obtain *very ML*, *more or less ML*, *L-ML*, and *ML-M*. Finally, we select the most similar among the four fuzzy sets to $A_{i,e,f}$ and assign the corresponding label to $A_{i,e,f}$. Once the fuzzy sets of all the rules have been examined, we again compute the similarity between each fuzzy set and the current reference terms in order to associate the most appropriate label with each fuzzy set. To generate the labels associated with the fuzzy sets shown in Figure 13.4, we have used a threshold $\tau=0.5$. Note that no further reference term has been added.

To interpret the rules, we follow this procedure: for each pattern $\mathbf{z}_n=[\mathbf{x}_i,\mathbf{x}_j,d_{i,j}]$ in the training set, we feed as input the values of the coordinates of $\mathbf{x}_i$ and $\mathbf{x}_j$ to the TS model and measure the activation degree of each rule. We aim to discover whether there exists a relation between the activation of a rule and the values of dissimilarity. Table 13.5 shows, for each rule, the mean value $\bar{d}_{i,j}$ of dissimilarity $d_{i,j}$ of the pairs $(\mathbf{x}_i,\mathbf{x}_j)$ of patterns of the training set that activate this rule more than the other rules. This association between rules and dissimilarity values helps us interpret the meaning of the rules. From rule r_4, for instance, we can deduce that if the abscissa and the ordinate of the first point are, respectively, *MH* and *MH*, and the abscissa and the ordinate of the second point are, respectively, *ML* and *M*, then the dissimilarity is high. This rule can be easily verified by observing the data-set in Figure 13.2.

We note that rules are activated by pairs of points with either high or low dissimilarity. Indeed, the mean value of dissimilarity is close to 0 or 1. This means that the antecedents of the rules determine regions of the plane which contain points belonging either to the same class or to different classes. This observation confirms the results shown in Table 13.3: using 15 % of points in the training pool, we achieved 82.6 % of correct classification.

13.3 RELATIONAL CLUSTERING

Let $Q=[\mathbf{x}_1,\ldots,\mathbf{x}_M]$ be the data-set. Once the MLP has been trained or the TS has been generated and optimized, we compute the dissimilarity value between each possible pair $(\mathbf{x}_i,\mathbf{x}_j)$ of patterns in the data-set Q. Such dissimilarity values are provided as an $M\times M$ relation matrix $D=[d_{i,j}]$. The value $d_{i,j}$ represents the extent to which $\mathbf{x}_i$ is dissimilar to $\mathbf{x}_j$. Thus, the issue of partitioning patterns described through a set of meaningful features is transformed into the issue of partitioning patterns described through the values of their reciprocal relations. This issue is tackled by *relational clustering* in the literature. One of the most popular relational clustering algorithms is the sequential agglomerative hierarchical nonoverlapping clustering algorithm, which generates clusters by sequentially merging pairs of clusters which are the closest to each other at each step (Sneath and Sokal, 1973). Another well-known relational clustering algorithm partitions the data-set around a fixed number of representative objects, denoted *medoids*. The medoids are chosen from the data-set in such a way that the sum of the intra-cluster dissimilarity is minimized (Kaufman and Rousseeuw, 1987, 1990). Two versions of this algorithm aimed at handling large data-sets were proposed by Kaufman and Rousseeuw (1987) and by Ng and Han (1994), respectively. The aforementioned algorithms generate crisp clusters. As we are interested in finding a fuzzy partition of the data-set, in the following we discuss fuzzy relational clustering algorithms. The most popular examples of fuzzy relational clustering are the fuzzy nonmetric model (FNM, Roubens, 1978), the assignment prototype model (AP, Windham, 1985), the relational fuzzy C-means (RFCM, Hathaway, Davenport, and Bekdek, 1989), the non-Euclidean relational fuzzy C-means (NERFCM, Hathaway, and Bezdek, 1994), the fuzzy analysis (FANNY, Kaufman, and Rousseeuw, 1990), the fuzzy C-medoids (FCMdd, Krishnapuram, Joshi, Nasraoni, and Yi, 2001), and fuzzy relational data clustering (FRC, Davé, and Sen, 2002). All these algorithms assume (at least) that $D = [d_{i,j}]$ is a positive, irreflexive, and symmetric fuzzy square binary dissimilarity relation, i.e., $\forall i,j \in [1..M]$, $d_{i,j} \geq 0$, $d_{i,i} = 0$, and $d_{i,j} = d_{j,i}$. Unfortunately, the relation D produced by the two models may be neither irreflexive nor symmetric, thus making the existing fuzzy relational clustering algorithms theoretically not applicable to this relation. Actually, as shown by Corsini, Lazzerini, and Marcelloni (2002, 2004), these algorithms can be applied, but their convergence to a reasonable partition is not guaranteed (see, for instance, Corsini, Lazzerini, and Marcelloni, 2005). Indeed, in some data-sets used in our experiments, we observed that these algorithms tend to converge to a partition with completely superimposed fuzzy sets, that is, each object belongs to all clusters with equal membership value. To overcome this difficulty, we suggested transforming a relational clustering problem into an object clustering problem (Corsini, Lazzerini, and Marcelloni, 2005).

The basic idea of our approach arises from the definition of relational clustering algorithm itself: a relational clustering algorithm groups together objects that are "closely related" to each other, and "not so closely" related to objects in other clusters. Given a set of M patterns, and a square binary relation matrix $D = [d_{i,j}]$, with i, j in $[1..M]$, two patterns $\mathbf{x}_i$ and $\mathbf{x}_j$ should belong to the same cluster if the two vectors of the M strengths of relation between, respectively, $\mathbf{x}_i$ and all the patterns in the data-set Q, and $\mathbf{x}_j$ and all the patterns in Q, are close to each other. The two vectors correspond to the rows D_i and D_j of the matrix D. As the relation strengths are real numbers, the two vectors D_i and D_j can be represented as points in the metric space $\Re^M$. The closeness between D_i and D_j can be computed by using any metric defined in $\Re^M$; for instance, we could adopt the Euclidean or the Mahalanobis distance. Then, patterns $\mathbf{x}_i$ and $\mathbf{x}_j$ have to be inserted into the same cluster if and only if the distance between D_i and D_j is small (with respect to the distances between D_i (resp. D_j) and all the other row vectors). Based on this observation, the problem of partitioning M patterns, which are described by relational data, moves to the problem of partitioning M object data D_k, $k = 1..M$, in the metric space $\Re^M$. Thus, any clustering algorithm applicable to object data can be used. In particular, as proposed by Corsini, Lazzerini, and Marcelloni (2005, 2006), where the resulting clustering algorithm has been named ARCA, we can use the classical FCM. In the experiments, we used $m = 2$ and $\varepsilon = 0.001$, where ε is the maximum difference between corresponding membership values in two subsequent iterations. Moreover, we implemented the FCM algorithm in an efficient way in terms of both memory requirement and computation time, thanks to the use of the technique described by Kolen and Hutcheson (2002).

We executed ARCA with C ranging from two to five and chose the optimal number of clusters based on the Xie–Beni index. Tables 13.6 and 13.7 show the percentage of correctly classified points in the five experiments when $C = 2$ for the synthetic dataset and $C = 3$ for the Iris data-set, respectively. Here, the second and fourth columns indicate the percentage of correctly classified points for dissimilarity relations extracted by, respectively, the MLP and the TS, and the third and fifth columns the corresponding partition coefficients. The partition coefficient (PC) is defined as the average of the squared membership degrees. PC essentially measures the distance the partition U is from being crisp by assessing the fuzziness in the rows of U. PC varies in the interval $\left[\frac{1}{C}, 1\right]$. Empirical studies show that maximizing PC leads to a good interpretation of data. Thus, the closer PC is to one, the better the partition is. As expected, the percentage of correctly classified points increases with the increase of

Table 13.6 Percentage of correctly classified points of the synthetic data-set in the five experiments.

	TS system		MLP system	
Training pool	Correctly classified points	Partition coefficient	Correctly classified points	Partition coefficient
5%	84.4% ± 6.5%	0.84 ± 0.07	87.1% ± 2.8%	0.83 ± 0.10
10%	87.5% ± 5.4%	0.89 ± 0.05	88.9% ± 2.8%	0.86 ± 0.07
15%	93.7% ± 3.5%	0.90 ± 0.04	93.8% ± 1.5%	0.88 ± 0.04
20%	94.1% ± 2.9%	0.92 ± 0.02	94.6% ± 1.5%	0.91 ± 0.03
25%	97.0% ± 1.8%	0.94 ± 0.03	97.3% ± 1.3%	0.92 ± 0.02

points in the training pool. Just for small percentages of points in the training pool, the combinations MLP–ARCA and TS–ARCA are able to trace the boundaries of the classes conveniently. The quality of the approximation improves when the points of the training pool are a significant sample of the overall data-set. The tables show that the class shape is almost correctly identified just with 5 % of the points of the data-set. Note that, as reported in Tables 13.1–13.4, the MLP and the TS are able to output only 70.1 % and 69.6 % of correct dissimilarity values for the synthetic data-set, and 81.2 % and 80.5 % for the Iris data-set, when trained with training pools containing the same percentage of points. Finally, the high values of the partition coefficient highlight that the partition determined by the relational clustering algorithm is quite good.

Tables 13.8 and 13.9 show the number of clusters (in the form (mean ± standard deviation)) in the five experiments for, respectively, the synthetic and Iris data-sets when using the TS. It can be observed that the percentage of trials in which the number of clusters is equal to the number of classes increases very quickly (up to 100 %) with the increase of the percentage of points in the training pool.

As shown in Tables 13.6 and 13.7, ARCA achieves very interesting results and is characterized by a certain stability. As comparison, we applied some of the most popular fuzzy clustering algorithms to the same relations extracted by the TS and we observed a strong dependence of the results on the initial partition and on the fuzzification constant m. In several trials, we found out that the algorithms converge to a partition composed completely superimposed fuzzy sets. Anyway, since ARCA adopts the Euclidean distance, it suffers from the well-known curse of dimensionality problems: when the dimensionality increases, distances between points become relatively uniform, thus making the identification of clusters practically impossible. Actually, the curse of dimensionality problems could arise because the dimension of the space is equal to the number of objects in the data-set. Thus, for very large data-sets, we should adopt distance functions more suitable for high-dimensional spaces in place of the Euclidean distance. We did not adopt this solution in the examples simply because it was not strictly necessary. We performed, however, some experiments with the version of FCM proposed by Klawonn and Keller (1999), which adopts the cosine distance in place of the Euclidean distance. We used large dissimilarity relations

Table 13.7 Percentage of correctly classified points of the Iris data-set in the five experiments.

	TS system		MLP system	
Training pool	Correctly classified points	Partition coefficient	Correctly classified points	Partition coefficient
5%	89.8% ± 5.5%	0.74 ± 0.08	90.8% ± 4.4%	0.78 ± 0.09
10%	92.5% ± 4.6%	0.86 ± 0.04	91.3% ± 3.7%	0.91 ± 0.07
15%	94.4% ± 3.0%	0.91 ± 0.04	94.1% ± 2.4%	0.88 ± 0.05
20%	95.2% ± 2.1%	0.93 ± 0.04	95.6% ± 2.7%	0.90 ± 0.05
25%	95.8% ± 1.6%	0.92 ± 0.03	96.0% ± 1.3%	0.91 ± 0.04

Table 13.8 Number of clusters in the five experiments (synthetic data-set).

Training pool	Number of clusters	Percentage of trials with number of clusters equal to number of classes
5%	2.1 ± 0.3	90%
10%	2.0 ± 0.0	100%
15%	2.0 ± 0.0	100%
20%	2.0 ± 0.0	100%
25%	2.0 ± 0.0	100%

$(10\,000 \times 10\,000)$ created artificially. We verified that, also in this case, the results obtained by the two versions of FCM are comparable. For instance, we generated a data-set composed of three clusters using uniform random distribution of points over three nonoverlapping circles centered in (700, 400), (400, 900), and (1000, 900), with radius equal to 530. The three clusters are composed of 3606, 3733, and 3606 points, respectively. Then, we generated a dissimilarity relation $(10\,945 \times 10\,945)$ using the Euclidean distance. We executed the FCM algorithm with the fuzzification coefficient m and the termination error ε equal to 1.4 and 0.01, respectively. We obtained 100 % classification rate for both the versions of FCM, with a partition coefficient equal to 0.95 and 0.98 for the version with the Euclidean distance and for the version with the cosine distance, respectively.

To further verify the validity of ARCA, we applied a well-known density-based algorithm, named OPTICS (Ankerst, Breuing, Kriegel, and Sander, 1999), to the dissimilarity relation produced by the TS. OPTICS is an extension of DBSCAN (Ester, Kriegel, Sander, and Xu, 1996), one of the best known density-based algorithms. DBSCAN defines a cluster to be a maximum set of density-connected points, which means that every core point in a cluster must have at least a minimum number of points (*MinPts*) within a given radius (*Eps*). DBSCAN assumes that all points within genuine clusters can be reached from one another by traversing a path of density-connected points and that points across different clusters cannot. DBSCAN can find arbitrarily shaped clusters if the cluster density can be determined beforehand and the cluster density is uniform. DBSCAN is very sensitive to the selection of *MinPts* and *Eps*. OPTICS reduces this sensitivity by limiting it to *MinPts*. To perform clustering, density-based algorithms assume that points within clusters are "density reachable" and points across different clusters are not. Obviously, the cluster shape depends on the concept of "density reachable" that, in its turn, depends on the definition of dissimilarity. Thus, we cannot consider adopting density-based algorithms to solve the initial problem, that is, to determine the most suitable dissimilarity measure and therefore the most suitable cluster shape. As an example, let us consider the data-set shown in Figure 13.6 (XOR problem). The points belong to two different classes: each class is composed of two compact clusters located on the opposite corners of a square, respectively.

A density-based clustering process performed in the feature space is not able to detect the correct structure, unless a specific proximity measure is defined. Indeed, the OPTICS algorithm finds four different clusters, i.e., it achieves 50 % classification rate. Figure 13.7 shows the output of the OPTICS algorithm.

Table 13.9 Number of clusters in the five experiments (Iris data-set).

Training pool	Number of clusters	Percentage of trials with number of clusters equal to number of classes
5%	2.5 ± 0.5	50%
10%	2.8 ± 0.6	60%
15%	3.3 ± 0.5	70%
20%	2.9 ± 0.3	100%
25%	3.0 ± 0.0	100%

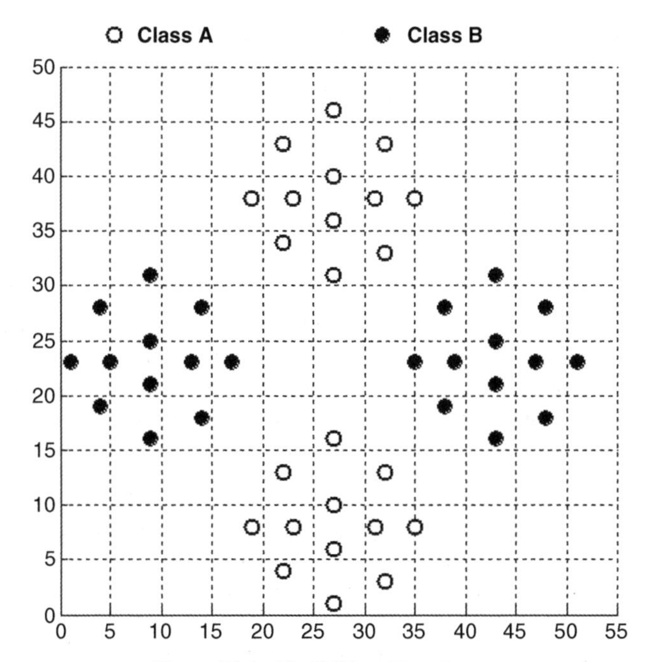

Figure 13.6 The XOR problem data-set.

On the contrary, our approach achieves 96.7 % ± 2.6 % classification rate using 20 % of points as training pool. Furthermore, we achieve a better classification rate than OPTICS even with 5 % points in the training pool. This example shows that our approach does not depend on the distribution of data and therefore on the concept of spatial density. Our method is certainly more time-consuming, but it has been introduced to solve clustering problems that are not automatically solvable with density-based clustering algorithms.

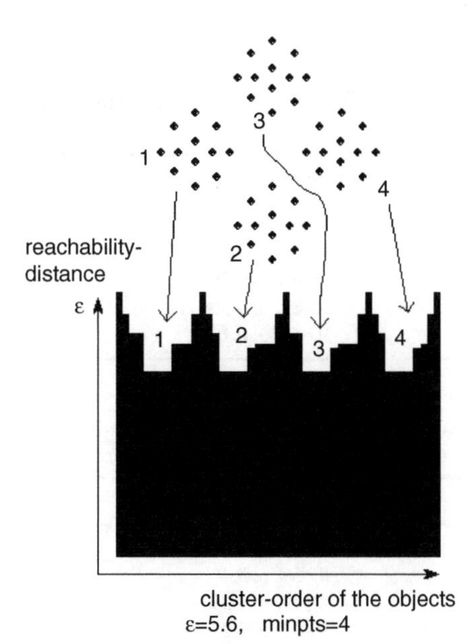

Figure 13.7 Output of the OPTICS algorithm.

On the other hand, since some density-based algorithms do not require distances but rather generic dissimilarity measures to determine the closeness of points, we can adopt OPTICS to cluster data described by the dissimilarity relations produced by the MLP and the TS. We performed different experiments and verified that the performance of OPTICS and ARCA are quite similar. In the XOR example, for instance, OPTICS achieves 95.6 % $\pm$ 2.3 % classification rate using 20 % of points as training pool.

13.4 EXPERIMENTAL RESULTS

In this section, we briefly discuss some results obtained by applying the combination TS–ARCA to some well-known data-sets provided by the University of California (UCI, 2006), namely the Wisconsin Breast Cancer (WBC) data-set, the wine data-set, and the Haberman's Survival (HS) data-set. We discuss only the TS approach because, as shown in Section 13.2, it is characterized by more interesting features.

The WBC data-set consists of 699 patterns belonging to two classes: 458 patterns are members of the "benign" class and the other 241 patterns are members of the "malignant" class. Each pattern is described by nine features: clump thickness, uniformity of cell size, uniformity of cell shape, marginal adhesion, single epithelial cell size, bare nuclei, bland chromatin, normal nucleoli, and mitoses. Since 16 patterns have a missing value, we decided to use only 683 patterns in our experiments.

The Wine data-set contains the chemical analysis of 178 wines grown in the same region in Italy but derived from three different cultivars, which represent the classes. As known in the literature (Setnes and Roubos, 2000), only some of the 13 features are effective for classification. Thus, we performed a feature selection based on the correlation between classes and features, and selected the following four features: total phenols, flavanoids, color intensity, and OD280/OD315 of diluted wines.

The HS data-set contains 306 cases from a study on the survival of patients who had undergone surgery for breast cancer. The three attributes represent the age of the patient, the year of the operation, the number of the positive axillary nodes. The two classes represent the survival status after 5 years. In this data-set, features have a low correlation with classes and therefore the data-set is quite difficult for clustering algorithms.

We carried out the same experiments described in previous sections. Tables 13.10–13.12 show the percentage of correctly classified points of the WBC, Wine, and HS data sets in the five experiments. We

Table 13.10 Percentage of correctly classified points of the WBC data-set in the five experiments.

Training pool	Correctly classified points	Partition coefficient
5%	95.9% $\pm$ 0.5%	0.94 $\pm$ 0.03
10%	96.1% $\pm$ 0.3%	0.97 $\pm$ 0.00
15%	96.8% $\pm$ 0.7%	0.96 $\pm$ 0.00
20%	96.8% $\pm$ 0.3%	0.95 $\pm$ 0.01
25%	97.1% $\pm$ 0.1%	0.96 $\pm$ 0.01

Table 13.11 Percentage of correctly classified points of the wine data-set in the five experiments.

Training pool	Correctly classified points	Partition coefficient
5%	83.7% $\pm$ 4.6%	0.84 $\pm$ 0.05
10%	85.5% $\pm$ 3.4%	0.88 $\pm$ 0.03
15%	89.7% $\pm$ 3.2%	0.91 $\pm$ 0.02
20%	91.3% $\pm$ 3.2%	0.93 $\pm$ 0.02
25%	94.1% $\pm$ 1.4%	0.95 $\pm$ 0.02

Table 13.12 Percentage of correctly classified points of the HS data-set in the five experiments.

Training pool	Correctly classified points	Partition coefficient
5%	$87.0\% \pm 2.9\%$	0.80 ± 0.02
10%	$88.1\% \pm 3.0\%$	0.80 ± 0.05
15%	$90.0\% \pm 3.4\%$	0.83 ± 0.04
20%	$88.7\% \pm 4.7\%$	0.84 ± 0.05
25%	$90.6\% \pm 5.0\%$	0.83 ± 0.04

can observe that the percentages of correct classifications are just quite high with training pools composed of only 5 % of patterns. These results compare favorably with several classification techniques proposed in the literature. Since our method is not a classification method because we do not suppose to know the labels of the classes, but rather some similarities between patterns, the results prove the effectiveness of the combination of learning algorithms and relational clustering algorithms.

13.5 CONCLUSIONS

Object clustering algorithms generally partition a data-set based on a dissimilarity measure expressed in terms of some distance. When the data distribution is irregular, for instance in image segmentation and pattern recognition where the nature of dissimilarity is conceptual rather than metric, distance functions may fail to drive the clustering algorithm correctly. Thus, the dissimilarity measure should be adapted to the specific data-set. For this reason, we have proposed extracting the dissimilarity relation directly from a few pairs of patterns of the data-set with known dissimilarity values. To this aim, we have used two different techniques: a multilayer perceptron with supervised learning and a Takagi–Sugeno fuzzy system. We have discussed and compared the two approaches with respect to generalization capabilities, computational overhead, and capability of explaining intuitively the dissimilarity relation. We have shown that the TS approach provides better characteristics than the MLP approach.

Once the dissimilarity relation has been generated, the partitioning of the data-set is performed by a fuzzy relational clustering algorithm, denoted ARCA, recently proposed by the authors. Unlike well-known relational clustering algorithms, this algorithm can manage the dissimilarity relations generated by the MLP and the TS, which are neither irreflexive nor symmetric. The experiments performed on some real data-sets have shown the good qualities of our approach. In particular, we have observed that just using a significantly low percentage of known dissimilarities, our method is able to cluster the data-sets almost correctly.

REFERENCES

Abonyi, J., Babuška, R., and Szeifert, F. (2002) 'Modified Gath–Geva fuzzy clustering for identification of Takagi–Sugeno fuzzy models', *IEEE Trans. on Systems, Man and Cybernetics-Part B*, **32**, 612–621.

Angelov, P.P. and Filev, D.P. (2004) 'An approach to online identification of takagi-sugeno fuzzy models'. *IEEE Trans. on Systems, Man, and Cybernetics-Part B: Cybernetics*, **34**, 484–498.

Ankerst, M., Breunig, M., Kriegel, H.-P. and Sander, J. (1999) 'Optics: ordering points to identify the clustering structure'. *Proc. ACM Int. Conf. Management of Data* (ACM SIGMOD '99), Philadelphia, PA, USA, pp. 49–60.

Babuška, R. (1996) *Fuzzy Modeling and Identification* PhD dissertation, Delft University of Technology, Delft, The Netherlands.

Bezdek, J.C. (1981) *Pattern Recognition with Fuzzy Objective Function Algorithms*. New York, NY, USA, Plenum.

Bezdek, J.C., Keller, J., Krisnapuram, R. and Pal, N.R. (1999) *Fuzzy Model and Algorithms for Pattern Recognition and Image Processing*. Kluwer Academic Publishing, Boston, MA, USA.

Chang, H. and Yeung, D.Y. (2005) 'Semi-supervised metric learning by kernel matrix adaptation', *Proc. International Conference on Machine Learning and Cybernetics* (ICMLC '05), Guangzhou, China, Vol. 5, pp. 3210–3215.

Corsini, P., Lazzerini, B. and Marcelloni, F. (2002) 'Clustering based on a dissimilarity measure derived from data'. *Proc. Knowledge-Based Intelligent Information and Engineering Systems* (KES '02), Crema, Italy, 885–889.

Corsini, P., Lazzerini, B. and Marcelloni, F. (2004) 'A fuzzy relational clustering algorithm based on a dissimilarity measure extracted from data', *IEEE Trans. on Systems, Man, and Cybernetics – Part B*, **34**, pp. 775–782.

Corsini, P., Lazzerini, B. and Marcelloni, F. (2005) 'A new fuzzy relational clustering algorithm based on the fuzzy C-means algorithm'. *Soft Computing*, **9**, 439–447.

Corsini, P., Lazzerini, B. and Marcelloni, F. (2006) 'Combining supervised and unsupervised learning for data clustering'. *Neural Computing and Applications*, **15**, 289–297.

Davé R.N. and Sen, S. (2002) 'Robust fuzzy clustering of relational data'. *IEEE Trans. on Fuzzy Systems*, **10**, 713–727.

Ester, M., Kriegel, H.-P., Sander, J. and Xu, X. (1996) 'A density-based algorithm for discovering clusters in large spatial databases'. *Proc. Int. Conf. Knowledge Discovery and Data Mining* (KDD '96), Portland, OR, USA, pp. 226–231.

Gath, I. and Geva, A.B. (1989) 'Unsupervised optimal fuzzy clustering'. *IEEE Trans. on Pattern Analysis and Machine Intelligence*, **11**, 773–781.

Gustafson, D.E. and Kessel, W.C. (1979) 'Fuzzy clustering with fuzzy covariance matrix'. *Advances in Fuzzy Set Theory and Applications* (M.M. Gupta, R.K. Ragade and R.R Yager eds), North Holland, Amsterdam, The Netherlands, pp. 605–620.

Hathaway, R.J. and Bezdek, J.C. (1994) 'NERF C-means: non-Euclidean relational fuzzy clustering'. *Pattern Recognition*, **27**, 429–437.

Hathaway, R.J., Davenport, J.W. and Bezdek, J.C. (1989), 'Relational duals of the c-means clustering algorithms', *Pattern Recognition*, **22**, 205–212.

Haykin, S. (1999) *Neural Networks: A Comprehensive Foundation* (2nd Edition). Prentice Hall.

Hertz, T., Bar-Hillel, A. and Weinshall, D. (2004) 'Learning distance functions for image retrieval'. *Proc. IEEE Computer Society Conference on Computer Vision and Pattern Recognition* (IEEE CVPR '04), Washington, DC, USA, Vol. II, pp. 570–577.

Jacobs, D.W., Weinshall, D. and Gdalyahu, Y. (2000) 'Classification with nonmetric distances: image retrieval and class representation'. *IEEE Trans. on Pattern Analysis and Machine Intelligence*, **22**, 583–600.

Jain, A.J. and Flynn, P.J. (eds) (1993) *Three dimensional Object Recognition Systems*. Elsevier Science, New York, USA.

Jain, A.K., Murty, M.N. and Flynn, P.J. (1999) 'Data clustering: a review'. *ACM Computing Surveys*, **31**, 265–323.

Jarvis, R.A. and Patrick, E.A. (1973) 'Clustering using a similarity method based on shared near neighbors'. *IEEE Trans. on Computers*, **C-22**, 1025–1034.

Kamgar-Parsi, B. and Jain, A.K. (1999) 'Automatic aircraft recognition: toward using human similarity measure in a recognition system'. *Proc. IEEE Computer Society Conference on Computer Vision and Pattern Recognition* (IEEE CVPR '99), Fort Collins, CO, USA, pp. 268–273.

Kaufman, J. and Rousseeuw, P.J. (1987) 'Clustering by means of medoids'. *Statistical Data Analysis Based on the L1 Norm* (Y. Dodge ed.), North Holland/Elsevier, Amsterdam, The Netherlands, pp. 405–416.

Kaufman, J. and Rousseeuw, P.J. (1990) '*Findings Groups in Data: An Introduction to Cluster Analysis* John Wiley & Sons, Ltd, Brussels, Belgium.

Klawonn, F. and Keller, A. (1999) 'Fuzzy clustering based on modified distance measures'. *Advances in Intelligent Data Analysis* (D.J. Hand, J.N. Kok and M.R. Berthold, eds), Springer, Berlin, Germany, pp. 291–301.

Kolen, J.F. and Hutcheson, T. (2002) 'Reducing the time complexity of the fuzzy C-means algorithm'. *IEEE Trans. on Fuzzy Systems*, **10**, 263–267.

Krishnapuram, R., Joshi, A., Nasraoui, O. and Yi, L. (2001) 'Low-complexity fuzzy relational clustering algorithms for web mining'. *IEEE Trans. on Fuzzy Systems*, **9**, 595–607.

Lange, T., Law, M.H.C., Jain, A.K. and Buhmann, J.M. (2005), 'Learning with constrained and unlabelled data'. *Proc. IEEE Computer Society Conference on Computer Vision and Pattern Recognition* (IEEE CVPR '05), San Diego, CA, USA, pp. 731–738.

Latecki, L.J. and Lakamper, R. (2000) 'Shape similarity measure based on correspondence of visual parts'. *IEEE Trans. on Pattern Analysis and Machine Intelligence*, **22**, 1185–1190.

Law, M.H., Topchy, A. and Jain, A.K. (2005) 'Model-based clustering with probabilistic constraints'. *Proc. SIAM International Conference on Data Mining* (SDM '05), Newport Beach, CA, USA, pp. 641–645.

Makrogiannis, S., Economou, G. and Fotopoulos, S. (2005) 'A region dissimilarity relation that combines feature-space and spatial information for color image segmentation'. *IEEE Trans. on Systems, Man and Cybernetics - Part B*, **35**, 44–53.

Michalski, R., Stepp, R.E. and Diday, E. (1983) 'Automated construction of classifications: conceptual clustering versus numerical taxonomy'. *IEEE Trans. on Pattern Analysis and Machine Intelligence*, **5**, 396–409.

Ng, R.T. and Han, J. (1994) 'Efficient and effective clustering methods for spatial data mining'. *Proc. International Conference on Very Large Data Bases* (VLDB '94), Santiago, Chile, pp. 144–155.

Pal, N.R. and Bezdek, J.C. (1995) 'On cluster validity for the fuzzy C-means model'. *IEEE Trans. on Fuzzy Systems*, **3**, 370–379.

Pedrycz, W. (2005) *Knowledge-based clustering: from Data to Information Granules*. John Wiley and Sons, Inc., Hoboken, NJ, USA.

Pedrycz, W. and Waletzky, J. (1997) 'Fuzzy clustering with partial supervision'. *IEEE Trans. on Systems, Man and Cybernetics – Part B: Cybernetics*, **27**, 787–795.

Pedrycz, W., Loia, V. and Senatore, S. (2004) 'P-FCM: a proximity-based fuzzy clustering'. *Fuzzy Sets and Systems*, **148**, 21–41.

Pedrycz, W. *et al.* (2001) 'Expressing similarity in software engineering: a neural model'. *Proc. Second International Workshop on Soft Computing Applied to Software Engineering* (SCASE '01), Enschede, The Netherlands.

Rezaee, M.R., Lelieveldt, B.P.F. and Reiber, J.H.C. (1998) A new cluster validity index for the fuzzy *C*-mean. *Pattern Recognition Letters*, **18**, 237–246.

Roubens, M. (1978) 'Pattern classification problems and fuzzy sets'. *Fuzzy Sets and Systems*, **1**, 239–253.

Santini, S. and Jain, R. (1999) 'Similarity measures'. *IEEE Trans. on Pattern Analysis and Machine Intelligence*, **21**, 871–883.

Setnes, M. and Roubos, H. (2000) 'GA-fuzzy modeling and classification: complexity and performance'. *IEEE Trans. on Fuzzy Systems*, **8**, 509–522.

Sneath, P.H. and Sokal, R.R. (1973) *Numerical Taxonomy – The Principles and Practice of Numerical Classification*. W.H. Freeman & Co., San Francisco, CA, USA.

Su, M.C. and Chou, C.H. (2001) 'A Modified version of the *K*-means algorithm with a distance based on cluster symmetry'. *IEEE Trans. of Pattern Analysis and Machine Intelligence*, **23**, 674–680.

Sugeno, M. and Yasukawa, T. (1993) 'A fuzzy logic-based approach to qualitative modeling'. *IEEE Trans. on Fuzzy Systems*, **1**, 7–31.

Takagi, T. and Sugeno, M. (1985) 'Fuzzy identification of systems and its application to modeling and control'. *IEEE Trans. on Systems, Man, and Cybernetics*, **15**, 116–132.

Tsang, I.W., Cheung, P.M. and Kwok, J.T. (2005) 'Kernel relevant component analysis for distance metric learning'. *Proc. IEEE International Joint Conference on Neural Networks* (IEEE IJCNN '05), Montréal, QB, Canada, **2**, pp. 954–959.

UCI Machine Learning Database Repository, http://www.ics.uci.edu/~mlearn/MLSummary.html, 2006.

Valentin, D., Abdi, H., O'Toole, A.J. and Cottrell, G.W. (1994) 'Connectionist models of face processing: a survey'. *Pattern Recognition*, **27**, 1208–1230.

Windham, M.P. (1985) 'Numerical classification of proximity data with assignment measures'. *Journal of Classification*, **2**, 157–172.

Wright, A.H. (1991) 'Genetic algorithms for real parameter optimization', *Foundations of Genetic Algorithms*, (G.J. Rawlins ed.); San Mateo, CA, USA, Morgan Kaufmann, pp. 205–218.

Xie, X.L. and Beni, G. (1991) 'A validity measure for fuzzy clustering'. *IEEE Trans. on Pattern Analysis and Machine Intelligence*, **13**, 841–847.

Xing, E.P., Ng, A.Y., Jordan, M.I. and Russell, S. 2003) 'Distance metric learning, with application to clustering with side-information'. *Advances in Neural Information Processing Systems* (S. Thrun, Becker, S. and K. Obermayer, (eds), Cambridge, MA, USA, **15**, pp. 505–512.

Xu, R. and Wunsch, D. II (2005) 'Survey of Clustering Algorithms'. *IEEE Trans. on Neural Networks*, **16**, 645–678.

Yang, M.S. and Wu, K.L. (2004) 'A similarity-based robust clustering method'. *IEEE Trans. on Pattern Analysis and Machine Intelligence*, **26**, 434–448.

14

Simultaneous Clustering and Feature Discrimination with Applications

Hichem Frigui

Multimedia Research Laboratory, CECS Department, University of Louisville, Louisville, USA

14.1 INTRODUCTION

The problem of selecting or weighting the best subset of features constitutes an important part of the design of good learning algorithms for real word tasks. Irrelevant features can degrade the generalization performance of these algorithms significantly. As a result, several methods have been proposed for feature selection and weighting [1, 20, 31, 37, 30]. In feature selection, the task's dimensionality is reduced by completely eliminating irrelevant features. This amounts to assigning binary relevance weights to the features (1 for relevant and 0 for irrelevant). Feature weighting is an extension of the selection process where the features are assigned continuous weights which can be regarded as degrees of relevance. Because it provides a richer feature relevance representation, continuous weighting tends to outperform feature selection from an accuracy point of view in tasks where some features are useful but less important than others.

Most feature weighting (and selection) methods assume that feature relevance is invariant over the task's domain, and hence learn a single set of weights for the entire data-set. This assumption can impose unnecessary and pernicious constraints on the learning task when the data is made of different categories or classes. If the data is already labeled (supervised learning), then it is possible to learn a different set of weights for each class. On the other hand, if the data is unlabeled (unsupervised learning), then existing feature weighting algorithms cannot be used to learn cluster-dependent feature weights. The classical approach in this case is to determine a single subset of features or feature weights for the entire unlabeled data prior to clustering. However, by ignoring the existence of different sub-structures in the data-set, which require different subsets of feature weights, the performance of any clustering procedure can be severely degraded. Hence, it is clear that the clustering and feature selection/weighting steps are coupled, and applying these steps in sequence can degrade the performance of the learning system. In fact, even if

Advances in Fuzzy Clustering and its Applications Edited by J. Valente de Oliveira and W. Pedrycz

the data is labeled according to several known classes, it is preferable to model each complex (non-convex) class by several simple sub-classes or clusters, and to use a different set of feature weights for each cluster.

To illustrate the need for different sets of feature weights for different clusters, we consider the unsupervised learning problem of segmenting the color image shown in Figure 14.1(a). In this problem, the pixels of the image must be categorized into meaningful clusters corresponding to different homogeneous regions of the image without any prior knowledge about its contents. In order to be categorized, the pixels are first mapped to feature vectors as will be explained in Section 14.5. Figure 14.1(b) and (c) show the segmentation results with different feature subsets using the fuzzy C-means (FCM) algorithm [2], when the number of clusters is fixed to four. The FCM, like other clustering algorithms, has no provision for cluster dependent feature weights. Any given feature must either be used (completely relevant) or ignored (irrelevant) for *all* clusters. Figure 14.1(b) shows the segmentation results when the x and y position features are not used during clustering. It can be seen that some clusters suffer from a fragmentation phenomenon. That is, instead of being clean and compact, these clusters are scattered around the image. Figure 14.1(c) shows the results when the x and y position features are used. As expected, the clusters obtained are more compact. However, except for the tiger cluster, the remaining clusters do not correspond to homogeneous regions in the original image. This is because, for example for the dirt region, the y-position feature is relevant while the x-position feature is irrelevant since this region forms a narrow strip that extends over the entire width of the image. For the grass region, both the x- and y- position features are irrelevant because grass can be found in almost all areas of the image. We conclude that in this case, the x- and y-position features are both useful, and thus relevant for delineating clusters corresponding to the compact regions of the image such as the tiger. However, that is not the case for noncompact regions such as the background. The remaining features cannot be easily visualized as in the case of position. However, their relevance is also expected to vary across the different regions of the image because of the inherent differences in their color and texture properties. This simple example shows how feature relevance can vary widely within the domain of a data-set.

In light of the above discussion, it is only natural to conclude that ideally, clustering and feature selection should be performed simultaneously. In this chapter, we describe an approach, called simultaneous clustering and attribute discrimination (SCAD)[12, 13], that performs clustering and feature weighting *simultaneously*. When used in conjunction with a supervised or unsupervised learning system, SCAD offers several advantages. First, it uses continuous feature weighting, hence providing a much richer feature relevance representation than feature selection. Second, SCAD learns the feature relevance representation of each cluster independently and in an unsupervised manner. Moreover, SCAD can adapt to the variations that exist within a data-set by categorizing it into distinct clusters, which are allowed to overlap because of the use of fuzzy membership degrees. Based on the SCAD approach, we present two clustering algorithms, called SCAD-1 and SCAD-2 respectively. These two algorithms achieve the same goal, however, they minimize different objective functions. We also present a coarse version of SCAD ($SCAD_c$) that avoids over-fitting when the dimensionality of the feature space is high. Instead of learning a weight for each feature, we divide the features into logical subsets and learn a relevance weight

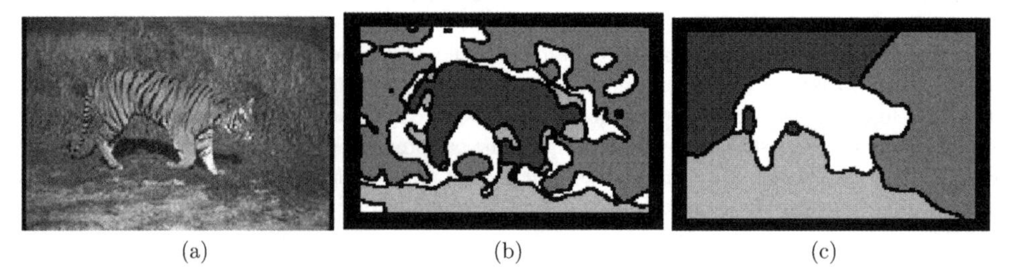

Figure 14.1 (a) Original color image, (b) results of the FCM with no position features, (c) results of the FCM with position features.

for each subset. Finally, because the number of categories in a data-set is not always known a priori, we present an extension of SCAD that can determine the optimal number of clusters using competitive agglomeration.

The rest of the chapter is organized as follows. In Section 14.2, we review existing approaches to feature selection/weighting, and prototype-based clustering algorithms. In Section 14.3, we introduce SCAD-1 and SCAD-2. In Section 14.4, we describe the coarse SCAD , and in Section 14.5, we extend SCAD to the case where the number of clusters is unknown. In Sections 14.6, 14.7 and 14.8, we illustrate the performance of SCAD when applied to the problems of image segmentation, text document categorization, and to learn associations between visual features and textual keywords. Finally, Section 14.9 contains the summary conclusions.

14.2 BACKGROUND

14.2.1 Feature Selection and Feature Weighting

Feature selection and weighting techniques generally rely on a criterion function and a search strategy. The criterion function is used to decide whether one feature subset is better than another, while the search strategy determines feature subset candidates. Depending on the criterion function used, there are two types of feature selection/weighting methods: the *wrapper* and *filter* approaches [18]. The wrapper approach relies on feedback from the performance algorithm, such as classifier accuracy, to learn feature weights or to decide whether a particular feature subset is superior to another. The filter approach optimizes a classifier independent criterion function. The wrapper approach tends to perform better, however it can cause overfitting [21]. Moreover, it should only be applied in combination with classifiers of low complexity to limit its computational cost.

Feature selection methods have exploited several search strategies. The most rudimentary strategy, exhaustive search, considers $2^n - 1$ (where n is the maximum number of features) possible feature subsets, and is impractical for large n. As a result, other optimized search strategies such as forward selection and backward selection [14] can be used. These approaches are computationally feasible. Unfortunately, they are only appropriate for binary weighting and are prone to yielding sub-optimal solutions [18]. Other strategies include random mutation hill climbing [33] and parallel search [26].

Feature selection/weighting can either be applied to the entire data-set to obtain a single set of features or weights, or to each class independently to obtain a different set of features or weights for each class. It is also possible to use the entire data and the class labels to learn a different set of weights for each class [22, 33]. In general, class dependent feature selection/weighting is superior to the approach yielding a single set of features/weights. Unfortunately, class dependent feature selection/weighting cannot be applied when the data is unlabeled.

14.2.2 Prototype-based Clustering

Let $\mathbf{X} = \{\mathbf{x}_j \mid j = 1, \ldots, N\}$ be a set of N feature vectors in an n-dimensional feature space. Let $\mathbf{B} = (\beta_1, \ldots, \beta_c)$ represent a C-tuple of prototypes each of which characterizes one of the C clusters. Each β_i consists of a set of parameters. Let u_{ij} represent the grade of membership of feature point $\mathbf{x}_j$ in cluster β_i. The $C \times N$ matrix $\mathbf{U} = [u_{ij}]$ is called a constrained fuzzy C-partition matrix if it satisfies the following conditions [2]

$$u_{ij} \in [0, 1] \ \forall i; \qquad 0 < \sum_{j=1}^{N} u_{ij} < N \ \forall i, j; \qquad \sum_{i=1}^{C} u_{ij} = 1 \ \forall j. \tag{14.1}$$

The problem of fuzzily partitioning the feature vectors into C clusters can be formulated as the minimization of an objective function $J(\mathbf{B}, \mathbf{U}; \mathbf{X})$ of the form

$$J(\mathbf{B}, \mathbf{U}; \mathbf{X}) = \sum_{i=1}^{C} \sum_{j=1}^{N} (u_{ij})^m d_{ij}^2, \tag{14.2}$$

subject to the constraint in (14.1). In (14.2), $m \in [1, \infty)$ is a weighting exponent called the fuzzifier, and d_{ij}^2 represents the distance from feature point $\mathbf{x}_j$ to prototype β_i. Minimization of (14.2) with respect to $\mathbf{U}$ subject to (14.1) yields [2]:

$$u_{ij} = \frac{1}{\sum_{k=1}^{C} \left(d_{ij}^2 / d_{kj}^2 \right)^{\frac{1}{m-1}}}. \tag{14.3}$$

Minimization of (14.2) with respect to $\mathbf{B}$ varies according to the choice of the prototypes and the distance measure. For example, in the fuzzy C-means (FCM) algorithm, the Euclidean distance is used, and each of the prototypes is described by the cluster center $\mathbf{c}_i$, which may be updated in each iteration using [2]:

$$\mathbf{c}_i = \frac{\sum_{j=1}^{N} u_{ij}^m \mathbf{x}_j}{\sum_{j=1}^{N} u_{ij}^m}. \tag{14.4}$$

14.2.3 Competitive Agglomeration

The objective function in (14.2), which is essentially the sum of (fuzzy) intra-cluster distances, has a monotonic tendency with respect to the number of clusters, C, and has the minimum value of zero when $C = N$. Therefore, it is not useful for the automatic determination of the "optimum" number of clusters, and C has to be specified a priori. The competitive agglomeration (CA) algorithm [11, 10] overcomes this drawback by adding a second regularization term to prevent over-fitting the data with too many prototypes. The CA algorithm starts by partitioning the data into a large number of small clusters. As the algorithm progresses, adjacent clusters compete for data points, and clusters that lose in the competition gradually vanish. The CA algorithm minimizes the following objective function

$$J_A(\mathbf{B}, \mathbf{U}; \mathbf{X}) = \sum_{i=1}^{C} \sum_{j=1}^{N} (u_{ij})^2 d_{ij}^2 - \alpha \sum_{i=1}^{C} \left[\sum_{j=1}^{N} u_{ij} \right]^2, \tag{14.5}$$

subject to the constraints in (14.1). It should be noted that the number of clusters C in (14.5) is dynamically updated in the CA. The first term in (14.5) controls the shape and size of the clusters and encourages partitions with many clusters, while the second term penalizes solutions with a large number of clusters and encourages the agglomeration of clusters. When both terms are combined and α is chosen properly, the final partition will minimize the sum of intra-cluster distances, while partitioning the data into the smallest possible number of clusters. It can be shown [10] that the membership update equation for (14.5) is given by

$$u_{ij} = u_{ij}^{FCM} + u_{ij}^{Bias}, \tag{14.6}$$

where

$$u_{ij}^{FCM} = \frac{1/d_{ij}^2}{\sum_{k=1}^{C} 1/d_{kj}^2}, \tag{14.7}$$

and

$$u_{ij}^{Bias} = \frac{\alpha}{d_{ij}^2} \left(N_i - \overline{N}_j \right). \tag{14.8}$$

In (14.8),

$$N_i = \sum_{j=1}^{N} u_{ij},$$

is the cardinality of cluster i, and

$$\overline{N}_j = \frac{\sum_{k=1}^{C} 1/d_{kj}^2 N_k}{\sum_{k=1}^{C} 1/d_{kj}^2},$$

is a weighted average of the cardinalities of all clusters. The first term in (14.6) is the membership term in the FCM algorithm [2] (see Equation (14.3)) which takes into account only the relative distances of the feature point to all clusters. The second term is a signed bias term which allows good clusters to agglomerate and spurious clusters to disintegrate.

The value of α needs to be initially small to encourage the formation of small clusters. Then, it should be increased gradually to promote agglomeration. After a few iterations, when the number of clusters becomes close to the "optimum," the value of α should again decay slowly to allow the algorithm to converge. In [10], it is recommended to update α in every iteration (t) using

$$\alpha(t) = \eta(t) \frac{\sum_{i=1}^{C} \sum_{j=1}^{N} (u_{ij}^{(t-1)})^2 (d_{ij}^2)^{(t-1)}}{\sum_{i=1}^{C} \left[\sum_{j=1}^{N} u_{ij}^{(t-1)} \right]^2}, \tag{14.9}$$

where

$$\eta(t) = \begin{cases} \eta_0 \, e^{-|t_0 - t|/\tau} & \text{if } t > 0 \\ 0 & \text{if } t = 0. \end{cases} \tag{14.10}$$

In (14.10), η_0 is the initial value, τ is a time constant and t_0 is the iteration number at which η starts to decrease. The superscript $(t-1)$ is used on u_{ij} and d_{ij}^2 to denote their values in the previous iteration, $(t-1)$. The default values for $(\eta_0,\ \tau,\ t_0)$ are (1, 10, 20).

14.3 SIMULTANEOUS CLUSTERING AND ATTRIBUTE DISCRIMINATION (SCAD)

The Simultaneous Clustering and Attribute Discrimination (SCAD) algorithm [12, 13] is designed to search for the optimal prototype parameters, **B**, and the optimal set of feature weights, **V**, simultaneously. Each cluster i is allowed to have its own set of feature weights $\mathbf{V}_i = [v_{i1}, \cdots, v_{in}]$. We present two versions of SCAD: SCAD-1 tries to balance between the two terms of a compound objective function in order to determine the optimal attribute relevance weights, while SCAD-2 minimizes a single term criterion.

14.3.1 Simultaneous Clustering and Attribute Discrimination – Version 1 (SCAD-1)

The SCAD-1 algorithm minimizes the following objective function:

$$J_1(\mathbf{C}, \mathbf{U}, \mathbf{V}; \mathbf{X}) = \sum_{i=1}^{C} \sum_{j=1}^{N} u_{ij}^m \sum_{k=1}^{n} v_{ik} d_{ijk}^2 + \sum_{i=1}^{C} \delta_i \sum_{k=1}^{n} v_{ik}^2, \tag{14.11}$$

subject to the constraint on fuzzy memberships, u_{ij} in (14.1), and the following constraint on the feature weights:

$$v_{ik} \in [0, 1] \ \forall\, i,\ k; \quad \text{and} \quad \sum_{k=1}^{n} v_{ik} = 1,\ \forall\, i. \tag{14.12}$$

In (14.11), v_{ik} represents the relevance weight of feature k in cluster i, and d_{ijk} is given by

$$d_{ijk} = |x_{jk} - c_{ik}|, \tag{14.13}$$

where x_{jk} is the kth feature value of data point $\mathbf{x}_j$, and c_{ik} is the kth component of the ith cluster center vector. In other words, d_{ijk} is the projection of the displacement vector between feature point $\mathbf{x}_j$ and the ith class center ($\mathbf{c}_i$) along the kth dimension.

The objective function in (14.11) has two components. The first one, which is similar to the FCM objective function [2], is the sum of feature-weighted Euclidean distances to the prototypes, additionally weighted by constrained memberships. This component allows us to obtain compact clusters. From a feature-relevance point of view, this term is minimized when, in each cluster, only one feature is completely relevant, while all other features are irrelevant. The second component in Equation (14.11) is the sum of the squared feature weights. The global minimum of this component is achieved when all the features are equally weighted. When both components are combined as in (14.11), and the coefficients δ_i are chosen properly, the final partition will minimize the sum of intra-cluster weighted distances, where the weights are optimized for each cluster.

To optimize J_1, with respect to **V**, we use the Lagrange multiplier technique, and obtain

$$J_1(\mathbf{\Lambda}, \mathbf{V}) = \sum_{i=1}^{C}\sum_{j=1}^{N}(u_{ij})^m \sum_{k=1}^{n} v_{ik}d_{ijk}^2 + \sum_{i=1}^{C}\delta_i \sum_{k=1}^{n} v_{ik}^2 - \sum_{i=1}^{C}\lambda_i\left(\sum_{k=1}^{n} v_{ik} - 1\right),$$

where $\Lambda = [\lambda_1, \cdots, \lambda_c]^t$. Since the rows of **V** are independent of each other, we can reduce the above optimization problem to the following C independent problems:

$$J_{1i}(\lambda_i, \mathbf{V}_i) = \sum_{j=1}^{N}(u_{ij})^m \sum_{k=1}^{n} v_{ik}(x_{jk} - c_{ik})^2 + \delta_i \sum_{k=1}^{n} v_{ik}^2 - \lambda_i\left(\sum_{k=1}^{n} v_{ik} - 1\right),$$

for $i = 1, \cdots, C$, where $\mathbf{V}_i$ is the ith row of **V**. By setting the gradient of J_{1i} to zero, we obtain

$$\frac{\partial J_{1i}(\lambda_i, \mathbf{V}_i)}{\partial \lambda_i} = \left(\sum_{k=1}^{n} v_{ik} - 1\right) = 0, \tag{14.14}$$

and

$$\frac{\partial J_{1i}(\lambda_i, \mathbf{V}_i)}{\partial v_{ik}} = \sum_{j=1}^{N}(u_{ij})^m d_{ijk}^2 + 2\delta_i v_{ik} - \lambda_i = 0. \tag{14.15}$$

Solving (14.14) and (14.15) for v_{ik}, we obtain

$$v_{ik} = \frac{1}{n} + \frac{1}{2\delta_i}\sum_{j=1}^{N}(u_{ij})^m\left[\frac{||\mathbf{x}_j - \mathbf{c}_i||^2}{n} - d_{ijk}^2\right]. \tag{14.16}$$

The first term in (14.16), $(1/n)$, is the default value if all attributes are treated equally, and no feature discrimination is performed. The second term is a bias that can be either positive or negative. It is positive for compact features where the projected distance along the corresponding dimension is, on average, less than the average projected distance values along all the dimensions. In other words, if an attribute is very compact, compared to the other attributes, for most of the points that belong to a given cluster (high u_{ij}), then it is considered to be very relevant for that cluster.

The choice of δ_i in Equation (14.11) is important to the performance of SCAD-1 since it reflects the importance of the second term relative to the first term. If δ_i is too small, then the first term dominates, and only one feature in cluster i will be maximally relevant and assigned a weight of one, while the remaining features are assigned zero weights. On the other hand, if δ_i is too large, then the second term will dominate, and all features in cluster i will be relevant, and assigned equal weights of $1/n$. Hence, the values of δ_i

should be chosen such that both terms are of the same order of magnitude. This can be accomplished by updating δ_i in iteration, t, using

$$\delta_i^{(t)} = K \frac{\sum_{j=1}^{N} \left(u_{ij}^{(t-1)}\right)^m \sum_{k=1}^{n} v_{ik}^{(t-1)} \left(d_{ijk}^{(t-1)}\right)^2}{\sum_{k=1}^{n} \left(v_{ik}^{(t-1)}\right)^2}. \tag{14.17}$$

In (14.17), K is a constant and the superscript $(t-1)$ is used on u_{ij}, v_{ik} and d_{ijk} to denote their values in iteration $(t-1)$.

It should be noted that depending on the values of δ_i, the feature relevance values v_{ik} may not be confined to [0,1]. This inequality constraint could be added to the objective function in (14.11), and the Karush–Kuhn–Tucker (KKT) Theorem [25] could be used to derive the necessary conditions. A much simpler and practical heuristic, that proved almost as effective as the KKT optimization conditions, is to simply set negative values to zero and to clip values that are greater than one to one. However, if this constraint is violated very often, then it is an indication that the value of δ is too small and it should be increased (increase K).

To minimize J_1 with respect to $\mathbf{U}$, we rewrite the objective function in (14.11) as

$$J_1(\mathbf{C}, \mathbf{U}, \mathbf{V}; \mathbf{X}) = \sum_{i=1}^{C} \sum_{j=1}^{N} u_{ij}^m \tilde{d}_{ij}^2 + \sum_{i=1}^{C} \delta_i \sum_{k=1}^{n} v_{ik}^2, \tag{14.18}$$

where $\tilde{d}_{ij}^2 = \sum_{k=1}^{n} v_{ik} d_{ijk}^2$ is the weighted Euclidean distance. Since the second term in (14.18) does not depend on u_{ij} explicitly, the update equation of the memberships is similar to that of the FCM (see Equation (14.3)), i.e.,

$$u_{ij} = \frac{1}{\sum_{k=1}^{C} \left(\tilde{d}_{ij}^2 / \tilde{d}_{kj}^2\right)^{1/m-1}}. \tag{14.19}$$

To minimize J_1 with respect to the centers, we fix $\mathbf{U}$ and $\mathbf{V}$, and set the gradient to zero. We obtain

$$\frac{\partial J}{\partial c_{ik}} = -2 \sum_{j=1}^{N} (u_{ij})^m v_{ik} (x_{jk} - c_{ik}) = 0. \tag{14.20}$$

Solving (14.20) for c_{ik}, we obtain

$$c_{ik} = \frac{v_{ik} \sum_{j=1}^{N} (u_{ij})^m x_{jk}}{v_{ik} \sum_{j=1}^{N} (u_{ij})^m}. \tag{14.21}$$

There are two cases depending on the value of v_{ik}:

- *Case 1*: $v_{ik} = 0$. In this case, the kth feature is completely irrelevant relative to the ith cluster. Hence, regardless of the value of c_{ik}, the values of this feature will not contribute to the overall weighted distance computation. Therefore, in this situation, any arbitrary value can be chosen for c_{ik}. In practice, we set $c_{ik} = 0$.
- *Case 2*: $v_{ik} \neq 0$. For the case when the kth feature has some relevance to the ith cluster, Equation (14.21) reduces to

$$c_{ik} = \frac{\sum_{j=1}^{N} (u_{ij})^m x_{jk}}{\sum_{j=1}^{N} (u_{ij})^m}.$$

To summarize, the update equation for the centers is

$$c_{ik} = \begin{cases} 0 & \text{if } v_{ik} = 0, \\ \frac{\sum_{j=1}^{N} (u_{ij})^m x_{jk}}{\sum_{j=1}^{N} (u_{ij})^m} & \text{if } v_{ik} > 0. \end{cases} \tag{14.22}$$

The SCAD-1 algorithm is summarized below.

Simultaneous Clustering and Attribute Discrimination: SCAD-1

Fix the number of clusters C;
Fix m, $m \in [1, \infty)$*;*
Initialize the centers;
Initialize the relevance weights to $1/n$*;*
Initialize the fuzzy partition matrix **U***;*
REPEAT
  Compute d_{ijk}^2 *for* $1 \le i \le C$, $1 \le j \le N$, *and* $1 \le k \le n$*;*
  Update the relevance weights v_{ik} *by using (14.16);*
  Update the partition matrix $\mathbf{U}^{(k)}$ *by using (14.19);*
  Update the centers by using (14.22);
  Update δ_i *by using (14.17);*
UNTIL *(centers stabilize);*

14.3.2 Simultaneous Clustering and Attribute Discrimination – Version 2(SCAD-2)

The SCAD-2 algorithm omits the second term from its objective function by incorporating a discriminant exponent, q, and minimizing

$$J_2(\mathbf{B}, \mathbf{U}, \mathbf{V}; \mathbf{X}) = \sum_{i=1}^{C} \sum_{j=1}^{N} (u_{ij})^m \sum_{k=1}^{n} (v_{ik})^q d_{ijk}^2, \tag{14.23}$$

subject to (14.1) and (14.12).

To optimize J_2 with respect to **V**, we use the Lagrange multiplier technique, and obtain

$$J_2(\mathbf{\Lambda}, \mathbf{V}) = \sum_{i=1}^{C} \sum_{j=1}^{N} (u_{ij})^m \sum_{k=1}^{n} (v_{ik})^q d_{ijk}^2 - \sum_{i=1}^{C} \lambda_i \left(\sum_{k=1}^{n} v_{ik} - 1 \right),$$

where $\mathbf{\Lambda} = [\lambda_1, \cdots, \lambda_c]^{\mathbf{t}}$ is a vector of Lagrange multipliers corresponding to the C constraints in (14.12). Since the rows of **V** are independent of each other and d_{ijk}^2 is independent of **V**, we can reduce the above optimization problem to the following C independent problems:

$$J_{2i}(\lambda_i, \mathbf{V}_i) = \sum_{j=1}^{N} (u_{ij})^m \sum_{k=1}^{n} (v_{ik})^q d_{ijk}^2 - \lambda_i \left(\sum_{k=1}^{n} v_{ik} - 1 \right) \quad \text{for } i = 1, \cdots, C,$$

where $\mathbf{V}_i$ is the ith row of **V**. By setting the gradient of J_{2i} to zero, we obtain

$$\frac{\partial J_{2i}(\lambda_i, \mathbf{V}_i)}{\partial \lambda_i} = \left(\sum_{k=1}^{n} v_{ik} - 1 \right) = 0, \tag{14.24}$$

and

$$\frac{\partial J_i(\lambda_i, \mathbf{V}_i)}{\partial v_{it}} = q \; (v_{it})^{(q-1)} \sum_{j=1}^{N} (u_{ij})^m d_{ijt}^2 - \lambda_i = 0. \tag{14.25}$$

Equation (14.25) yields

$$v_{it} = \left[\frac{\lambda_i}{q \; \sum_{j=1}^{N} (u_{ij})^m d_{ijt}^2} \right]^{1/(q-1)}. \tag{14.26}$$

Substituting (14.26) back into (14.24), we obtain

$$\sum_{k=1}^{n} v_{ik} = [\lambda_i/q]^{1/(q-1)} \sum_{k=1}^{n} \left[\frac{1}{\sum_{j=1}^{N} (u_{ij})^m d_{ijk}^2} \right]^{1/(q-1)} = 1.$$

Thus,

$$[\lambda_i/q]^{1/(q-1)} = \frac{1}{\sum_{k=1}^{n} \left[\frac{1}{\sum_{j=1}^{N} (u_{ij})^m d_{ijk}^2)} \right]^{1/(q-1)}}.$$

Substituting this expression back in (14.26), we obtain

$$v_{it} = \frac{\left[\frac{1}{\sum_{j=1}^{N} (u_{ij})^m d_{ijt}^2} \right]^{1/(q-1)}}{\sum_{k=1}^{n} \left[\frac{1}{\sum_{j=1}^{N} (u_{ij})^m d_{ijk}^2} \right]^{1/(q-1)}}. \tag{14.27}$$

Simplifying (14.26) and using k to represent the dimension, we obtain

$$v_{ik} = \frac{1}{\sum_{t=1}^{n} \left(\tilde{D}_{ik}/\tilde{D}_{it} \right)^{1/(q-1)}}, \tag{14.28}$$

where $\tilde{D}_{ik} = \sum_{j=1}^{N} (u_{ij})^m d_{ijk}^2$, can be interpreted as a measure of dispersion of the ith cluster along the kth dimension, and $\sum_{t=1}^{n} \tilde{D}_{it}$ can be viewed as the total dispersion of the ith cluster (taking all dimensions into account). Hence, v_{ik} is inversely related to the ratio of the dispersion along the kth dimension to the total dispersion for the ith cluster. This means that the more compact the ith cluster is along the kth dimension (smaller $\tilde{D}_{ik}$), the higher will the relevance weight, v_{ik}, be for the kth feature.

Equation (14.28) may be likened to the estimation and use of a covariance matrix in an inner-product norm-induced metric in various clustering algorithms [17, 8]. In fact, if d_{ijk}^2 is defined to be the Euclidean distance between x_{jk} and c_{ik}, then $\tilde{D}_{ik}$ becomes a measure of the fuzzy variance of the ith cluster along the kth dimension, and SCAD-2 becomes similar to the Gustafson-Kessel (GK) algorithm [17] with a diagonal covariance matrix. However, the estimation of a covariance matrix relies on the assumption that the data has a multivariate Gaussian distribution. On the other hand, SCAD-2 is free of any such assumptions when estimating the feature weights. This means that SCAD-2 can incorporate more general dissimilarity measures.

The role of the attribute weight exponent, q, can be deduced from Equation (14.28), and is subject to the following theorem.

Theorem 14.1

$$\lim_{q \to 1^+} v_{ik} = \begin{cases} 1 & \text{if } \tilde{D}_{ik} = \min_{t=1}^{n} \tilde{D}_{it}, \\ 0 & \text{otherwise.} \end{cases}$$

Proof (see Appendix).

Theorem 14.1 implies that as q approaches 1, v_{ik} tends to take binary values. This case is analogous to the winner-take-all situation where the feature along which the ith cluster is the most compact gets all the relevancy ($v_{ik} = 1$), while all other attributes get assigned zero relevance, and hence do not contribute to the distance or center computations. On the other hand, when q approaches infinity, it can easily be shown that $v_{ik} = 1/n$. This means that all attributes share the relevancy equally. This is equivalent to the situation where no feature selection/weighting takes place. For the case where q takes finite values in $(1, \infty)$, we obtain weights that provide a moderate level of feature discrimination. For this reason, we will refer to q as a "discrimination exponent."

To minimize J_2 with respect to **U**, we follow similar steps and obtain

$$u_{ij} = \frac{1}{\sum_{k=1}^{C} \left[\frac{\tilde{d}_{ij}^2}{\tilde{d}_{kj}^2} \right]^{1/(m-1)}}, \tag{14.29}$$

where

$$\tilde{d}_{ij}^2 = \sum_{k=1}^{n} (v_{ik})^q d_{ijk}^2. \tag{14.30}$$

To minimize J_2 with respect to the centers, we fix **U** and **V**, and set the gradient to zero. We obtain

$$\frac{\partial J}{\partial c_{ik}} = -2 \sum_{j=1}^{N} (u_{ij})^m (v_{ik})^q (x_{jk} - c_{ik}) = 0. \tag{14.31}$$

Equation (14.31) is similar to (14.20), and its solution yields the same equation to update the centers as in SCAD-1.

The SCAD-2 algorithm is summarized below.

Simultaneous Clustering and Attribute Discrimination: SCAD-2

Fix the number of clusters C;
Fix the fuzzifier m, $m \in [1, \infty)$*;*
Fix the discrimination exponent q, $q \in [1, \infty)$*;*
Initialize the centers and the fuzzy partition matrix **U***;*
Initialize all the relevance weights to $1/n$*;*
REPEAT
Compute d_{ijk}^2 *for* $1 \le i \le C$, $1 \le j \le N$, *and* $1 \le k \le n$*;*
Update the relevance weights matrix **V** *by using (14.28);*
Compute $\tilde{d}_{ij}^2$ *by using (14.30);*
Update the partition matrix **U** *by using (14.29);*
Update the centers by using (14.22);
UNTIL (*centers stabilize*);

We now illustrate the performance of SCAD-1 and SCAD-2 on a simple data-set. Table 14.1 contains the coordinates of 20 points of two synthetic Gaussian clusters with $(\mu_1, \Sigma_1) = ([0,0]^T, \mathbf{I}_2)$ and $(\mu_2, \Sigma_2) = ([5,5]^T, \mathbf{I}_2)$. In this example, m, K (for SCAD-1) and q (for SCAD-2) were set to 2.0, and the centers and the fuzzy partition matrix were initialized by running the FCM algorithm for two iterations. SCAD-1 converged after four iterations, and SCAD-2 converged after five iterations giving the results displayed in Tables 14.2 and 14.3 respectively. Since both features are relevant, SCAD-1 and SCAD-2 assigned high weights for both features, and the estimated center coordinates are close to those of the actual centers.

To demonstrate the ability of these algorithms to cluster and identify relevant features, we increase the number of features to four by adding two irrelevant features to each cluster which are highlighted in Table 14.4. The first two features of the first cluster are uniformly distributed in the intervals [0,20] and [0,10] respectively. Features two and four of the second cluster are uniformly distributed in the intervals [0,10] and [0,5] respectively. A traditional feature selection algorithm can only discriminate against the second feature since it is irrelevant for both clusters. Clustering the remaining three features will not provide a compact description of each cluster. SCAD-1 converged after 10 iterations and SCAD-2 converged after five iterations, and their results are displayed in Tables 14.5 and 14.6 respectively. The first feature of the first cluster is correctly identified as irrelevant by both algorithms. The second feature of the first cluster is

Table 14.1 Two two-dimensional Gaussian clusters.

Cluster # 1		Cluster # 2	
x_1	x_2	x_1	x_2
−0.33	1.11	4.66	6.11
−2.02	−0.73	2.97	4.26
−0.33	0.72	4.66	5.72
−0.25	0.04	4.74	5.04
−1.08	−0.37	3.91	4.62
0.15	−0.36	5.15	4.63
−1.22	0.11	3.77	5.11
1.80	1.43	6.80	6.43
−1.48	−0.70	3.51	4.29
−0.87	1.02	4.12	6.02
−0.21	−0.45	4.78	4.54
−0.28	1.06	4.71	6.06
0.45	0.16	5.45	5.16
−2.29	1.98	2.70	6.98
0.84	−0.68	5.84	4.31
1.49	1.61	6.49	6.61
−0.23	0.31	4.76	5.31
−0.46	−0.82	4.53	4.17
−1.58	−1.09	3.41	3.90
0.72	1.27	5.72	6.27

identified as irrelevant by SCAD-2 (relevance=0.05), but was assigned a larger weight by SCAD-1 (relevance=0.23) because it has a relatively smaller dynamic range. Feature four of the second cluster was not identified as totally irrelevant by both algorithms. This is because it has a dynamic range that is close to that of the actual features, and therefore this feature will be treated as almost equally important. Notice, however, that this feature was judged by SCAD-2 to be more irrelevant compared with the remaining features than SCAD-1.

We have used SCAD-1 and SCAD-2 to cluster several other data-sets, and we have observed that they have similar behavior. Both algorithms succeed in identifying the relevant and irrelevant features, and assign similar weights. In the rest of this chapter, we will extend only SCAD-1 to the case of high-dimensional feature space and to the case where the number of clusters is unknown. The extension of SCAD-2 follows similar steps.

Table 14.2 Results of SCAD-1 on the data-set in Table 14.1.

	Cluster # 1		Cluster # 2	
Features	x_1	x_2	x_1	x_2
Centers	−0.4	0.24	4.65	5.27
Relevance weights	0.49	0.51	0.48	0.52

Table 14.3 Results of SCAD-2 on the data-set in Table 14.1

	Cluster # 1		Cluster # 2	
Features	x_1	x_2	x_1	x_2
Centers	−0.37	0.27	4.64	5.28
Relevance weights	0.43	0.57	0.43	0.57

Table 14.4 Two four-dimensional clusters.

Cluster # 1				Cluster # 2			
x_1	x_2	x_3	x_4	x_1	x_2	x_3	x_4
19.00	**2.09**	−0.33	1.11	4.66	**2.13**	6.11	**0.28**
4.62	**3.79**	−2.02	−0.73	2.97	**6.43**	4.26	**1.76**
12.13	**7.83**	−0.33	0.72	4.66	**3.20**	5.72	**4.06**
9.71	**6.80**	−0.25	0.04	4.74	**9.60**	5.04	**0.04**
17.82	**4.61**	−1.08	−0.37	3.91	**7.26**	4.62	**0.69**
15.24	**5.67**	0.15	−0.36	5.15	**4.11**	4.63	**1.01**
9.12	**7.94**	−1.22	0.11	3.77	**7.44**	5.11	**0.99**
0.37	**0.59**	1.80	1.43	6.80	**2.67**	6.43	**3.01**
16.42	**6.02**	−1.48	−0.70	3.51	**4.39**	4.29	**1.36**
8.89	**0.50**	−0.87	1.02	4.12	**9.33**	6.02	**0.99**
12.30	**4.15**	−0.21	−0.45	4.78	**6.83**	4.54	**0.07**
15.83	**3.05**	−0.28	1.06	4.71	**2.12**	6.06	**3.73**
18.43	**8.74**	0.45	0.16	5.45	**8.39**	5.16	**2.22**
14.76	**0.15**	−2.29	1.98	2.74	**6.28**	6.98	**4.65**
3.52	**4.98**	0.84	−0.68	5.84	**1.33**	4.31	**2.33**
8.11	**7.67**	1.49	1.61	6.49	**2.07**	6.61	**2.09**
18.70	**9.70**	−0.23	0.31	4.76	**6.07**	5.31	**4.23**
18.33	**9.90**	−0.46	−0.82	4.53	**6.29**	4.17	**2.62**
8.20	**7.88**	−1.58	−1.09	3.41	**3.70**	3.90	**1.01**
17.87	**4.38**	0.72	1.27	5.72	**5.75**	6.27	**3.36**

Table 14.5 Results of SCAD-1 on the data-set in Table 14.4.

	Cluster # 1				Cluster # 2			
Features	x_1	x_2	x_3	x_4	x_1	x_2	x_3	x_4
Centers	**13.06**	**5.56**	−0.32	0.22	4.67	**5.17**	5.19	**2.08**
Relevance Weights	**0.00**	**0.23**	0.38	0.40	0.28	**0.16**	0.29	**0.27**

Table 14.6 Results of SCAD-2 on the data-set in Table 14.4.

	Cluster # 1				Cluster # 2			
Features	x_1	x_2	x_3	x_4	x_1	x_2	x_3	x_4
Centers	**12.72**	**5.39**	−0.40	0.26	4.62	**5.26**	5.26	**2.03**
Relevance Weights	**0.02**	**0.05**	0.40	0.53	0.32	**0.06**	0.42	**0.20**

14.4 CLUSTERING AND SUBSET FEATURE WEIGHTING

Learning a relevance weight for *each* feature may not be appropriate for high-dimensional data and may lead to over-fitting. This is because the generated clusters tend to have a few relevant features and may not reflect the actual distribution of the data. In this section, we present a coarse approach to the feature weighting problem. Instead of learning a weight for each feature, we divide the set of features into logical subsets, and learn a weight for each feature subset. The partition of the features into subsets is application dependent. For instance, in image segmentation, we can have a subset for the color features, another subset for the shape features, and a third subset for the texture features.

In the following, we assume that the n features have been partitioned into K subsets: $FS_1, FS_2, \cdots, FS_K$, and that each subset, FS_s, includes k_s features. Let d_{ij}^s be the partial distance between data vector $\mathbf{x}_j$ and cluster i using the sth feature subset. Note that we do not require that d_{ij}^s be the Euclidean distance. Moreover, different distance measures could be used for different feature subsets. For instance, the L_p norm, Mahalanobis distance, and fuzzy t-norm and t-conorm could be used for different subsets. We only require the different measures be normalized to yield values in the same dynamic range.

Let v_{is} be the relevance weight for feature subset FS_s with respect to cluster i. The total distance, D_{ij}, between $\mathbf{x}_j$ and cluster i is then computed by aggregating the partial degrees of similarities and their weights. In the following, we derive the equations for the simple case of a weighted average operator. That is, we let

$$D_{ij}^2 = \sum_{s=1}^{K} v_{is} (d_{ij}^s)^2. \tag{14.32}$$

Other aggregation operators such as the ordered weighted averaging operator (OWA) [38], and the fuzzy integral [16, 19] can be integrated into this approach.

The coarse SCAD algorithm (SCAD$_c$) minimizes

$$J = \sum_{i=1}^{C} \sum_{j=1}^{N} u_{ij}^m \sum_{s=1}^{K} v_{is} (d_{ij}^s)^2 + \sum_{i=1}^{C} \delta_i \sum_{s=1}^{K} v_{is}^2, \tag{14.33}$$

subject to (14.1), and

$$v_{is} \in [0, 1] \ \forall \ i, \ s; \quad \text{and} \quad \sum_{s=1}^{K} v_{is} = 1, \ \forall \ i. \tag{14.34}$$

To optimize J, with respect to $\mathbf{V}$, we use the Lagrange multiplier technique, follow similar steps to those outlined in Section 14.3.1, and obtain

$$v_{is} = \frac{1}{K} + \frac{1}{2\delta_i} \sum_{j=1}^{N} (u_{ij})^m \left[D_{ij}^2 / K - (d_{ij}^s)^2 \right]. \tag{14.35}$$

The first term in (14.35), $(1/K)$, is the default value if all K feature subsets are treated equally, and no discrimination is performed. The second term is a bias that can be either positive or negative. It is positive for compact feature subsets where the partial distance is, on average, less than the total distance (normalized by the number of feature subsets). If a feature subset is compact, compared with the other subsets, for most of the points that belong to a given cluster (high u_{ij}), then it is very relevant for that cluster.

To minimize J with respect to $\mathbf{U}$, we rewrite the objective function in (14.33) as

$$J = \sum_{i=1}^{C} \sum_{j=1}^{N} u_{ij}^m D_{ij}^2 + \sum_{i=1}^{C} \delta_i \sum_{s=1}^{K} v_{is}^2 \tag{14.36}$$

Since the second term in (14.36) does not depend on u_{ij} explicitly, the update equation of the memberships is similar to that of the FCM (see Equation (14.3)), i.e.,

$$u_{ij} = \frac{1}{\sum_{k=1}^{C} \left(D_{ij}^2 / D_{kj}^2 \right)^{\frac{1}{m-1}}}. \tag{14.37}$$

Minimization of J with respect to the prototype parameters depends on the choice of the partial distance measures d_{ij}^s. Since the partial distances are treated independent of each other (i.e., disjoint feature subsets), and since the second term in (14.33) does not depend on prototype parameters explicitly, the objective function in (14.33) can be decomposed into K independent problems:

$$J_s = \sum_{i=1}^{C} \sum_{j=1}^{N} u_{ij}^m v_{is} (d_{ij}^s)^2 \quad \text{for } s = 1, \cdots, K. \tag{14.38}$$

Each J_s would be optimized with respect to a different set of prototype parameters. For instance, if d_{ij}^s is a Euclidean distance, minimization of J_s would yield the update equation for the centers of subset s (same as FCM). Also, if d_{ij}^s is the weighted Mahalanobis distance proposed in [17] (or in [15]), minimization of J_s would yield the update equations for the centers and covariance matrices of subset s.

Finally, we should note that if each feature subset includes only one feature, SCAD_c reduces to SCAD-1. Thus, SCAD_c can be viewed as a generalization of SCAD-1.

14.5 CASE OF UNKNOWN NUMBER OF CLUSTERS

The objective function of SCAD_c (Equation (14.33)) can be easily combined with the objective function of the competitive agglomeration (CA) [10] algorithm (Equation (14.5)). The algorithm that minimizes the resulting objective function would inherit the advantages of the CA and the SCAD_c algorithms. This algorithm, called SCAD_c–CA, minimizes the following objective function

$$J = \sum_{i=1}^{C}\sum_{j=1}^{N}(u_{ij})^2 D_{ij}^2 + \sum_{i=1}^{C}\delta_i \sum_{s=1}^{K} v_{is}^2 - \alpha \sum_{i=1}^{C}\left[\sum_{j=1}^{N} u_{ij}\right]^2. \tag{14.39}$$

The additional third term in (14.39) does not depend on v_{is} explicitly. Thus, minimization of (14.39) with respect to $\mathbf{V}$ yields the same update equation for v_{is} as in SCAD_c (see Equation (14.35)). Moreover, since the second term in (14.39) does not depend on u_{ij}, minimization of (14.39) with respect to the u_{ij} yields the same membership update equation as the CA [10]. That is

$$u_{ij} = u_{ij}^{\text{FCM}} + u_{ij}^{\text{Bias}}, \tag{14.40}$$

where

$$u_{ij}^{\text{FCM}} = \frac{1/D_{ij}^2}{\sum_{k=1}^{C} 1/D_{kj}^2}, \tag{14.41}$$

and

$$u_{ij}^{\text{Bias}} - \frac{\alpha}{D_{ij}^2}\left(\sum_{j=1}^{N} u_{ij} - \frac{\sum_{k=1}^{C} 1/D_{kj}^2 N_k}{\sum_{k=1}^{C} 1/D_{kj}^2}\right).$$

14.6 APPLICATION 1: COLOR IMAGE SEGMENTATION

In this section, we illustrate the ability of SCAD to perform clustering and learn cluster-dependent feature weighting by using it to segment color images. First, we use a small set of features and use SCAD to partition the image into homogeneous regions and learn a relevance weight for each feature in each region. Then, we increase the number of features and use SCAD_c–CA to learn relevance weights for subsets of features.

14.6.1 Segmentation with SCAD

We start by mapping each pixel in the original image to a feature vector consisting of color, texture and position features. These are the same features used by Carson *et al.* in their Blobword content-based image retrieval system [6]. In the following, we will first give a brief description of these features, then show the results of using SCAD–CA to segment several images.

14.6.1.1 Feature Data Extraction

Texture Features. First, the image $I(x, y)$ is convolved with Gaussian smoothing kernels $G_\sigma(x, y)$ of several scales, σ, as follows:

$$M_\sigma(x, y) = G_\sigma(x, y) * (\nabla I(x, y))(\nabla I(x, y))^t.$$

(a) (b)

Figure 14.2 (a) Original color image, (b) results of SCAD-1.

Then, the following three features are computed at each pixel location [6]:

(1) *Polarity*: $p = \frac{|E_+ - E_-|}{E_+ + E_-}$, where E_+ and E_- represent the number of gradient vectors in the window $G_\sigma(x, y)$ that are on the positive and negative sides of the dominant orientation, respectively. For each pixel, an optimal scale value is selected such that it corresponds to the value where polarity stabilizes with respect to scale. Let p^* be the polarity at the selected scale.
(2) *Anisotropy*: $a = 1 - \lambda_2/\lambda_1$, where λ_1 and λ_2 are the eigenvalues of $M_\sigma(x, y)$ at the selected scale.
(3) *Normalized texture contrast*: $c = 2(\sqrt{\lambda_1 + \lambda_2})^3$.

Since the anisotropy and polarity are meaningless in regions of low contrast, these two features are scaled by the contrast value to yield the texture feature vector $[ac, p^*c, c]$.
Color Features. The three color features are the L*a*b* coordinates of the color image computed after smoothing the image with a Gaussian kernel at the selected optimal scale. Note that smoothing is performed to avoid over-segmentation of regions due to local color variations.
Position Features. The (x, y) coordinates of each pixel are used to reduce over-segmentation of some regions, and to obtain smoother regions.

14.6.1.2 Feature Data Clustering

The first color image to be segmented is shown in Figure 14.2(a). This is the same image that was used to illustrate the need for adaptive relevance weights in Section 14.1 (See Figure 14.1(a)). As discussed in the introduction, the FCM succeeds in delineating the compact cluster corresponding to the tiger, but fails to correctly delineate the different regions in the background. This is because the FCM is unable to assign cluster dependent feature relevance weights. When the number of clusters is also fixed to four, SCAD-1 succeeds in simultaneously segmenting the image into four meaningful clusters as shown in Figure 14.2(b), and in determining appropriate relevance weights for each cluster, as displayed in Table 14.7. Since a total of eight features were used in this experiment, the resulting feature weights should be compared to $1/8 \approx 0.12$, for the purpose of judging their relevancy. As it can be seen, the x-position feature is found to be irrelevant ($v = 0.014$) for the "dirt" cluster (cluster #4 in Figure 14.2(b)), while the y-position ($v = 0.102$) is deemed relevant for this cluster. This is because the "dirt" region forms a narrow

Table 14.7 Feature relevance weights of the segmented objects in Figure 14.2(b).

	Color features			Texture features			Position features	
Clusters	C_1	C_2	C_3	T_1	T_2	T_3	x	y
Cluster # 1	0.194	0.107	0.207	0.029	0.208	0.039	0.074	0.141
Cluster # 2	0.142	0.156	0.038	0.301	0.232	0.108	0.009	0.013
Cluster # 3	0.117	0.073	0.417	0.131	0.046	0.084	0.042	0.091
Cluster # 4	0.048	0.069	0.295	0.207	0.093	0.173	0.014	0.102

strip that extends over the entire width of the image. For the "grass" cluster (cluster #2 in Figure 14.2(b)), both the x and y features are found to be irrelevant because this region is scattered around the entire image. Similarly, the color and texture features receive unequal weights depending on the cluster. This is because the regions are not all characterized by the same features. For instance, the normalized texture contrast for the tiger region (cluster #1) was determined to be practically irrelevant ($v = 0.039$) because this feature has a large dynamic range. In fact, the value of this feature varies from small (in the uniformly textured areas) to large. Also, for the tiger cluster, the first texture feature, ac, was found to be practically irrelevant ($v = 0.029$). This means that the anisotropy feature ($a = 1 - \lambda_2/\lambda_1$) is irrelevant. This is because the orientation of the stripes varies significantly in some parts of the tiger cluster. In other words, some locations of this cluster have a dominant orientation ($\lambda_1 \gg \lambda_2$), while others do not.

In the next experiment, we compare the performance of CA and SCAD-1–CA to illustrate the importance of the simultaneous clustering and feature weighting mechanism when the number of clusters is unknown. The six color images in Figure 14.3(a) are to be segmented into an unknown number of

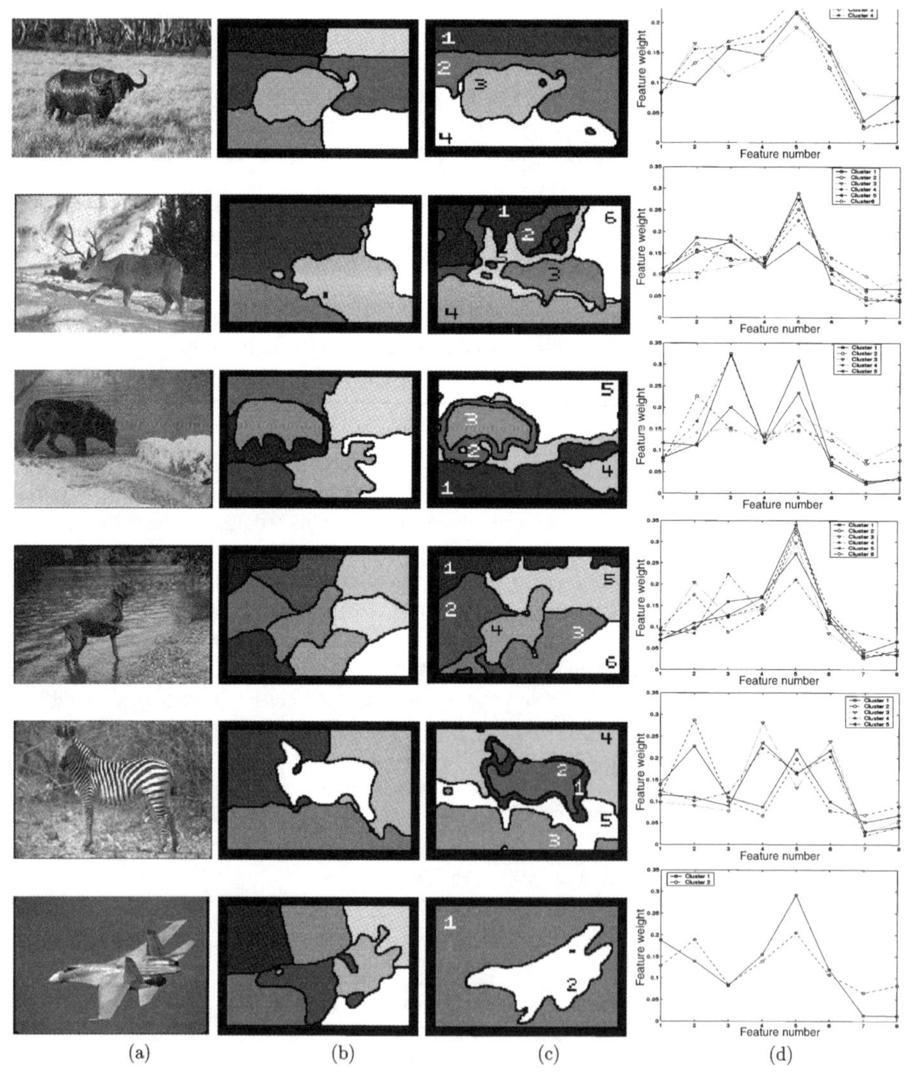

Figure 14.3 Image segmentation with CA and SCAD-1–CA: (a) original color images, (b) CA segmentation results, (c) SCAD-2–CA segmentation results, and (d) feature relevance weights for the regions labeled in (c).

clusters. Both CA and SCAD-1–CA were initialized by first dividing the image into a grid of 16 subimages. Then, the average of each feature within the *i*th subimage is computed to yield the corresponding initial feature value of the *i*th prototype. For the agglomeration coefficient, we set $(\eta_0, \tau, k_0) = (1.0, 10, 20)$. Figure 14.3(b), displays the segmentation results using CA, and Figure 14.3(c) displays the segmentation results using SCAD-1–CA. The identified regions (i.e., clusters) are enumerated in Figure 14.3(c), and their feature relevance weights are plotted in Figure 14.3(d). As can be seen, the feature relevance weights differ not only from one feature to another, but also from one cluster to another.

As in the previous experiment, we notice that the SCAD-1–CA performs better than CA, and it is able to yield meaningful clusters corresponding to both the objects in the foreground and the different areas of the background. This is because, unlike CA, SCAD-2–CA is capable of assigning distinct feature weights to different clusters. For example, the position features *x* and *y* are useful to delineate clusters corresponding to compact regions such as the animals in some pictures. Hence, both algorithms are able to delineate these clusters. However, the *x*-position feature is irrelevant for most of the areas in the background such as the sky or the grass regions. Without an unsupervised feature weighting mechanism, the CA is unable to recognize this selective irrelevance. The same phenomenon can also be observed if an object in the foreground is too large. For example, the plane image in the last row of Figure 14.3(b) gets over-segmented by CA into two clusters. In contrast, SCAD-1–CA was capable of segmenting all the foreground and background regions in all the images in Figure 14.3 into an appropriate number of clusters. These examples clearly demonstrate the effectiveness of SCAD-1–CA in color image segmentation.

14.6.2 Segmentation with $SCAD_c$

In this section, we increase the number of features and use $SCAD_c$–CA to learn relevance weights for subsets of features. We increase the number of features from eight to 14 by adding six features that were used in the SIMPLIcity [36] content-based image retrieval (CBIR) system.

14.6.2.1 Feature Extraction and Grouping

- FS_1 *Texture Features*. This subset of features includes the polarity, anisotropy and normalized texture contrast described in Section 14.6.1.1.
- FS_2 *L*a*b* Color Features*. The L*a*b* coordinates of the color image computed after smoothing the image with a Gaussian kernel at the selected optimal scale are used as a subset of three color features.
- FS_3 *LUV Color Features*. The color image is transformed to the LUV color space, and the three coordinates of each pixel in this space are used as a feature subset.
- FS_4 *Wavelet Texture Features*. First, we apply a one-level Daubechies-4 wavelet transform to the L component of the image. Then, three features were extracted from the HL, LH and HH bands. Each feature corresponds to the average coefficients in a 2×2 block (see [36] for more details).
- FS_5 *Horizontal Position Feature*. The *x* coordinate of each pixel is used as a position feature subset.
- FS_6 *Vertical Position Feature*. The *y* coordinate of each pixel is used as another position feature subset. Notice that the *x* and *y* position features are treated as two independent feature sets. This is because it is typical to find regions where one of these position features is important while the other is not.

14.6.2.2 Feature Data Clustering

The extracted features are normalized in the range [0,1], and the Euclidean distance is used for each feature subset. The initial number of clusters is fixed at 25. Figure 14.4(a) shows the original color image to be segmented. Figure 14.4(b) shows the segmentation results using CA when the *x* and *y* position features are not used. In this case, the CA fails to identify compact regions. For instance, there are two clusters that combine pixels from the animal and the background. Figure 14.4(c) displays the results of CA when

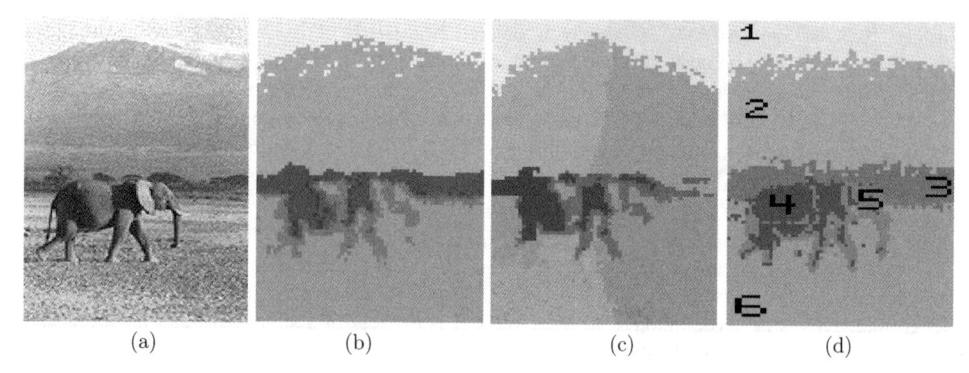

Figure 14.4 (a) Original Image, (b) results of CA without position features, (c) results of CA with position features, and (d) results of SCAD$_c$–CA.

Table 14.8 Feature relevance weights of the segmented objects in Figure 14.4(d).

Clusters	FS_1	FS_2	FS_3	FS_4	FS_5	FS_6
1	0.208	0.195	0.196	0.199	0.202	0.000
2	0.213	0.203	0.198	0.197	0.189	0.000
3	0.209	0.176	0.152	0.213	0.251	0.000
4	0.159	0.070	0.000	0.247	0.298	0.227
5	0.074	0.156	0.113	0.210	0.249	0.198
6	0.200	0.196	0.184	0.219	0.201	0.000

position features were used. In this case, not all clusters correspond to different homogeneous regions in the image. This is because most of the regions that make up the background tend to be noncompact. Figure 14.4(d) shows the segmentation results of SCAD$_c$–CA, which are clearly better than the CA results. This is because SCAD$_c$–CA is able to assign cluster-dependent weights that can adjust feature relevancy to the individual regions of the images. These feature weights are displayed in Table 14.8 (the cluster numbers listed in the table correspond to the annotated regions in Figure 14.4(d)). These weights should be compared with the default value ($1/K = 0.167$) to quantify their relevancy. For instance, FS_6 (y-position) is completely irrelevant for regions 1,2,3 and 6. This is because these regions extend over the entire width of the image. Region 4 is another instance where weights vary significantly: the color features (FS_2 and FS_3) are either weakly relevant (0.07) or completely irrelevant (due to the shaded areas on the animal). Notice also that the two texture (or color) feature sets are not equally relevant for the same regions (e.g., region 5). This fact illustrates the need for multiple feature sets to segment generic images.

Figure 14.5 displays the segmentation results obtained by SCAD$_c$–CA on three more images. As in the previous example, the relevance weights for the different feature subsets are not equally important for the different clusters and have similar behavior to those listed in Table 14.8.

14.7 APPLICATION 2: TEXT DOCUMENT CATEGORIZATION AND ANNOTATION

14.7.1 Motivation

One of the important tasks performed as part of many text mining and information retrieval systems is clustering. Clustering can be used for finding the nearest neighbors of a document efficiently [5], for improving the precision or recall in information retrieval systems [35, 24], for aid in browsing a collection

Figure 14.5 Segmentation of diverse color images: (a) original images, (b) segmented images.

of documents [7], for the organization of search engine results [39] and lately for the personalization of search engine results [29].

Most document clustering approaches work with the vector-space model, where each document is represented by a vector in the term-space. This vector consists of the keywords important to the document collection. For instance, the respective term or word frequencies (TF) [23] in a given document can be used to form a vector model for this document. In order to discount frequent words with little discriminating power, each word can be weighted based on its inverse document frequency (IDF) [23, 29] in the document collection. It is expected that the distribution of words in most real document collections can vary drastically from one cluster of documents to another. Hence, relying solely on the IDF for keyword selection can be inappropriate and can severely degrade the results of the clustering and/or any other learning tasks that follow it. For instance, a group of "News" documents and a group of "Business" documents are expected to have different sets of important keywords. Note that if the documents have already been manually preclassified into distinct categories, then it would be trivial to select a different set of keywords for each category based on IDF. However, for large, dynamic and *unlabeled* document collections, such as those on the World Wide Web, this manual classification is impractical. Hence, the need for automatic or unsupervised classification/clustering that can handle categories that differ widely in their best keyword sets.

Selecting and weighting subsets of keywords in text documents is similar to the problem of clustering and feature weighting in pattern recognition. Thus, this task can be accomplished by SCAD. Using SCAD to categorize text documents offers the following advantages compared to existing document clustering techniques. First, its *continuous* term weighting provides a much richer feature relevance representation than binary feature selection. This is especially true when the number of keywords is large. For example, one would expect the word "playoff" to be more important than the word "program" to distinguish a group of "sports" documents. Secondly, a given term is not considered *equally* relevant in *all* categories: For instance, the word "film" may be more relevant to a group of "entertainment" related documents than to a group of "sports" documents.

14.7.2 Adapting SCAD to Document Clustering

In Section 14.3, SCAD was derived for data lying in some Euclidean space, and relied on the Euclidean distance. For the special case of text documents, it is well known that the Euclidean distance is not appropriate. This is due mainly to the high dimensionality of the problem, and to the fact that two documents may not be considered similar if keywords are missing in both documents. In this section, we adapt SCAD to cluster text documents and to learn *dynamic category-dependent* keyword set weights. Since only one logical set of features is used (term frequency), we do not divide the features into subsets, and we use the standard SCAD.

The SCAD's criterion function can be modified to incorporate other dissimilarity measures. The only constraint is the ability to decompose the dissimilarity measure across the different attribute directions. An appropriate similarity for text document clustering is the cosine measure [23]. Let $\mathbf{x}_i = (x_{i1}, \cdots, x_{in})$ be the vector representation of document i, where x_{ij} is the frequency of the jth term in this document. The cosine similarity measure between two documents $\mathbf{x}_i$ and $\mathbf{x}_j$ is given by:

$$S(\mathbf{x}_i, \mathbf{x}_j) = \frac{\sum_{k=1}^{n} x_{ik} \times x_{jk}}{\sqrt{\sum_{k=1}^{n} x_{ik}^2}\sqrt{\sum_{k=1}^{n} x_{jk}^2}}. \tag{14.42}$$

To incorporate this measure into SCAD, we define the dissimilarity between document $\mathbf{x}_j$ and the ith cluster center vector as:

$$\tilde{D}_{ij} = \sum_{k=1}^{n} v_{ik} D_{ij}^k, \tag{14.43}$$

where

$$D_{ij}^k = \frac{1}{n} - (x_{jk} \cdot c_{ik}). \tag{14.44}$$

$\tilde{D}_{ij}$ is a weighted aggregate sum of cosine-based distances along the individual dimensions, c_{ik} is the kth component of the ith cluster center vector and $\mathbf{V} = [v_{ik}]$ is the relevance weight of keyword k in cluster i. Note that the individual products are not normalized in (14.43) because it is assumed that the data vectors are normalized to unit length before they are clustered, and that all cluster centers are normalized after they are updated in each iteration.

14.7.3 Categorization Results

We illustrate the performance of SCAD on text documents collected from the World Wide Web from several preclassified categories. Students were asked to collect 50 distinct documents from each of the following categories: news, business, entertainment, and sports. Thus, the entire collection consists of 200 documents. The documents' contents were preprocessed by eliminating stop words and stemming words to their root source. Then the inverse document frequencies (IDF) [23] of the terms were computed and sorted in descending order so that only the top 200 terms were chosen as final keywords. Finally, each document was represented by the vector of its document frequencies, and this vector was normalized to unit length. We should note here that the class labels were not used during the clustering process. They were used only to validate the final partition.

SCAD converged after 27 iterations, resulting in a partition that closely resembles the distribution of the documents with respect to their respective categories. The class distribution is shown in Table 14.9. Table 14.10 lists the six most relevant keywords for each cluster. As can be seen, the terms with the highest

Table 14.9 Distribution of the 50 documents from each class into the four clusters.

	Cluster 1 (business)	Cluster 2 (entertainment)	Cluster 3 (news)	Cluster 4 (sports)
Class 1	48	1	1	0
Class 2	7	31	5	7
Class 3	2	1	47	0
Class 4	0	0	3	47

Table 14.10 Term relevance for the top six relevant words in each cluster.

Cluster # 1		Cluster # 2		Cluster # 3		Cluster # 4	
$v_{1(k)}$	$w_{(k)}$	$v_{2(k)}$	$w_{(k)}$	$v_{3(k)}$	$w_{(k)}$	$v_{4(k)}$	$w_{(k)}$
0.029	compani	0.031	film	0.016	polic	0.025	game
0.016	percent	0.012	star	0.011	govern	0.015	season
0.011	share	0.010	week	0.010	state	0.010	plai
0.010	expect	0.008	dai	0.009	offici	0.009	york
0.008	market	0.008	peopl	0.009	nation	0.009	open
0.008	stock	0.008	open	0.009	sai	0.009	run

feature relevance weights in each cluster reflected the general topic of the category winning the majority of the documents that were assigned to the cluster. Thus, these cluster-dependent relevant keywords could be used to provide a *short summary* for each cluster and to *annotate* documents automatically.

The partition of the documents of class 2 showed most of the error in assignment due to the mixed nature of some of the documents therein. For example, by looking at the excerpts (shown below) from documents from Class 2 (*entertainment*) that were assigned to Cluster 1 with relevant words relating to *business* as seen in Table 14.10, one can see that these documents are hard to classify into one category, and that the keywords present in the documents in this case have mislead the clustering process. However, in the case of document 78, the fuzzy membership values in the business and entertainment clusters do not differ much, indicating a document related to several topic classes simultaneously. This illustrates the advantage of *fuzzy* memberships in text clustering.

Excerpt from Document 70 (memberships $u_{0j} = 0.853, u_{1j} = 0.140, u_{2j} = 0.005, u_{3j} = 0.003$):

LOS ANGELES (Reuters) - Ifilm and Pop.com, the would-be Web site backed by film makers Steven Spielberg, Ron Howard and other Hollywood moguls, have ended talks to merge, according to an e-mail sent to Ifilm employees on Friday. ... "The companies will continue to enjoy many overlapping shareholder and personal relationships," the memo said. Industry observers said the founders of Pop.com, which has never aired a single show or launched its Web site, are looking for a graceful exit strategy out of the venture, which has been plagued by infighting and uncertainty about the company's direction and business plan ...

Excerpt from Document 78 (memberships $u_{0j} = 0.319, u_{1j} = 0.252, u_{2j} = 0.232, u_{3j} = 0.197$):

... The Oxford-based quintet's acclaimed fourth release, "Kid A," opened at No. 1 with sales of 207,000 copies in the week ended Oct. 8, the group's Capitol Records label said Wednesday. The tally is more than four times the first-week sales of its previous album.

The last Stateside No. 1 album from the U.K was techno act Prodigy's "The Fat of the Land" in July 1997. That very same week, Radiohead's "OK Computer" opened at No. 21 with 51,000 units sold. It went on to sell 1.2 million copies in the United States ...

14.8 APPLICATION 3: BUILDING A MULTI-MODAL THESAURUS FROM ANNOTATED IMAGES

In this section, we illustrate the application of SCAD to learn associations between low-level visual features and keywords. We assume that a collection of images is available and that each image is globally annotated. The objective is to extract representative visual profiles that correspond to frequent homogeneous regions, and to associate them with keywords. These labeled profiles would be used to build a multi-modal thesaurus that could serve as a foundation for hybrid navigation and search algorithms in content-based image retrieval (CBIR) [27, 34]. This application involves two main steps. First, each image is coarsely segmented into regions, and visual features are extracted from each region. Second, the regions of all training images are grouped and categorized using SCAD$_c$–CA. As a result, we obtain

Figure 14.6 Examples of image-level annotations that refer to different and not segmented regions in the image.

clusters of regions that share subsets of relevant features. Representatives from each cluster and their relevant visual and textual features would be used to build a thesaurus.

14.8.1 Feature Extraction and Representation

We assume that we have a large collection of training images and that each image is annotated by a few keywords. We do not assume that the annotation is complete or accurate. For instance, the image may contain many objects, and we do not have a one-to-one correspondence between objects and words. This scenario is very common as images with annotations are readily available, but images where the regions themselves are labeled are rare and difficult to obtain. Moreover, we do not know which specific visual attributes best describe the keywords. Figure 14.6 displays three images annotated at the image level. Some keywords, such as "grass," can be clearly associated with color features. Others, such as "house," may be associated with shape features. Other words may be associated with any combination of color, texture, and shape features. This information, if it could be learned, would improve the efficiency of image annotation and hybrid searching and browsing.

First, each training image needs to be segmented into homogeneous regions based on color and/or texture features. It is not required to have an accurate segmentation as subsequent steps can tolerate missing and over-segmented regions. Then, each region would be described by visual features such as color, texture, shape and a set of keywords. Let $\{f_{j1}^{(i)}, \cdots, f_{jk_j}^{(i)}\}$ be a k_j dimensional vector that encodes the jth visual feature set of region R_i of a given image. For the keywords, we use the standard vector space model with term frequencies as features [32]. Let $\{w_1, w_2, \cdots, w_p\}$ be the representation of the keywords describing the given image (not region-specific). An image that includes n regions $(R_1, \ldots, R_n)$ would be represented by n vectors of the form:

$$\underbrace{f_{11}^{(i)}, \cdots, f_{1k_1}^{(i)}}_{\text{visual feat 1 of } R_i}, \cdots, \underbrace{f_{C1}^{(i)}, \cdots, f_{Ck_C}^{(i)}}_{\text{visual feat C of } R_i}, \underbrace{w_1, \cdots, w_p}_{\text{Keywords}}, \quad i = 1 \cdots n.$$

Figure 14.7 illustrates the image representation approach. We should note here that since the keywords are not specified per region, they get duplicated for each region representation. The assumption is that if word w describes a given region R_i then a subset of its visual features would be present in many instances across the image database. Thus, an association rule among them could be mined. On the other hand, if none of the words describe R_i, then these instances would not be consistent and will not lead to strong associations.

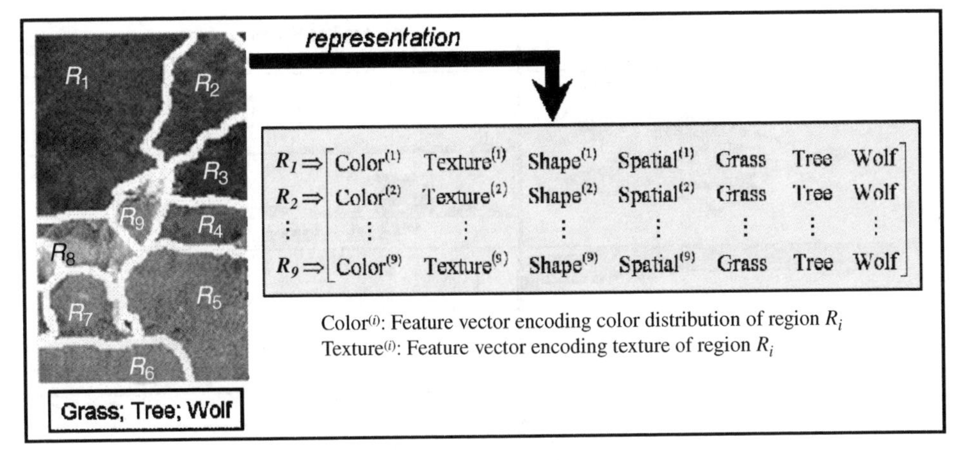

Figure 14.7 Representation of visual and textual features.

14.8.2 Categorization and Thesaurus Construction

A subset of the Corel image collection is used to illustrate the ability of SCAD to learn associations between multimodal features and to construct a multimodal thesaurus. We used a total of 1331 images, where each image is labeled by one to six keywords. The keywords provide a global description of the image and are not explicitly associated with specific regions. A total of 24 words were used. First, each image is coarsely segmented as described in Section 14.6.1. Segmentation of all the images resulted in 6688 regions. Second, each region is characterized by one color, one texture and one textual feature set. The color feature consists of a 64 bin RGB histogram. The texture feature consists of one global 5-Dim edge histogram [28]. The textual feature set consists of a 24-Dim vector that indicates the presence/absence of each keyword. Each feature set is normalized such that its components sum to one.

The 6688 feature vectors with the three feature subsets were clustered by $SCAD_c$ into $C = 100$ clusters of homogeneous regions and the relevant feature sets for each cluster were identified. Figure 14.8 displays some regions that were assigned to one of the clusters. For each region, we show the keywords that were used to annotate the original images from which the region was extracted. As can be seen, not all words are valid. However, some of the words (in this case "Sky") would be more consistent across all the regions

Figure 14.8 Sample regions from the "Sky" cluster.

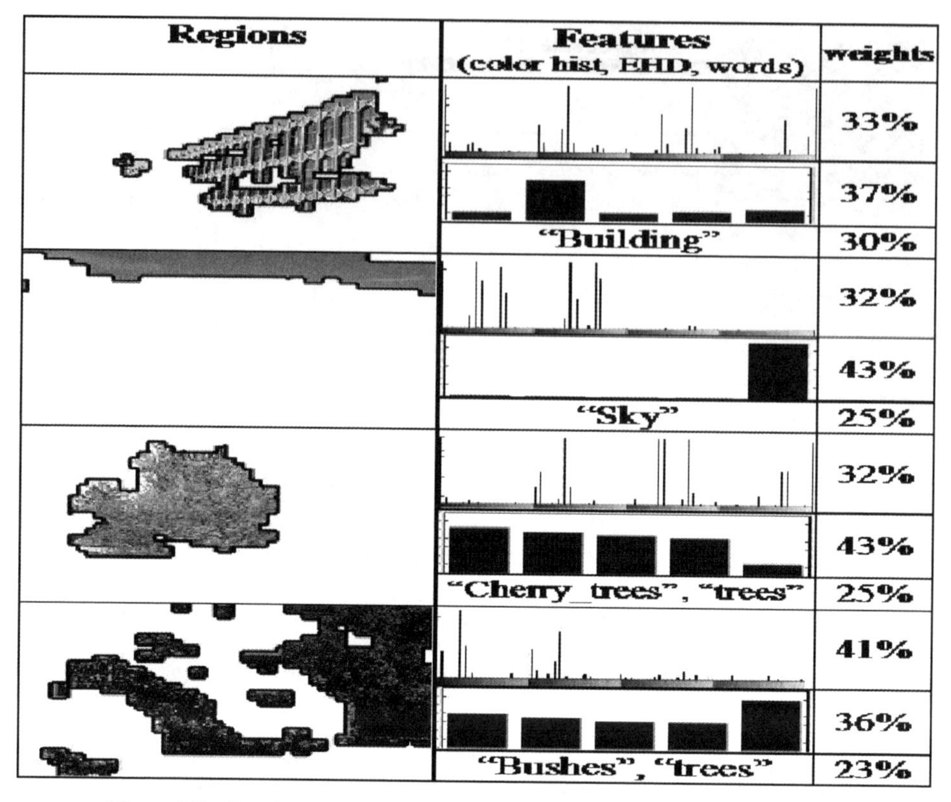

Figure 14.9 Samples of clusters' representatives with their visual and textual features.

of the cluster. Consequently, these words will be the dominant terms in the textual feature set. The visual features of this cluster consist of a *bluish* color and *smooth* texture.

After categorizing all the regions, the dominant keywords, the representative visual features and the feature relevance weights of each cluster would be used to create one entry in the thesaurus. Figure 14.9 displays samples of clusters' representatives. For each representative (i.e., cluster center), we show the image of the closest region, the color feature components (color histogram), the texture feature components (edge histogram where the five components indicate the proportion of horizontal, vertical, diagonal, anti-diagonal, and non-edge pixels in the region), and one or two dominant components (keywords) of the textual features. We also show the relevance weight of each feature subset. The second profile corresponds to the cluster sampled in Figure 14.8. From this profile, we can deduce that: "if the color is *bluish* and the texture is *smooth*, then the label is *sky*." Moreover, texture is the most consistent and thus the most important feature of this profile. For other profiles, such as the fourth sample in Figure 14.9, color is the most important feature.

14.8.3 Image Annotation

The constructed multimodal thesaurus could be used to annotate new test images as follows. Given a test image, we first segment it into homogeneous regions (as described in Section 14.6.1). Then for each region, R_k, we extract its color feature, R_k^{c}, its texture feature R_k^{t}, and compare it to the clusters' representatives using

$$D_i = v_{i\mathrm{c}} dist\left(R_k^{\mathrm{c}}, c_i^{\mathrm{c}}\right) + v_{i\mathrm{t}} dist\left(R_k^{\mathrm{t}}, c_i^{\mathrm{t}}\right), \quad i = 1, \cdots, C. \tag{14.45}$$

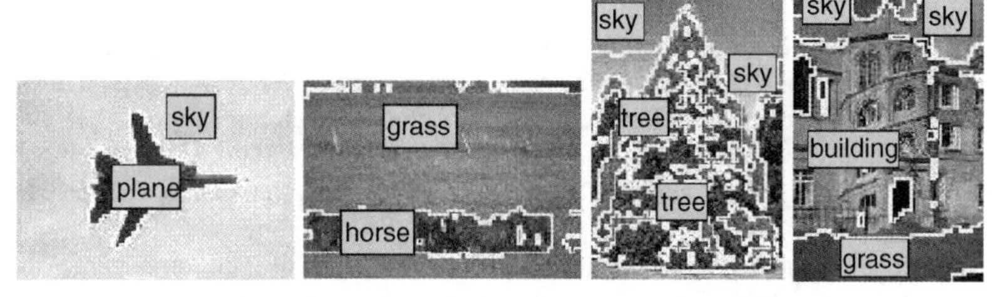

Figure 14.10 Samples of segmented images that were labeled correctly.

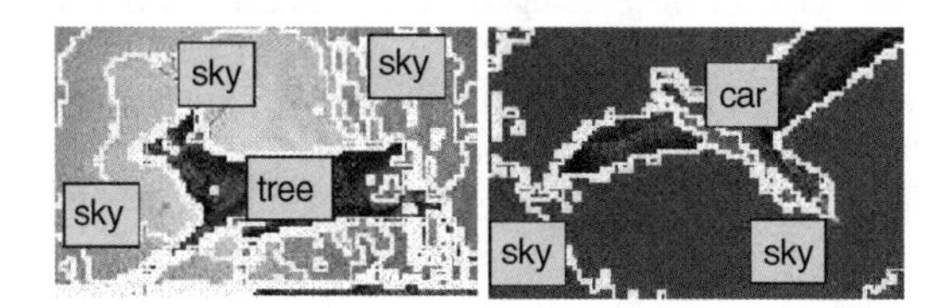

Figure 14.11 Samples of segmented images with some regions labeled incorrectly.

In (14.45), c_i^{c} and c_i^{t} are the center of the color and texture feature subsets of the ith entry of the constructed thesaurus, and v_{ic}, v_{it} are their respective relevance weights. Based on the distances D_i and the distribution of the clusters, several ways could be used to annotate R_k and assign a confidence value to each label. Here, we show the results using a simple approach that selects one word from the closest cluster only.

Figure 14.10 displays samples of test images that were segmented and labeled correctly. The first two images have good segmentation, and a correct label was assigned to each region. The last two images were over-segmented as both the sky and the tree were split into two regions. In this case, both regions of the same object were assigned the same correct label.

Figure 14.11 displays typical cases where this approach can fail. For the first image, the "deer" region was labeled incorrectly as "tree." This is because in the training image collection, there are several instances of trees that have the same visual features (brown color with small edges in all directions). The second image has an object (bird) that was not included in the training data, and the word "bird" was not included in the thesaurus. Since the adopted labeling approach did not assign a confidence value to each selected label, it does not have the option of rejecting the label.

14.9 CONCLUSIONS

In this chapter, we presented an approach that performs clustering and feature weighting simultaneously. When used as part of an unsupervised learning system, SCAD-1 and SCAD-2 can categorize the unlabeled data while determining the best feature weights within each cluster. Both SCAD-1 and SCAD-2 minimize one objective function for the optimal prototype parameters and feature weights for each cluster. This optimization is done iteratively by dynamically updating the prototype parameters and the feature weights. This makes the proposed algorithms computationally simple and efficient. Moreover, since the objective functions of SCAD-1 and SCAD-2 are based on that of the FCM, they inherit most of the advantages of FCM-type clustering algorithms. In fact, the SCAD approach can easily benefit from the advances and improvements that led to several K means and fuzzy C-means variants in the data mining and pattern recognition communities. In particular, the techniques developed to handle noise [11], to determine the number of clusters [10], to cluster very large data-sets [3, 9] and to improve initialization [4].

We have also presented SCAD$_{\mathrm{c}}$, a generalization of SCAD that learns feature relevance weights for subsets of features. If each extracted attribute is considered a feature subset (with one feature), then

$SCAD_c$ reduces to SCAD. There are two main advantages of using $SCAD_c$ as opposed to using SCAD. First, if there are too many features, then learning a relevance weight for *each* one may cause overfitting as clusters tend to be small and characterized by few relevant features. Second, $SCAD_c$ allows the use of different distance measures tailored for the different feature subsets. For instance, one could use the Mahalanobis distance for texture features, L_1 norm for the position features, cosine similarity (or fuzzy t-norm and t-conorm) for histogram-based features, etc. The only requirement is that the different distances should be normalized to yield values in the same dynamic range.

The SCAD paradigm was validated with three applications that involve clustering large data-sets with different types of features. In the image segmentation application, we have shown that learning cluster-dependent feature relevance weights improves the partition of the feature space and results in more homogeneous regions. In the text document categorization, we have shown that in addition to improving the quality of the categorization, the relevant features (keywords) could be used to provide a brief summary of each collection. In the third application, we used SCAD to learn associations between visual features of image regions and textual keywords. These associations are based on the identified relevant features from each modality, and were used to construct a multimodal thesaurus.

APPENDIX 14A.1

Theorem 1 $\lim_{q \to 1^+} v_{ik} = \begin{cases} 1 & \text{if} \tilde{D}_{ik} = \min_{t=1}^{n} \tilde{D}_{it} \\ 0 & \text{otherwise.} \end{cases}$

Proof

The relevance weight of feature k in cluster i is defined as

$$v_{ik} = \frac{1}{\sum_{t=1}^{n} (\tilde{D}_{ik}/\tilde{D}_{it})^{1/(q-1)}}, \qquad (14.46)$$

Since the exponent $\frac{1}{q-1} \to \infty$ as $q \to 1^+$, the individual terms in the denominator have the following tendency

$$\lim_{q \to 1^+} (\tilde{D}_{ik}/\tilde{D}_{it})^{1/(q-1)} = \begin{cases} \infty & \text{if } \tilde{D}_{ik} > \tilde{D}_{it}, \\ 0 & \text{if } \tilde{D}_{ik} < \tilde{D}_{it}, \\ 1 & \text{if } \tilde{D}_{ik} = \tilde{D}_{it}. \end{cases}$$

Therefore, two cases arise:

- *Case 1*:

$$\exists\, t \neq k \mid \tilde{D}_{ik} > \tilde{D}_{it} \Longrightarrow \tilde{D}_{ik} \neq \min_{t=1}^{n} \tilde{D}_{it}.$$

The denominator in (14.46) becomes infinite, and

$$\lim_{q \to 1^+} v_{ik} = 0.$$

- *Case 2*:

$$\nexists\, t \neq k \mid \tilde{D}_{ik} > \tilde{D}_{it} \Longrightarrow \tilde{D}_{ik} = \min_{t=1}^{n} \tilde{D}_{it}.$$

Hence,

$$\begin{aligned} v_{ik} &= \frac{1}{(\tilde{D}_{ik}/\tilde{D}_{it})^{1/(q-1)} + \sum_{t=1, t \neq k}^{n} (\tilde{D}_{ik}/\tilde{D}_{it})^{1/(q-1)}} \\ &= \frac{1}{1 + \sum_{t=1, t \neq k}^{n} (\tilde{D}_{ik}/\tilde{D}_{it})^{1/(q-1)}}. \end{aligned}$$

Since $\lim_{q\to 1^+}(\tilde{D}_{ik}/\tilde{D}_{it})^{1/(q-1)} = 0$ for all $t \neq k$,

$$\lim_{q\to 1^+} v_{ik} = 1.$$

Therefore, we conclude that

$$\lim_{q\to 1^+} v_{ik} = \begin{cases} 1 & \text{if} \tilde{D}_{ik} = \min_{t=1}^{n} \tilde{D}_{it}, \\ 0 & \text{otherwise.} \end{cases}$$

ACKNOWLEDGEMENTS

This work was supported in part by the National Science Foundation Career Award No. IIS-0514319, by an Office of Naval Research award number N00014-05-10788, and by a grant from the Kentucky Science and Engineering Foundation as per Grant agreement \#KSEF-148-502-05-153 with the Kentucky Science and Technology Corporation. The views and conclusions contained in this document are those of the authors and should not be interpreted as representing the official policies, either expressed or implied, of the Office of Naval Research or the U. S. Government.

REFERENCES

[1] Almuallim, H. and Dietterich, T.G. (1991) 'Learning with many irrelevant features'. In *Ninth National Conference on Artificial Intelligence*, pp. 547–552.

[2] Bezdek, J.C. (1981) *Pattern Recognition with Fuzzy Objective Function Algorithms*. Plenum Press, New York, USA.

[3] Bradley, P.S., Fayyad, U.M., and Reina, C. (1998) 'Scaling clustering algorithms to large databases'. In *Knowledge Discovery and Data Mining*, pp. 9–15.

[4] Bradley, P.S. and Fayyad, U.M. (1998) 'Refining initial points for K-Means clustering'. In *Proceedings of the 15th International Conference on Machine Learning*, pp. 91–99. Morgan Kaufmann, San Francisco, CA, USA.

[5] Buckley, C. and Lewit, A.F. (1985) 'Optimizations of inverted vector searches'. In *SIGIR'85*, pp. 97–110.

[6] *Carson, C. et al.* (1999) 'Blobworld: a system for region-based image indexing and retrieval'. In *Third International Conference on Visual Information Systems*. Springer.

[7] Cutting, D.R., Karger, D.R., Pedersen, J.O. and Tukey, J.W. (1992) 'Scatter/gather: a cluster-based approach to browsing large document collections'. In *SIGIR'92*, pp. 318–329.

[8] Dempster, A.P., Laird, N.M. and Rubin, D.B. (1977) 'Maximum Likelihood from incomplete data via the EM algorithm', *Journal of the Royal Statistical Society Series B*, **39**(1): 1–38.

[9] Farnstrom, F., Lewis, J. and Elkan, C. (2000) 'Scalability for clustering algorithms revisited', *SIGKDD Explorations*, **2**(1): 51–57.

[10] Frigui, H. and Krishnapuram, R. (1997) 'Clustering by competitive agglomeration'. *Pattern Recognition*, **30**(7): 1223–1232.

[11] Frigui, H. and Krishnapuram, R. (1999) 'A robust competitive clustering algorithm with applications in computer vision'. *IEEE Transactions on Pattern Analysis and Machine Intelligence*, **21**(5): 450–465.

[12] Frigui, H. and Nasraoui, O. (2000) 'Simultaneous clustering and attribute discrimination'. In *Proceedings of the IEEE Conference on Fuzzy Systems*, pp. 158–163, San Antonio, Texas.

[13] Frigui, H. and Nasraoui, O. (2004) 'Unsupervised learning of prototypes and attribute weights'. *Pattern Recognition Journal*, **37**:567–581.

[14] Fukunaga, K. (1990) *Statistical Pattern Recognition*. Academic Press, San Diego, California, USA.

[15] Gath, I. and Geva, A.B. (1989) 'Unsupervised optimal fuzzy clustering'. *IEEE Transactions on Pattern Analysis and Machine Intelligence*, **11**(7): 773–781.

[16] Grabisch, M., Murofushi, T. and Sugeno, M. (1992) 'Fuzzy measure of fuzzy events defined by fuzzy integrals', *Fuzzy Sets and Systems*, **50**:293–313.

[17] Gustafson, E.E. and Kessel, W.C. (1979) 'Fuzzy clustering with a fuzzy covariance matrix'. In *IEEE CDC*, pp. 761–766, San Diego, California, USA.

[18] John, G., Kohavi, R. and Pfleger, K. (1994) 'Irrelevant features and the subset selection problem'. In *Eleventh International Machine Learning Conference*, pp. 121–129.
[19] Keller, J., Gader, P. and Hocaoglu, K. (2000) 'Fuzzy integrals in image processing and recognition'. In (M. Grabisch, Murofushi, T. and Sugeno, M., eds) *Fuzzy Measures and Integrals*. Springer-Verlag.
[20] Kira, K. and Rendell, L.A. (1992) 'The feature selection problem: traditional methods and a new algorithm'. In *Tenth National Conference on Artificial Intelligence*, pp. 129–134.
[21] Kohavi, R. and Sommerfield, D. (1995) 'Feature subset selection using the wrapper model: overfitting and dynamic search space topology' In *First International Conference on Knowledge Discovery and Data Mining*, pp. 192–197.
[22] Kononenko, I. (1994) 'Estimation attributes: analysis and extensions of relief'. In *European Conference on Machine Learning*, pp. 171–182.
[23] Korfhage, R.R. (1997) *Information Storage and Retrieval*. John Wiley & Sons, Ltd, Chichester, UK.
[24] Kowalski, G. (1997) *Information retrieval systems-theory and implementations*. Kluwer Academic Publishers.
[25] Kuhn, H.W. and Tucker, A.W. (1950) 'Nonlinear programming'. In (J. Neyman, ed.) *Proceedings of the Second Berkeley Symposium on Mathematical Statistics and Probability*, pp. 481–492.
[26] Lee, M.S. and Moore, A.W. (1994) 'Efficient algorithms for minimizing cross validation error'. In (W. Cohne and H. Hirsh, eds) *Machine Learning: Proceedings of the Eleventh International Conference*, pp. 190–198.
[27] Lu, Ye. *et al.* (2000) 'A unified framework for semantics and feature based relevance feedback in image retrieval systems'. In *ACM Multimedia*, pp. 31–37.
[28] Manjunath, B.S., Salembier, P. and Sikora, T. (2002) *Introduction to MPEG 7: Multimedia Content Description Language*. John Wiley & Sons, Ltd, Chichester, UK.
[29] Mladenic, D. (1999) 'Text learning and related intelligent agents'. *IEEE Expert*, July.
[30] Pedrycz, W. and Vukovich, G. (2002) 'Feature analysis through information granulation and fuzzy sets'. *Pattern Recognition*, **35**(4):825–834.
[31] Rendell, L.A. and Kira, K. (1992) 'A practical approach to feature selection'. In *International Conference on Machine Learning*, pp. 249–256.
[32] Salton, G. and McGill, M.J. (1983) *An Introduction to Modern Information Retrieval*. McGraw-Hill.
[33] Skalak, D. (1994) 'Prototype and feature selection by sampling and random mutation hill climbing algorithms', In *Eleventh International Machine Learning Conference (ICML-94)*, pp. 293–301.
[34] Smeulders, A.W.M. *et al.* (2000) 'Content based image retrieval at the end of the early years'. *IEEE Transactions on Pattern Analysis and Machine Intelligence*, **22**(12):1349–1380.
[35] VanRijsbergen, C. J. (1989) *Information Retrieval*. Buttersworth, London, UK.
[36] Wang, J., Li, J. and Wiederhold, G. (2001) 'SIMPLIcity: semantics-sensitive integrated matching for picture LIbraries'. *IEEE Transactions on Pattern Analysis and Machine Intelligence*, **23**:947–963.
[37] Wettschereck, D., Aha, D.W. and Mohri, T. (1997) 'A review and empirical evaluation of feature weighting methods for a class of lazy learning algorithms'. *Artificial Intelligence Review*, **11**:273–314.
[38] Yager, R.R. (1988) 'On ordered weighted averaging aggregation operators in multicriteria decision making'. *IEEE Trans. systems, Man, and Cybernetics*, **18**(1):183–190.
[39] Zamir, O., Etzioni, O., Madani, O. and Karp, R.M. (1997) 'Fast and intuitive clustering of web documents'. In *KDD'97*, pp. 287–290.

Part IV

Real-time and Dynamic Clustering

15

Fuzzy Clustering in Dynamic Data Mining – Techniques and Applications

Richard Weber

Department of Industrial Engineering, University of Chile, Santiago, Chile

15.1 INTRODUCTION

During the past 20 years or so we have witnessed the successful performance of data mining systems in many companies and organizations worldwide. Most of these systems are, however, static in the sense that they do not explicitly take into consideration a changing environment. If we want to improve such systems and/or update them during their operation the dynamics of the data should be considered.

Since data mining is just one step of an iterative process called knowledge discovery in databases (KDD, Han, and Kamber, 2001), the incorporation of dynamic elements could also affect the other steps. The entire process consists basically of activities that are performed before doing data mining (such as selection, preprocessing, transformation of data (Famili, Shen, Weber, and Simoudis, 1997)), the actual data mining part and subsequent steps (such as interpretation and evaluation of results). Clustering techniques are used for data mining if the task is to group similar objects in the same classes (segments) whereas objects from different classes should show different characteristics (Jain, Murty, and Flynn, 1999). Such clustering approaches could be generalized in order to treat different dynamic elements, such as dynamic objects and/or dynamic classes as shown below. Incorporating fuzzy logic into clustering algorithms offers various advantages, especially for the treatment of dynamic elements occurring in the problem as will be seen in this chapter.

Section 15.2 reviews the literature related to dynamic clustering using crisp as well as fuzzy approaches. Section 15.3 presents recently developed methods for dynamic fuzzy clustering. Applications of these techniques are shown in Section 15.4. Future perspectives and conclusions are provided in Section 15.5.

15.2 REVIEW OF LITERATURE RELATED TO DYNAMIC CLUSTERING

In data mining various methods have been proposed in order to find interesting information in databases. Among the most important ones are decision trees, neural networks, association rules, and clustering

Advances in Fuzzy Clustering and its Applications Edited by J. Valente de Oliveira and W. Pedrycz

methods (Han and Kamber, 2001). Since the analyzed objects in many real-world domains depend on time-evolving phenomena, the related dynamic aspects have to be explicitly taken into account. Model updating is one way to address the problem caused by continuously receiving new data.

For each of the above-mentioned data mining methods updating can be done in different ways and several approaches have been proposed in literature:

- *Decision trees*: Various techniques for incremental learning and tree restructuring (see, for example, Utgoff (1997)) as well as the identification of concept drift (see, for example, Black and Hickey (1999)), have been proposed.
- *Neural networks*: Updating is often used in the sense of re-learning or improving the network's performance by presenting new examples.
- *Association rules*: Raghavan and Hafez (2000) have developed systems for dynamic data mining for association rules.
- *Clustering*: In this section, we describe in more detail approaches for dynamic data mining using clustering techniques that can be found in literature.

One of the first dynamic cluster methods has been presented in Diday (1973) where groups of objects called "samplings" adapt over time and evolve into interesting clusters. In order to structure the following considerations dynamic clustering approaches can be characterized regarding the nature of input data, i.e., whether the respective input data is static or dynamic. We call input data static if no time-dependent variation is considered; in the opposite case we speak about dynamic input data. Several clustering systems have been proposed that treat static input data but are using dynamic elements during classifier design, i.e., dynamic adaptations of the respective algorithm are performed while applying it to a set of static input data. "Chameleon" is such a system, which uses hierarchical clustering where the merging decision on each hierarchy dynamically adapts to the current cluster characteristics (Karypis, Han, and Kumar 1999).

Other than in hierarchical clustering, objective function-based clustering methods such as, for example, C-means and fuzzy C-means, need to specify the number of clusters (here c) before running the respective algorithm. The determination of an appropriate number of clusters has been the subject of several investigations (see, for example, Bezdek, Keller, Krishnapuram, and Pal (1999)). "Dynamic partitional clustering using evolution strategies" is an approach where the cluster number is optimized during runtime (Lee, Ma, and Antonsson, 2001) using evolutionary algorithms. Adaptive fuzzy clustering (AFC) estimates dynamically the respective parameters during the classifier construction (see, for example, Krishnapuram and Kim (1999)).

The clustering methods mentioned above are using dynamic elements during their application to a set of static input data. Next, we will analyze the situation of dynamic input data distinguishing between the following two cases. On one hand, clustering can be performed at a certain point of time explicitly taking into consideration the previous development of feature values. In this case dynamic clustering means clustering of feature trajectories instead of real-valued feature vectors. On the other hand, we want to understand how class structures may change over time and are interested in updating a classifier when new data becomes available. In the following, both cases will be studied in more detail.

In situations where current feature values are not sufficient in order to explain the underlying phenomenon it may be interesting to analyze the respective feature trajectories. This is the case, for instance, in medicine, where patients' conditions depend not only on the current values of, for example, blood pressure but also on their development during the relevant past. Another example could be stock price prediction, where a stock's current value is not sufficient in order to predict future price development. For such a prediction we would be interested in the chart exhibiting the respective stock prices of the relevant past.

An approach where trajectories of feature values are clustered is Matryoshka (Li, Biswas, Dale, and Dale, 2002), which is based on a hidden Markov model (HMM) for temporal data clustering. Given as input objects described by temporal data (trajectories) it determines the optimal number of classes where each class is characterized as an HMM and an assignment of objects to classes. The research area of data mining where objects are described by dynamic data (trajectories) is also called temporal data mining (Antunes and Oliveira, 2001).

For the case of dynamically changing classes fuzzy clustering has been used, for example, as a preprocessing tool for a rule-based fuzzy system. This approach iterates between fuzzy clustering and tuning of the fuzzy rule base "until the number of the data belonging to a class that are misclassified into another class does not exceed the prescribed number" (Abe, 1998).

A dynamic evolving neural-fuzzy inference system (DENFIS) has been used for dynamic time series prediction (Kasabov and Song, 2002). Based on an evolving clustering method (ECM) a first-order Takagi–Sugeno-type fuzzy rule set for prediction is created dynamically.

A prototype-based clustering algorithm called dynamic data assigning assessment (DDAA) has been proposed in Klawonn and Georgieva (2006). It is based on the noise clustering technique and finds good single clusters one by one and at the same time separates noisy data. A framework for dynamic fuzzy clustering that explicitly integrates the process of monitoring the analyzed phenomenon is presented in Angstenberger (2001).

Fuzzy logic has also proved its potential in clustering of data streams. Such a data stream can be defined as a continuously increasing sequence of time-stamped data. The problem of clustering evolving data streams is described, for example, in Chapter 16 of this book. The authors show that in such applications fuzzy clustering offers particular advantages since changes of the respective classes are frequently smooth rather than abrupt.

15.3 RECENT APPROACHES FOR DYNAMIC FUZZY CLUSTERING

Table 15.1 provides a taxonomy of various situations where we are faced with dynamic input data in the case of clustering. In the case of static classes we apply clustering only at one point of time; no updating of class structures will be considered. If additionally objects are static, i.e., described by real-valued feature vectors containing just current values of the features used, we have the classical case of clustering. If we want to consider explicitly the development of feature values previous to the clustering step we are assigning dynamic objects to static classes. In this case the feature vectors consist of trajectories instead of real values.

In the case of dynamic classes the class structure should be updated iteratively. Should the respective feature vectors be composed of current feature values, we are updating with static objects. Assigning dynamic objects to dynamic classes is the case of class updating with feature trajectories.

The following sections present fuzzy clustering approaches developed for the cases of "static classes"/"dynamic objects" and "dynamic classes"/"static objects." It will be shown that working with fuzzy logic provides particular advantages in dynamic clustering since the respective membership values represent a strong tool in order to capture changing environments. The following developments are based on fuzzy C-means (FCM) that is described, for example, in (Bezdek, Keller, Krishnapuram, and Pal, 1999). Here we use the following notation:

c: number of classes
n: number of objects

Table 15.1 Clustering approaches for dynamic data mining.

	Static classes	Dynamic classes
Static objects	Classical case: clustering of real-valued feature vectors	Clustering of a set of changing feature vectors
Dynamic objects	Clustering of feature trajectories	Clustering of a set of changing feature trajectories

p: number of features describing each object
$\mathbf{x}_i$: feature vector of object i, $i = 1, \ldots, n$
$\mathbf{v}_j$: center of class j, $j = 1, \ldots, c$
μ_{ij}: degree of membership of object i to class j, $i = 1, \ldots, n; j = 1, \ldots, c$

15.3.1 Assigning Dynamic Objects to Static Classes

In many real-world applications current feature values are not sufficient in order to explain the underlying phenomenon. As has been seen already in Section 15.2, in such situations we may be interested in the development of these values over time so that features become trajectories.

15.3.1.1 Approaches for Handling Trajectories in Clustering

Incorporating trajectories into data mining systems can be done during the transformation step of the KDD-process and then applying classical clustering methods or directly within accordingly modified methods. Transforming original trajectories into real-valued feature vectors maintaining the dynamic information contained in these trajectories can be done by determining, for example, mean values, variance, derivatives, and others. We have to keep in mind, however, that reducing the original trajectory to a set of real values in general means a loss of information. The functional fuzzy C-means (FFCM) described next works with the entire trajectories using a modified distance measure.

15.3.1.2 Description of the Functional Fuzzy c-means (FFCM)

In order to be able to cluster dynamic objects, we need a distance measure between two vectors where each component is a trajectory (function) instead of a real number. Functional fuzzy C-means is a fuzzy clustering algorithm where the respective distance is based on the similarity between two trajectories that is determined using membership functions. Alternative approaches to determine the similarity between two functions are wavelets (see, for example, Angers (2002)) and time warping (see, for example, Keogh (2002)).

The FFCM is a generalization of standard fuzzy C-means (FCM). In each iteration FCM uses the following formula in order to calculate the membership value $\mu_{i,j}$ of object i to class j:

$$\mu_{i,j} = \frac{1}{\sum_{k=1}^{c} \left(\frac{d(\mathbf{X}_i, \mathbf{V}_j)}{d(\mathbf{X}_i, \mathbf{V}_k)} \right)^{\frac{2}{m-1}}}$$

where $\mathbf{x}_i$ is the feature vector of object i; $\mathbf{v}_j$ and $\mathbf{v}_k$ are the class centers of the classes j and k, respectively; c is the number of classes; $m \in (1, \infty)$ is a parameter determining the degree of fuzziness of the generated clusters.

The expression $d(\mathbf{x}, \mathbf{y})$ used in the above formula determines the distance between two vectors $\mathbf{x}$ and $\mathbf{y}$ in the feature space. In the case of feature vectors containing real numbers the Euclidean distance can be applied for this calculation. The main idea of the FFCM is to generalize the calculation of the distance between a pair of feature vectors containing real numbers to the calculation of the distance between a pair of feature vectors containing their trajectories. The latter calculation can be performed using the concept of a membership function. The idea presented below can be applied, however, to any data mining technique where the distance between objects is required.

FFCM determines the distance between two objects applying the following five steps:

(1) A fuzzy set A "approximately zero" with membership function μ is defined (Figure 15.1(a))
(2) The degree of membership $\mu(f(x))$ of an arbitrary function $f(x)$ to the fuzzy set A is calculated for every point x. These degrees of membership can be interpreted as (pointwise) similarities of the function f to the zero function (Figure 15.1 (b)).

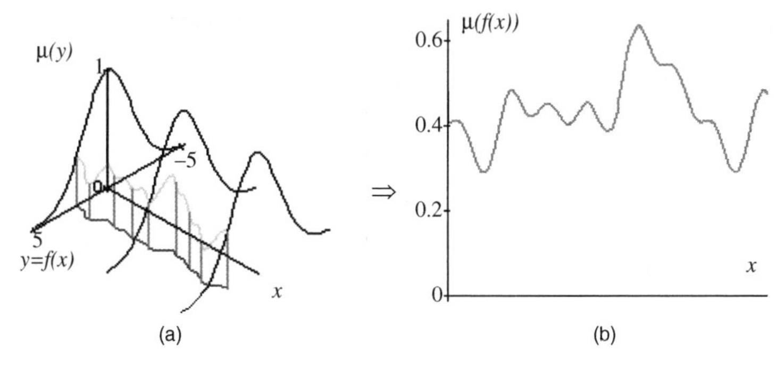

Figure 15.1 (a) The fuzzy set "approximately zero"($\mu(y)$), the function $f(x)$, and the resulting point wise similarity $\mu(f(x))$, Projection of the point wise similarity into the plane $(x, \mu(f(x)))$.

(3) $\mu(f(x))$ is transformed into a real number expressing the overall degree of being zero by using specific transformations (e.g., γ-operator, fuzzy integral (Zimmermann, 2001)).
(4) The similarity measure defined by steps (1) to (3) is invariant with respect to the addition of a function, i.e., $s(f, g) = s(f + h, g + h)$ holds for all functions f, g, and h, (Joentgen, Mikenina, Weber, and Zimmermann, 1999). This allows us to calculate the similarity between an arbitrary pair of functions f and g by adding the function $h := -g$ and determining the similarity of $f - g$ to the zero-function.
(5) Finally, the similarity $s(f, g)$ is transformed to a distance $d(f, g)$ by: $d(f, g) := 1/s(f, g) - 1$.

Applying this new distance measure between functions FFCM works as FCM determining classes of dynamic objects. The respective class centers are composed of the most representative trajectories in each class (see Figure 15.2).

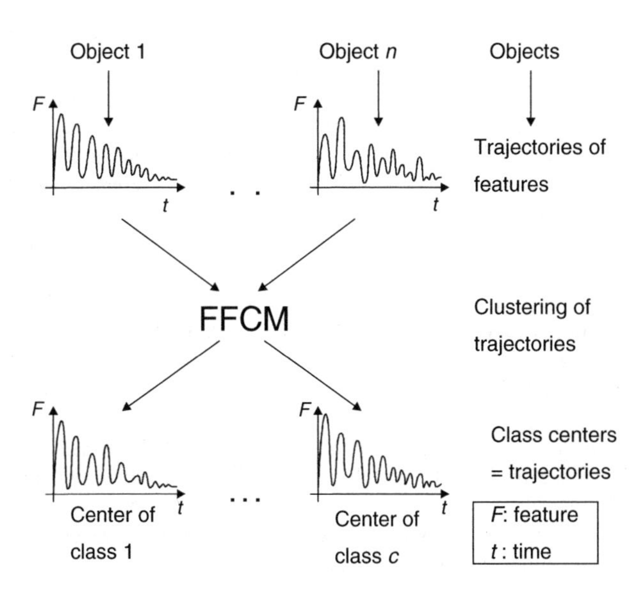

Figure 15.2 Functional fuzzy C-means (FFCM).

15.3.2 Assigning Static Objects to Dynamic Classes

We present a methodology for dynamic data mining based on fuzzy clustering that assigns static objects to "dynamic classes," i.e., classes with changing structure over time. It starts with a given classifier and a set of new objects, i.e., objects that appeared after the creation of the current classifier. The period between the creation of a classifier and its update is called a **cycle**. The length of such a cycle depends on the particular application, e.g., we may want to update buying behavior of customers in a supermarket once a year whereas a system for dynamic machine monitoring should be updated every 5 minutes. Next, we present a global view of our approach before we provide details of each of its steps.

15.3.2.1 Global View of the Proposed Methodology

The methodology presented next does not assume identifiable objects, i.e., we do not need an identifier for each object. This would be the case, for example, in segmentation of bank customers where we can identify each customer and his/her activities (see, for example, Weber (1996)). A possible application with non-identifiable objects would be segmentation of supermarket customers where we do not have personalized information.

Possible changes of a classifier's structure we may want to detect in each cycle are:

- creation of new classes;
- elimination of classes;
- movement of classes in the feature space.

The following five steps are applied in order to detect changes in the class structure and realize the corresponding modifications.

Step I: Identify objects that represent changes. We first want to know if the new objects can be explained well by the given classifier or not. In other words we want to identify objects that ask for possible changes in the classifier structure because they are not well classified. If there are many objects that represent such possible changes we proceed with Step II, otherwise we go immediately to Step III.

Step II: Determine changes of class structure. Here we want to decide if we have to create new classes in order to improve the classifier for the new objects or if it is sufficient to just move the existing classes. If Step I identifies "many new objects" representing changes we have to create a new class, otherwise we just move the existing classes in the feature space.

Step III: Change the class structure. We perform the changes according to the results of Steps I and II ((a) move or (b) create classes).

Step III(a): Move classes. We update the position of the existing class centers based on the information provided by the new objects and knowing that they do not ask for new classes.

Step III(b): Create classes. If we know that classes have to be created, we first determine an appropriate class number. Then we apply fuzzy C-means with the new class number to the available data-set.

Step IV: Identify trajectories of classes. We identify trajectories of the classes from previous cycles in order to decide if they received new objects. Classes that did not receive new objects during several cycles should be eliminated.

Step V: Eliminate unchanged classes. According to the result of Step IV we eliminate classes that did not receive new objects during an "acceptable period."

Figure 15.3 exhibits a general view of the proposed methodology; see (Gespo and Weber 2005).

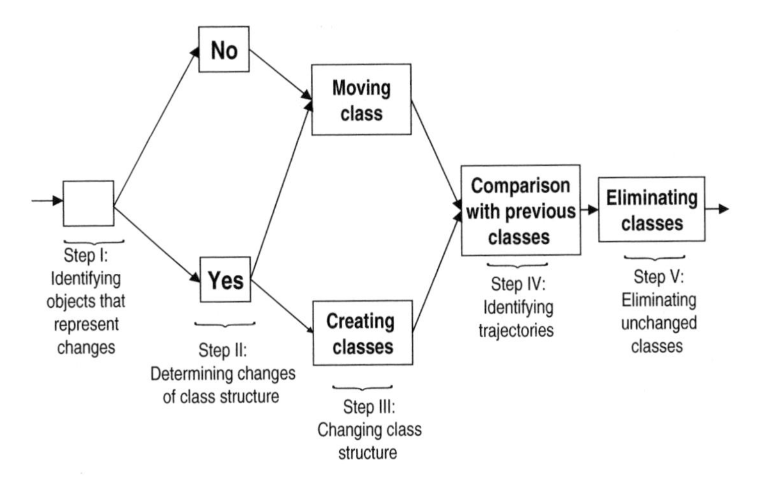

Figure 15.3 General view of our methodology for dynamic fuzzy clustering considering changing class structures.

15.3.2.2 Detailed Description of a Cycle of the Proposed Methodology

During a cycle m new objects have appeared. Let $k = n+1, \ldots, n+m$ be the index of the m new objects.

Step I: Identify objects that represent changes. The goal of this step is to identify those objects that are not well classified by the given classifier. For this reason we need the distances between each pair of the currently existing class centers. This pairwise distance is:

$$d(\mathbf{v}_i, \mathbf{v}_j)\ \forall i \neq j, i, j \in \{1, \ldots, c\}.$$

Additionally we need the distance between the new object k and the center $\mathbf{v}_i$ from the current class structure. This distance is:

$$\hat{d}_{ik} = \hat{d}(\mathbf{x}_k, \mathbf{v}_i), \qquad i \in \{1, \ldots, c\} \qquad k \in \{n+1, \ldots, n+m\}.$$

Finally we apply the given classifier to the m new objects and obtain

$$\hat{\mu}_{ik} = \text{membership of new object } k \text{ to class } i \qquad i \in \{1, \ldots, c\} \qquad k \in \{n+1, \ldots, n+m\}.$$

Based on this preliminary work we identify objects that represent changes of the given classifier. In this step we just want to know if it is adequate to create a new class or if class movement is sufficient. Therefore, we want to identify objects that are "not classified well by the existing classifier" and "far away from the current classes." The following two conditions are applied in order to detect such objects.

Condition 1: $|\hat{\mu}_{ik} - 1/c| \leq \alpha \quad \forall k \in \{n+1, \ldots, n+m\} \quad \forall i \in \{1, \ldots, c\}$.
Given a value $\alpha \geq 0$, Condition 1 determines those objects that have a membership value close to the inverse of the class number ($1/c$). A new object k that has all its membership values close to $1/c$ cannot be classified satisfactorily.

Parameter α can be determined applying one of the following strategies:

- It can be fixed context-dependently if the respective knowledge exists.
- If we know the correct class of some objects in a given cycle, we can determine α dynamically depending on the rate of correct classification of objects. For example, if the rate of correct classification is high in one cycle, α should be closer to zero in the following cycle in order to generate less changes of the classifier structure.

Condition 2: $\hat{d}_{ik} > \min\{d(v_i, v_j)\} \quad \forall k \in \{n+1, \ldots, n+m\} \quad \forall i \neq j \in \{1, \ldots, c\}$.

Condition 2 determines if a new object k is "far away from the current classes." We assume this to be the case, if its distance to each class i is larger than the minimal distance between two classes i and j.

Based on these two conditions we define:

$$1_{IC}(x_k) = \begin{cases} 1 & x_k \text{ fulfills Conditions 1 and 2,} \\ 0 & \text{else.} \end{cases}$$

In other words: $1_{IC}(x_k)$ has value 1 if and only if object k cannot be classified well by the current classifier.

If $\sum_{k=n+1}^{n+m} 1_{IC}(\mathbf{x}_k) = 0$ we proceed with Step III(a), otherwise we go to Step II.

Step II: Determine changes of class structure. Given that at least one object asks for a change of the classifier structure we now want to check if we need a new class or if moving the existing classes is sufficient. To do so, we apply the following criterion:

$$\frac{\sum_{k=n+1}^{n+m} 1_{IC}(\mathbf{x}_k)}{m} \geq \beta, \quad \text{with a parameter } \beta \quad 0 \leq \beta \leq 1.$$

If the relation between new objects that represent changes and the total number of new objects (m) is above a predefined threshold β we create new classes (III(b)). Otherwise we just move the existing classes (III(a)).

Parameter β can be determined applying one of the following strategies:

- It can be fixed context-dependently if the respective knowledge exists.
- If the correct class of the objects in a given cycle is known, β can be determined as the rate of correct classification of objects. This way we have an adaptive parameter setting for β in the following cycle, i.e., if many objects are classified correctly in one cycle it needs more new objects that represent a possible change in order to create a new class in the following cycle.

Step III: Change the class structure. Depending on the result of Step II we may want to create new classes or just move the existing ones.

Step III(a) Move classes. There are basically two options for class movement:

- Applying the underlying clustering algorithm (in our case fuzzy C-means) with previous and new objects (without changing the number of classes).
- Determining "class centers" representing only the new objects and combining them with the previous class centers.

The second option of moving classes combines the centers of the existing classes with centers representing the new objects belonging to the same class, respectively. For this reason we define the indicator function of a new object k for class i:

$$1_{C_i}(\mathbf{x}_k) = \begin{cases} 1 & \text{object } k \text{ is assigned to class } i, \\ 0 & \text{else.} \end{cases}$$

We assign an object to a class if it has its highest membership value in this class. For each class i we determine the class centers representing only the new objects of this class by:

$$v_i^* = \frac{\sum_{k=n+1}^{n+m} (1 - 1_{IC}(\mathbf{x}_k))(\hat{\mu}_{ik})^m \mathbf{x}_k}{\sum_{k=n+1}^{n+m} (1 - 1_{IC}(\mathbf{x}_k))(\hat{\mu}_{ik})^m} \quad 1 \leq i \leq c.$$

Combining these "centers of the new objects" with the previous centers we determine the new class centers by:

$$\hat{v}_i = (1 - \lambda_i)v_i + \lambda_i v_i^*,$$

where the weight λ_i indicates the proportion of new objects assigned to class i:

$$\lambda_i = \frac{\sum_{k=n+1}^{n+m} [1_{C_i}(\mathbf{x}_k) \cdot (1 - 1_{IC}(\mathbf{x}_k)) \cdot \hat{\mu}_{ik}]}{\sum_{j=1}^{n} (1_{C_i}(\mathbf{x}_j) \cdot \mu_{ij}) + \sum_{k=n+1}^{n+m} [1_{C_i}(\mathbf{x}_k) \cdot (1 - 1_{IC}(\mathbf{x}_k)) \cdot \hat{\mu}_{ik}]}.$$

Step III(b) Create classes. If Step II tells us that we have to create one or more new classes since many new objects cannot be assigned properly to the existing classes, we first have to determine an adequate new number of classes (e.g., c^{new}). For this purpose we apply the concept presented by Li and Mukaidono (1995), which is called *structure strength* and is based on the idea that "the knowledge of a part allows us to guess easily the rest of the whole." A loss function $L(c)$ is defined as within-group sum-of-squared error (WGSS) for a given cluster solution:

$$L(c) = \sum_{k=1}^{N} \sum_{i=1}^{c} u_{ik} {}^* d_{ik}^2,$$

where c = number of classes, N = number of objects, u_{ik} = degree of membership of object k to class i, and d_{ik} = distance between object k and class center i.

Based on this loss function the number c of classes is determined as follows:

$$\begin{aligned}&S(c) = \text{structure strength} = \\ &\alpha^*(\text{effectiveness of classification}) + (1-\alpha)^* \ (\text{accuracy of classification}) = \\ &\alpha^* \log(N/c) + (1-\alpha)^* \log(L(1)/L(c)).\end{aligned}$$

Li and Mukaidono (1995) suggest that measuring the effectiveness by $\log(N/c)$, i.e., a classification with less classes, is more effective. They propose measuring the accuracy by the term $\log(L(1)/L(c))$, i.e. a classification with more classes is more accurate. $L(1)$ is the variance of the entire data set and α is the weight between effectiveness and accuracy. The authors suggest $\alpha = 0.5$ in the case of an unbiased estimation, i.e., effectiveness and accuracy should have the same weight. The value c, which maximizes $S(c)$, is supposed to be an adequate class number. Using an adequate new number of classes (e.g., c^{new}) for all objects we continue with our basic clustering algorithm (here fuzzy C-means) in order to determine the best c^{new} classes representing all objects.

Step IV: Identify trajectories. Having performed necessary movements and/or creations of classes we now have to check if there are classes that should be eliminated. As preparation for the elimination step (Step V), we identify the development of each class during previous cycles based on its trajectory and a counter c_i^t for class i in cycle t. Here we have the following two cases:

- class i has been created in cycle $t-1$. In this case we set its counter $c_i^t = 1$.
- class i is the result of moving a certain class j in cycle $t-1$. In this case we set: $c_i^t = c_j^{t-1} + 1$.

Step V:Eliminate unchanged classes. The idea of *eliminating* a class can be stated in the following way: "A class has to be eliminated if it did not receive new objects during a *long* period." What "*long* period" means has to be defined.

In Step IV we identified for each class i in cycle t its counter (c_i^t), i.e., the number of cycles it has been active. We define a maximum number of cycles a class could stay active without receiving new objects (here T cycles). If a class does not receive new objects for T cycles, it will be eliminated. In the proposed methodology we use for all classes the same threshold value T that has to be specified by the user. In certain applications it could be interesting to use different threshold values for the existing classes.

It should be mentioned that eliminating a class does not mean this class is forgotten completely. When eliminating a class, the respective information will be kept in a separate memory. If in a later cycle in Step III(b) (create classes) a new class will be generated that is very similar to a previously eliminated one, we obtain additional information from the periodically entering data. This way we can detect "cycles" in our data structure if, for example, a class is eliminated and newly created with certain periodicity (e.g., seasonal customer behavior in a supermarket).

15.4 APPLICATIONS

Below we present applications of the methods described in Section 15.3. First we will show the potential of dynamic fuzzy clustering for strategic planning where functional fuzzy C-means (FFCM) will be used to cluster scenarios as one step of scenario planning. Then we explain our methodology for class updating using an illustrative example with simulated data. Finally, we show how dynamic fuzzy clustering can improve traffic management by dynamic identification of traffic states.

15.4.1 Scenario Planning

One goal of scenario planning is to investigate possible future developments (Schoemaker, 1995). In order to cover almost all possibilities it is desirable to analyze as many different scenarios as possible. On the other hand the complexity of the analysis grows as the number of scenarios increases that often limits their consideration. Here clustering of scenarios (trajectories) offers the advantage of complexity reduction while maintaining explicitly the information contained in these trajectories.

The process of scenario planning can be subdivided into the three main steps of analysis, forecasting, and synthesis. The analysis step consists of an adequate problem definition and studying the considered phenomenon. During the forecasting step many base scenarios are constructed, which are then aggregated to few final scenarios in the synthesis step. This aggregation step is where dynamic fuzzy clustering has been used providing final scenarios as the respective class centers. This way dynamic fuzzy clustering offers a powerful tool to support strategic planning based on scenario analyses.

The subsequent analysis of different scenarios regarding the price of crude oil until the end of 2020 will show the potential of the algorithm FFCM as presented in Section 15.3. Scenario analysis has already been used successfully for oil price prediction (see, for example, Austvik (1992)) but without applying clustering for scenario aggregation.

First, based on expert knowledge and simulation 151 different base scenarios have been generated (see, for example, Hofmeister *et al.* (2000) for more details). Each base scenario consists of trajectories of 16 features such as oil price, growth of economy, political stability, oil reserves, demand and supply, among others, over a period of 104 quarters. FFCM clustered these 151 base scenarios into four classes (aggregated scenarios) where each class center is represented by a set of 16 trajectories (features). Figure 15.4 displays the four most important features for these aggregated scenarios. The aggregated scenarios have been interpreted by an expert as will be shown next.

Scenario 1: Recession. In this scenario supply is higher than demand that leads first to lower oil prices resulting on one hand in a slightly lower solidarity between OPEC-countries and on the other hand in higher world-wide demand. Over-capacities on the supply side will lead to reduced production rates and consequently to higher oil prices increasing income and solidarity among OPEC-countries.

Scenario 2: Innovation. Higher economic growth in industrialized nations leads to steadily increasing oil prices providing even more incentives for rationalization investments in order to decrease oil-dependence. Political stability among OPEC-countries increases first due to higher income but goes

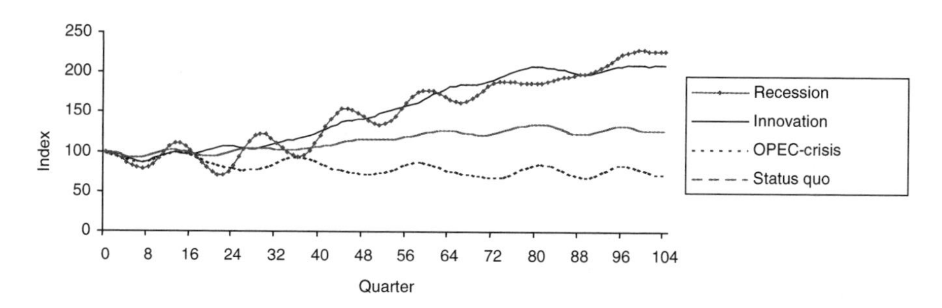

Economy growth

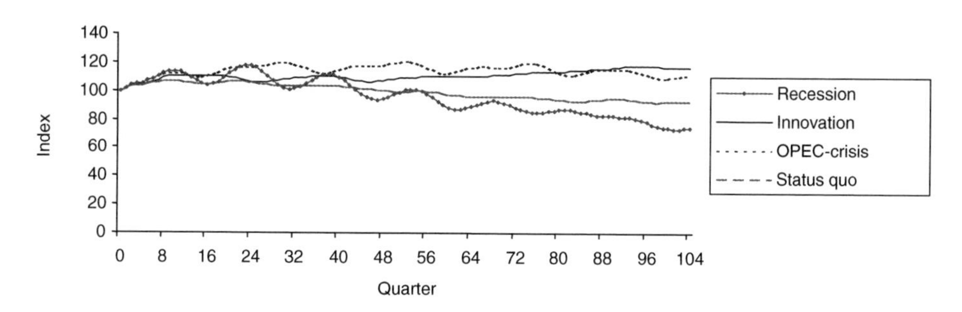

Political stability

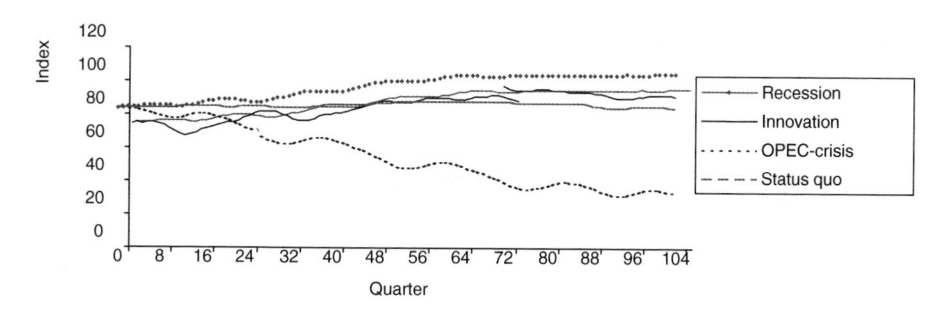

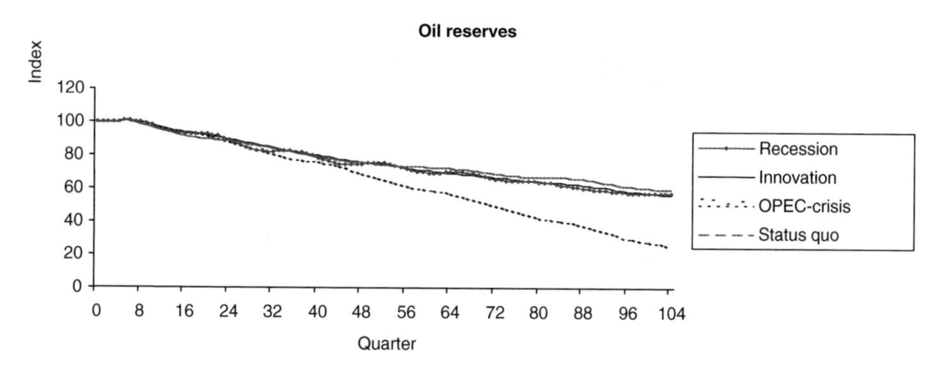

Figure 15.4 Development of the four most important features for the aggregated scenarios.

down at the end of the considered period due to stagnation on the demand side. Less cooperation finally leads to higher supply that further reduces oil prices.

Scenario 3: OPEC crisis. If solidarity among OPEC-countries decreases drastically – this scenario will be called OPEC-crisis – production will increase but demand stays basically unchanged, which results in lower prices and consequently less income for the associated countries. This will reduce solidarity among them even more.

Scenario 4: Status quo. In this scenario the oil price decreases slightly that offers incentives for innovations but takes pressure from investments aimed at rationalization. As a consequence energy dependence in general, and participation of oil in the energy mix in particular, remains almost constant. Increasing demand for crude oil can be satisfied since the oil-producing countries will increase their production in order to maintain their income. After some quarters the higher economic growth reduces oil reserves leading to slightly higher prices.

By identifying the most representative aggregated scenarios from a large set of base scenarios dynamic fuzzy clustering provides insight into the complex phenomena and supports this way strategic planning.

15.4.2 Application of Dynamic Fuzzy Clustering to Simulated Data

Here we apply our methodology for dynamic clustering to a simulated data set in order to illustrate each of its steps (movement of class centers, eliminating classes, creating classes).

15.4.2.1 Description of Data Set Used and Initial Solution

An initial solution has been determined using 500 artificially generated objects. Each object is described by two features whose values are normally distributed with the following mean values: (0, 15), (8, 35), (15, 0), and (15, 20), respectively. Figure 15.5 shows the initial data set.

We applied fuzzy C-means with $c = 4$ classes and $m = 2$. Table 15.2 presents the respective cluster solution.

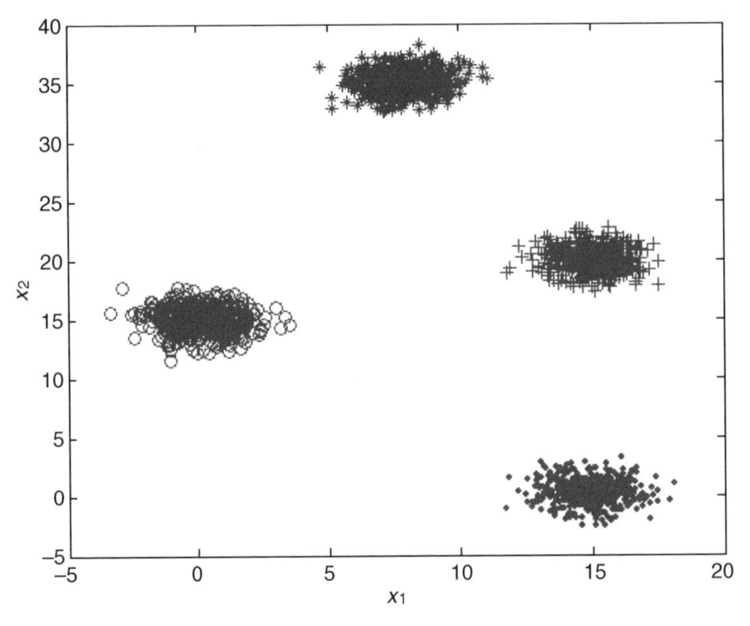

Figure 15.5 Initial data set of simulated data.

Table 15.2 Initial cluster solution of simulated data set.

Class	Variables X_1	X_2	Number of objects
1	14.98	0.23	500
2	0.19	15.03	500
3	8.01	35.01	500
4	15.05	20.08	500
			Total: 2000

15.4.2.2 Applying Dynamic Clustering to Static Objects

We chose the following parameters:

- Step I: in order to identify objects that represent changes we used $\alpha = 0.05$.
- Step II: with $\beta = 0.2$ we created a new class if the number of objects representing changes (result from Step I) is larger than 20 % of the total number of new objects.
- In Step V we eliminated a class if it did not receive new objects for $T = 2$ periods.

In the first cycle 600 new objects arrive as shown in Figure 15.6. Conditions 1 and 2 of Step I indicate that no new object represents changes of the classifier structure. We go immediately to Step III and move the centers of Classes 1, 3, and 4 that received 200 new objects each. Results are shown in Table 15.3. The last three columns of this table contain binary variables indicating if the respective change has been performed.

In the second cycle 500 new objects arrive. The entire set of objects is shown in Figure 15.7. Applying our methodology in Cycle 2 we obtain Table 15.4. Since 200 out of 500 new objects (40 %) represent

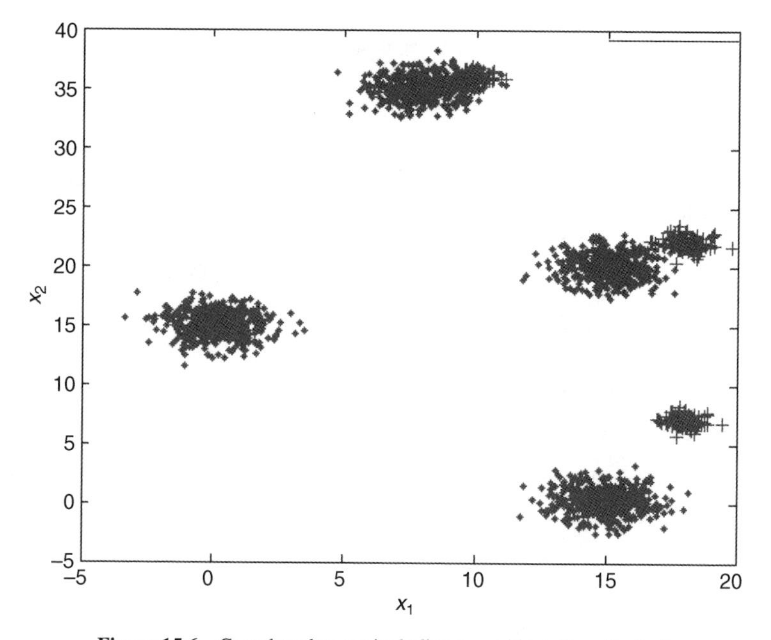

Figure 15.6 Complete data-set including new objects from Cycle 1.

Table 15.3 Result after first cycle.

Class	Variables X_1	X_2	Number of objects	Class moved	Class created	Class eliminated
1	15.55	1.65	700	1	0	0
2	0.19	15.03	500	0	0	0
3	8.70	35.54	700	1	0	0
4	15.72	20.36	700	1	0	0
			Total: 2600			

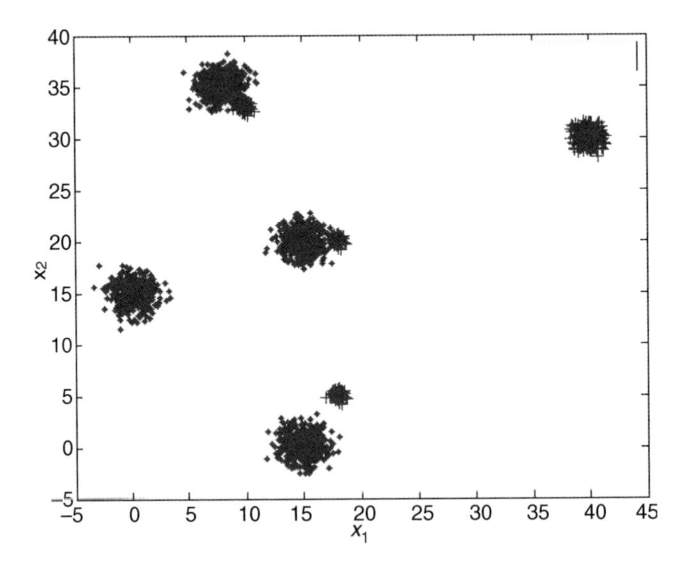

Figure 15.7 Complete data set including new objects from Cycle 2.

changes, we created a new class (Class 5). Additionally, centers of classes 1, 3, and 4 have been moved. In Step V we eliminated Class 2 in Cycle 2 because it did not receive new objects during $T = 2$ cycles.

In the third cycle 600 new objects arrive leading to the data set as shown in Figure 15.8 (objects belonging to the previously eliminated Class 2 are not shown). Applying our methodology in the third cycle we get the result as shown in Table 15.5. Analyzing explicitly the performed changes, for example,

Table 15.4 Result after second cycle.

Class	Variables X_1	X_2	Number of objects	Class moved	Class created	Class eliminated
1	15.83	2.03	800	1	0	0
2	—	—	0	0	0	1
3	8.86	35.24	800	1	0	0
4	16.05	20.39	800	1	0	0
5	39.98	29.96	200	0	1	0
:			Total: 2600			

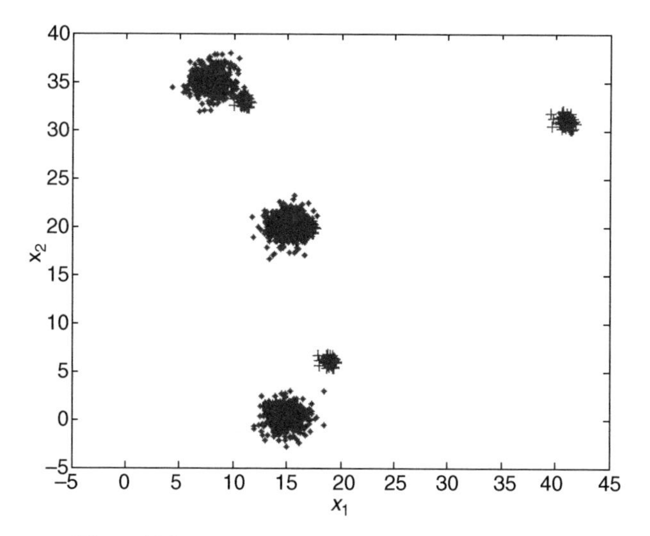

Figure 15.8 Data set including new objects from Cycle 3.

elimination of Class 2 and creation of Class 5, could provide further insights into the behavior of the observed system.

15.4.3 Dynamic Traffic State Identification

Traffic data has been collected between June 21, 1999 and October 30, 2000 on a German highway (Bastian, Kirschfink, and Weber, 2001). We analyzed the traffic behavior of the 66 Mondays in this period. According to experts in traffic management, Mondays typically show a different pattern compared with other days. Figure 15.10 shows the number of vehicles passing one particular sensor. The x-axis represents time (0 – 24 hours) and the y-axis represents number of vehicles per hour.

The raw data is collected in intervals of 5 minutes. Applying the Fourier transform to the original time series and representing it using the first 12 coefficients of the Fourier transform has led to the approximation shown as the smooth line in Figure 15.9.

In this application, we try to learn from the measured data how traffic behavior changed over time. The objects of the respective analysis are the time series representing each day's traffic flow described by a 12-dimensional feature vector containing the respective Fourier coefficients. From our data set we used the first 33 Mondays in order to obtain the initial solution with three traffic states. Then we updated this

Table 15.5 Cluster solution in Cycle 3.

	Variables					
Class	X_1	X_2	Number of objects	Class moved	Class created	Class eliminated
1	16.16	2.44	900	1	0	0
2	—	—	0	0	0	0
3	9.09	35.00	900	1	0	0
4	16.15	20.33	900	1	0	0
5	40.16	30.16	500	1	0	0
:			Total: 3200			

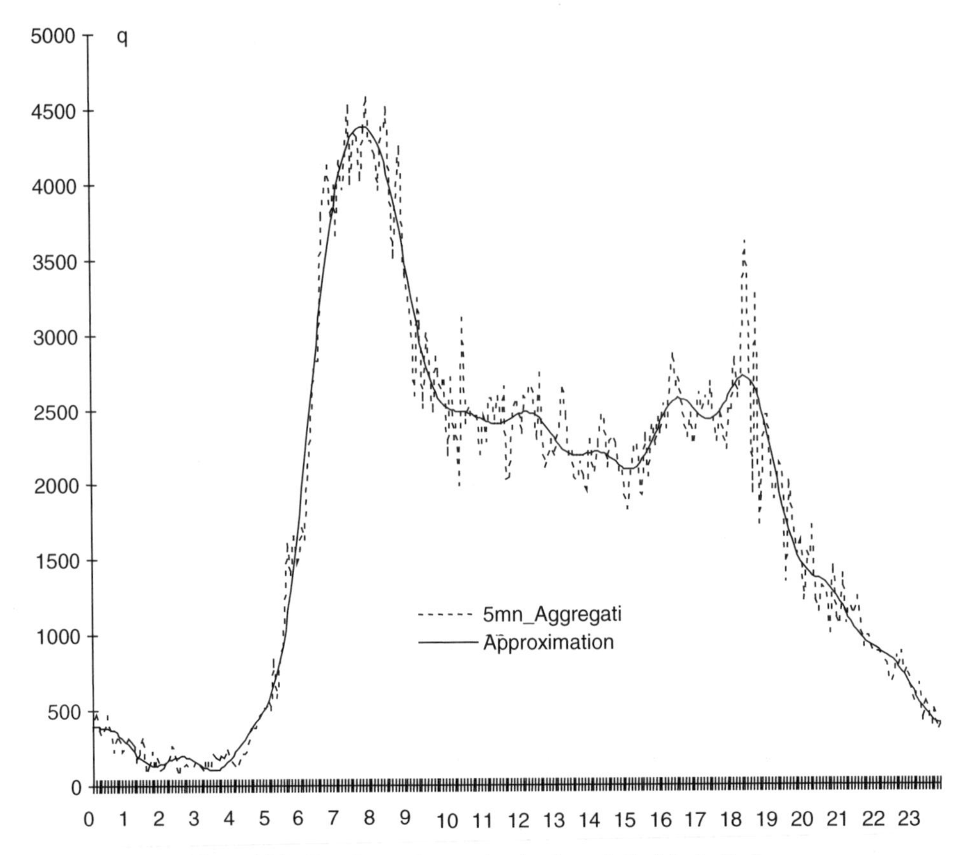

Figure 15.9 Raw data and approximation (smooth line) for traffic flow.

solution each month, by adding four new objects (four new Mondays) in each iteration leading to eight iterations.

Figure 15.10 shows (a) the approximations corresponding to the class centers of the initial structure and (b) after eight iterations. During the first seven iterations the number of classes did not change, class centers just moved slightly. As can be seen, one class has been eliminated in Iteration 8. Revising the

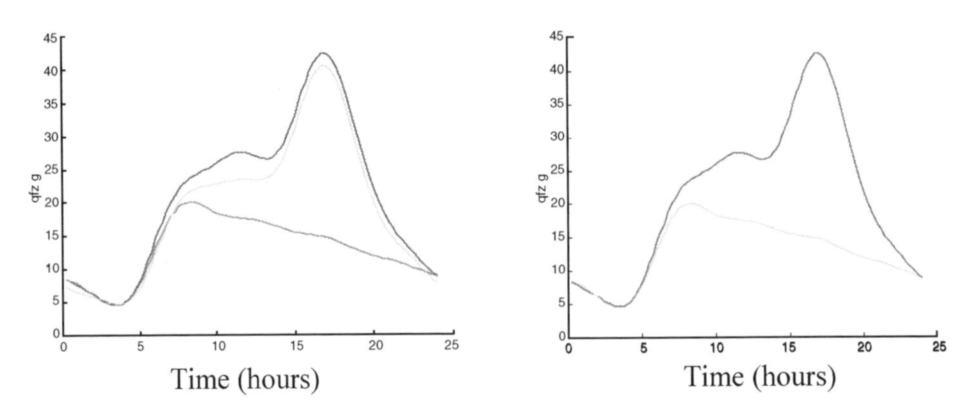

Figure 15.10 (a) Initial representation of class centers and (b) representation of class centers after eight iterations.

available objects and consulting the experts this change could be confirmed for the last period analyzed (October, 2000).

15.5 FUTURE PERSPECTIVES AND CONCLUSIONS

In this chapter, different systems for dynamic fuzzy clustering have been presented. There is, however, still a lot of work to be done in order to better understand the way "dynamic elements" could be modeled adequately in complex real-world applications. Next, we will give some hints on future research directions regarding dynamic fuzzy clustering.

The use of more sophisticated and context-dependent membership functions could improve the distance function proposed in Section 15.3.1 and used within the functional fuzzy C-means (FFCM). It would also be interesting to use the idea of alternative clustering algorithms instead of fuzzy C-means in order to cluster trajectories.

The methodology for class updating presented in Section 15.3.2 does not require identifiable objects. In applications where these objects can be identified (e.g., dynamic segmentation of bank customers), we have additional information at hand, which could be used to develop more powerful updating methodologies. Again, it would also be interesting to use alternative clustering algorithms instead of fuzzy C-means as the basic technique within the proposed methodology. First experiments using semi-supervised fuzzy C-means gave promising results since the combination of objects with known as well as unknown class membership makes special sense in dynamic settings.

So far we dealt with dynamic objects and/or dynamic classes. Another interesting issue of dynamic clustering that has not received proper attention yet is dynamic feature selection. The underlying idea is to understand the changing importance of features used for clustering. A similar updating strategy for dynamic feature selection for regression analysis has been presented, for example, in REG-UP (Guajardo, Weber, and Miranda 2006).

As has been shown in this chapter dynamic clustering offers huge potential for future research as well as improving existing data mining systems. Fuzzy logic provides special advantages when it comes to incorporating dynamic elements into clustering solutions since the respective membership values provide strong tools for treating time-dependent information, such as updating class structures. The methods presented in Section 15.3 and some of their applications as presented in Section 15.4 underline this potential. A lot of work still needs to be done in order to understand better how to develop the respective methods for dynamic clustering and their possible applications in real-world problems.

ACKNOWLEDGEMENT

This project was supported by Conicyt (FONDECYT Project number: 1040926) and the Nucleus Millennium Science on Complex Engineering Systems (www.sistemasdeingenieria.cl). Figure 15.2 is reprinted from (Joentgen, Mikenina, Weber, Zimmermann 1999) with permission from Elsevier. Figure 15.3, 15.9, and 15.10 are reprinted from (Crespo, Weber 2005) with permission from Elsevier.

REFERENCES

Abe, S. (1998) 'Dynamic Cluster Generation for a Fuzzy Classifier with Ellipsoidal Regions'. *IEEE Transactions on Systems, Man, and Cybernetics—Part B: Cybernetics* **28**, No. 6, 869–876

Angers, J.-F. (2002) Curves comparison using wavelet. Working Paper CRM-2833, Départment de mathématiques et de statistique; Université de Montréal; Montreal, Quebec, February 2002

Angstenberger, L. (2001) *Dynamic Fuzzy Pattern Recognition with Application to Finance and Engineering.* Springer-Verlag, Heidelberg, Germany.

Antunes, C. M., and Oliveira, A. L. (2001) 'Temporal data mining: an overview'. *Workshop on Temporal Data Mining, (KDD2001)*, San Francisco, September 2001, 1–13

Austvik, O. G. (1992) 'Limits to Oil Pricing; Scenario Planning as a Device to Understand Oil Price Developments'. *Energy Policy* **20**, 11, 1097–1105

Bastian, M., Kirschfink, H., and Weber, R. (2001) TRIP: 'Automatic TRaffic State Identification and Prediction as Basis for Improved Traffic Management Services'. *Proceedings of the Second Workshop on Information Technology*, Cooperative Research between Chile and Germany, 15–17 January 2001, Berlin, Germany.

Bezdek, J. C., Keller, J., Krishnapuram, R., and Pal, N. R. (1999) *Fuzzy Models and Algorithms for Pattern Recognition and Image Processing.* Kluwer, Boston, London, Dordrecht.

Black, M., and Hickey, R. J. (1999) 'Maintaining the performance of a learned classifier under concept drift'. *Intelligent Data Analysis* **3**, 6, 453–474

Diday, E. (1973) 'The dynamic cluster method in non-hierarchical clustering'. *International Journal of Parallel Programming* **2**, 1, 61–88

Famili, A., Shen, W.-M., Weber, R., and Simoudis, E. (1997) 'Data preprocessing and intelligent data analysis'. *Intelligent Data Analysis* **1**, 1, 3–23

Guajardo, J., Miranda, J., and Weber, R. (2005) 'A hybrid forecasting methodology using feature selection and support vector regression'. *Fifth International Conference on Hybrid Intelligent Systems HIS 2005*, 6–9 of November 2005, Rio de Janeiro, Brazil, 341–346

Han, J., and Kamber M. (2001) *Data Mining: Concepts and Techniques.* San Francisco: Morgan Kaufmann Publishers.

Hofmeister, P. et al. (2000) Komplexitätsreduktion in der Szenarioanalyse mit Hilfe dynamischer Fuzzy-Datenanalyse. *OR-Spektrum* **22**, 3, 403–420 (in German).

Jain, A.K., Murty, M.N., and Flynn, P.J. (1999) 'Data clustering: a review'. *ACM Computing Surveys*, **31**, 3, 264–323

Joentgen, A., Mikenina, L., Weber, R., and Zimmermann, H.-J. (1999) 'Dynamic fuzzy data analysis based on similarity between functions'. *Fuzzy Sets and Systems* **105**, 1, 81–90.

Karypis, G., Han, E.-H., and Kumar, V. (1999) Chameleon: 'A hierarchical Clustering Algorithm Using Dynamic Modeling'. *IEEE Computer, Special Issue on Data Analysis and Mining* **32**, 8, 68–75

Kasabov, N.K., and Song, Q. (2002) DENFIS: 'Dynamic evolving neural-fuzzy inference system and its application for time-series prediction'. *IEEE Transactions on Fuzzy Systems* **10**, 2, 144–154

Keogh, E. (2002) 'Exact indexing of dynamic time warping'. *Proc. of 28th International Conference on Very Large Data Bases*, Hong Kong, China, 406–417

Klawonn, F., and Georgieva, O. (2006) 'Identifying single clusters in large data sets'. In (J. Wang, ed.) *Encyclopedia of Data Warehousing and Mining.* Idea Group, Hershey, 582–585

Krishnapuram, R., and Kim, J. (1999) 'A note on the Gustafson–Kessel and adaptive fuzzy clustering algorithms'. *IEEE Transactions on Fuzzy Systems* **7**, 4, 453–461

Lee, C.-Y., Ma, L., and Antonsson, E. K. (2001) 'Evolutionary and adaptive synthesis methods'. In *Formal Engineering Design Synthesis* (E. K. Antonsson and J. Cagan, eds), Cambridge University Press, Cambridge, U.K., 270–320.

Li, C., Biswas, G., Dale, M., and Dale, P. (2002) 'Matryoshka: a HMM based temporal data clustering methodology for modeling system dynamics'. *Intelligent Data Analysis*, **6**, 3, 281–308

Li, R.-P., and Mukaidono, M. (1995) 'A maximun-entropy approach to fuzzy clustering', in *Proc. Int. Join Conf. 4th IEEE Int. Conf. Fuzzy/2nd Int. Fuzzy Eng. Symp.* (FUZZ/IEEE-IFES), Yokohama, Japan, March 1995, 2227–2232.

Raghavan V., and Hafez A. (2000) 'Dynamic data mining'. In (R. Loganantharaj, G. Palm and M. Ali, eds), *Intelligent Problem Solving – Methodologies and Approaches: Proc. of Thirteenth International Conference on Industrial Engineering Applications of AI & Expert Systems.* New York: Springer, 220–229.

Schoemaker, P.J.H. (1995) 'Scenario planning: a tool for strategic thinking', *Sloan Management Review*, Winter, 25–40

Utgoff, P. E. (1997) 'Decision tree induction based on efficient tree restructuring'. *Machine Learning* **29** (1) 5–44.

Weber, R. (1996) 'Customer segmentation for banks and insurance groups with fuzzy clustering techniques'. In (J. F. Baldwin, ed.) *Fuzzy Logic*, John Wiley and Sons, Ltd, Chichester, 187–196

Zimmermann, H.-J. (2001) *Fuzzy Set Theory - and Its Applications.* 4th ed. Kluwer Academic Publishers, Boston, Dordrecht, London.

16
Fuzzy Clustering of Parallel Data Streams

Jürgen Beringer and Eyke Hüllermeier

Fakultät für Informatik, Otto-von-Guericke-Universität, Magdeburg, Germany

16.1 INTRODUCTION

In recent years, so-called *data streams* have attracted considerable attention in different fields of computer science such as database systems, data mining, or distributed systems. As the notion suggests, a data stream can roughly be thought of as an ordered sequence of data items, where the input arrives more or less continuously as time progresses [16, 13, 6]. There are various applications in which streams of this type are produced such as network monitoring, telecommunication systems, customer click streams, stock markets, or any type of multi-sensor system.

A data stream system may constantly produce huge amounts of data. To illustrate, imagine a multi-sensor system with 10 000 sensors each of which sends a measurement every second of time. Regarding aspects of data storage, management, and processing, the continuous arrival of data items in multiple, rapid, time-varying, and potentially unbounded streams raises new challenges and research problems. Indeed, it is usually not feasible to simply store the arriving data in a traditional database management system in order to perform operations on that data later on. Rather, stream data must generally be processed in an online manner in order to guarantee that results are up-to-date and that queries can be answered with only a small time delay. The development of corresponding *stream processing systems* is a topic of active research [3].

In this chapter we consider the problem of clustering data streams. Clustering is one of the most important and frequently used data analysis techniques. It refers to the grouping of objects into homogeneous classes or groups and is commonly seen as a tool for discovering structure in data. In our context, the goal is to maintain classes of data streams such that streams within one class are similar to each other in a sense to be specified below. Roughly speaking, we assume a large number of evolving data streams to be given, and we are looking for groups of data streams that evolve similarly over time. Our focus is on time-series data streams, which means that individual data items are real numbers that can be thought of as a kind of measurement. There are numerous applications for this type of data analysis such as clustering of stock rates.

Advances in Fuzzy Clustering and its Applications Edited by J. Valente de Oliveira and W. Pedrycz

Apart from its practical relevance, this problem is also interesting from a methodological point of view. In particular, the aspect of efficiency plays an important role. First, data streams are complex, extremely high-dimensional objects making the computation of similarity measures costly. Secondly, clustering algorithms for data streams should be adaptive in the sense that up-to-date clusters are offered at any time, taking new data items into consideration as soon as they arrive. In this chapter we develop techniques for clustering data streams that meet these requirements. More specifically, we develop an efficient online version of the fuzzy C-means clustering algorithm. The efficiency of our approach is mainly due to a scalable online transformation of the original data which allows for a fast computation of approximate distances between streams.

The remainder of this chapter is organized as follows. Section 16.2 provides some background information, both on data streams and on clustering. The maintenance and adequate preprocessing of data streams is addressed in Section 16.3. Section 16.4 covers the clustering of data streams and introduces an online version of the fuzzy C-means algorithm. Section 16.5 discusses quality and distance measures for (fuzzy) cluster models appropriate for the streaming setting. Finally, experimental results are presented in Section 16.6.

16.2 BACKGROUND

16.2.1 The Data Stream Model

The *data stream model* assumes that input data are not available for random access from disk or memory, such as relations in standard relational databases, but rather arrive in the form of one or more continuous data streams. The stream model differs from the standard relational model in the following ways [1]:

- The elements of a stream arrive incrementally in an "online" manner. That is, the stream is "active" in the sense that the incoming items trigger operations on the data rather than being sent on request.
- The order in which elements of a stream arrive are not under the control of the system.
- Data streams are potentially of unbounded size.
- Data stream elements that have been processed are either discarded or archived. They cannot be easily retrieved unless they are stored in memory, which is typically small relative to the size of the stream. (Stored/condensed information about past data is often referred to as a *synopsis*, see Figure 16.1.)
- Due to limited resources (memory) and strict time constraints, the computation of exact results will usually not be possible. Therefore, the processing of stream data commonly produces *approximate* results [4].

16.2.2 Clustering

Clustering refers to the process of grouping a collection of objects into classes or "clusters" such that objects within the same class are *similar* in a certain sense, and objects from different classes are

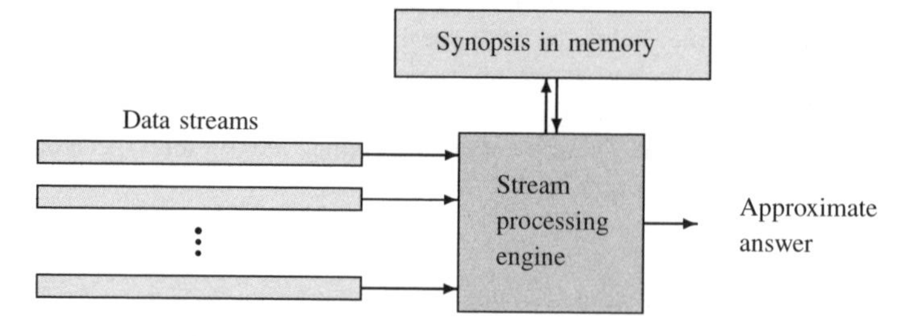

Figure 16.1 Basic structure of a data stream model.

dissimilar. Clustering algorithms proceed from given information about the similarity between objects, for example in the form of a *proximity matrix*. Usually, objects are described in terms of a set of measurements from which similarity degrees between pairs of objects are derived, using a kind of similarity or distance measure.

One of the most popular clustering methods is the so-called K-means algorithm [19]. This algorithm starts by guessing K cluster centers and then iterates the following steps until convergence is achieved:

- clusters are built by assigning each element to the closest cluster center;
- each cluster center is replaced by the mean of the elements belonging to that cluster.

K-means usually assumes that objects are described in terms of quantitative attributes, i.e., that an object is a vector $x \in \Re^n$. Dissimilarity between objects is defined by the Euclidean distance, and the above procedure actually implements an iterative descent method that seeks to minimize the variance measure ("within cluster" point scatter)

$$\sum_{k=1}^{K} \sum_{x_i, x_j \in C_k} ||x_i - x_j||^2, \tag{16.1}$$

where C_k is the kth cluster. In each iteration, the criterion (16.1) is indeed improved, which means that convergence is assured. Still, it is not guaranteed that the global minimum will be found, i.e., the final result may represent a suboptimal local minimum of (16.1).

Fuzzy (C-means) clustering is a generalization of standard (K-means) clustering that has proved to be useful in many practical applications.[1] In standard clustering, each object is assigned to one cluster in an unequivocal way. In contrast to this, in fuzzy clustering an object x may belong to different clusters at the same time, and the degree to which it belongs to the ith cluster is expressed in terms of a *membership degree* $u_i(x)$. Consequently, the boundary of single clusters and the transition between different clusters are usually "smooth" rather than abrupt.

The fuzzy variant of K-means clustering seeks to minimize the following objective function [2]:

$$\sum_{i=1}^{n} \sum_{j=1}^{K} ||x_i - c_j||^2 \, (u_{ij})^m, \tag{16.2}$$

where $u_{ij} = u_j(x_i)$ is the membership of the ith object x_i in the jth cluster, and c_i is the jth center. In the commonly employed *probabilistic* version of fuzzy C-means (FCM), it is assumed that

$$\sum_{j=1}^{K} u_{ij} = \sum_{j=1}^{K} u_j(x_i) = 1 \tag{16.3}$$

for all x_i [20]. The constant $m > 1$ in (16.2) is called the *fuzzifier* and controls the overlap ("smoothness") of the clusters (a common choice is $m = 2$).

Minimizing (16.2) subject to (16.3) defines a constrained optimization problem.[2] The clustering algorithm approximates an optimal (or at least locally optimal) solution by means of an iterative scheme that alternates between recomputing the optimal centers according to

$$c_j = \frac{\sum_{i=1}^{n} x_i \, (u_{ij})^m}{\sum_{i=1}^{n} (u_{ij})^m}$$

and membership degrees according to

$$u_{ij} = \left(\sum_{\ell=1}^{K} \left(\frac{||x_i - c_j||}{||x_i - c_\ell||} \right)^{2/(m-1)} \right)^{-1}.$$

[1] We adhere to the common practice of using the term fuzzy C-means (FCM) instead of fuzzy K-means. However, for reasons of coherence, we will go on denoting the number of clusters by K, also in the fuzzy case.

[2] As most clustering problems, the standard K-means problem is known to be NP-hard (see, for example, [11]).

16.2.3 Related Work

Stream data mining [10] is a topic of active research, and several adaptations of standard statistical and data analysis methods to data streams or related models have been developed recently (for example, [7, 28]). Likewise, several online data mining methods have been proposed (for example, [21, 8, 25, 5, 15, 12]). In particular, the problem of clustering in connection with data streams has been considered in [17, 9, 24]. In these works, however, the problem is to cluster the elements of one individual data stream, which is clearly different from our problem, where the objects to be clustered are the streams themselves rather than single data items in then. To the best of our knowledge, the problem in this form has not been addressed in the literature before.

There is a bunch of work on time series data mining in general and on clustering time series in particular [23]. Even though time series data mining is of course related to stream data mining, one should not overlook important differences between these fields. Particularly, time series are still static objects that can be analyzed offline, whereas the focus in the context of data streams is on dynamic adaptation and online data mining.

16.3 PREPROCESSING AND MAINTAINING DATA STREAMS

The first question in connection with the clustering of (active) data streams concerns the concept of distance or, alternatively, similarity between streams. What does similarity of two streams mean, and why should they, therefore, fall into one cluster?

Here, we are first of all interested in the qualitative, time-dependent evolution of a data stream. That is to say, two streams are considered similar if their evolution over time shows similar characteristics. As an example consider two stock rates both of which continuously increased between 9:00 a.m. and 10:30 a.m. but then started to decrease until 11:30 a.m.

To capture this type of similarity, we shall simply derive the Euclidean distance between the normalization of two streams (a more precise definition follows below). This measure satisfies our demands since it is closely related to the (statistical) *correlation* between these streams. In fact, there is a simple linear relationship between the correlation of normalized time series (with mean 0 and variance 1) and their (squared) Euclidean distance. There are of course other reasonable measures of similarity for data streams or, more specifically, time series [18], but Euclidean distance has desirable properties and is commonly used in applications.

16.3.1 Data Streams and Sliding Windows

The above example (clustering of stock rates) already suggests that one will usually not be interested in the entire data streams, which are potentially of unbounded length. Instead, it is reasonable to assume that recent observations are more important than past data. Therefore, one often concentrates on a *time window*, that is a subsequence of a complete data stream. The most common type of window is a so-called *sliding window* that is of fixed length and comprises the w most recent observations (Figure 16.2). A more general approach to taking the relevancy of observations into account is that of *weighing*. Here, the idea is to associate a weight in the form of a real number to each observation such that more recent observations receive higher weights.

When considering data streams in a sliding window of length w, a stream (resp. the relevant part thereof) can formally be written as a w-dimensional vector $X = (x_0, x_1, \ldots, x_{w-1})$, where a single observation x_i is simply a real number. As shown in Figure 16.3, we further partition a window into m blocks (basic windows) of size v, which means that $w = m \cdot v$ (Table 16.1 provides a summary of notation):[3]

$$X = (\underbrace{x_0, x_1, \ldots, x_{v-1}}_{B_1} \mid \underbrace{x_v, x_{v+1}, \ldots, x_{2v-1}}_{B_2} \mid \ldots \mid \underbrace{x_{(m-1)v}, x_{(m-1)v+1}, \ldots, x_{w-1}}_{B_m}).$$

[3]Typical values as used in our experiments later on are $w = 2048$ and $v = 128$.

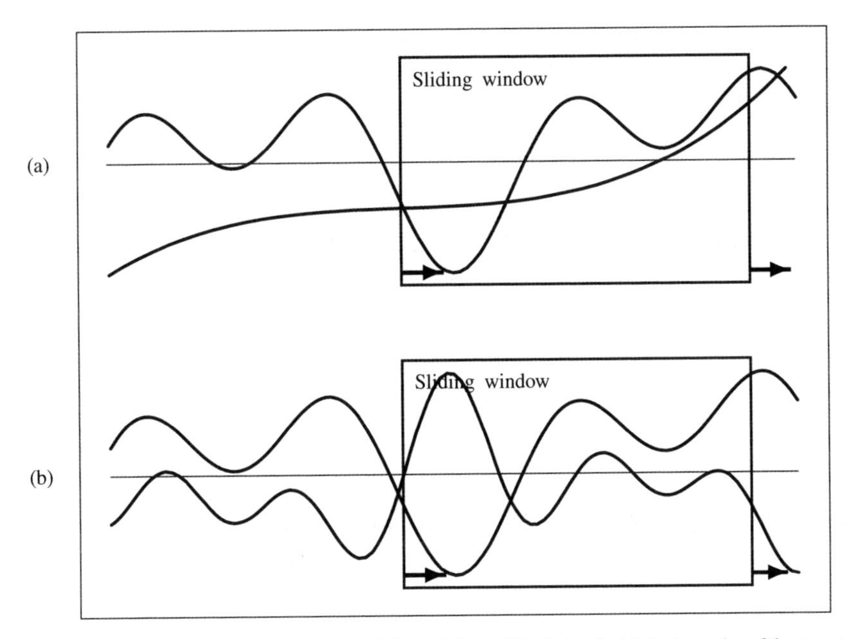

Figure 16.2 Data streams are compared within a sliding window of fixed size. In (a) the behavior of the two streams is obviously quite different. In (b) the two streams are similar to some extent.

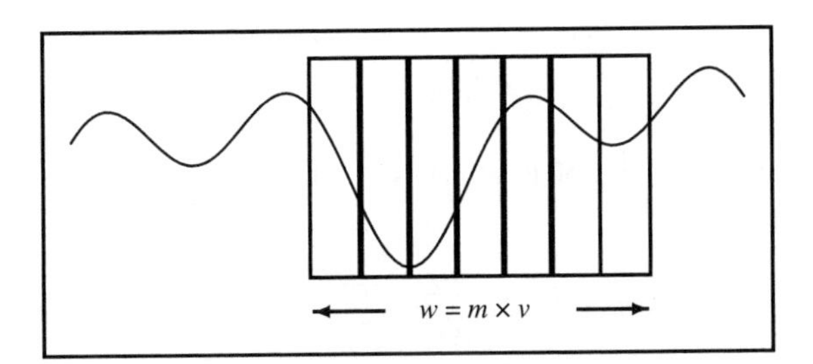

Figure 16.3 A window of length w is divided into m blocks of size v.

Table 16.1 Notation.

Symbol	Meaning
X	Data stream
X^n	Normalized data stream
x_i	Single observation
x_{ij}	$(j+1)$th element of block B_i
w	Window length
v	Length of a block
m	Number of blocks in a window
c	Weighing constant
V	Weight vector
$\bar{x}$	Mean value of a stream
s	Standard deviation of a stream

Data streams will then be updated in a "block-wise" manner each time v new items have been observed. This approach gains efficiency since the number of necessary updates is reduced by a factor of v. On the other hand, we tolerate the fact that the clustering structure is not always up-to-date. However, since the delay is at most one block size, this disadvantage is limited at least for small enough blocks. Apart from that, one should note that a small number of observations can change a stream but only slightly, hence the clustering structure in the "data stream space" will usually not change abruptly.

We assume data items to arrive synchronously, which means that all streams will be updated simultaneously. An update of the stream X, in this connection also referred to as X^{old}, is then accomplished by the following shift operation, in which B_{m+1} denotes the *entering* block:

$$\begin{array}{llll} X^{\text{old}} : & B_1 & |B_2|B_3|\ldots|B_{m-1}|B_m & \\ X^{\text{new}} : & & B_2|B_3|\ldots|B_{m-1}|B_m|B_{m+1} & \end{array}. \tag{16.4}$$

Finally, we allow for an exponential weighing of observations (within a window). The weight attached to observation x_i is defined by c^{w-i-1}, where $0 < c \leq 1$ is a constant. We denote by V the weight vector $(c^{w-1}, c^{w-2} \ldots c^0)$ and by $V \odot X$ the coordinate-wise product of V and a stream X:

$$V \odot X \stackrel{\text{df}}{=} (c^{w-1}x_0, c^{w-2}x_1 \ldots c^0 x_{w-1}).$$

16.3.2 Normalization

Since we are interested in the *relative* behavior of a data stream, the original streams have to be normalized in a first step. By normalization one usually means a linear transformation of the original data such that the transformed data has mean 0 and standard deviation 1. The corresponding transformation simply consists of subtracting the original mean and dividing the result by the standard deviation. Thus, we replace each value x_i of a stream X by its normalization

$$x_i^n \stackrel{\text{df}}{=} \frac{x_i - \bar{x}}{s}. \tag{16.5}$$

Considering in addition the weighing of data streams, $\bar{x}$ and s become the *weighted* average and standard deviation, respectively:

$$\begin{aligned} \bar{x} &= \frac{1-c}{1-c^w} \cdot \sum_{i=0}^{w-1} x_i \cdot c^{w-i-1}, \\ s^2 &= \frac{1-c}{1-c^w} \cdot \sum_{i=0}^{w-1} (x_i - \bar{x})^2 \cdot c^{w-i-1} \\ &= \frac{1-c}{1-c^w} \cdot \sum_{i=0}^{w-1} (x_i)^2 \cdot c^{w-i-1} - (\bar{x})^2. \end{aligned}$$

As suggested above, $\bar{x}$ and s^2 are updated in a block-wise manner. Let X be a stream and denote by x_{ij} the $(j+1)$th element of the ith block B_i. Particularly, the exiting block leaving the current window ("to the left") is given by the first block $B_1 = (x_{10}, x_{11}, \ldots, x_{1,v-1})$. Moreover, the new block entering the window ("from the right") is $B_{m+1} = (x_{m+1,0}, x_{m+1,1}, \ldots, x_{m+1,v-1})$. We maintain the following quantities for the stream X:

$$Q_1 \stackrel{\text{df}}{=} \sum_{i=0}^{w-1} x_i \cdot c^{w-i-1}, \quad Q_2 \stackrel{\text{df}}{=} \sum_{i=0}^{w-1} (x_i)^2 \cdot c^{w-i-1}.$$

Likewise, we maintain for each block B_k the variables

$$Q_1^k \stackrel{\text{df}}{=} \sum_{i=0}^{v-1} x_{ki} \cdot c^{v-i-1}, \quad Q_2^k \stackrel{\text{df}}{=} \sum_{i=0}^{v-1} (x_{ki})^2 \cdot c^{v-i-1}.$$

An update via shifting the current window by one block is then accomplished by setting

$$Q_1 \leftarrow Q_1 \cdot c^v - Q_1^1 \cdot c^w + Q_1^{m+1},$$
$$Q_2 \leftarrow Q_2 \cdot c^v - Q_1^2 \cdot c^w + Q_2^{m+1}.$$

16.3.3 Discrete Fourier Transform

Another preprocessing step replaces the original data by its discrete fourier transform (DFT). As will be explained in more detail in Section 16.5, this provides a suitable basis for an efficient approximation of the distance between data streams and, moreover, allows for the elimination of noise. The DFT of a sequence $X = (x_0, \ldots, x_{w-1})$ is defined by the DFT coefficients

$$\mathrm{DFT}_f(X) \stackrel{\mathrm{df}}{=} \frac{1}{\sqrt{w}} \sum_{j=0}^{w-1} x_j \cdot \exp\left(\frac{-i2\pi fj}{w}\right), \qquad f = 0, 1, \ldots, w-1,$$

where $i = \sqrt{-1}$ is the imaginary unit.

Denote by X^n the normalized stream X defined through values (16.5). Moreover, denote by V the weight vector $(c^{w-1}, \ldots, c^0)$ and recall that $V \odot X^n$ is the coordinate-wise product of V and X^n, i.e., the sequence of values

$$c^{w-i-1} \times \frac{x_i - \bar{x}}{s}.$$

Since the DFT is a linear transformation, which means that

$$\mathrm{DFT}(\alpha X + \beta Y) = \alpha \mathrm{DFT}(X) + \beta \mathrm{DFT}(Y)$$

for all $\alpha, \beta \geq 0$ and sequences X, Y, the DFT of $V \odot X^n$ is given by

$$\begin{aligned} \mathrm{DFT}(V \odot X^n) &= \mathrm{DFT}\left(V \odot \frac{X - \bar{x}}{s}\right) \\ &= \mathrm{DFT}\left(\frac{V \odot X - V \odot \bar{x}}{s}\right) \\ &= \frac{\mathrm{DFT}(V \odot X) - \mathrm{DFT}(V) \odot \bar{x}}{s}. \end{aligned}$$

Since $\mathrm{DFT}(V)$ can be computed in a preprocessing step, an incremental derivation is only necessary for $\mathrm{DFT}(V \odot X)$.

16.3.4 Computation of DFT Coefficients

Recall that the exiting and entering blocks are $B_1 = (x_{10}, x_{11}, \ldots, x_{1,v-1})$ and $B_{m+1} = (x_{m+1,0}, x_{m+1,1}, \ldots, x_{m+1,v-1})$, respectively. Denote by $X^{\mathrm{old}} = B_1|B_2|\ldots|B_m$ the current and by $X^{\mathrm{new}} = B_2|B_3|\ldots|B_{m+1}$ the new data stream. Without taking weights into account, the DFT coefficients are updated as follows [28]:

$$\mathrm{DFT}_f \leftarrow e^{\frac{i2\pi fv}{w}} \cdot \mathrm{DFT}_f + \frac{1}{\sqrt{w}} \left(\sum_{j=0}^{v-1} e^{\frac{i2\pi f(v-j)}{w}} x_{m+1,j} - \sum_{j=0}^{v-1} e^{\frac{i2\pi f(v-j)}{w}} x_{1,j} \right).$$

In connection with our weighing scheme, the weight of each element of the stream must be adapted as well. More specifically, the weight of each element x_{ij} of X^{old} is multiplied by c^v, and the weights of the new elements $x_{m+1,j}$, coming from block B_{m+1}, are given by c^{v-j-1}, $0 \leq j \leq v-1$. Noting that $\mathrm{DFT}_f(c^v \cdot X^{\mathrm{old}}) = c^v \mathrm{DFT}_f(X^{\mathrm{old}})$ due to the linearity of the DFT, the DFT coefficients are now modified as follows:

$$\mathrm{DFT}_f \leftarrow e^{\frac{i2\pi fv}{w}} \cdot c^v \mathrm{DFT}_f + \frac{1}{\sqrt{w}} \left(\sum_{j=0}^{v-1} e^{\frac{i2\pi f(v-j)}{w}} c^{v-j-1} x_{m+1,j} - \sum_{j=0}^{v-1} e^{\frac{i2\pi f(v-j)}{w}} c^{w+v-j-1} x_{1,j} \right).$$

Using the values

$$\beta_k^f \stackrel{\text{df}}{=} \sum_{j=0}^{v-1} e^{\frac{i2\pi f(v-j)}{w}} c^{v-j-1} x_{k,j}, \tag{16.6}$$

the above update rule simply becomes

$$\text{DFT}_f \leftarrow e^{\frac{i2\pi fv}{w}} \cdot c^v \text{DFT}_f + \frac{1}{\sqrt{w}}(\beta_{m+1}^f - c^w \beta_1^f).$$

As can be seen, the processing of one block basically comes down to maintaining the Q- and β-coefficients. The time complexity of the above update procedure is therefore $O(nvu)$. Moreover, the procedure needs space $O(nmu + nv)$: for each stream, the β-coefficients have to be stored for each block plus the complete last block (u is the number of DFT coefficients used for representing a stream, see Section 16.3.5 below).

16.3.5 Distance Approximation and Smoothing

We now turn to the problem of computing the Euclidean distance

$$||X - Y|| = \left(\sum_{i=0}^{w-1} (x_i - y_i)^2 \right)^{1/2}$$

between two streams X and Y (resp. the distance $||V \odot X^n - V \odot Y^n||$ between their normalized and weighted versions) in an efficient way. A useful property of the DFT is the fact that it preserves Euclidean distance, i.e.,

$$||X - Y|| = ||\text{DFT}(X) - \text{DFT}(Y)||. \tag{16.7}$$

Furthermore, the most important information is contained in the first DFT coefficients. In fact, using only these coefficients within the inverse transformation (which recovers the original signal X from its transform $\text{DFT}(X)$) comes down to implementing a low-pass filter and, hence, to using DFT as a smoothing technique (see Figure 16.4 for an illustration).

Therefore, a reasonable idea is to approximate the distance (16.7) by using only the first $u \ll w$ rather than all of the DFT coefficients and, hence, to store the values (16.7) only for $f = 0, \ldots, u-1$. More specifically, since the middle DFT coefficients are usually close to 0, the value

$$\left(2 \sum_{f=1}^{u-1} (\text{DFT}_f(X) - \text{DFT}_f(Y)) \overline{(\text{DFT}_f(X) - \text{DFT}_f(Y))} \right)^{1/2}$$

is a good approximation to (16.7). Here, we have used that $\text{DFT}_{w-f+1} = \overline{\text{DFT}_f}$, where $\overline{\text{DFT}_f}$ is the complex conjugate of DFT_f. Moreover, the first coefficient DFT_0 can be dropped, as for real-valued sequences the first DFT coefficient is given by the mean of that sequence, which vanishes in our case (recall that we normalize streams in a first step).

The above approximation has two advantages. First, by filtering noise we capture only those properties of a stream that are important for its characteristic time-dependent behavior. Second, the computation of the distance between two streams becomes much more efficient due to the related dimensionality reduction.

16.4 FUZZY CLUSTERING OF DATA STREAMS

The previous section has presented an efficient method for computing the (approximate) pairwise Euclidean distances between data streams in an incremental way. On the basis of these distances, it is

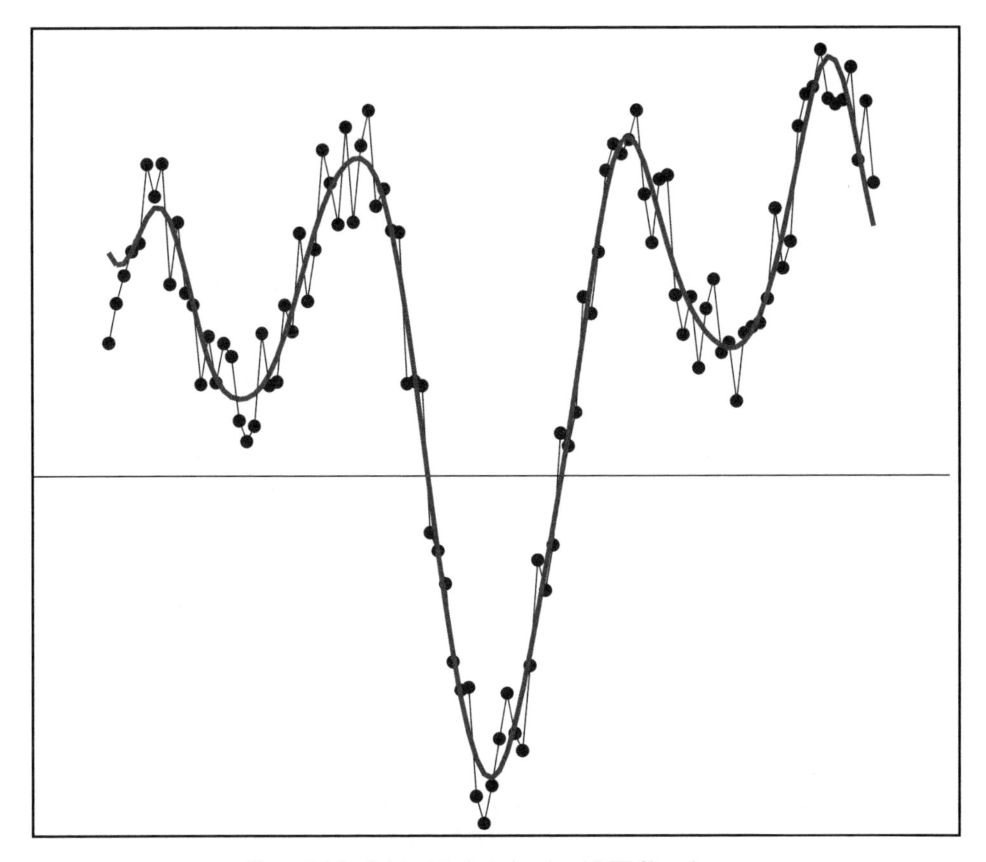

Figure 16.4 Original (noisy) signal and DFT-filtered curve.

possible in principle, to apply any clustering method. In this section, we propose an incremental version of the fuzzy C-means algorithm, subsequently referred to as FCM-DS (for fuzzy C-means on data streams).

For the following reasons, FCM appears to be especially suitable in our context. First, K-means is an iterative procedure that can be extended to the online setting in a natural way. Secondly, the *fuzzy* variant of this algorithm is strongly advised, since data streams are evolving objects. As a consequence, the clustering structure will change over time, and typically doing so in a "smooth" rather than an abrupt manner.

To illustrate the second point, suppose that the objects to be clustered are continuously moving points in a two-dimensional space.[4] Figure 16.5 shows snapshots of a dynamic clustering structure at different time points. At the beginning, there is only one big cluster. However, this cluster begins to divide itself into three small clusters, two of which are then again combined into a single cluster. Thus, there are time points where the structure definitely consists of one (first picture), two (sixth picture), and three (fourth picture) clusters. In-between, however, there are intermediate states, for which it is neither possible to determine the number of clusters nor to assign an object to one cluster in an unequivocal way.

Our incremental FCM method works as shown in Figure 16.6. The standard FCM algorithm is run on the current data streams. As soon as a new block is available for all streams, the current streams are updated by the shift operation (16.4). FCM is then simply continued. In other words, the cluster model of

[4]This example just serves as an illustration. Even though the objects might in principle be thought of as data streams over an extremely short window ($w = 2$), their movement is not typical of data streams.

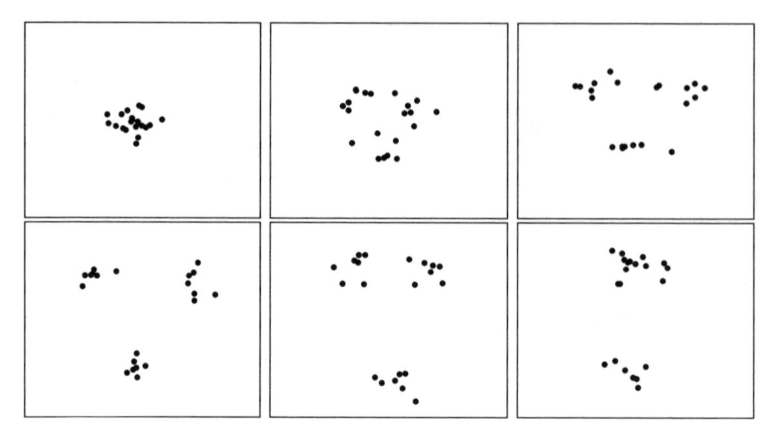

Figure 16.5 Snapshots of a dynamic clustering structure at different time points (upper left to lower right).

the current streams is taken as an initialization for the model of the new streams. This initialization will usually be good or even optimal since the new streams will differ from the current ones but only slightly.

An important additional feature that distinguishes FCM-DS from standard FCM is an incremental adaptation of the cluster number K. As already mentioned above, such an adaptation is very important in the context of our application where the clustering structure can change over time. Choosing the right number K is also a question of practical importance in standard K-means, and a number of (heuristic) strategies have been proposed. A common approach is to look at the cluster dissimilarity (the sum of distances between objects and their associated cluster centers) for a set of candidate values $K \in \{1, 2, \ldots, K_{\max}\}$. The cluster dissimilarity is obviously a decreasing function of K, and one expects that this function will show a kind of irregularity at K^*, the optimal number of clusters: The benefit of increasing the cluster number K will usually be large if $K < K^*$ but will be comparatively small later on. This intuition has been formalized, for example, by the recently proposed *gap statistic* [26].

Unfortunately, the above strategy is not practicable in our case as it requires the consideration of too large a number of candidate values. Instead, we pursue a local adaptation process that works as follows. In each iteration phase, one test is made in order to check whether the cluster model can be improved by increasing or decreasing K^*, the current (hopefully optimal) cluster number. By iteration phase we mean the phase between the entry of new blocks, i.e., while the streams to be clustered remain unchanged. Further, we restrict ourselves to adaptations of K^* by ± 1, which is again justified by the fact that the clustering structure will usually not change abruptly. In order to evaluate a cluster model, we make use of a quality measure (validity function) $Q(\cdot)$ that will be introduced in Section 16.5 below. Let $Q(K)$ denote this quality measure for the cluster number K, i.e., for the cluster model obtained for this number. The optimal cluster number is then updated as follows:

$$K^* \leftarrow \arg\max\{Q(K^* - 1), Q(K^*), Q(K^* + 1)\}.$$

```
1. Initialize K cluster centers at random
2. Repeat
3.      Assign membership degrees of each stream to the cluster centers
4.      Replace each center by the center of its associated fuzzy cluster
5.      If a new block is complete:
6.           Update the streams and pairwise distances
7.           Update the optimal cluster number K
```

Figure 16.6 Incremental version of the K-means algorithm for clustering data streams.

Intuitively, going from K^* to $K^* - 1$ means that one of the current clusters has disappeared, for example, because the streams in this cluster have become very similar to the streams in a neighboring cluster. Thus, $Q(K^* - 1)$ is derived as follows. One of the current candidate clusters is tentatively removed, which means that each of its elements is re-assigned to the closest cluster (center) among the remaining ones (note that different elements might be assigned to different clusters). The quality of the cluster model thus obtained is then computed. This is repeated K times, i.e., each of the current clusters is removed by way of trial. The best cluster model is then chosen, i.e., $Q(K^* - 1)$ is defined by the quality of the best model.

Going from K^* to $K^* + 1$ assumes that an additional cluster has emerged, for example, because a homogeneous cluster of streams has separated into two groups. To create this cluster we complement the existing K^* centers by one center that is defined by a randomly chosen object (stream). The probability of a stream being selected is reasonably defined as an increasing function of the stream's distance from its cluster center. In order to compute $Q(K^* + 1)$, we try out a fixed number of randomly chosen objects and select the one that gives the best cluster model.

16.5 QUALITY MEASURES

16.5.1 Fuzzy Validity Function

Regarding the evaluation of a cluster model in terms of a measure $Q(\cdot)$, several proposals can be found in literature. Unfortunately, most of these measures have been developed for the non-fuzzy case. Indeed, validity functions of that kind might still be (and in fact often are) employed, namely, by mapping a fuzzy cluster model onto a crisp one first (i.e., assigning each object to the cluster in which it has the highest degree of membership) and deriving the measure for this latter structure afterwards. However, this approach can of course be criticized as it comes with a considerable loss of information. On the other hand, many of the non-fuzzy measures can be adapted to the fuzzy case in a natural way.

Validity functions typically suggest finding a trade off between intra-cluster and inter-cluster variability (see Tables 16.2 and 16.3, respectively, for some examples), which is of course a reasonable principle. Besides, our application gives rise to a number of additional requirements:

(a) Since the number K of clusters is only changed locally by ± 1, i.e., in the style of hill-climbing, our adaptation procedure might get stuck in local optima. Consequently, the convexity (resp. concavity) of the validity function is highly desirable. That is, $Q(K)$ should be maximal (resp. minimal) for the

Table 16.2 Measures of intra-cluster variability. C_k denotes the kth cluster and c_k its center.

Standard variant	Fuzzy variant	Complexity
$\max_k \max_{x,y \in C_k} \|\|x - y\|\|$	$\max_k \max_{i,j} u_{ik}^m u_{jk}^m \|\|x_i - x_j\|\|^2$	$O(Kn^2)$
$\max_k \max_{x \in C_k} \|\|x - c_k\|\|$	$\max_k \max_i u_{ik}^m \|\|x_i - c_k\|\|^2$	$O(Kn)$
$\sum_k \max_{x,y \in C_k} \|\|x - y\|\|$	$\sum_k \max_{i,j} u_{ik}^m u_{jk}^m \|\|x_i - x_j\|\|^2$	$O(Kn^2)$

Table 16.3 Measures of intra-cluster variability. C_k denotes the kth cluster and c_k its center.

Standard (= fuzzy) variant	Complexity
$\min_{k,\ell} \min_{x \in C_k, y \in C_\ell} \|\|x - y\|\|^2$	$O(K^2n^2)$
$\sum_k \min_\ell \min_{x \in C_k, y \in C_\ell} \|\|x - y\|\|^2$	$O(K^2n^2)$
$\sum_k \min_\ell \|\|c_k - c_\ell\|\|^2$	$O(K^2)$
$\min_{k,\ell} \|\|c_k - c_\ell\|\|^2$	$O(K^2)$

optimal number K^* of clusters and decrease (resp. increase) in a monotonic way for smaller and larger values (at least within a certain range around K^*). Unfortunately, most existing measures do not have this property and instead show a rather irregular behavior.

(b) As already explained above, to adapt the cluster number K, we provisionally consider two alternative structures that we obtain, respectively, by removing and adding a cluster. Both candidate structures, however, are not fully optimized with regard to the objective function (16.2). In fact, this optimization only starts *after* the apparently optimal structure has been selected. In order to avoid this optimization to invalidate the previous selection, the validity measure $Q(\cdot)$ should harmonize well with the objective function (16.2).

(c) Finally, since the validity function is frequently evaluated in our application, its computation should be efficient. This disqualifies measures with a quadratic complexity such as, for example, the maximal distance between two objects within a cluster.

A widely used validity function is the so-called Xie–Beni index or separation [27], which is defined as

$$\frac{\frac{1}{n}\sum_{i=1}^{n}\sum_{k=1}^{K} u_{ik}^m ||x_i - c_k||^2}{\min_{k,\ell} ||c_k - c_\ell||^2}. \tag{16.8}$$

As most validity measures do, (16.8) puts the intra-cluster variability (numerator) in relation to the inter-cluster variability (denominator). In this case, the latter is simply determined by the minimal distance between two cluster centers. Obviously, the smaller the separation, the better the cluster model.

Since the nominator of (16.8) just corresponds to the objective function (16.2), the Xie–Beni index looks quite appealing with regard to point (b) above. Moreover, it is also efficient from a computational point of view. Still, point (a) remains problematic, mainly due to the minimum in the denominator.

To remedy this problem, we replace the minimum by a summation over all (pairwise) cluster dissimilarities, with smaller dissimilarities having a higher weight than larger ones. Simply defining the dissimilarity between two clusters by the distance between the corresponding centers is critical, however, since it neglects the variability (size) of these clusters. Therefore, we define the variability of a cluster in terms of the average (squared) distance from the center,

$$V_k \overset{\text{df}}{=} \frac{\sum_i u_{ik} ||x_i - c_k||^2}{\sum_i u_{ik}}$$

and the dissimilarity between two clusters as

$$D(C_k, C_\ell) \overset{\text{df}}{=} \frac{||c_k - c_\ell||^2}{V_k + V_\ell}.$$

These dissimilarities are aggregated by means of

$$\frac{1}{K(K-1)} \sum_{1 \le k < \ell \le K} \frac{1}{D(C_k, C_\ell)}, \tag{16.9}$$

thereby putting higher weight on smaller dissimilarities. Replacing the denominator in (16.8) by (16.9), we thus obtain

$$\sum_{1 \le k < \ell \le K} \frac{1}{D(C_k, C_\ell)} \cdot \sum_{i=1}^{n} \sum_{k=1}^{K} ||x_i - c_k||^2 \, u_{ik}^m. \tag{16.10}$$

It is of course not possible to prove the concavity of (16.10) in a formal way. Still, our practical experience so far has shown that it satisfies our requirements in this regard very well and compares favorably with alternative measures. Corresponding experimental results are omitted here due to reasons of space.

The only remaining problem concerns clusters that are unreasonably small. To avoid such clusters, we add a penalty of M/k for every cluster having less than three elements with membership of at least $1/2$ (M is a very high, implementation-dependent constant).

16.5.2 Similarity Between Cluster Models

A validity function $Q(\cdot)$ as introduced above measures the quality of a single cluster model. What we still need (in the experimental section below) is a measure of similarity (distance) for comparing two alternative structures (fuzzy partitions), say, $\mathcal{X} = \{C_1 \dots C_k\}$ and $\mathcal{Y} = \{C'_1 \dots C'_\ell\}$. In literature, such measures are known as *relative* evaluation measures.

Intuitively, a partition $\mathcal{X} = \{C_1 \dots C_k\}$ is similar to a partition $\mathcal{Y} = \{C'_1 \dots C'_\ell\}$ if, for each cluster in $\mathcal{X}$, there is a similar cluster in $\mathcal{Y}$ and, vice versa, for each cluster in $\mathcal{Y}$, there is a similar cluster in $\mathcal{X}$. Formalizing this idea, we can write

$$S(\mathcal{X},\mathcal{Y}) = s(\mathcal{X},\mathcal{Y}) \otimes s(\mathcal{Y},\mathcal{X}), \tag{16.11}$$

where $s(\mathcal{X},\mathcal{Y})$ denotes the similarity of $\mathcal{X}$ to $\mathcal{Y}$ (in the above sense) and vice versa $s(\mathcal{Y},\mathcal{X})$ the similarity of $\mathcal{Y}$ to $\mathcal{X}$:

$$s(\mathcal{X},\mathcal{Y}) = \bigotimes_{i=1\dots k} \bigoplus_{j=1\dots \ell} s(C_i, C'_j), \tag{16.12}$$

where $\otimes$ is a t-norm (modeling a logical conjunction), $\oplus$ a t-conorm (modeling a logical disjunction), and $s(C_i, C'_j)$ denotes the similarity between clusters C_i and C'_j. Regarding the latter, note that C_i and C'_j are both fuzzy subsets of the same domain (namely, all data streams), so that we can refer to standard measures for the similarity of fuzzy sets. One such standard measure is

$$s(C_i, C'_j) = \frac{|C_i \cap C'_j|}{|C_i \cup C'_j|} = \frac{\sum_q \min(u_{q,i}, u'_{q,j})}{\sum_q \max(u_{q,i}, u'_{q,j})},$$

where u_{qi} and u'_{qj} denote, respectively, the membership of the qth data stream in the clusters C_i and C'_j.

As one potential drawback of (16.12) let us mention that it gives the same influence to every cluster, regardless of its size. That is, the degree

$$s_i = \bigoplus_{j=1\dots \ell} s(C_i, C'_j) \tag{16.13}$$

to which there is a cluster in $\mathcal{Y}$ similar to C_i has the same influence for every i, regardless of the size of cluster C_i. Thus, one might think of weighting (16.13) by the relative size $w_i = |C_i|/n$, where n is the number of objects (data streams), that is, to replace (16.13) in (16.12) by $m(w_i, s_i)$. This comes down to using a *weighted* t-norm aggregation instead of a simple one [22]:

$$s(\mathcal{X},\mathcal{Y}) = \bigotimes_{i=1\dots k} m\left(w_i, \bigoplus_{j=1\dots \ell} s(C_i, C'_j)\right). \tag{16.14}$$

In the experimental section below, we shall employ two different versions of (16.14)}, with $\otimes$, $\oplus$, $m(\cdot)$ defined as follows:

$$a \otimes b = \min(a,b), \quad a \oplus b = \max(a,b), \quad m(w,s) = \max(1-w,s) \tag{16.15}$$

$$a \otimes b = ab, \quad a \oplus b = a + b - ab, \quad m(w,s) = s^w. \tag{16.16}$$

Again, we refrain from a more detailed discussion of the pros and cons of the above similarity measure. It should be noted, however, that there are of course other options, and that alternative measures can indeed be found in literature. Anyway, as a reasonable feature of (16.11), note that it is a normalized measure between 0 and 1, where the latter value is assumed for perfectly identical structures. This property is often violated for fuzzifications of standard (relative) evaluation measures such as, for example, those based on the comparison of coincidence matrices.

16.6 EXPERIMENTAL VALIDATION

A convincing experimental validation of FCM-DS as introduced above is difficult for several reasons. First, the evaluation of clustering methods is an intricate problem anyway, since an objectively "correct

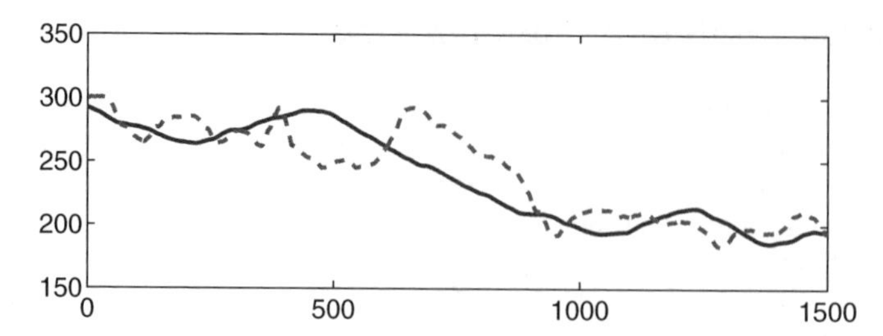

Figure 16.7 Example of a prototypical data stream (solid line) and a distorted version (dashed line).

solution" in the sense of a real clustering structure does not usually exist, at least not in the case of real-world data. Moreover, the performance of a clustering method strongly depends on the choice of the data-set (selective superiority problem). In fact, most methods give good results for a particular type of data but otherwise perform poorly. Secondly, since FCM-DS is the first method for clustering complete data streams, there are no alternative methods to compare with. Thirdly, real-world streaming data is currently not available in a form that is suitable for conducting systematic experiments. Therefore, we decided to carry out experiments with synthetic data. As an important advantage of synthetic data let us note that it allows for experiments to be conducted in a *controlled* way and, hence, to answer specific questions concerning the performance of a method and its behavior under particular conditions.

Synthetic data was generated in the following way. First, a *prototype* $p(\cdot)$ is generated for each cluster. This prototype is either predefined in terms of a specific (deterministic) function of time, or is generated as a stochastic process defined by means of a second-order difference equation:

$$\begin{aligned} p(t+\Delta t) &= p(t) + p'(t+\Delta t) \\ p'(t+\Delta t) &= p'(t) + u(t), \end{aligned} \tag{16.17}$$

$t = 0, \Delta t, 2\Delta t \ldots$. The $u(t)$ are independent random variables, uniformly distributed in an interval $[-a, a]$. Obviously, the smaller the constant a is, the smoother the stochastic process $p(\cdot)$ will be. The elements that (should) belong to the cluster are then generated by "distorting" the prototype, both horizontally (by stretching the time axis) and vertically (by adding noise). More precisely, a data stream $x(\cdot)$ is defined by

$$x(t) = p(t + h(t)) + g(t),$$

where $h(\cdot)$ and $g(\cdot)$ are stochastic processes that are generated in the same way as the prototype $p(\cdot)$.[5] Figure 16.7 shows a typical prototype together with a distortion $x(\cdot)$.

Of course, the above data generating process seems to be quite convenient for the K-means method. It should be stressed, therefore, that our experiments are not intended to investigate the performance of K-means itself (by now a thoroughly investigated algorithm with known advantages and disadvantages). Instead, our focus is more on the extensions that have been proposed in previous sections. More specifically, we are interested in the performance of our adaptation scheme for the cluster number K, the advantages of using a fuzzy instead of a crisp cluster method in dynamic environments, and the trade off between efficiency and quality in connection with the data preprocessing in FCM-DS.

16.6.1 First Experiment

In a first experiment, we investigated the ability of FCM-DS to adapt to a changing number of clusters. To this end, we varied the number of artificial clusters in the data generating process: starting with two

[5]However, the constant a that determines the smoothness of a process can be different for $p(\cdot), h(\cdot)$, and $g(\cdot)$.

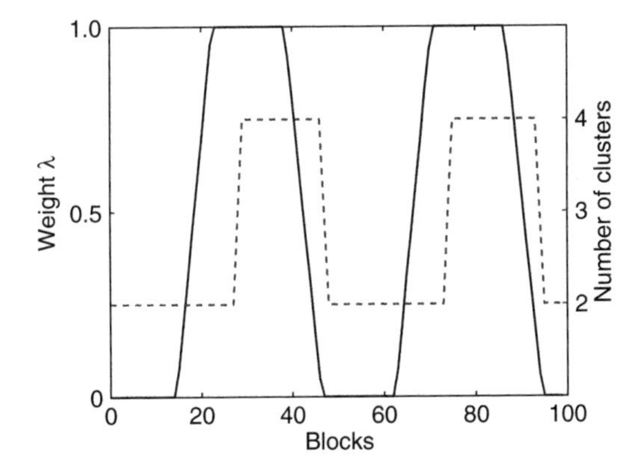

Figure 16.8 Weight of the parameter λ (solid line) and number of clusters (dashed line) in the first experiment.

clusters, the number of clusters was repeatedly doubled to four (in a "smooth" way) and later again reduced to two. Technically, this was accomplished as follows. The overall number of 100 data streams is divided into four groups. The first and second group are represented by the prototype $p_1(t) = \sin(t)$ and $p_2(t) = 1 - \sin(t)$, respectively. The third group is characterized by the prototype $p_3(t) = (1 - \lambda)p_1(t) + \lambda \sin(t + \pi/2)$, where $\lambda \in [0, 1]$ is a parameter. Likewise, the fourth group is characterized by the prototype $p_4(t) = (1 - \lambda)p_2(t) + \lambda(1 - \sin(t + \pi/2))$. As explained above, all streams were generated as distortions of their corresponding prototypes, using the values 0.04 and 0.5, respectively, as a smoothness parameter for the processes $h(\cdot)$ and $g(\cdot)$. The original streams were compressed using 100 DFT coefficients.

As can be seen, for $\lambda = 0$, the third (fourth) and the first (second) group form a single cluster, whereas the former moves away from the latter for larger values of λ, and finally constitutes a completely distinct cluster for $\lambda = 1$. The parameter λ is changed from 0 to 1 and back from 1 to 0 in a smooth way within a range of eight blocks (the block size is 512 data points).

Figure 16.8 shows the value of λ and the number of clusters generated by FCM-DS as a function of time, that is, for each block number. As can be seen, our approach correctly adapts the number of clusters, but of course with a small delay. We obtained qualitatively very similar results with other numbers of clusters and other data generating processes.

16.6.2 Second Experiment

The second experiment is quite similar to the first one. This time, we simulated a scenario in which some data streams move between two clusters. Again, these two clusters are represented, respectively, by the prototypes $p_1(t) = \sin(t)$ and $p_2(t) = 1 - \sin(t)$. Additionally, there are two streams that are generated as distortions of the convex combination $(1 - \lambda)p_1 + \lambda p_2$, where $\lambda \in [0, 1]$.

Figure 16.9 shows the value of λ and the (average) membership degree of the two streams in the second cluster. As can be seen, the membership degrees are again correctly adapted with a small delay of time.

We repeated the same experiment, this time using five instead of only two streams that move between the two clusters. The results, shown in Figure 16.10, are fairly similar, with one notable exception: since the number of moving streams is now higher (than three), FCM-DS creates an additional cluster in between. This cluster suddenly emerges when the streams are relatively far away from the first cluster and disappears when they come close enough to the second cluster. The degree of membership in the intermediate cluster, again averaged over the moving elements, is shown by the additional solid line in Figure 16.10.

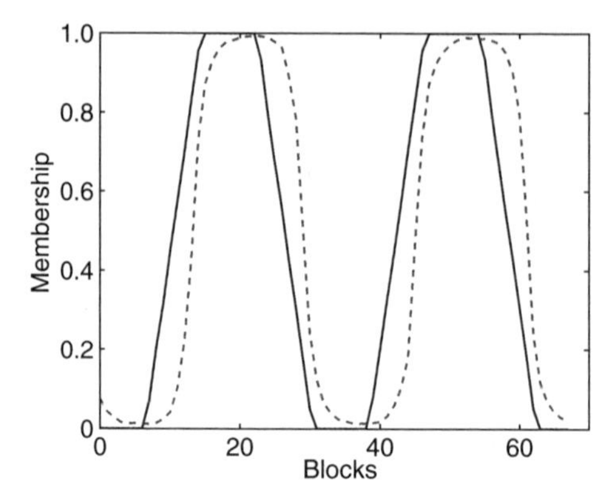

Figure 16.9 Weight of the parameter λ (solid line) and degree of membership of the moving streams in the second cluster (dashed line).

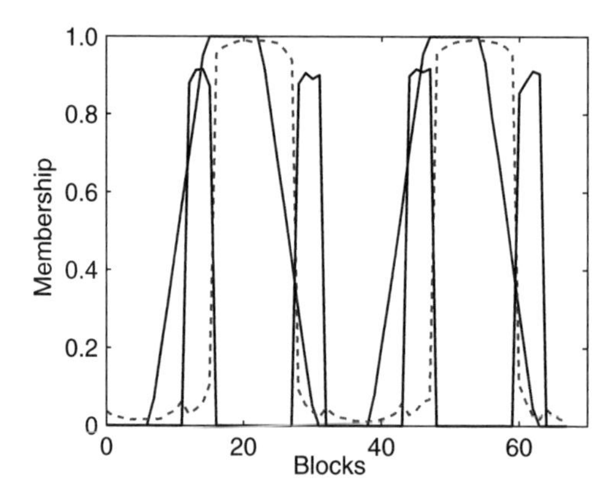

Figure 16.10 Weight of the parameter λ and degree of membership of the moving streams in the second cluster (dashed line) and the intermediate cluster (solid line).

16.6.3 Third Experiment

The purpose of the third experiment was to study the scalability of FCM-DS, that is, the trade off between efficiency and quality of data stream clustering. To this end, we have conducted experiments with 100 streams, generated as distortions of six prototypes (16.17). The parameter a was again set, respectively, to 0.04, 0.04, and 0.5 for the processes $p(\cdot)$, $h(\cdot)$, and $g(\cdot)$. The window size and block size were set to 2048 and 128, respectively.

As an efficiency parameter we have measured the time needed in order to process one block, that is, to preprocess the data and to update the cluster model. Figure 16.11 shows the mean processing time together with the standard deviation for different numbers of DFT coefficients. Moreover, we have plotted the average processing time and standard deviation for the original streams. As was to be expected, the time complexity increases as an approximately linear function of the number of DFT coefficients. The critical number of coefficients is around 600. That is, when using 600 or more DFT coefficients, the preprocessing of the data will no longer pay off.

As can also be seen in Figure 16.11, the processing time for one block is < 1 second. Thus, the system can process streams with an arrival rate of $\approx$ 150 elements per second. When approximating the original streams with 100 DFT coefficients, an arrival rate of $\approx$ 1000 per second can be handled. As an aside, we

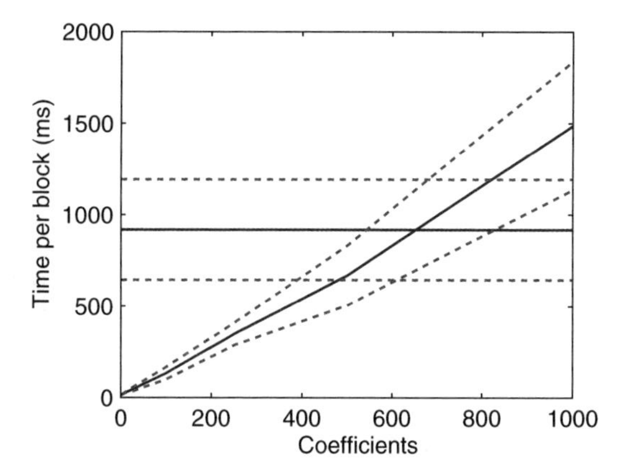

Figure 16.11 Mean processing time per block (solid) $\pm$ standard deviation (dashed) for different numbers of DFT coefficients (increasing curve), average processing time $\pm$ standard deviation for the original streams (horizontal lines).

note that the memory requirements are not critical in this application, as the number of data streams is still manageable.

Apart from efficiency aspects, we were interested in the quality of the results produced by FCM-DS. More specifically, we aimed at comparing the cluster models obtained by our online method, which preprocesses data streams first and clusters them in a low-dimensional space afterward, with the models that would have been obtained by clustering the *original* data streams. In fact, recall that in FCM-DS, distances between data streams are computed after having transformed the original streams from a w-dimensional space to a u-dimensional space, where $u \ll w$. In other words, a fuzzy partition is derived, not for the actual streams, but only for *approximations* thereof. On the one hand, this makes clustering in an online scenario more efficient. On the other hand, clustering in a space of lower dimension might of course come along with a quality loss, in the sense that the fuzzy partition in this space is different from the partition in the original space.

To compare the cluster models obtained, respectively, for the uncompressed and compressed data streams, we used the two versions (16.15) and (16.16) of the similarity measure (16.14). More specifically, data streams were clustered over a time horizon of 200 blocks, with and without DFT, and the mean similarity (16.14) was derived. This was repeated 10 times. Table 16.4 shows the averages over these 10 runs and the corresponding standard deviations. As can be seen, the results are quite satisfying in

Table 16.4 Average similarity (and standard deviation) for cluster models derived with and without transformation of data streams, for different numbers of DFT coefficients.

# coeff.	Similarity (16.15)	Similarity (16.16)
2	0.72867 (0.00341)	0.95853 (0.00680)
5	0.73267 (0.00432)	0.96988 (0.00293)
10	0.74191 (0.00726)	0.97900 (0.00387)
25	0.86423 (0.00988)	0.98814 (0.00388)
50	0.93077 (0.01116)	0.99384 (0.00366)
100	0.93721 (0.01454)	0.99273 (0.00482)
250	0.95006 (0.00811)	0.99509 (0.00341)
500	0.95266 (0.01421)	0.99532 (0.00330)
750	0.93768 (0.01651)	0.99367 (0.00378)
1000	0.94329 (0.01002)	0.99525 (0.00303)

Table 16.5 Average similarity (and standard deviation) for cluster models derived with and without transformation of data streams. In the latter case, the cluster number was assumed to be known.

# coeff.	Similarity (16.15)	Similarity (16.16)
2	0.78258 (0.00639)	0.99192 (0.00206)
5	0.79732 (0.00031)	0.99743 (0.00002)
10	0.80665 (0.00113)	0.99893 (0.00013)
25	0.83063 (0.00420)	0.99242 (0.00098)
50	0.81600 (0.00543)	0.98403 (0.00251)
100	0.81269 (0.00557)	0.98110 (0.00208)
250	0.81171 (0.00730)	0.98267 (0.00336)
500	0.80855 (0.00369)	0.98207 (0.00204)
750	0.81145 (0.00669)	0.98236 (0.00184)
1000	0.80992 (0.01129)	0.98323 (0.00353)

the sense that the similarity degrees are reasonably high. Moreover, the more DFT coefficients are used, the better the results become, even though good performance is already achieved for comparatively few coefficients.

We repeated the same experiment, this time using the correct number of $K = 6$ clusters for the reference model. That is, to cluster the original, uncompressed streams, we did not try to estimate (adapt) the best number of clusters and instead assumed this number to be known. The results, shown in Table 16.5, are rather similar to those from the first experiment, with an important exception: the quality is no longer a monotone increasing function of the number of DFT coefficients. Instead, the optimal number of coefficients is around 25, and using more exact approximations of the original streams deteriorates rather than improves the cluster models. This interesting phenomenon can be explained as follows. Using a small number of DFT coefficients comes along with a strong smoothing effect of the original streams, and this apparently helps to reveal the real structure in the cluster space. Stated differently, by filtering noise, the data streams are moved closer to their prototypes and, hence, the complete structure closer to the "ground truth." Seen from this point of view, data preprocessing does not only increase efficiency but also improves the quality of results.

16.7 CONCLUSIONS

In this chapter, we have addressed the problem of clustering data streams in an online manner. To this end, we have developed FCM-DS, an adaptable, scalable online version of the fuzzy C-means algorithm. Data streams can be perceived as moving objects in a very high-dimensional data space, the clustering structure of which is subject to continuous evolution. Therefore, a fuzzy approach appears particularly reasonable.

Another key aspect is that FCM-DS is an efficient preprocessing step that includes an incremental computation of the distance between data streams, using a DFT approximation of the original data. In this way, it becomes possible to cluster thousands of data streams in real time.

In order to investigate the performance and applicability of FCM-DS in a systematic way, we have performed experiments with synthetic data. The results of these experiments have shown that FCM-DS achieves an extreme gain in efficiency at the cost of an acceptable loss in quality. Depending on one's point of view and on the assumptions on the data generating process, it can even be argued that preprocessing can improve the quality by removing noise in the original streams.

Going beyond the relatively simple K-means approach by trying out other clustering methods is a topic of ongoing and future work. In this respect, one might think of extensions of K-means, such as Gath–Geva clustering [14], as well as alternative methods such as, for example, self-organizing maps. Likewise, other techniques might be tested in the preprocessing step of our framework, especially for the online approximation of data streams.

We conclude the chapter by noting that FCM-DS does not necessarily produce the end product of a data mining process. Actually, it can be considered as transforming a set of data streams into a new set of streams the elements of which are cluster memberships. These "cluster streams" can be analyzed by means of other data mining tools.

The Java implementation of FCM-DS is available for experimental purposes and can be downloaded, along with documentation, from the following address: wwwiti.cs.uni-magdeburg.de/iti_dke.

REFERENCES

[1] Babcock, B. *et al.*. 'Models and issues in data stream systems'. In *Proceedings of the Twenty-first ACM SIGACT-SIGMOD-SIGART Symposium on Principles of Database Systems, June 3-5, Madison, Wisconsin, USA*, pages 1–16. ACM, 2002.

[2] Bezdek, J.C. *Pattern Recognition with Fuzzy Objective Function Algorithm.* Plenum Press, New York, 1981.

[3] Cherniack, M. *et al.*. 'Scalable distributed stream processing'. In *Proceedings, C.I.DR–03: First Biennial Conference on Innovative Database Systems*, pages 257–268, Asilomar, CA, 2003.

[4] Considine, J., Li, F., Kollios, G. and Byers, J.W. 'Approximate aggregation techniques for sensor databases'. In *ICDE–04: 20th IEEE International Conference on Data Engineering*, 00:449, 2004.

[5] Cormode, G. and Muthukrishnan, S. 'What's hot and what's not: tracking most frequent items dynamically'. In *Proceedings of the 22nd ACM SIGMOD-SIGACT-SIGART Symposium on Principles of Database Systems*, pages 296–306. ACM Press, 2003.

[6] Das, A., Gehrke, J. and Riedewald, M. 'Approximate join processing over data streams'. In *Proceedings of the 2003 ACM SIGMOD International Conference on Management of Data*, pages 40–51. ACM Press, 2003.

[7] Datar, M., Gionis, A., Indyk, P. and Motwani, R. 'Maintaining stream statistics over sliding windows'. In *Proceedings of the Annual, A.C.M-SIAM Sympsium on Discrete Algorithms (SODA 2002)*, pages 635–644, 2002.

[8] Datar, M. and Muthukrishnan, S. 'Estimating rarity and similarity over data stream windows'. In *Algorithms - ESA 2002*, pages 323–334. Springer, 2002.

[9] Domingos, P. and Hulten, G. 'A general method for scaling up machine learning algorithms and its application to clustering'. In *Proceedings of the Eighteenth International Conference on Machine Learning*, pages 106–113, Williamstown, MA, 2001.

[10] Domingos, P. and Hulten, G. 'A general framework for mining massive data streams'. *Journal of Computational and Graphical Statistics*, **12**(4):945–949, 2003.

[11] Drineas, P. *et al.*. 'Clustering large graphs via the singular value decomposition'. *Machine Learning*, **56**:9–33, 2004.

[12] Gaber, M.M. Krishnaswamy, S. and Zaslavsky, A. 'Cost-efficient mining techniques for data streams'. In *Proceedings of the Second Workshop on Australasian Information Security, Data Mining and Web Intelligence, and Software Internationalisation*, pages 109–114. Australian Computer Society, Inc., 2004.

[13] Garofalakis, M., Gehrke, J. and Rastogi, R. 'Querying and mining data streams: you only get one look'. In *Proceedings of the 2002 ACM SIGMOD International Conference on Management of Data*, pages 635–635. ACM Press, 2002.

[14] Gath, I. and Geva, A. 'Unsupervised optimal fuzzy clustering'. *IEEE Transactions on Pattern Analysis and Machine Intelligence*, **11**(7):773–781, 1989.

[15] Giannella, C. *et al.*. 'Mining frequent patterns in data streams at multiple time granularities'. In, Kargupta, H., A. Joshi, Sivakumar, K., and Y. Yesha, editors, *Next Generation Data Mining*, pages 105–124. AAAI/MIT, 2003.

[16] Golab, L. and Tamer, M. 'Issues in data stream management'. *SIGMOD Rec.*, **32**(2):5–14, 2003.

[17] Guha, S., Mishra, N., Motwani, R. and O'Callaghan, L. 'Clustering data streams'. In *IEEE Symposium on Foundations of Computer Science*, pages 359–366, 2000.

[18] Gunopulos, D. and Das, G. 'Time series similarity measures and time series indexing'. In *Proceedings of the 2001 ACM SIGMOD International Conference on Management of Data*, page 624, Santa Barbara, California, 2001.

[19] Hartigan, J.A. *Clustering Algorithms.* John Wiley & Sons, Inc., New York, 1975.

[20] Höppner, F., Klawonn, F., Kruse, F. and Runkler, T. *Fuzzy Cluster Analysis.* John Wiley & Sons, Ltd, Chichester, 1999.

[21] Hulten, G., Spencer, L. and Domingos, P. 'Mining time-changing data streams'. In *Proceedings of the Seventh ACM SIGKDD International Conference on Knowledge Discovery and Data Mining*, pages 97–106. ACM Press, 2001.

[22] Kaymak, U. and van Nauta Lemke, H.R. 'A sensitivity analysis approach to introducing weight factors into decision functions in fuzzy multicriteria decision making'. *Fuzzy Sets and Systems*, **97**(2):169–182, 1998.

[23] Keogh, E. and Kasetty, S. 'On the need for time series data mining benchmarks: A survey and empirical demonstration'. In *Eighth, A.C.M. SIGKDD International Conference on Knowledge Discovery and Data Mining*, pages 102–111, Edmonton, Alberta, Canada, July 2002.

[24] O'Callaghan, L. *et al.*. 'Streaming-data algorithms for high-quality clustering', 00:0685, 2002.

[25] Papadimitriou, S., Faloutsos, C. and Brockwell, A. 'Adaptive, hands-off stream mining'. In *Twentyninth International Conference on Very Large Data Bases*, pages 560–571, 2003.

[26] Tibshirami, R., Walther, G. and Hastie, T. 'Estimating the number of clusters in a data set via the gap statistic'. *Journal of the Royal Statistical Society – Series B*, **63**(2):411–423, 2001.

[27] Xie, X.L. and Beni, G.A. 'Validity measure for fuzzy clustering'. *IEEE Transactions on Pattern Analysis and Machine Intelligence*, **3**(8):841–846, 1991.

[28] Zhu, Y. and Shasha, D. Statstream: 'Statistical monitoring of thousands of data streams in real time'. *Analyzing Data Streams - VLDB 2002*, pages 358–369, Springer, 2002.

17
Algorithms for Real-time Clustering and Generation of Rules from Data

Dimitar Filev[1] and Plamen Angelov[2]

[1]*Ford Motor Company, Dearborn, Michigan, USA*
[2]*Department of Communication Systems, UK*

17.1 INTRODUCTION

"We are drowning in information and starving for knowledge"

R.D. Roger

Data clustering by unsupervised learning has been around for quite some time (Kohonen, 1988). It has found important applications to:

- group unlabeled data;
- neural and fuzzy modeling and real-time modeling (Specht, 1991; Wang, 1994);
- novelty detection and fault isolation techniques (Neto and Nehmzow, 2004; Simani, Fantuzzi, and Patton, 2002);
- learning of the structure and parameters of fuzzy and neuro-fuzzy models (Yager and Filev, 1993; Chiu, 1994);
- feature selection for a successive classification (Kecman, 2001), etc..

Due to the fact that providing a classification label, reinforcement, or an error feedback is not a prerequisite for this technique, it is inherently very appropriate for online and real-time applications. This advantage can be exploited to design powerful algorithms for fuzzy (and neuro-fuzzy) model structure identification in real-time. When combining the computational simplicity of unsupervised clustering with the recursive online learning of models consequents' parameters very powerful schemes for real-time generation of interpretable rules from data can be developed as discussed later.

In this chapter, we present two main approaches to computationally efficient real-time clustering and generation of rules from data. The first one stems from the mountain/subtractive clustering approach

Advances in Fuzzy Clustering and its Applications Edited by J. Valente de Oliveira and W. Pedrycz

(Yager and Filev, 1993) while the second one is based on the k-nearest neighbors (k-*NN*) (Duda and Hart, 1973) and self-organizing maps (SOM) (Kohonen, 1988). They both address a common challenge that the advanced manufacturing, robotic, communication and defense systems face currently, namely the ability to have a *higher level of adaptation* to the environment and to the changing data patterns. Thus, the changes in the data pattern are reflected not just by the adaptation/adjustment of the parameters (as it is according to the conventional adaptive systems theory (Astrom and Wittenmark, 1997) applicable mostly for linear processes), but they are also reflected by a more substantial change of the model/system structure in the data space. In this way, significant changes of the object inner state (appearance or development of a fault, new physiological state in biological and living organisms (Konstantinov and Yoshida, 1989), different regime of operation (Murray-Smith and Johansen, 1997)), or of the environment (reaction to a sudden change or treat) can be appropriately reflected in a new model/system structure. At the same time the adaptation of the system structure is *gradual*. Two of the basic methods for real-time clustering the input/output data, which are presented in this chapter, play a crucial role in this process of *model structure evolution*.

The two methods that address this problem have different origins but they both have commonality in the procedure and the principal logic. The common features include:

- they are noniterative (no search is involved in the clustering) and thus they are one-pass (incremental);
- they have very low memory requirements, because recursive calculations are used thus they are computationally nonexpensive and applicable to real-time problems;
- the number of clusters/groups is *not* predefined (as it is in the fuzzy C-means known as FCM (Bezdek, 1974) etc. and thus they both are *truly* unsupervised according to the definition given in Gath and Geva (1989);
- they can start "*from scratch*" from the very first data sample assumed (temporarily) to be the first center of a cluster;
- changes of the cluster number and parameters (position of the center, zone of influence or spread) are *gradual*, not abrupt.

The main differences between these two methods are in the mechanism of the *evolution* (gradual change) of the cluster structure (which is the basis of the model structure if we consider a fuzzy rule-based or neuro-fuzzy model generated using these clusters). The first approach is based on the recursive calculation of the value called *potential*. Potential calculated at a data point is a function of accumulated proximity that represents the *density* of the data surrounding this data point. Therefore, this approach is called *density-based* real-time clustering. The second difference between the two approaches is that the first (density-based) one is also prototype-based. That means the data points/samples are used as prototypes of cluster centers. The second approach is *mean-based* – the centers of clusters are located at the *mean*, which does not coincide with any data point, in general.

Density-based clustering stems from the mountain clustering method (Yager and Filev, 1993) where the so-called *mountain function* was calculated at vertices of a grid; in the version of this approach called subtractive clustering (Chiu, 1994) the *potential* was calculated for each data sample. Both mountain and subtractive clustering approaches are available in the Matlab® toolbox Fuzzy Logic as *subclust* (Fuzzy Logic Toolbox, 2001). Neither of them requires the number of clusters to be prespecified, but they both are offline and multipass in the sense that the same data samples are processed many times. Density-based real-time clustering considered later in this chapter is an online, one-pass, noniterative extension of mountain/subtractive clustering and has been applied to classification, rule-base generation, prediction, and novelty detection (Angelov and Filev, 2004).

In this chapter, an alternative approach called *minimum-distance-based* clustering is considered that is also online, one-pass, noniterative, computationally efficient, starts "from scratch," is fully unsupervised, and is applying a gradual change to the cluster structure. There are a number of representatives of the *minimum-distance-based* clustering, all of which revolve around the idea of incrementally adding of new clusters whenever the new data does not fit into the existing cluster structure. The new cluster center is assumed to be the new data sample and the level of adequacy of the existing cluster structure is usually measured by the *minimum-distance* of the new data sample to the centers of already existing clusters using a threshold. Examples of this type of clustering are rule-based guided adaptation (Filev, Larsson, and Ma, 2000), evolving SOM (Kasabov, 2001), neural gas networks (Fritzke, 1995), recourse allocation networks (Plat, 1991), adaptive resonance theory neural networks, ART (Carpenter, Grossberg, and Reynolds, 1991), etc.. Minimum-distance-based clustering does not use the potential as a measure based on which new clusters are formed. Therefore, these approaches usually form a large number of clusters that one needs to 'prune' later (Huang, Saratchandran, and Sundarajan, 2005). Density-based clustering methods also use the minimum-distance between the new point and the cluster centers but this is just an auxiliary condition that helps the decision to form a new cluster or *to replace* an existing cluster center (Angelov and Filev, 2004).

Both approaches can be used as a first stage of a more thorough optimal clustering. In this case, they will determine in a computationally efficient way the number of candidate cluster centers. More interestingly, they can be used as a first stage generating interpretable linguistic fuzzy rules (or equivalently neuro-fuzzy systems) from raw data in real-time (Angelov and Filev, 2004; Filev and Tardiff, 2004). This has been done for the so-called Takagi–Sugeno (TS) type fuzzy rule-based models, controllers, classifiers, and for simplified Mamdani-type fuzzy models that can also be seen as zero-order TS models. Due to the proven similarity of TS models with radial-basis function (RBF) neural networks the results are equally applicable to neuro-fuzzy systems. The identification of the TS model (and thus of the neuro-fuzzy system) in real-time is, therefore, performed in two stages. Note that both stages take place during one time step (the time instant between reading the present and the next data sample) similarly to the adaptation and prediction in conventional adaptive systems (Astrom and Wittenmark, 1997). Usually consequents of the TS models are assumed to be linear, however, one can also use zero-order consequents (singletons) and thus generate a model that is close to the Mamdani-type fuzzy model (it is called in this work simplified Mamdani model, sM model). Such a model is more suitable for classification and some control applications because it is fully linguistic (Filev and Tardiff, 2005).

In this chapter, a comparative analysis of the two ways to address the problem of real-time clustering is made and the basic procedures for real-time rule generation from data using these two approaches are given. Examples of application of these approaches to real problems are also presented.

17.2 DENSITY-BASED REAL-TIME CLUSTERING

17.2.1 Potential and Scatter Definitions

The aim of the clustering is to group the data into clusters in such a way that commonality/closeness (in a generic sense) between members of the same cluster is higher than the dissimilarity/distance between members of different clusters (Duda and Hart, 1973). Yager and Filev (1993) have introduced the so-called mountain function as a measure of *spatial density* around vertices of a grid:

$$M_1(v_i) = \sum_{j=1}^{N} e^{-\alpha\|v_i - z_j\|} \tag{17.1}$$

where α is a positive constant; M_1 *is* the mountain function calculated at the ith vertex v_i during the pass one; N is the total number of data points/samples that is assumed to be available before the algorithm starts; $\|\bullet\|$ denotes the (Euclidean) distance between the points used as arguments; z_j is the current data sample/point.

Obviously, a vertex surrounded by many data points/samples will have a high value for this function and, conversely, a vertex with no neighboring data sample/point will have a low value for the mountain function. It should be noted that this is the function used for the first pass through the data only. For the subsequent passes through the data the function is defined by subtraction of the value proportional to the peak value of the mountain function (Yager and Filev, 1993). The mountain function is *monotonic, normal* $(0 < M \leq 1)$, and is *inversely proportional to the sum of the distances* between the vertex and all other data points. In a very similar way the subtractive clustering approach uses so-called *potential*, which is defined as:

$$P_1(z_i) = \sum_{j=1}^{N} e^{-\alpha\|z_i - z_j\|^2} \tag{17.2}$$

where $P_1(z_i)$ denotes the *potential* of the data point/sample (z_i) calculated at the first pass through the data.

In a similar way, a data point/sample with many points/samples in its neighborhood will have a high value of potential, while a remote data point/sample will have a low value of potential. By definition, the *potential* is also a *monotonic, normal* function with value $P = 1$ for the extreme case when all the data points/samples coincide $(z_i = z_j; \forall i; \forall j)$, i.e., when the distances between all data points are 0 $((\| z_i - z_j \|^2 = 0; \forall i; \forall j))$. Both mountain and subtractive clustering, however, were designed for offline (batch) data processing.

Real-time density-based clustering, called *e-clustering* (from evolvable), also uses the concept of *potential* as a density measure; thus vetting the data points/samples that are candidates to be a cluster center. However, because of the online nature of the data processing only the current data point/sample is available at a particular time instant. This requires a recursive way to calculate the potential. One possible such function is the approximation in Taylor series of the exponential function used in (17.1) and (17.2), which, in fact, is the Cauchy function (Angelov and Filev, 2004):

$$P_k(z_k) = \frac{1}{1 + \frac{1}{p(k-1)} \sum_{j=1}^{p} \sum_{i=1}^{k-1} (z_i - z_k)^2} \tag{17.3}$$

where $P_k(z_k)$ denotes the potential of the data point (z_k) calculated at time k starting from $k = 2$ and p is the dimensionality of the data space $(z \in R^p)$.

In the type of density-based clustering called *simpl_eTS* so-called *scatter* was used instead of *potential*. *Scatter* was defined as a weighted accumulated proximity measure (Angelov and Filev, 2005):

$$S_k(z_k) = \frac{1}{p(k-1)} \sum_{j=1}^{p} \sum_{i=1}^{k-1} (z_i - z_k)^2. \tag{17.4}$$

In fact, the scatter is proportional (instead of reciprocal) to the sum of squared distance projections. Therefore, the logic of generating clusters in Angelov and Filev (2005) was also inverted. The best candidate to be a cluster center is the point with the *lower* scatter (instead of the point with the *higher* potential). The range of possible values of the scatter measured at a certain point is obviously [0;1] with 0 meaning all of the data samples coincide (which is extremely improbable) and 1 meaning that all of the data points are on the vertices of the hypercube formed as a result of the normalization of the data.

17.2.2 Data Normalization

It should be mentioned that the input/output data is assumed to be normalized. Normalization can be done based on the range of the data or based on the mean and standard deviation (Hastie, Tibshirani, and Friedman, 2001). The ranges of the variables ($[\underline{x}_i, \overline{x}_i]$ for the inputs and $[\underline{y}_i, \overline{y}_i]$ for the outputs, where $\underline{x}_i = \min_{i=1}^{n}(x_i); \overline{x}_i = \max_{i=1}^{n}(x_i); \underline{y}_i = \min_{i=1}^{m}(y_i); \overline{y}_i = \max_{i=1}^{m}(y_i)$) can be suggested for a process, but in real-time they may change. Therefore, they need to be updated in real-time, which is straightforward. If normalization based on the mean, z_{kj}^M, and standard deviation, σ_{kj}^2, is used then mean and standard deviation will also need to be updated recursively. This can be done using the following relations (Angelov and Filev, 2005):

$$z_{kj}^M = \frac{(k-1)z_{(k-1)j}^M + z_{kj}}{k}; z_{1j}^M = 0 \tag{17.5}$$

$$\sigma_{kj}^2 = \frac{k-1}{k}\sigma_{(k-1)j}^2 + \frac{\left(z_{kj} - z_{kj}^M\right)^2}{k-1}; \sigma_{1j}^2 = 0. \tag{17.6}$$

Then the normalized input/output value is:

$$|z|_k = \frac{z_{kj} - z_{kj}}{\sigma_{kj}}. \tag{17.7}$$

In order to simplify the notations we assume in the next steps of the algorithm that the input/output vector z has already been normalized.

17.2.3 Procedure of the Density-based Clustering

The process of forming new clusters in real-time *density-based* clustering can start either from a set of predefined cluster centers based on available a priori knowledge or, which is more attractive, it can also start "*from scratch*" adopting the first available data point/sample as a center of a cluster ($z_1^* = z_1$). In cases when it starts from an initial set of cluster centers they are being further refined. The potential of the first cluster center when starting "*from scratch*" is set to the highest possible value, that is $P_1(z_1) = 1$ (respectively, the scatter if we use (17.4) is set to the lowest possible value, that is $S_1(z_1) = 0$). The procedure continues further by a loop that ends only when there are no new data to be read. This loop includes the following steps:

(1) calculate the potential, $P_k(z_k)$) (or respectively the scatter, $S_k(z_k)$, of the current data point/sample;
(2) update the potential (or respectively the scatter) of the previously existing cluster centers affected by adding the new data point/sample), $P_k(z^*)$ (or respectively, $S_k(z^*)$);
(3) compare $P_k(z_k)$ with $P_k(z^*)$or respectively $S_k(z_k)$ with $S_k(z^*)$ and take one of the following actions and *form a new* cluster center, z^{*R+1} around the new data point, z_k; if certain condition (1) (to be specified later) is satisfied;
(4) in the case when a new cluster is added, check if it *includes* any of the previously existing cluster centers (the precise formulation of includes will be specified later); if this is the case, remove/delete previously existing centers for which this is true.

In more detail, step (1) includes the *recursive* calculation of the potential (or respectively) the scatter of the new data point/sample. If expression (17.3) is used the potential can be calculated recursively in a similar way as it is detailed in Angelov and Filev (2004), which results in:

$$P_k(z_k) = \frac{p(k-1)}{p(k-1)(a_k+1) - 2c_k + b_k}, \tag{17.8}$$

where the following notations have been used:

$$b_k = b_{k-1} + a_{k-1}; b_1 = 0;\ a_k = \sum_{j=1}^{p} (z_k^j)^2; b_k = \sum_{i=1}^{k-1}\sum_{j=1}^{p} (z_i^j)^2 \tag{17.9}$$

$$c_k = \sum_{j=1}^{p} z_k^j f_k^j; f_k^j = \sum_{i=1}^{k-1} z_i^j\ ;\ f_k^j = f_{k-1}^j + z_{k-1}^j; f_1^j = 0. \tag{17.10}$$

To summarize, during step (1), the potential of the newly read data sample/point is calculated using (17.8) where quantities b_k and f_k^j are recursively accumulated sums computed from (17.9)–(17.10) starting from 0 values.

If, alternatively, we use scatter, the (17.4) recursive expression can be derived (Angelov and Filev, 2005) in a similar way.

In step (2) we calculate recursively the adjustment to the potential (or respectively scatter) of the existing cluster centers due to the distorted data density by adding a new data sample/point. The expression for the update of the potential of the existing cluster centers can be derived in a similar way as in Angelov and Filev (2004) for the expression (17.3). We then have

$$P_k(z^*) = \frac{(k-1)P_{k-1}(z^*)}{(k-2) + P_{k-1}(z^*) + \frac{1}{p}P_{k-1}(z^*)\sum_{j=1}^{p}(z^* - z_{k-1})^2}. \tag{17.11}$$

One can find a similar expression for the case when scatter is used (17.4).

In step (3) one can compare the values calculated in (17.8) and (17.11). Let us denote the number of clusters formed at the moment of time k by R. Each one of them will have different density (expressed by its potential/scatter respectively). We are interested in the spatial difference that the new data point brings (Angelov and Zhou, 2006). If the new data sample/point has a potential that is not represented by the existing centers we assume that this new data sample brings valuable new information:

$$(\Delta P_i > 0) \quad \text{OR} \quad (\Delta P_i < 0); \forall i; i = [1, R] \tag{17.12}$$

where $\Delta P_i = P_k(z_k) - P_k(z^{i*})$.

In the first case we continue the procedure without changing the cluster structure, because we assume that this data sample/point can be represented well by the existing cluster centers. If (17.12) is satisfied, we assume that the current data point/sample brings valuable new information that is *not* represented so far by the currently existing clusters. Therefore, we form a new cluster around this point/sample *unless* the following condition holds:

$$\min_{i=1}^{R} \| x_k - x^{*i} \| < T \tag{17.13}$$

where T denotes a threshold; Filev and Tseng (2006) suggest values for T in the range of [0.15; 0.3], Angelov and Zhou (2006) suggest a value of 1/3; both assume that the data is normalized.

If the condition (17.13) holds then the current data point/sample brings valuable new information in terms of data density (17.12) and the old center is inside the zone of influence of the newly formed rule. Therefore, the old center has been removed (Angelov and Zhou, 2006). In a similar way, one can use the scatter; in this case the logic is inverted, (see Angelov and Filev, 2005).

The procedure described above underlines the *density-based on-line clustering* approach that ensures a *gradually evolving* cluster structure by upgrading and modifying it. This approach is noniterative, incremental, and thus computationally very efficient (it has very low memory requirements and is a noniterative, single-pass procedure). This approach can be used for real-time classification, prediction, and control applications. An application to landmark recognition and novelty detection by autonomous robotic systems is presented in Section 17.4.

17.3 FSPC: REAL-TIME LEARNING OF SIMPLIFIED MAMDANI MODELS

Density-based clustering is a convenient tool for real-time identification of similar data patterns that are characterized with strong input/output correlation and can be represented as combinations of multiple (linear) input/output models. Systems with fast dynamics, multiple, and frequently changing operating modes are the typical area of application of density-based clustering. The main objective of the system model in this case is to cover the transitions between the operating modes adequately. The Takagi–Sugeno model provides the formal structure for accomplishing this task by representing such nonlinear systems as a continuously switching mixture of piecewise linear models.

For a different class of systems that are dominated by fewer operating modes and rarely changing steady states, the steady state representation is more critical than the transitions between the operating modes. This is the case of the typical industrial systems that are designed to operate predominantly in certain operating modes where the transitions between the modes are rather exceptions. The so-called simplified Mamdani (sM) model (Yager and Filev, 1994) – a collection of *IF-THEN* rules with fuzzy predicates and deterministic consequents – is the mathematical framework for representing systems with characteristics described above. By combining the rules and applying a fuzzy reasoning mechanism we obtain a rule-based process model mapping the relationship between the input(s) and output(s). The assumed fuzziness of the predicates provides the model with a capability to cover the transitions between the steady states through interpolation. For this type of system we consider a special type of distance-based clustering that is inspired by the statistical process control (SPC) – a well-established method for process variability monitoring in industry.

SPC is a methodology for identifying variations in monitored output variables that are due to actual input changes rather than process noise. Although, it is sometimes presented as a control tool, classical SPC is in reality a diagnostic means signaling a *change in the distribution* of the monitored variable that is due to a special (assignable) cause, i.e., a change of the input variable(s). In general, the SPC methodology is not based on the idea of an input–output model. However, the concept of SPC can be applied to identify clusters in the output domain that are associated with certain steady states; it can also be used to induce the corresponding clusters in the input domain. We use these input/output data clusters to learn online the rule antecedent and consequent parameters of the sM system model.

When the output (monitored process variable) is in statistical control, i.e., the output y is within the process control limits (usually set at $\pm 3\sigma_{yi}$, where σ_{yi} is the standard deviation of the output) we associate its current mean y_i^M with the current vector of the inputs (special causes) x_i^M. This situation is covered by the rule:

$$IF \quad (x \text{ is close to } x_i^M) \quad THEN \quad (y \text{ is } y_i^M), \tag{17.14}$$

where x_i^M and y_i^M are the means of the input and output variables and are obtained directly from the data; *close* is a linguistic quantifier that is not explicitly defined (we assume a Gaussian type fuzzy set for its definition)

$$close = e^{-\frac{(x-x_i^M)^2}{2\sigma_{xi}^2}} \tag{17.15}$$

and σ_{xi} is the standard deviation of the inputs that correspond to the output values within the ith set of process control limits

The fuzzy rule (17.14) clusters all input/output pairs that are associated with output values y satisfying the SPC process control condition:

$$|y - y_i^M| < 3\sigma_{yi}. \tag{17.16}$$

If the output takes values that are out of the process control limits this is an indication of a substantial change of the input. Additional out of control indicators include a trend over time, or multiple points below or above the mean. In the cases when the process is out of statistical control the process control limits are recalculated resulting in a new mean value y_j^M. Denoting the corresponding input parameters x_j^M and σ_{xj} we obtain a new rule:

$$IF \quad (x \text{ is } close \text{ to } x_j^M) \; THEN \; (y \text{ is } y_j^M)$$

that summarizes the new set of input values (assignable causes) resulting in a set of output values within the jth process control range, etc.. By combining the rules we identify a rule-based process model mapping the relationship between the special causes and the monitored process variables. The simplified reasoning method provides an analytical input–output model:

$$\hat{y} = \sum_{i=1}^{R} \tau_i y_i^M \Big/ \sum_{i=1}^{R} \tau_i \tag{17.17}$$

where

$$\tau_i = e^{-\frac{(x-x_i^M)^2}{2\sigma_{xi}^2}}$$

is the degree of firing (firing level) of the ith rule predicate by the current input x and R is the number of segments of the SPC characteristic with different process control limits.

The SPC provides a formal mechanism for clustering the output data into regions that are characterized with similar system behavior. These regions are used to guide the determination of corresponding clusters in the input domain. We call the method of learning sM fuzzy models that is inspired by the SPC methodology the fSPC learning. The fSPC learning is driven by a distance measure with a threshold that corresponds to the $+/-3\sigma$ process control limit (17.16) that is used in the SPC technique.

The concept of SPC based fuzzy modeling is illustrated on the system represented by the input/output data-sets shown in Figure 17.1. The SPC chart of the output (assuming a subgroup size of four) along with

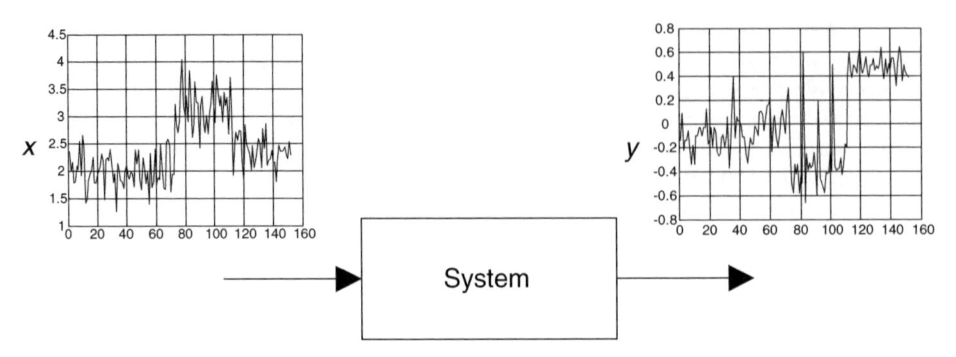

Figure 17.1 Input and output data-sets.

the corresponding partitioning of the input and output data is presented in Figure 17.2. We can see that the SPC chart identifies three output clusters characterized with different values of the mean and control limits. These regions determine three related clusters in the input and output domain.

From the perspective of system modeling we can consider the areas where the process is under statistical control as steady states related to a certain operating mode. In these states the system output can be approximated with a high probability with its mean y_i^M. Each of those states will correspond to an individual rule of the fuzzy model.

For MIMO systems x and y are n and m dimensional vectors, respectively. In this case the statistical control condition is defined through the T^2 *Hotelling* statistics of the m dimensional output vector (Ryan, 1989):

$$T_i^2 = \left(y - y_i^M\right)^T S_{yi}^{-1} \left(y - y_i^M\right) \tag{17.18}$$

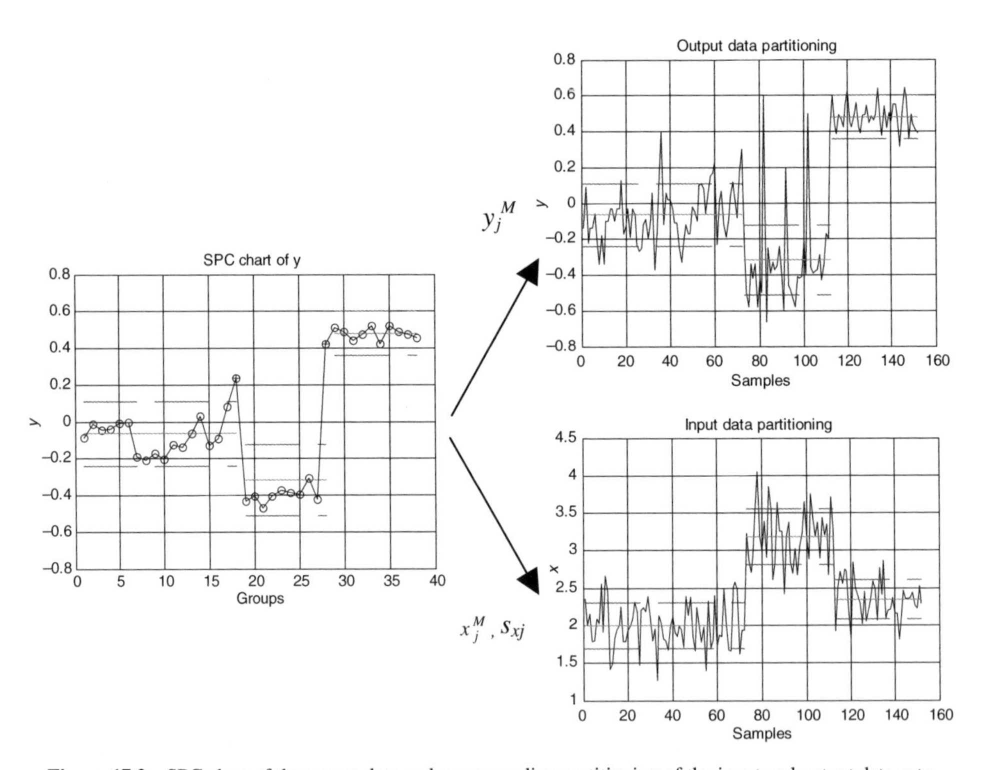

Figure 17.2 SPC chart of the output data and corresponding partitioning of the input and output data-sets.

where y_i^M is the mean, S_{yj} is the output covariance (single sample group size is assumed since exponential moving average SPC is considered). The output vector is considered to be in statistical control when the T^2 is upper bounded by the chi-squared distribution

$$T_i^2 < \chi^2_{m,\beta}, \tag{17.19}$$

where $\chi^2_{m,\beta}$ is the $(1-\beta)$th value of the chi-squared distribution with p degrees of freedom and β is the probability of a false alarm, e.g., $\chi^2_{2,0.0027} = 11.8290$, $\chi^2_{3,0.0027} = 14.1563$, while $\chi^2_{1,0.0027} = 9$ corresponds to the well-known $\pm 3\sigma$ limit rule for the case of a single output. Condition (17.19) covers the single output case (for $m = 1$).

For multiple inputs the antecedent fuzzy set changes as follows:

$$close = e^{-\frac{\left(x - x_i^M\right)^T S_{xi}^{-1}\left(x - x_i^M\right)}{2}} \tag{17.20}$$

where x_i^M is the vector of input means and S_{xi} is the input covariance matrix corresponding to y_i^M.

We apply the exponentially weighted moving average (EWMA) SPC technique to derive a recursive algorithm for learning fuzzy models that is suitable for online modeling. We assume that the sample groups contain a single output value, therefore replacing the expression for the process mean by its EWMA counterpart:

$$y_i^M(k) = \alpha y_i^M(k-1) + (1-\alpha)y(k), \tag{17.21}$$

where $\alpha, \alpha \in (0,1)$ is the exponentially forgetting parameter that controls the weight of the new observations and $y_i^M(k)$ is the mean associated with the ith local output cluster.

Alternatively, we derive an EWMA version of the expression for output covariance. In order to calculate the Hotteling statistics T^2 and verify statistical control condition (17.19) we need the inverse of the covariance $S_{yi}^{-1}(k)$ rather than the actual covariance matrix $S_{iy}(k)$:

$$S_{yi}(k) = S_{yi}(k-1) + (y(k) - y_i^M(k))(y(k) - y_i^M(k))^T = S_{yi}(k-1) + \Delta y_i(k)\Delta y_i^T(k)$$

where $\Delta y_i(k) = y(k) - y_i^M(k)$. Including the effect of exponential forgetting we can rewrite the above expression for updating the covariance matrix as a weighted sum of the current covariance matrix and the current update:

$$S_{yi}(k) = \alpha S_{yi}(k-1) + (1-\alpha)\Delta y_i(k)\Delta y_i^T(k).$$

By applying the matrix inversion lemma (Astrom and Wittenmark, 1997)

$$(A + BCD)^{-1} = A^{-1} - A^{-1}B(C^{-1} + DA^{-1}B)^{-1}DA^{-1}$$

(assuming nonsingular square matrices A, C, and $C^{-1} + DA^{-1}B$) on the inverse covariance matrix $Q_{yi}(k) = S_{yi}^{-1}(k)$:

$$Q_{yi}(k) = (\alpha S_{yi}(k-1) + (1-\alpha)\Delta y_i(k)\Delta y_i^T(k))^{-1}$$

$$Q_{yi}(k) = \alpha^{-1}Q_{yi}(k-1) - \alpha^{-1}(1-\alpha)Q_{yi}(k-1)\Delta y(k)$$
$$(1 + \alpha^{-1}(1-\alpha)\Delta y_i^T(k)Q_{yi}(k-1)\Delta y_i(k))^{-1}\alpha^{-1}\Delta y_i^T Q_{yi}(k-1))$$

we obtain a recursive expression for updating the inverse process covariance matrix $Q_{yi}(k) = S_{yi}^{-1}(k)$:

$$Q_{yi}(k) = (I - (1-\alpha)Q_{yi}(k-1)\Delta y(k)(1 + (1-\alpha)\Delta y_i^T(k)Q_{yi}(k-1)\Delta y_i(k)/\alpha)^{-1})\Delta y_i^T Q_{yi}(k-1))/\alpha \tag{17.22}$$

where $Q_{yi}(0)$ is a diagonal matrix with sufficiently large elements.

Expressions (17.21) and (17.22) recursively estimate the ith output cluster center $y_i^M(k)$ and covariance matrix $S_{yi}^{-1}(k)$ (and inverse variance σ_{yi}^{-2} in the single output case).

Formally same expressions as (17.21) and (17.22) are applied to calculate the parameters x_i^M and $Q_{xi}(k) = S_{xi}^{-1}(k)$ of the rule antecedents:

$$x_i^M(k) = \alpha x_i^M(k-1) + (1-\alpha)x(k) \tag{17.23}$$

$$Q_{xi}(k) = Q_{xi}(k-1)(1-(1-\alpha)\Delta x_i(k)(1+(1-\alpha)\Delta x_i^T(k)Q_{xi}(k-1)\Delta x_i(k))^{-1}\Delta x_i^T Q_{xi}(k-1))/\alpha \tag{17.24}$$

where

$$\Delta x_i(k) = x(k) - x_i^M(k).$$

Condition (17.19) identifies whether the current output vector belongs to the same output cluster. If condition (17.19) is satisfied, the output vector $y(k)$ continues to belong to the current output cluster center y_i^M. The center and the covariance matrix of this particular cluster are updated according to expressions (17.21) and (17.22). Similarly the center and the covariance matrix of the corresponding ith cluster in the input – expressions (17.23) and (17.24) are updated. Failure of condition (17.19) indicates that the output vector no longer belongs to the current ith output cluster – the output vector is assigned to an existing cluster or a new cluster is initiated. If a previously identified output cluster center y_j^M, $j \in [1, i-1]$ satisfies (17.19), i.e., there exist an i^* such that

$$i^* = \arg\min[(y - y_j^*)^T S_{yj}^{-1}(y - y_j^*)] \; ; \quad j = [1, i-1] \tag{17.25}$$

then the output vector $y(k)$ is assigned to the particular output cluster center $y_{i^*}^M$. Consequently, the center and the covariance matrix of the i^*th output cluster are updated by applying (17.21) and (17.22). If none of the existing output cluster centers satisfies (17.19), i.e., $i^* = \emptyset$, then new output and input clusters are initiated.

The main steps of the recursive fSPC learning algorithm are summarized below:

(1) Check statistical process control condition (17.19).
- If (17.19) is not satisfied:
  - Check for closeness to a pre-existing output cluster (17.25)
  - If $i^* = \emptyset$
    - Increment i and start a new output cluster and input cluster, and a new rule.
    - Initialize new rule parameters y_i^M, Q_{yi}, x_i^M, and Q_{xi}, and set the sample counter at $k = 1$.
  - Otherwise, update the parameters of the i^*th rule
    - Set $i = i^*$.
    - Go to (2).
- Otherwise set the sample counter at $k = k + 1$. Go to (2).

(2) Update ith rule parameters y_i^M, Q_{yi}, Q_i^M, and Q_{xi}:
- Rule consequent mean y_i^M (17.21).
- Inverse output inverse covariance matrix Q_{yi} (17.22).
- Rule antecedent mean by y_i^M (17.23)
- Rule antecedent inverse covariance matrix by Q_{xi} (17.24).
- Go to (1).

17.4 APPLICATIONS

In this section practical applications of the proposed methods to landmark recognition and novelty detection in mobile robotics and to manufacturing automation are presented.

17.4.1 Application of Density-based Clustering to Landmark Recognition and Novelty Detection in Robotics

This experiment exemplifies the *density-based* real-time clustering approach presented in Section 17.2. The experiment has been motivated by the need to develop self-localization techniques to improve the navigation of mobile robots (due to wheel slippage so-called *dead reckoning* hampers the use of odometer readings for navigation (Neto and Nehmzow, 2004). Another motivation is the need to localize if maps or global positioning such as GPS are unavailable or unreliable, especially in a hostile environment (Durrant-White, 2001). As a first stage of our experiment, we have simulated the same experimental environment as in Nehmzow, Smithers, and Hallam (1991) using the computer-based robot simulator ARIA (ARCOS, 2005). The performance of the *density-based* clustering approach described in Section 17.2 was compared to that of the self-organizing maps (SOM) used by Nehmzow, Smithers, and Hallam (1991) and reported in (Zhou and Angelov, 2006). As a next step we consider a realistic indoor environment in a real office. Important differences comparing to the original experiment (Nehmzow, Smithers, and Hallam (1991) are the absence of pretraining, the recursive, and fully unsupervised nature of the approach (including the number of clusters). Due to the limited computational resources available to an autonomous mobile robot an effective algorithm must process in real-time large amounts of sensory data; therefore, a recursive real-time algorithm is highly desirable to cope with the memory and time limitations.

17.4.1.1 The Experiment

For our experiment we used an autonomous mobile robot, Pioneer-3DX (Figure 17.3) equipped with an on-board computer (Pentium III CPU, 256 MB RAM), camera, digital compass, sonar and bumper sensors, wireless connection for transmission of data to a desktop or laptop in real-time controlled from the on-board computer in a client-server mode using embedded microprocessor ARCOS (ARCOS, 2005).

Odometer and sonar transducers data were processed in real-time by the on-board computer that also runs the density-based evolvable clustering algorithm. The task is to identify the 'landmarks' while performing a routine 'wall following' behavior in a real office environment (office B-69 in InfoLab21 building on campus at Lancaster University, Lancaster, UK, Figure 17.4). Neither the number of the corners, their type (concave or convex), the distance between different corners, etc., is predefined. The robot has to identify all of the real corners correctly without any pretraining and any prior knowledge in that respect. The results are compared to the results of application of fixed structure, pretrained SOM by a mobile robot performing the same task reported in Nehmzow, Smithers, and Hallam (1991). The proposed approach demonstrated superiority in several aspects (Zhou and Angelov, 2006): higher recognition rate;

Figure 17.3 Pioneer DX autonomous mobile robot.

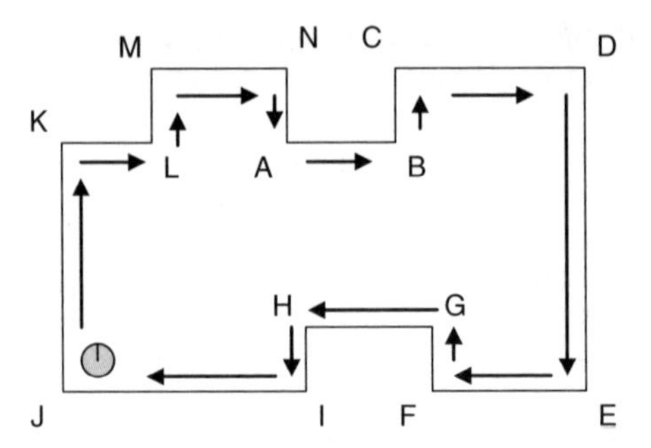

Figure 17.4 Experimental enclosure (office, B-69, InfoLab21, Lancaster).

higher degree of autonomy (density-based *e*Clustering used is fully unsupervised); higher flexibility (its structure is not fixed and can accommodate more neurons/rules/clusters if the environment has changed). In addition, an important feature of the proposed approach is that the information stored in the rule-base is fully linguistically transparent and interpretable.

The inputs include instructions to the robot motor to rotate right (90 degrees) or left (−90 degrees). These turning commands are taken along with the time duration or the distance between turnings. The generated clusters are, therefore, closely related to the corners. For example, if we have 16 corners in the office, ideally, there should be precisely 16 clusters corresponding to each corner. The data are not evenly distributed; some of the clusters formed in real-time are quite similar in terms of adjacent corner types and distances, for example corners 'A' and 'D' or 'M' or 'C' (Figure 17.4).

The clustering algorithm can be presented as a neuro-fuzzy system (Figure 17.5 and Equation (17.26).

R_1: **IF**(T_0 is *Right*) **AND** (t_0 is close to 0.08) **AND** (T_1 is *Left*) **AND** (t_1 is close to 0.25)
AND ($T_1$is *Left*)
THEN (Corner is A) (17.26)

....

R_R: ......

The input vector (Figure 17.6) comprises the time duration and heading/direction change in several consecutive turnings (Right or Left; + 90° or −90°;); $x =$ [*T_0(current turning);t_0 (time duration preceding the current turning); T_{-1}(previous turning);t_{-1} (time duration preceding the previous turning);*

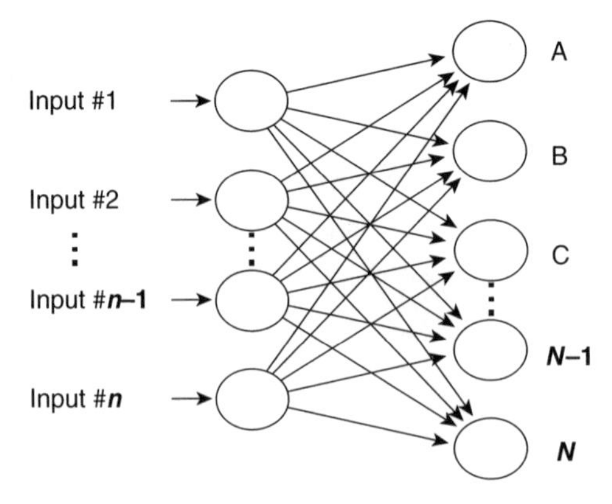

Figure 17.5 Density-based evolving clustering used for novelty (corner) detection.

Input vector 1	T_0	$\Delta D_{0\text{-}1}$	T_1	$\Delta D_{1\text{-}2}$	T_2
Input vector 2	T_0	$\Delta D_{0\text{-}1}$	T_1	$\Delta D_{1\text{-}2}$	
Input vector 3	T_0	$\Delta D_{0\text{-}1}$			

T_0: Current turning in degree.
$\Delta D_{0\text{-}1}$: Distance between two adjacent turnings, T_0 and T_1

Figure 17.6 Input vector.

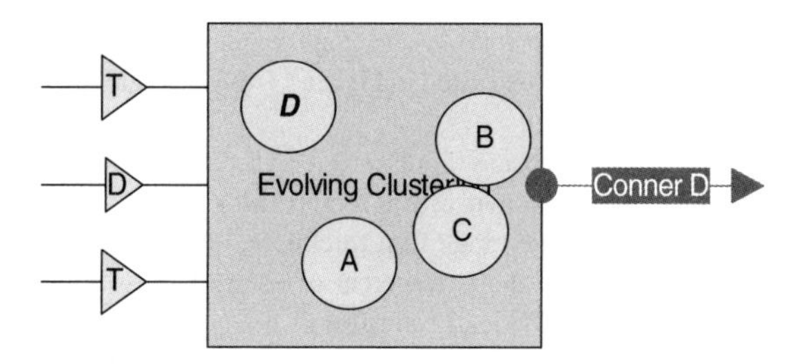

Figure 17.7 Density-based clustering of robot odometer and sonar data to identify corners/landmarks.

T_{-2}*(past turning))]*. $T_k, k = 0, -1, -2$ is a binary variable with value 0 corresponding to a *Right* turning and 1 corresponding to a *Left* turning.

Each cluster can be labeled to represent a distinctive landmark (corner) using, for example, alphabetic letters (Figure 17.7).

Note that new clusters were formed around most of the corners automatically (the detailed results are presented in Table 17.1), overwhelmingly using the second condition in (17.12):

$$\Delta P_i < 0. \tag{17.27}$$

The rationale is that the data points that correspond to the routine operation of the robot ('wall following' behavior) may have high potential because they are similar in terms of heading/direction due to absence of turning while each corner when it first appears is a 'strange', new point and its potential is very low.

After the robot makes one full run in an anticlockwise direction it was able to recognize successfully 10 out of the 16 real corners with the remaining corners incorrectly classified due to the close similarity between description of corners (as mentioned above). When comparing the results (Table 17.1) it should be noted the proposed method works without any prior knowledge and pretraining, while the method described in Nehmzow, Smithers, and Hallam (1991) is offline, needs pretraining, and a fixed number of 50 neurons (clusters) were used. It should be mentioned that the features that are selected for the input vector are critical to the result. Experiments were conducted with between two and five features (Zhou and Angelov, 2006). Uncertainties are related to the fact that two clusters can describe the same corner.

Table 17.1 Result comparison (T: turning, D: distance).

Experiment	Input vector	No. of corners	Correct	Uncertain	Missed	Clusters	Description
1	TDTDT	8	5	0	3	50	(Nehmzow, Smithers and Hallam (1991)
2	TDTDT	8	7	0	1	7	(Zhou and Angelov, 2006)
3	TDTDTD	16	10	5	1	15	Lab Environment, 16 corners

17.4.2 Application of Minimum-distance-based Clustering to Modeling of Automotive Paint Process

The *minimum-distance-based clustering fSPC* method for learning simplified Mamdani fuzzy models (sM) was applied to approximate the average film thickness on the vehicle body for a given combination of factors governing the process in automotive paint booths (Filev and Tardiff, 2004). These variables include: (i) fluid flow rates of the applicators (robotized bells and guns) – the main parameter directly affecting the film thickness; (ii) air down draft velocity, temperature, and humidity (air is continuously supplied to the booth to remove the paint overspray and to virtually separate individual sections of the booth). Since the effect of the air downdraft on the horizontal and vertical surfaces is different, alternative models are used to describe the relationships between the process variables and paint film thickness on the left and right vertical, and horizontal surfaces (Filev and Tardiff, 2004). We demonstrate the fSPC learning algorithm on a set of paint process input/output data representing the left vertical surfaces of 250 painted vehicles (the input /output data is presented in Figure 17.8; the actual scale and measurement units are omitted).

The fuzzy model maps the vector of process inputs $x =$ [FF_L (average fluid flow rate, left side); DDb (down draft, bell zone); Down Draft (down draft, reciprocator zone); T (air temperature); H (air humidity)] to process output $y =$ FT_L (average film thickness, left side). We denote the input vectors x and the output y. Similarly we denote the parameters of each of those models x^*, y^*, and S^*. This formally results in an evolving family of *If* ... *Then* rules of the type:

$$\textit{If} \quad x \textit{ is close to } x^* \textit{ Then } y \textit{ is } y^*.$$

Initially the model includes a single rule:

$$\textit{If} \quad \text{FF\_L} \approx 75 \textit{ and } \text{DDb} \approx 65 \textit{ and } \text{DDr} \approx 80 \textit{ and } T \approx 90 \textit{ and } H \approx 34 \textit{ Then } \text{FB\_L} \approx 0.67.$$

After the 38 data sample a new rule reflecting a well-defined cluster in the output domain is identified by the algorithm and added to the fuzzy model:

$$\textit{If } \text{FF\_L} \approx 88 \textit{ and } \text{DDb} \approx 60 \textit{ and } \text{DDr} \approx 63 \textit{ and } \text{T} \approx 80 \textit{ and } \text{H} \approx 17 \textit{ Then } \text{FB\_L} \approx 0.73.$$

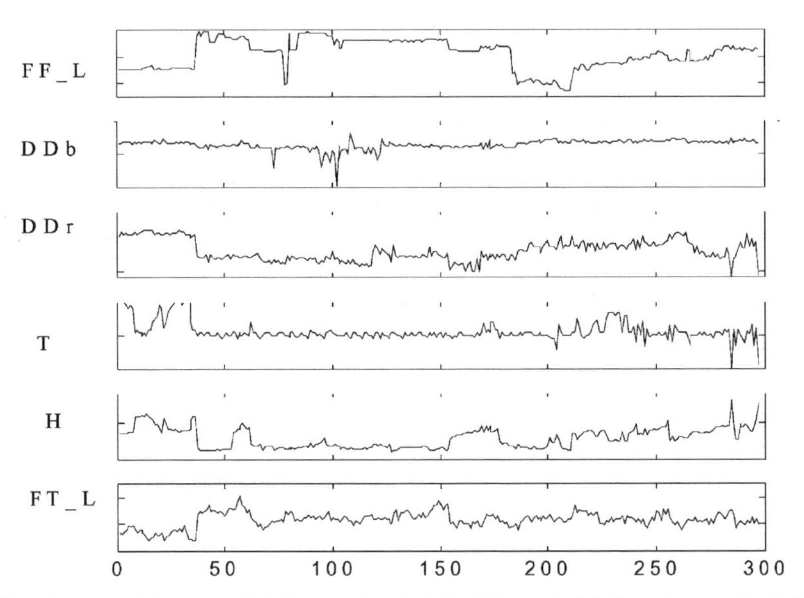

Figure 17.8 Inputs - FF_L (average fluid flow rate, left side); DDb (down draft, bell zone); Down Draft (down draft, reciprocator zone); T (air temperature); H (air humidity) and output - FT_L (average film thickness, left side).

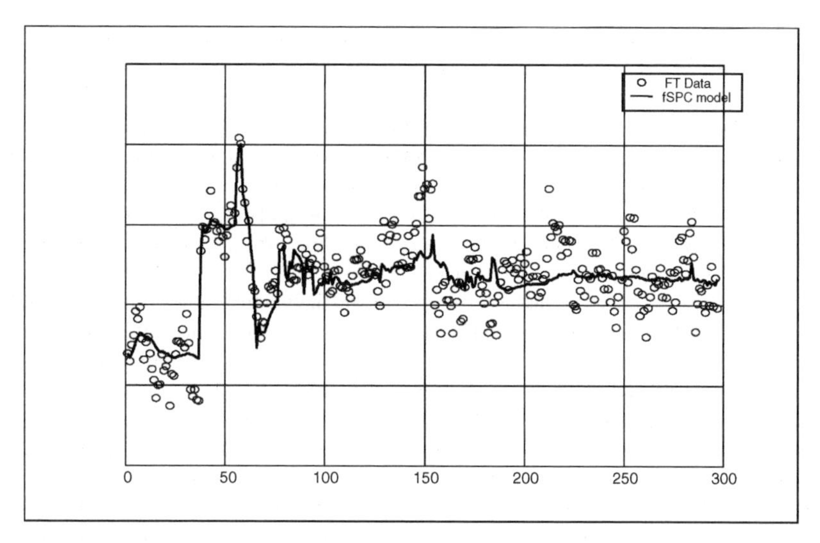

Figure 17.9 Measured film thickness vs. an fSPC based fuzzy model and an RBF neural network approximation.

After reading 150 data samples the model evolves to a set of three rules:

If FF_L $\approx$ 82 *and* DDb $\approx$ 61 *and* DDr $\approx$ 69 *and* T $\approx$ 82 *and* H $\approx$ 28 T*hen* FB_L $\approx$ 0.70
If FF_L $\approx$ 86 *and* DDb $\approx$ 61 *and* DDr $\approx$ 62 *and* T $\approx$ 80 *and* H $\approx$ 19 T*hen* FB_L $\approx$ 0.75
If FF_L $\approx$ 87 *and* DDb $\approx$ 62 *and* DDr $\approx$ 63 *and* T $\approx$ 80 *and* H $\approx$ 31 T*hen* FB_L $\approx$ 0.79.

At the end of the learning process rule parameters are further refined:

If FF_L $\approx$ 80 *and* DDb $\approx$ 64 *and* DDr $\approx$ 68 *and* T $\approx$ 82 *and* H $\approx$ 30T*hen* FB_L $\approx$ 0.70
If FF_L $\approx$ 84 *and* DDb $\approx$ 63 *and* DDr $\approx$ 64 *and* T $\approx$ 81 *and* H $\approx$ 25 T*hen* FB_L $\approx$ 0.75
If FF_L $\approx$ 85 *and* DDb $\approx$ 62*and* DDr $\approx$ 63 *and* T $\approx$ 80 *and* H $\approx$ 26 T*hen* FB_L $\approx$ 0.78.

The prediction of the fuzzy model is shown in Figure 17.9. The main advantage of the fSPC model is that it provides the means for a real time approximation of the input/output mapping and a straightforward interpretation of model parameters as a collection of rules that are extracted from the data and that describe the main operating modes of the process.

17.5 CONCLUSION

This chapter presents two approaches to real-time data clustering that can be used effectively for knowledge generation in the form of fuzzy rules. The first approach, density-based *e*Clustering stems from mountain/subtractive clustering and is suitable for nonstationary processes with relatively fast dynamics. It can be used for generation of fuzzy models in real-time from data, which was illustrated in the example of novelty detection and landmark recognition by a mobile robot in an unknown (office) environment.

The second approach stems from statistical control and is distance-based. It is suitable for generating fuzzy rules of simplified Mamdani (sM) type, which one can also treat as zero-order Takagi–Sugeno models. An example from real process control in the automotive industry was used to illustrate this approach.

Both approaches offer an online, an one-pass, a noniterative solution that is based on fuzzy recursive calculations and thus suitable for real-time applications. They both "learn from experience," do not need pretraining, and thus can start "from scratch."

The density-based *e*Clustering is a prototype-based approach, i.e., the cluster centers are some of the existing data points in the input/output data space. It also takes into account the whole accumulated history of the data and its spatial distribution. Thus it avoids creating a large number of cluster centers that would later need pruning. It is an effective basis for fuzzy rules antecedents generation from data in real-time. When combined with recursive least squares it is a very powerful tool for data space partitioning needed for fuzzy rule-based models antecedent part identification in real-time. New characteristics of cluster quality such as age and population have been introduced and discussed. The experimental results concern novelty detection and landmark recognition in mobile robotics where eClustering is used for unsupervised learning in an unknown environment.

The fSPC learning establishes a link between the SPC methodology – a well-known engineering tool for quality monitoring that is widely accepted in industry – and one of the main soft computing techniques – the concept of fuzzy rule base models. A simplified Mamdani (sM) fuzzy model is derived based on the information that is obtained through SPC monitoring. An output value that is out of the process control limits indicates that the system output is to be assigned to a different region in the output domain. The mean of the output region identifies the parameter of rule consequent. The output regions guide the creation of corresponding regions in the input domain. The traditional SPC technique is augmented with a procedure for calculation of the mean and the standard deviations / covariance matrices of the input regions – those parameters determine the rule antecedents. The fSPC learning algorithm can be implemented recursively and does not require multiple iterations compared to the traditional back-propagation-based algorithms for learning rule-based models. One of the main advantages of the fSPC approach is in the straightforward interpretability of the model. The model essentially consists of logical statements describing prototypical operating conditions that are associated with different modes of operation. The role of the fuzziness of the rule antecedents is to interpolate between these basic operating conditions through the normalized firing levels of the rules in. Model parameters have statistical meaning, are related to the SPC parameters, and can be easily calculated from the data without using specialized software packages. The fuzzy reasoning mechanism transforms the family of rules into an analytical model.

REFERENCES

Angelov, P. and Filev, D. 'An approach to on-line identification of evolving Takagi-Sugeno models', *IEEE Trans. on Systems, Man and Cybernetics, part B*, **34**, 2004, 484–498.

Angelov, P. and Zhou, X.-W. 'Evolving Fuzzy Systems from Data Streams in Real-Time', *2006 International Symposium on Evolving Fuzzy Systems*, 7–9 September, 2006, Ambelside, Lake District, UK, IEEE Press, pp.26–32, ISBN 0-7803-9719-3.

ARCOS (2005) 'ActivMedia's Robot Control & Operations Software', ActivMedia Robotics, LLC.

Astrom, K. and Wittenmark, B. (1997) *Computer Controlled Systems: Theory and Design*, 3rd edn. Prentice Hall, NJ, USA.

Bezdek, J. (1974) 'Cluster Validity with Fuzzy Sets', *Journal of Cybernetics*, **3** (3), 58–71Carpenter, G.A., Grossberg, S. and Reynolds, J.H. (1991). 'ARTMAP: Supervised real-time learning and classification of non-stationary data by a self-organizing neural network'. *Neural Networks*, **4**, 565–588.

Chiu, S.L. 'Fuzzy model identification based on cluster estimation'. *Journal of Intelligent and Fuzzy Systems*, **2**, 267–278, 1994.

Durrant-Whyte, H. *A Critical Review of the State-of-the-Art in Autonomous Land Vehicle Systems and Technology*, Sandia Report, SAND2001-3685, Nov. 2001.

Filev, D. and Tardiff, J. (2004) 'Fuzzy Modeling within the Statistical process control framework', *International Joint Conference on Neural Networks and International Conference on Fuzzy Systems*, IJCNN-FUZZ-IEEE, Budapest, Hungary, 25–29 July, 2004, pp

Filev, D.P., Larsson, T. and Ma, L. 'Intelligent Control, for automotive manufacturing-rule-based guided adaptation', in *Proc. IEEE Conf. IECON'00*, Nagoya, Japan, Oct. 2000, pp.283–288.

Filev, D. and Tseng, F. 'Novelty detection based machine health prognosis', *2006 International Symposium on Evolving Fuzzy Systems*, 7–9 September, 2006, Ambelside, Lake District, UK, IEEE Press, pp.181–186, ISBN 0-7803-9719-3.

Fritzke, B. (1995) 'A growing neural gas network learns topologies', In (Touretzky and Keen, eds.), *Advances in Neural Information Processing Systems*, vol.7, pp.625–632, MIT Press, Cambridge, MA, USAFuzzy Logic Toolbox (2001) User Guide, v.2, Mathworks Inc.

Gath, I. and Geva, A.B. (1989) 'Unsupervised optimal fuzzy clustering', *IEEE Transactions on Pattern Analysis and Machine Intelligence*, v.7, pp.773–781Hastie, T., Tibshirani, R. and Friedman, J. (2001) *The Elements of Statistical Learning: Data Mining, Inference, and Prediction*, Springer, 2001Huang, G.-B., Saratchandran, P. and Sundarajan, N. (2005) A generalized growing and pruning RBF (GGAP-RBF) neural network for function approximation, *IEEE Transaction on Neural Networks*, **16** (1) 57–67.

Kasabov, N. (2001) 'Evolving fuzzy neural networks for on-line supervised/unsupervised', knowledge-based learning, *IEEE Trans. SMC – part B, Cybernetics* **31**, 902–918.

Kecman, V. (2001) *Learning and Soft Computing*, MIT Press, Cambridge, MA, USAKohonen, T. (1988) *Self-Organization and Associative Memory*, Springer Verlag, Berlin, Heidelberg, New York.

Konstantinov, K. and Yoshida (1989) 'Physiological state control of fermentation processes', *Biotechnology and Bioengineering*, **33** (9), 1145–1156Murray-Smith, R. and Johansen, T.A. (Eds.) *Multiple Model Approaches to Modelling and Control*. Taylor and Francis, London, 1997Nehmzow, U., Smithers, T. and Hallam, J. (1991) Location Recognition in a Mobile Robot Using Self-Organising Feature Maps, in (G. Schmidt, ed.), *Information Processing in Autonomous Mobile Robots*, Springer Verlag.

Neto, H. V. and Nehmzow, U. (2004) 'Visual Novelty Detection for Inspection Tasks using Mobile Robots', In *Proc. of SBRN: 8th Brazilian Symposium on Neural Networks*, IEEE CS Press, São Luís, Brazil, ISBN 8589029042.

Plat, J. (1991) 'A resource allocation allocation network for function interpolation', *Neural Computation*, **3** (2) 213–225.

Ryan, T. (1989) *Statistical Methods for Quality Improvement*, John Wiley & Sons, Inc., New York, USA.

Shing, J. and Jang, R. (1993) 'ANFIS: Adaptive Network-based Fuzzy Inference Systems', *IEEE Transactions on Systems, Man & Cybernetics*, **23** (3), 665–685.

Simani, S., Fantuzzi, C., and Patton, R.J. (2002) *Model-based Fault Diagnosis in Dynamic Systems Using Identification Techniques*. Springer Verlag, Berlin Heidelberg.

Specht, D. F. (1991) 'A general regression neural network', *IEEE Trans. on Neural Networks*, 568–576Wang, L. (1994). *Adaptive fuzzy system and control: design and stability analysis*. Prentice Hall, Inc.

Yager, R.R. and Filev, D.P. (1993) 'Learning of Fuzzy Rules by Mountain Clustering', *Proc. of SPIE Conf. on Application of Fuzzy Logic Technology*, Boston, MA, USA, pp. 246–254.

Yager, R.R. and Filev, D.P. (1994) *Essentials of Fuzzy Modeling and Control*, John Wiley and Sons, Inc., NY, USA.

Zhou, X. and Angelov, P. (2006) 'Joint Landmark Recognition and Classifier Design', *Proc. World Congress on Computational Intelligence*, WCCI, Vancouver, Canada, July, 16–21, 2006, pp.6314–6321.

Part V
Applications and Case Studies

18

Robust Exploratory Analysis of Magnetic Resonance Images using FCM with Feature Partitions

Mark D. Alexiuk[2] and Nick J. Pizzi[1,2]

[1]*Institute for Biodiagnostics, National Research Council, Winnipeg, Canada*
[2]*Department of Electrical and Computer Engineering, University of Manitoba, Winnipeg, Canada*

18.1 INTRODUCTION

Exploratory data analysis (EDA) [1] differs from model based analysis in that aggressive limits are imposed on mathematical models of the data. Since models can mask data characteristics, this limitation is seen as a method to reveal the intrinsic properties of the data. When models are used in EDA, they are typically data-driven; model parameters are determined solely by data-set statistics. The main benefit of using EDA techniques is the possibility of detecting "important," yet unanticipated, features as is the case with functional magnetic resonance imaging (fMRI), where spatial images of brains are acquired over a time interval. Data-driven algorithms are ostensibly objective means to discover data structure since a priori structures are deprecated. The elicitation of intrinsic data organization is the purpose of unsupervised learning algorithms such as FCM [2].

With EDA, the imposition of a mathematical model on the data should be done only with strong justification. While it is not surprising that EDA methods often do not facilitate an augmented analysis using auxiliary mathematical models, justification for such models do, at times, exist. As an example, consider a data-set in which relationships between features are known. Most EDA techniques find it difficult to use this information to elicit overall structure. This is simply due to the lack of a mechanism to describe and integrate these relations. There are good reasons to group features. Since the collection of more sample features is usually not difficult, it is common for data-sets to have features acquired through different modalities, at different times and under different conditions. These features may exhibit different statistical properties, contain different levels of observational error and noise, and often have a de facto measurement or comparison

Advances in Fuzzy Clustering and its Applications Edited by J. Valente de Oliveira and W. Pedrycz

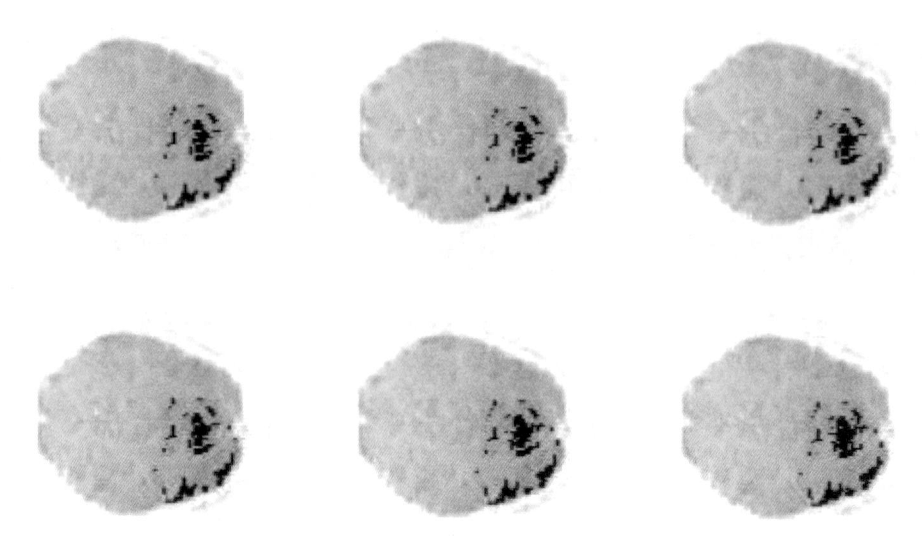

Figure 18.1 A sequence of FCMP-analyzed brain images with increasing spatial proximity relevance.

method associated with them. This is the driving force behind feature reduction techniques. Sometimes as little as less than 1% of features are used to discriminate biomedical data with high dimensionality [3,4].

It is apparent that a formalism is necessary to describe relationships between feature subsets and their combination with respect to a particular problem. This formalism may be expressed using FCM with feature partitions (FCMP), a novel variant of the FCM algorithm that exploits relations between features, which enables feature-specific processing and a ranking of feature subsets in terms of relative importance. For example, fMRI analysis often involves clustering along the temporal dimension (temporal partition) in order to group similar neural activations occurring over a period of time followed by a subsequent analysis of the spatial proximity (spatial partition) of the activations to one another. The first step strictly emphasizes temporal similarity to the exclusion of spatial proximity, while the second step conversely ranks their relative importance. However, if FCMP is used for fMRI analysis, these partitions may be treated jointly, that is, temporal similarity and spatial proximity are of similar importance and are evaluated simultaneously. This is illustrated in Figure 18.1, which is a sequence of FCMP-analyzed brain images with increasing (from top left to bottom right) relative importance of spatial proximity. Note the increasing number of similar neural activations (dark points) within the spatial area referred to as the visual cortex (right of centre region).

The next section describes the formalism that we have developed as well as our novel FCMP algorithm followed by a discussion of fMRI data-sets used to demonstrate the efficacy of this algorithm. Additional background theory for this chapter may be obtained from literature on time series [5,6,7], space–time analysis [8], and fuzzy concepts relating to spatio-temporal models [9,10,11,12,13,14].

18.2 FCM WITH FEATURE PARTITIONS

FCMP [15] is a generalization of FCM that provides a means of describing feature relationships and integrating feature subsets. Due to its general nature, many well-known clustering algorithms may be expressed as specializations of FCMP. The FCMP model also expresses several recent developments in cluster analysis [16,17]. Like FCM, FCMP is an iterative clustering algorithm based on an objective function. Its novelty lies in the definition of feature partitions and their advantageous integration. A feature partition is a formal mechanism to express relations between a single set of features and between sets of features.

A feature partition is a triple consisting of a metric, a weight and a set of feature indices. This triple describes how the features are grouped (the feature indices denote a distinct sub-group), how they relate (what metric is used to compare features), and how feature partitions relate to each (how are different sets of features integrated using weights). One benefit of making feature relations explicit in this manner is the continuous integration of information from a single feature partition into an existing clustering process.

18.2.1 Generalizing FCM

The formalized notation for data analysis follows that of Höppner, Klawonn, Kruse and Runkler [17] and is used to derive update equations for the FCM algorithm and has been extended to include feature partitions. Some definitions are repeated here for completeness. For data space $S, S \neq \phi$, and results space $P, |P| \geq 2, A(S,P)$ is the analysis space defined as the collection of mappings U from a specific data-set X to a possible solution Y. That is,

$$A(S,P) := \{f | U : X \rightarrow Y, X \subset S, X \neq \phi, Y \in P\}. \tag{18.1}$$

Analysis spaces are evaluated by an objective function $J, J : A(S,P) \rightarrow \mathbb{R}$. Partial derivatives are taken on J to determine maxima or minima and define update equations. Let J_x be the FCMP objective function with respect to a single sample, x. Membership update equations are determined by setting the partial derivative of the objective function to zero (where λ is a Lagrange multiplier).

$$\frac{\partial J_x}{\partial f_x} = 0, \frac{\partial J_x}{\partial \lambda} = 0. \tag{18.2}$$

We now extend the notation for the general FCMP clustering framework. Let the mapping $U : X \rightarrow Y \in A(S,P)$ denote a cluster partition (membership matrix) characterized by

$$\forall x \in X \sum_{v \in V} u_{xv} = 1, \ \forall v \in V \sum_{x \in X} u_{xv} > 0 \tag{18.3}$$

for sample $x \in X$ and centroid $v \in V$. For X and clusters V, the fuzzy objective function is

$$J(U) = \sum_{x \in X} \sum_{v \in V} u_{xv}^m d^2(x,v) \tag{18.4}$$

where d is a distance metric and m is the fuzzy exponent (coefficient of fuzzification).

Let G be the total number of sample features and n the number of samples. Let n_v denote the number of samples associated with cluster v. The number of partitions is $P = |P| \leq G$. (We use the terms sample and feature vector interchangeably.) Let an index I denote the columns (features) of the data-set. The decomposition of the feature vector into distinct feature partitions is now discussed. Consider a sample $x = [x_1 x_2 \ldots x_G]$ with G features. Define a partition of the G features as the P distinct feature subsets P,

$$P = \{p | p \neq \phi, \cup p = \{1, 2, \ldots, G\}, \forall p, q \in P, p \cap q = \phi\}. \tag{18.5}$$

One method of expressing the features in a partition is through a projection operator π. A projection operator transforms a vector (sample) onto a reference plane. The resulting projection may be a subset of the vector dimensions (sample features). (For instance, projections onto bases are common. In the case of three dimensions and unit basic vectors, $e_1 = [1\,0\,0], e_2 = [0\,1\,0], e_3 = [0\,0\,1]$, any vector in three dimensions may be viewed as additive components along the projection axes e_1, e_2, e_3. Using a transform such as principal component analysis, one can also define planes that are orthogonal but maximize sample variance on the projected plane.) The projection operator is beneficial for FCMP since it is used to index specific feature partitions. Let $\pi_i(x)$ denote the projection of vector x onto the standard basis e_i and extend this notation to allow for the union of projections, $\pi_i(x) = \cup_{i \in I} \pi_i(x)$. Now, the features for a sample x, associated with a particular feature partition $p \in P$, is represented as $\pi_p(x)$. Using the projection operator, the domain objective function is

$$J(U) = \sum_{x \in X} \sum_{v \in V} \sum_{p \in P} v_p f^m(\pi_p(x), \pi_p(v)) d^2(\pi_p(x), \pi_p(v)), \tag{18.6}$$

where $v = [v_1, v_2, \ldots, v_p]$ is a weighting factor denoting the relative importance between the feature partitions $p \in P$,

$$\sum_{p \in P} v_p = 1, \; \forall p, v_p > 0. \tag{18.7}$$

A shorthand device of denoting the projections of x onto the features in the subset p is $x' = \pi_p(x)(v' = \pi_p(v))$. Here p is implicit in $x'(v')$ and is determined in equations by v_p. This results in the simplified equation

$$J(U) = \sum_{x \in X} \sum_{v \in V} \sum_{p \in P} v_p u_{xk}^m d^2(x', k'). \tag{18.8}$$

In order to express algebraically the many types and combinations of metrics that may be used in a general cluster algorithm, the concept of distance is generalized. A metric over all features is replaced with a weighted sum of metrics over features partitions. Each partition $p \in P$ uses a, possibly unique, distance function d_p. They are combined in the cluster algorithm using the weighting parameter $v = [v_1 \ldots v_j \ldots v_p], \forall v_j, 0 < v_j < 1$. This results in a generalized distance

$$D(x, k) = \sum_{p \in P} v_p d_p^2(x', v'). \tag{18.9}$$

The membership update equation exchanges its distance metric with a weighted sum of distances on feature partitions

$$u_{xc} = \left(\sum_{v \in V} \frac{\sum_{p \in P} v_p d_p^2(x', c')}{\sum_{p \in P} v_p d_p^2(x', v')} \right)^{(m-1)} = \left(\sum_{v \in V} \frac{D(x, c)}{D(x, v)} \right)^{(m-1)}. \tag{18.10}$$

The centroid update equation remains unchanged

$$v = \frac{\sum_{x \in X} u_{xv}^m x}{\sum_{x \in X} u_{xk}^m x}. \tag{18.11}$$

Convergence criteria for FCMP are the same as for FCM and a proof of convergence for the algorithm is provided in [18].

18.2.2 FCMP Parameters

In addition to the FCM parameters, FCMP includes a non-empty set of feature partitions. FCMP uses the feature partitions to adapt to specific problems. When a feature partition collection contains only one feature partition it is proper to describe the algorithm as FCM. A feature partition is composed of a triple: a metric $\mu: \mathbb{R}^G \times \mathbb{R}^G \rightarrow \mathbb{R}^G$, a weight $v \in \mathbb{R}$, and a set of feature indices $P = \{p\}$. We denote a single feature partition as $\psi = \{\mu, v, p\}$ with $\Psi = \{\psi\}$ denoting the set of feature partitions. The metric used in FCMP is a composite of weighted distance metrics on the respective partition. Parameter initialization concerns the number of feature partitions to be used, the assignment of weights to rank partitions, and the assignment of metrics for each partition. Many methods may be used to select feature partitions including:

(1) *Entropy based grouping* – appropriate for partially supervised learning. Choose n best discriminating features. Partitions may be formed using the increasing, discriminatory ability of collections of features.
(2) *Statistical heuristics* – best for unsupervised learning. Rank the features by variance or other statistical properties.
(3) *Local domain* – useful for unsupervised or partially supervised learning. An example application is fMRI analysis. If there are time instances (temporal values), form a partition by selecting activated and unactivated epochs. Allow lag times for the activated epochs and form another subset.

Metrics may be chosen to ameliorate noise detected in the data or by external associations between features and metrics. For example, it is customary to use the Pearson correlation coefficient for fMRI time series analysis.

18.2.3 FCMP Specialization

Moving from an abstract (general) formulation to a concrete (specific) adaptation is known as specialization and eliminates algorithmic degrees of freedom from the abstract formulation. Specialization constrains the formula in a particular application of the algorithm. Expectations about the generalized performance of specializations should be tempered by the fact that specialization is essentially a data-set specific process. The casting of the abstract (or meta-) formulation to a particular data-set aims at local optima. For example, feature partitions determined to be optimal for one data-set have no necessary relation to optimal partitions in a different one. The selection of feature partitions tunes the algorithm to the current objective function. Algorithms are often multipurpose (general) and are applied to many types of data. An algorithm designed to meet a data-set specific objective function is often not available or feasible to implement. FCMP allows a general purpose algorithm, FCM, to exploit feature relations in the data and heuristics from external sources (say, an MRI technologist) by modifying the role of each feature and feature partition in the data-set.

An optimal feature partition integrates the distinct contributions that collections of features (partitions) encapsulate in order to maximize (optimize) an external sample-class label pairing. A partition aggregates features that have a bearing on, say, one heuristic that applies to one class of samples. This partition is minimal in that it contains only features related to this rule, and is generalizable in that it applies to all samples in the class. An optimal partition allows diverse and adumbrated structures in the data to be made manifest. For example, in fMRI, an optimal partition for a neural activation study of the visual cortex does the following: (a) defines a spatial region identifiable as the visual cortex where time courses have temporal similarity to the paradigm (a time course (TC), a time series of intensities for an image voxel, is more fully described in Section 18.3.3); (b) highlights other spatial regions where the TCs are correlated to the paradigm; (c) enables the analyst to dismiss quickly and correctly large portions of the data-set as irrelevant (in FCMP, noise TCs are also grouped spatially and may be rejected from further analysis upon an initial scan of the spatial partition image); (d) reduces the percentages of type I and type II errors (FCMP rejects outliers and includes proximal TCs that are also temporally similar); (e) facilitates the discovery of novelty (see (b) above); (f) allows general statements to be made about relationships in different domains and between domains (for instance, with fMRI, statements should be made about spatial, temporal and spatio-temporal properties as well as data-set properties on activated/unactivated epochs, and so on). Overall, the use of feature partitions should never degrade the analysis of a data-set. The cost of using FCMP is the lack of a priori knowledge as to the appropriate levels of integration between the partitions. A range of metrics and weights should be expected to be evaluated.

Several examples of optimal parameters are examined for specialization instances. The utility of the algebraic expression of FCMP compared to FCM will now be shown in the following areas: robust clustering, preprocessing, and partial supervision.

18.2.3.1 Robust Clustering

Robust clustering is designed to reduce the effects of outliers or noise on the centroids. This is done through preprocessing operations such as outlier detection tests. Outliers can then be eliminated from the data-set. Norms and metrics can also be used to reduce the impact of outliers. One such metric is the ε-tolerant metric, which considers as equal all sample pairs that are within an error tolerance, ε. Leski [19] presents the following approach to robust clustering. Define an ε-insensitive metric or norm where

$$|t|_{\varepsilon} = \left\{ \begin{array}{cc} 0 & \text{if } |t| \leq \varepsilon \\ |t| - \varepsilon & \text{if } |t| > \varepsilon \end{array} \right\}. \tag{18.12}$$

Then the cluster objective function is

$$J_\varepsilon = \sum_{i=1}^{c}\sum_{k=1}^{N}(u_{ik})^m |x_k - v_i|_\varepsilon. \tag{18.13}$$

This applies to the robust clustering algorithms ε-FCM, α-FCM, β-FCM [19] and can also be used with the fuzzy c-median. The general form of FCMP entails a simple specialization: one partition containing all features, the selected robust metric, and a feature partition weight of 1.

18.2.3.2 Preprocessing

Several preprocessing operations can be incorporated into the FCMP formalism such as principal component analysis (PCA), independent component analysis (ICA), feature selection, and small signal detection.

Designing an EDA system often requires manipulating the raw data before the clustering process begins. PCA is one common transformation and projects the samples onto axes of maximal variance. Preprocessing with PCA is incorporated by FCMP through using the eigenvalues λ of the data-set eigenvectors Λ as partition weights. In the notation of feature partitions, $\nu = \lambda$. The number of partitions corresponds to the number of principal components used. The number of components used is often determined by (a) a goal of accounting for at least a certain percentage of the total variance, or (b) isolating projection axes that discriminate between the samples. Each feature partition contains all the feature indices and requires that the samples be projected onto the principal components. Of the triple Ψ, only the metrics μ are left to define as parameters.

ICA defines independent components and a mixing matrix that describes how the original signals may have been combined. A whitening matrix, which makes the distribution Gaussian, is often used. The mixing matrix can be used to define a partition similar to that in the previous example using PCA. Each feature partition Ψ includes all feature indices ($p_i = \{1, \ldots, G\}, \forall\, i$) and the ICA mixing matrix is used to transform the data from the original feature space to the independent component space. Typically, the number of independent components is much less than the number of features.

Determining which features to use in a partition can be determined using an exhaustive search, heuristics or randomized selection. Each method results in a binary value for a feature (0 = not selected, 1 = selected). A binary vector defines the selection of features. This vector defines feature indices in the feature partition. Some methods, such as GMDH [20], generate new features by including feature products in combination. This new feature is the product of two or more existing features. The partition in both these cases is the feature index vector multiplied element-wise by the feature selection mask.

18.2.3.3 Partial Supervision

How can knowledge of class labels be exploited by FCMP? Samples with labels have an increased weight or cause the algorithm to switch from a robust metric to a regular metric. Changing metrics for samples with trusted labels increases the contribution of these samples to the centroid. Other methods can be used such that samples with labels will receive a higher weighting in the feature partition. Clusters that have samples with a variety of trusted labels can be split into clusters with samples having the same trusted label. This is similar to the partially supervised FCM of Pedrycz and Waletzky [21]. Let $\Omega = \{\omega_1, \omega_2, \ldots, \omega_c\}$ denote the set of C class labels. Let ω_0 denote the class assigned to outliers, classification rejections, and unknown and ambiguous samples. Let $\Omega^* = \Omega \cup \omega_0$ be the set of all possible class labels. Let $\omega(x)$ denote the class label associated with sample x. Define a data-set of unlabeled samples $X = \{x|\omega(x) = \omega_0\}$ and a data-set of labeled samples $Y = \{x|\omega(x) \in \Omega\}$. Consider the following metric μ_{XY}

$$\mu_{XY}(a,b) = \left\{ \begin{array}{lll} \mu_1(a,b) & \text{if} & a \in X, b \in Y \\ \mu_2(a,b) & \text{if} & a, b \in X \\ \mu_3(a,b) & \text{if} & a, b \in Y \end{array} \right\} \tag{18.14}$$

Label information is integrated by switching between different metrics. Alternately, consider feature subsets that are discriminatory with respect to the labeled samples. Let $\Gamma \in \{\gamma_1, \gamma_2, \ldots, \gamma_D\}$ be the set of discriminating features. Let p_1 be the feature indices in feature partition Ψ_1 that are all discriminatory, $p_1 = \{p|p \subset \Gamma\}$. Let feature partition Ψ_2 contain only indices of non-discriminatory features, $p_2 = \{p|p = \phi\}$. Let weights v_1, v_2 be associated with Ψ_1, Ψ_2 and assign more weight to the discriminating features. When this occurs, $v_1 > v_2$ and knowledge of class labels is effectively exploited.[1]

18.2.3.4 Small Signal Detection

The solution to small signal detection is similar to that of exploiting class labels. It is a question of selecting the appropriate metric. We define a probe as a finely-tuned function which determines the presence or absence of a localized signal. Probe localization may be spatial, temporal, in the frequency domain, a combination of domains (space–time, time–frequency, etc.), or may be defined by heuristics. Let us consider features acquired over a time interval: for example, a time series of n instances from an fMRI study. Denote activated epochs as E_A and the unactivated epochs as E_U. Define feature partitions Ψ_1, Ψ_2, with feature indices

$$\begin{aligned} p_1 &= \{p|p \in E_A\} \\ p_2 &= \{p|p \notin E_A\}. \end{aligned} \tag{18.15}$$

Assign weights such that $|p_1|v_1 > |p_2|v_2$ to augment the activated epochs. This method can also be used to discount features whose values have been contaminated. For labels that exhibit uncertainty, this method may also be applied.

Finally, consider a probe to be a thresholded metric or similarity function such that the output is binary signifying presence (1) or absence (0) of a phenomenon. One difficult problem in fMRI analysis is searching for small signals that are not linearly related to the paradigm. These small signals presumably reside in only a few percent of the TCs and will not form their own cluster with a general cluster algorithm. Detecting a cluster with a centroid highly correlated to the paradigm is insufficient to also identify TCs containing this small signal. However, a series of probes may be assembled to detect a variety of nonlinear small signals. When the signal is detected in the clustering process (say, after a cluster has defined a centroid sufficiently similar to the paradigm), the probes, being based on the centroid, will be able to detect the small signal. Heuristics can be devised to change metrics when the probe indicates that the small signal is present. In this manner, the clustering process of the entire data-set is combined with a search for small related signals.

18.3 MAGNETIC RESONANCE IMAGING

Magnetic resonance imaging (MRI) [22,23] is a noninvasive imaging modality. The ability of MRI to contrast various soft tissues has led to new imaging applications of the brain, abdominal organs, the musculo-skeletal system, breast, heart, and blood vessels. It is the de facto standard for biomedical imaging. Different echo sequences, magnetic strengths, goal-specific techniques (for example, contrast agents), and coils provide high resolution spatial images. FMRI examine blood flow intensity changes produced by a structured stimulus. A stimulus may be any physical change produced near, on or in the subject. Common stimuli include cognitive, visual, tactile, or auditory effects. The stimulus is applied over an interval of time (the activated epoch) and then the subject is allowed to rest (the unactivated epoch). Intensity values are scanned continuously over the alternating epoch pairs. That is, each epoch contains multiple scans. MRI studies generate voluminous amounts of data for each acquisition (in the scores of megabytes). Analysis tests for order and relation in the presence of multiple noise sources with the cognizance that novel information underlies the mass of measurements and may not yet be incorporated into mathematical models in common use. Since its commercial deployment, MRI/fMRI has been the methodology of choice for investigating the structure of the human body and its functional behavior. In particular, brain activation studies present challenges for standard pattern recognition and analysis techniques.

[1]Note that, depending on the relative number of features in and p_1 and p_2, even the case $v_1 < v_2$ may show improvement in terms of discrimination. This can occur when there are few discriminating features. In general, the formula is $|p_1|v_1 > |p_2|v_2$ for two classes.

18.3.1 Nuclear Magnetic Resonance

The phenomenon of nuclear magnetic resonance involves the interaction of static and oscillating magnetic fields. For a volume of tissue outside of a magnetic field, the spins of constituent protons are randomly distributed and yield a (near) null net field. Within a static magnetic field B_0, the protons precess around B_0. Perpendicular to the field, the spin orientations are still randomly distributed. Parallel to B_0, the coupling of the static field and the spin orientations produces the so-called Zeeman interaction, which exhibits an energy difference between parallel and anti-parallel spins. The lower energy orientation (parallel) has the larger proton population and is characterized as a Boltzmann distribution. The equilibrium between the parallel and anti-parallel spins is known as the induced magnetization M_0. Irradiation of an object in a static magnetic field by an oscillating magnetic field (where the frequency is matched to the precession frequency) rotates the magnetization of the sample into the transverse field. Magnetic resonance occurs between an external magnetic field and a nucleus with a nonzero magnetic moment (spin). Hydrogen (H) has spin 1/2. Hydrogen has a high gyro-magnetic ratio γ and is thus sensitive to magnetic fields. The presence of hydrogen in both blood and fat makes viable imaging of *in vivo* tissues.

18.3.2 Image Acquisition and Noise

MRI data acquisition occurs while the subject lies in a static magnetic field generated. Current, common magnetic field strengths range from 1 to 5 tesla. The application of short radio-frequency (r.f.) pulses to the magnetized tissue causes the tissue to absorb, and subsequently re-emit, the energy. Note that the frequency of the r.f. pulse must be matched to the energy difference between the spin up and spin down orientations (parallel and anti-parallel). Energy absorption at the resonant frequency is resonance absorption. The time between energy absorption and re-emission is known as the relaxation time. Two main relaxation times are used to define imaging intensities: T_1 and T_2. The T_1 (spin–lattice relaxation) time is the time necessary for the z component of M to return to 63 % of its original value following an r.f. pulse. It measures the rate at which spins return to their original configuration. The T_2 (spin–spin relaxation) time is the time required for the transverse component of M to decay to 37 % of its initial value via irreversible processes. The TCs activation levels acquired relate the intensities of the deoxygenated blood to the activity.
There are many compromising factors in MR imaging:

- Various echo sequences have associated noise concerns and known interactions with other noise sources.
- Motion artifacts – aperiodic motion leads to blurring of the moving tissue. Peristaltic motion is seen as noise over the stationary tissue. Periodic motion is manifested in ghost images. Flow artifacts effects depend on flow velocity.
- Misregistration (pixels misalignment) may occur due to machine drift.
- Coil noise; inhomogeneity in magnetic field.

18.3.3 Functional Magnetic Resonance Imaging

FMRI consists of a series of images that infer organ, typically brain, function through de-oxygenated blood flow intensities [24]. A pattern of activations is generated by having the subject repeat a task or receive stimuli. At each acquisition instance, the subject is considered either passive (resting or 0) or active (receiving stimulus or 1). The paradigm is the series of off–on activity states and is used as a reference for the intensity acquisitions. A TC is a time series of intensities for an image voxel. The neural regions stimulated in a study will include regions known to relate to the activity, novel (unanticipated) areas, and noisy voxels. It is common practice to employ EDA techniques in fMRI studies as mathematical models continue to develop. When developing strategies to reduce noise in the acquired signals, it is critical that the processing should not remove novel patterns and relations.

18.4 FMRI ANALYSIS WITH FCMP

The following applications of FCMP to problems using fMRI data demonstrate the versatility of a generalized clustering algorithm and the practicality of various specializations.

18.4.1 Adding Spatial Context to FCMP

Consider a partition of the sample features into spatial ($S = \{x, y, z\}$) and temporal features ($T = \{t_1, t_2, \ldots, t_n\}$) for n time instances. Using the previous notation, the feature partition P is

$$P = \{S, T\} = \{\{x, y, z\}, \{t_1, t_2, \ldots, t_n\}\} \tag{18.16}$$

and the respective partition weights are $v = \{v_S, v_T\}$. Denote distances (metrics) for the partitions as d_S and d_T. Substituting these specific values into the expanded formula, the FCMP objective function adapted for fMRI is

$$J = v_S \sum_{x \in X} \sum_{v \in V} u_{xk}^m d_S^2(x, k) + v_T \sum_{x \in X} \sum_{v \in V} u_{xk}^m d_T^2(x, k) - \lambda \left(\sum_{x \in X} \sum_{v \in V} u_{xk} - 1 \right). \tag{18.17}$$

For a particular sample x, the objective function is

$$J = v_S \sum_{v \in V} u_{xk}^m d_S^2(x, k) + v_T \sum_{v \in V} u_{xk}^m d_T^2(x, k) - \lambda \left(\sum_{v \in V} u_{xk} - 1 \right). \tag{18.18}$$

The membership update equation is

$$f(x, c) = \left[\sum_{v \in V} \frac{v_S d_S^2(x, c) + v_T d_T^2(x, c)}{v_S d_S^2(x, v) + v_T d_T^2(x, v)} \right]^{(m-1)}. \tag{18.19}$$

The interpretation of the objective function is this. Minimize the temporal distance of the sample to the time centroid and minimize the spatial distance to the spatial centroid based on the weights v_S and v_T. Spatial proximity and temporal similarity are considered at each iteration of the cluster process at specified levels of integration.

18.4.2 Parameters and Typical Use

Consider the remaining FCMP parameters with spatial context. The Euclidean metric is normally used with spatial features. A variety of metrics may be applicable to temporal features. In practice, a distance metric based on the Pearson correlation coefficient is often used for temporal features. Variations on the theme of temporal distance include considering only activated epochs (stimulus is applied), including anticipation and relaxation responses of the subject (this would consider portions of the unactivated epochs proceeding and following the stimulation epochs), or allocating increased weight to activated epochs.

A fuzzification exponent, m, close to but greater than 1 (typically, 1.1–1.3) tends to reduce the effects of noise. This is significant since common signal-noise (SNR) values for fMRI studies are 2–5 [25]. Algorithms that use cluster merging (for instance, EvIdent® [26]) often initialize the algorithm with more clusters than the analyst expects in the data-set (typically, 10–40).

18.4.3 Extending Spatial Context in FCMP

Additional information may be implicitly included in the FCMP functional by considering the spatial neighbors of a sample TC, x_i. Consider the effect of replacing x_i by a representative TC that shares both spatial proximity and signal similarity. Let γ denote a mapping from a given sample index i to the index of a neighbor j that best represents the neighborhood around x_i. For each x_i, form its set of neighbors, where the neighborhood is defined in terms of spatial and temporal similarity. The functional now has the form

$$J_\gamma(U,V;X) = \sum_{v \in V} \sum_{x \in X} u^m_{\gamma(x)v} D^2_{\gamma(x)v} \tag{18.20}$$

where $\gamma(x)$ is the neighborhood (spatial) representative for x_i. There are two additional degrees of freedom in this modification: (a) the definition of a neighborhood for a sample TC and (b) the selection of a representative TC from the neighborhood. This alternate formulation appends a term to the FCMP objective function

$$J_{m,\alpha,F}(U,V;X) = \sum_{k=1}^{K} \sum_{n=1}^{N} u^m_{nk} D^2_{nk} + \alpha \sum_{n=1}^{N} F(x_n, N_e(n)) \tag{18.21}$$

where $F(x_n, N_e(n))$ is a function of sample x_n and its neighborhood $N_e(n)$.

One possible implementation of F is a weighted sum of the neighborhood scatter matrix:

$$F = \beta_i \sum_{j=1}^{|N_e(i)|} (x_i - x_j)^T (x_i - x_j) = S_c(N_e(i)). \tag{18.22}$$

Such an objective function is said to contain spatial and temporal terms.

18.5 DATA-SETS

Both synthetic and authentic data-sets will be used. The synthetic data-sets that were generated are discussed first, followed by a discussion on the authentic data-sets.

Neural activation studies examine the spatial regions (volume elements or voxels) associated with paradigm-correlated TCs. Regions of interest are defined in space for time sequences of importance. Such a designation increases the importance of other spatially proximal TCs. While clustering algorithms define centroids, they can also define spatial partitions on the fMRI images, using the cluster membership values as voxel memberships in various partitions.

18.5.1 Spatially Separated Clusters with Noise

This synthetic fMRI data-set (synth1), based on a stimulus paradigm with various levels of noise added, demonstrates how feature partitions interact when noise degrades the feature partitions independently and when the partitions are correlated. Synth1 contains two feature partitions with specific noise distributions and comprises 100 degraded and 100 noisy samples (Figure 18.2). The first feature partition contains spatial coordinates (x, y); the second temporal intensities (TCs). The spatial data is composed of uniform sampling (radii and angle) in two circular areas ($\mu_1 = (0.1, 0.1), \mu_2 = (0.9, 0.9)$) with the same radii ($r_1 = r_2 = 0.1$). Temporal data is composed of degraded TCs and noise TCs. A stimulus defines the activation by the sequence [0 1 0 1 0 1 0 1 0 0]. A paradigm TC maps the stimulus to the number of sampling instants. Thus, successive time instances may belong to the same stimulus epoch. The degraded TCs are generated by adding noise to the paradigm TC at a specific SNR level. The noise TCs are uniform random values in [0,1]. The data-set is the concatenation of the spatial data to the temporal data (the degraded TCs are associated with μ_1; the noise TCs are associated with μ_2).

18.5.2 Visual Cortex Study

This "real-world" data-set (auth1) was acquired from a visual cortex study using a visual stimulus with a paradigm of [0 1 0 1 1 0]. The stimulus was presented to an individual and the neuron activations of a single slice were recorded. Figure 18.3 shows the mean intensity coronal image (regions of high average intensity appear lighter in the image). The data-set is composed of TCs with dimension

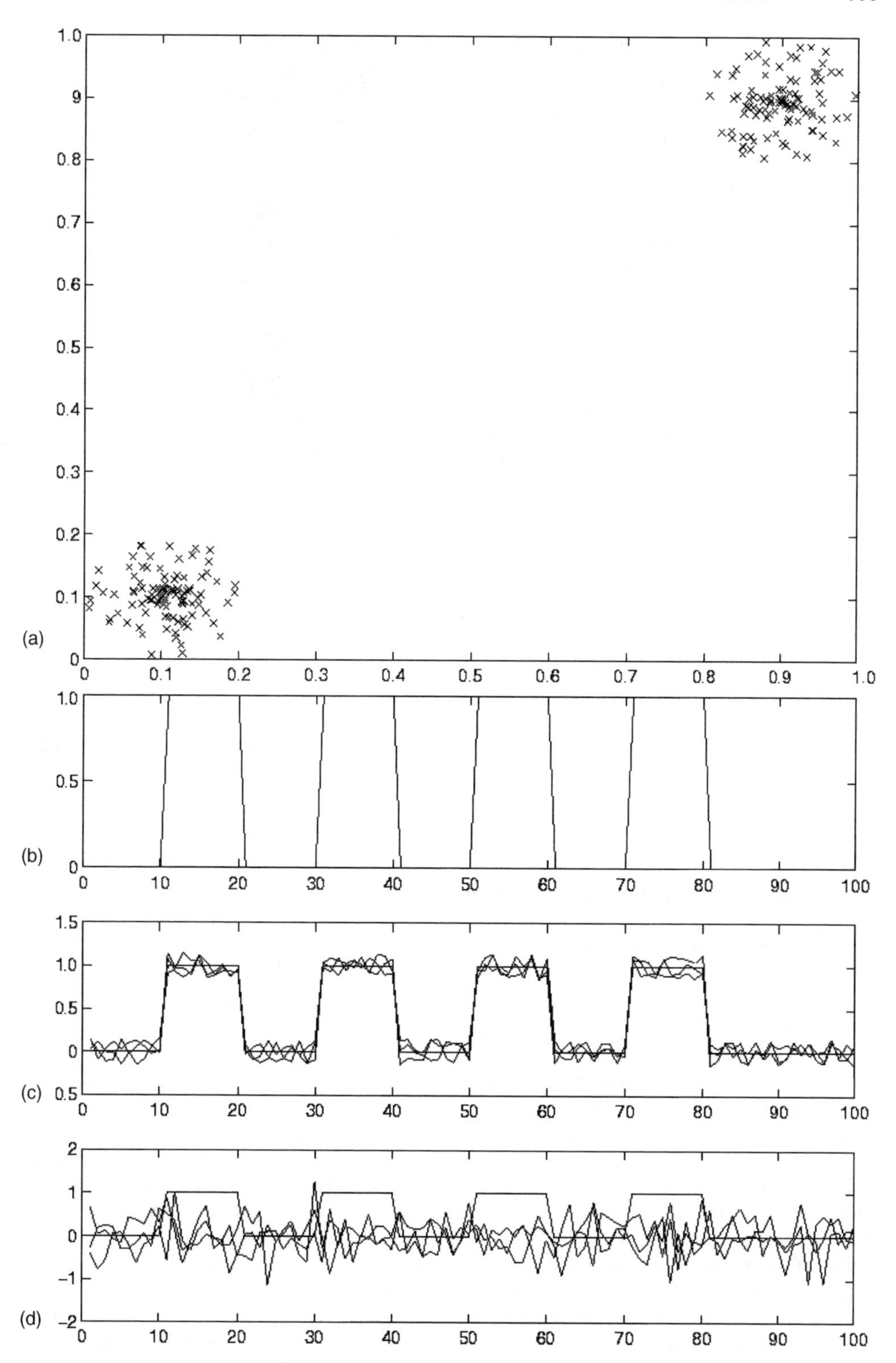

Figure 18.2 (a) Spatial distribution of synth1. Time courses: (b) paradigm, (c) degraded TCs, (d) noisy TCs.

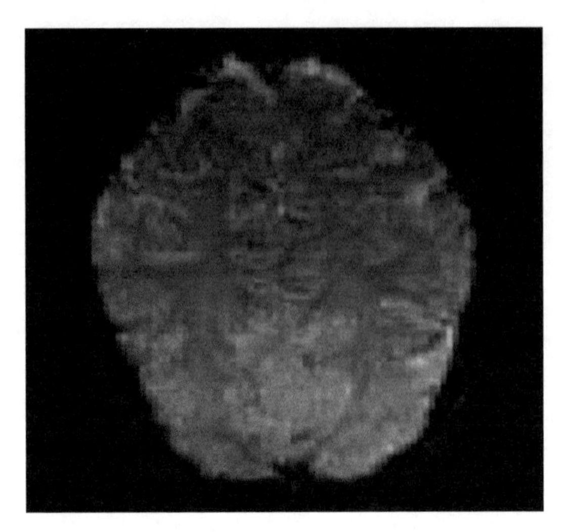

Figure 18.3 Mean intensity coronal image from visual cortex study. An example of single slice fMRI acquisition.

$128 \times 256 \times 1 \times 42(X, Y, Z, T)$. Figure 18.4 shows some typical TCs with above average intensity and above average correlation to the paradigm.

An examination of the histogram values in auth1 (Figure 18.5) shows that regions with TCs highly correlated to the paradigm will be small. (Any region growing operations in the region of interest will suffer loss of continuity unless they accept marginally correlated TCs that are nonetheless spatially proximal.) Examining the spatial distribution of correlated voxels, a correspondence is seen between the high average intensity voxels. Note that (a) the visual cortex has a significant number of highly correlated voxels, (b) some noise TCs (outside of the subject's body) have a significant correlation, and (c) a second separate region of correlated voxels exists, again, near the bottom of the image. Figure 18.6 shows the spatial distribution of TCs with Pearson correlation > 0.5.

Figure 18.7 shows the stimulus paradigm mapped to the 42 sampling instants, TCs in the visual cortex, and noise TCs.

18.6 RESULTS AND DISCUSSION

18.6.1 Proof of Concept

The proof of concept experiment uses auth1, FCM as a benchmark, and the minimum mean square error (MSE) of the centroid to the paradigm as the validation criterion. We compare FCMP to FCM over a range of SNR values for the degraded TCs (2, 5, 10, 20, 30 and 40). The spatial distance between the spatial cluster centres is 1 and Euclidean and Pearson distances are the metrics applied to each partition. The weights applied to each partition are 0.0, 0.1, 0.9, and 1.0.

For high SNR values, the feature partition weight contributed little additional enhancement in terms of MSE and correlation. As the SNR decreased, there is an optimal weighting pair for the spatial and temporal feature partitions. Several expected trends were confirmed by the experiments. As the number of clusters increases, the overall MSE decreases, regardless of SNR for the correlated TCs and the weight β_S. However, for an increase in the number of clusters, both FCM and FCMP were more likely to generate additional noise clusters. As the noise in the data-set increases, the minimum MSE of the resulting centroids increases. The following relation is noted between SNR, MSE and β_S: as $\beta_S \rightarrow 0$, the MSE depends increasingly on the SNR. That is, TCs with higher temporal SNR have lower values of β_S for the same MSE. As reliance on the low noise feature domain increased ($\beta_S \rightarrow 1$), the MSE decreases. This occurred for almost all cases; for the synthetic data-set at low SNR, several outlying points were

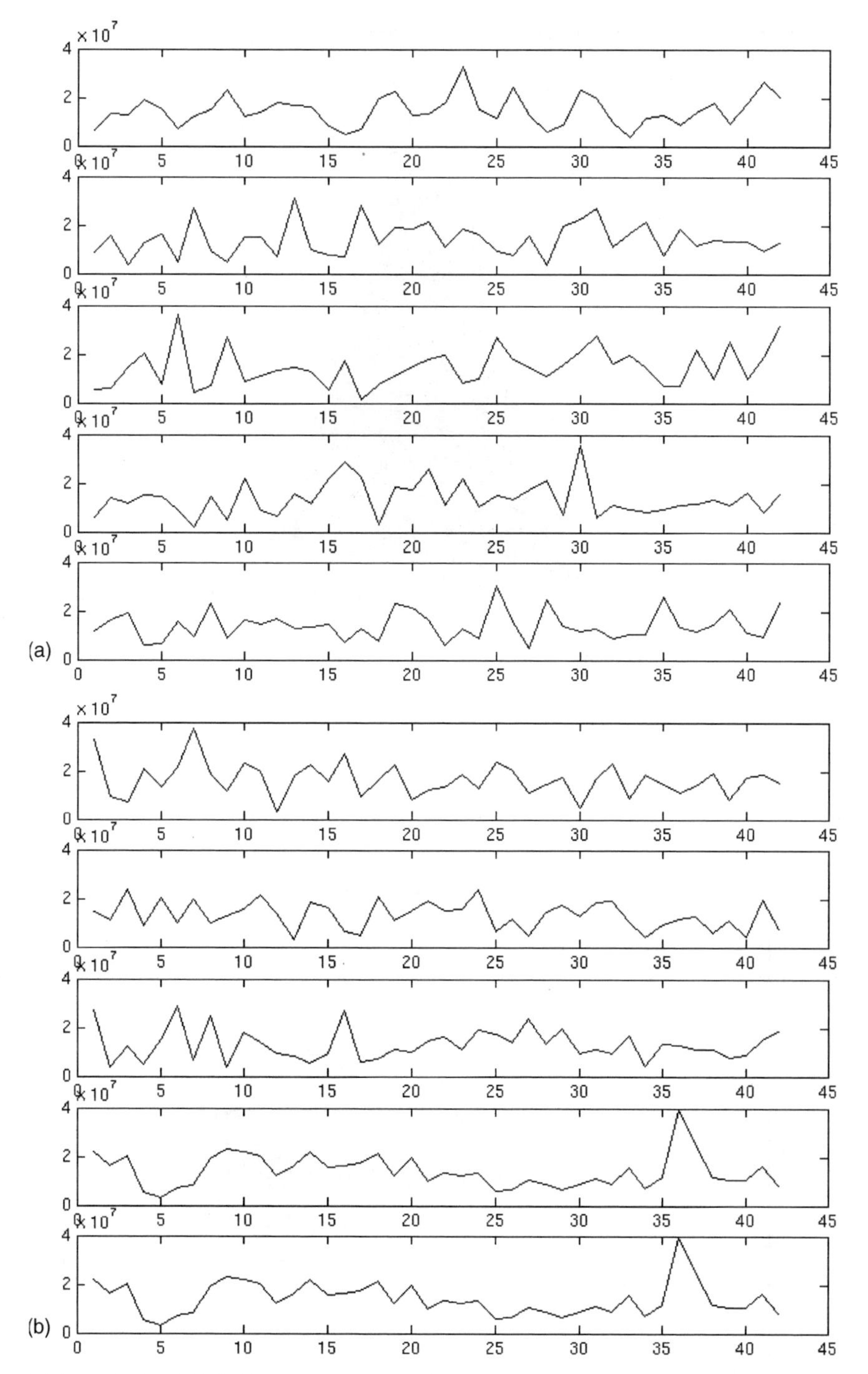

Figure 18.4 Typical auth1 TCs (a) with above average intensity and (b) above average correlation to paradigm.

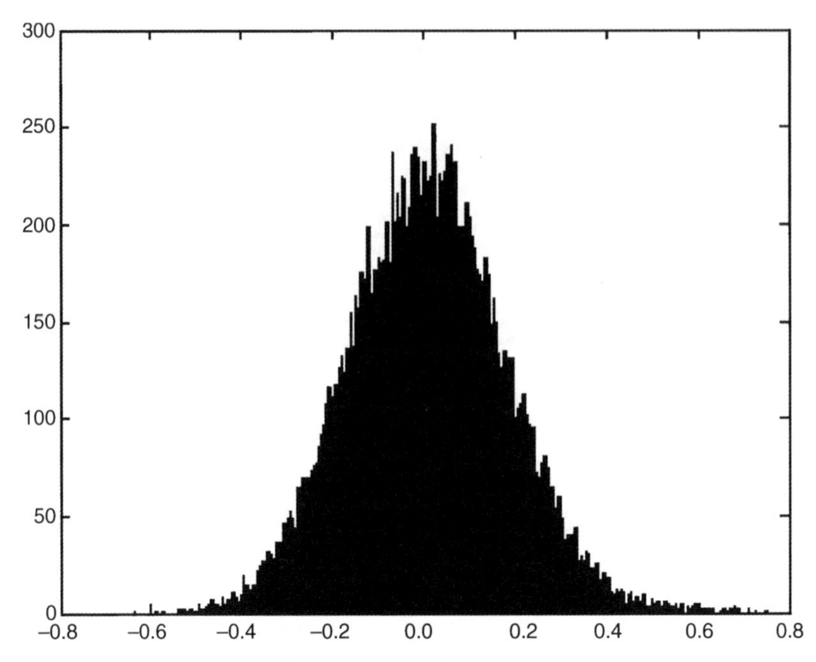

Figure 18.5 Histogram of correlation values.

discovered. This corresponds to a locally optimal value of data coordination, where the incorporation of degraded signals is beneficial. These inflection points in the general trends of decreasing MSE for increasing β_S were noted for mainly low SNR data-sets and occurred at $\beta_S = 0.9$ as shown in Figure 18.8. A range of SNR values reproduce this inflection; $\beta_S = 0.9$ seems optimal for these particular feature domains. The slope of the trends in Figure 18.8 suggest a critical SNR value at which β_S has a significant ameliorative effect (2–5 SNR).

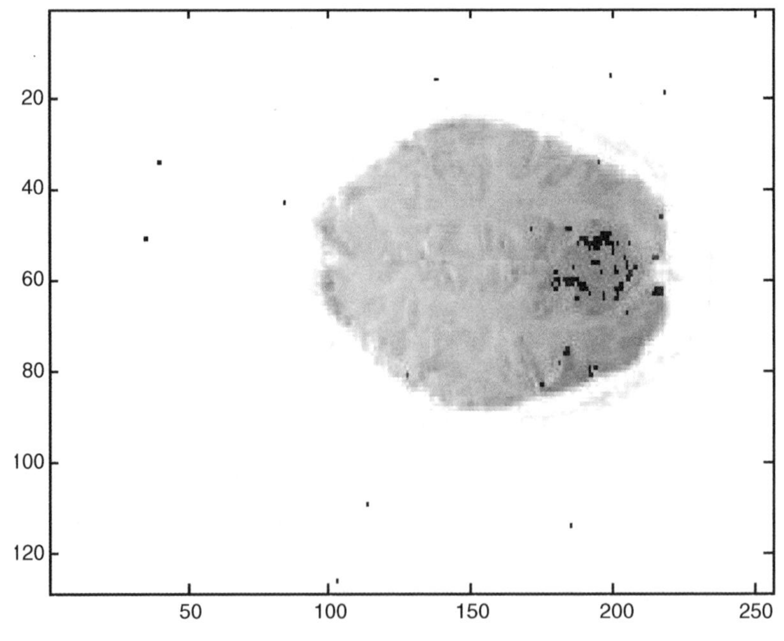

Figure 18.6 Spatial distribution of time courses with Pearson correlation >0.5.

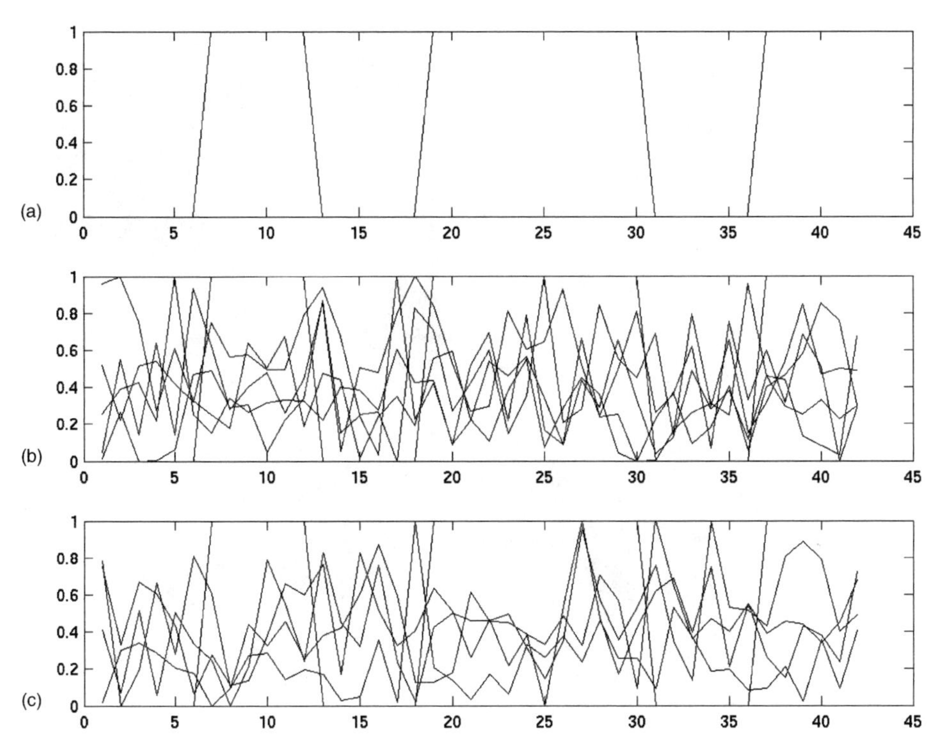

Figure 18.7 (a) Auth1 paradigm, (b) visual cortex TCs, (c) noisy TCs.

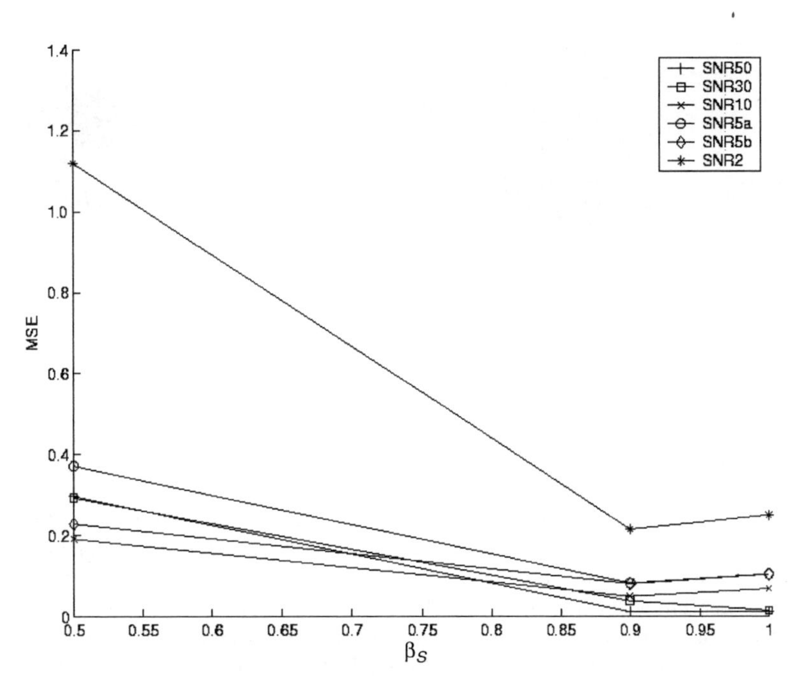

Figure 18.8 Inflection points for the spatial weighting term and MSE.

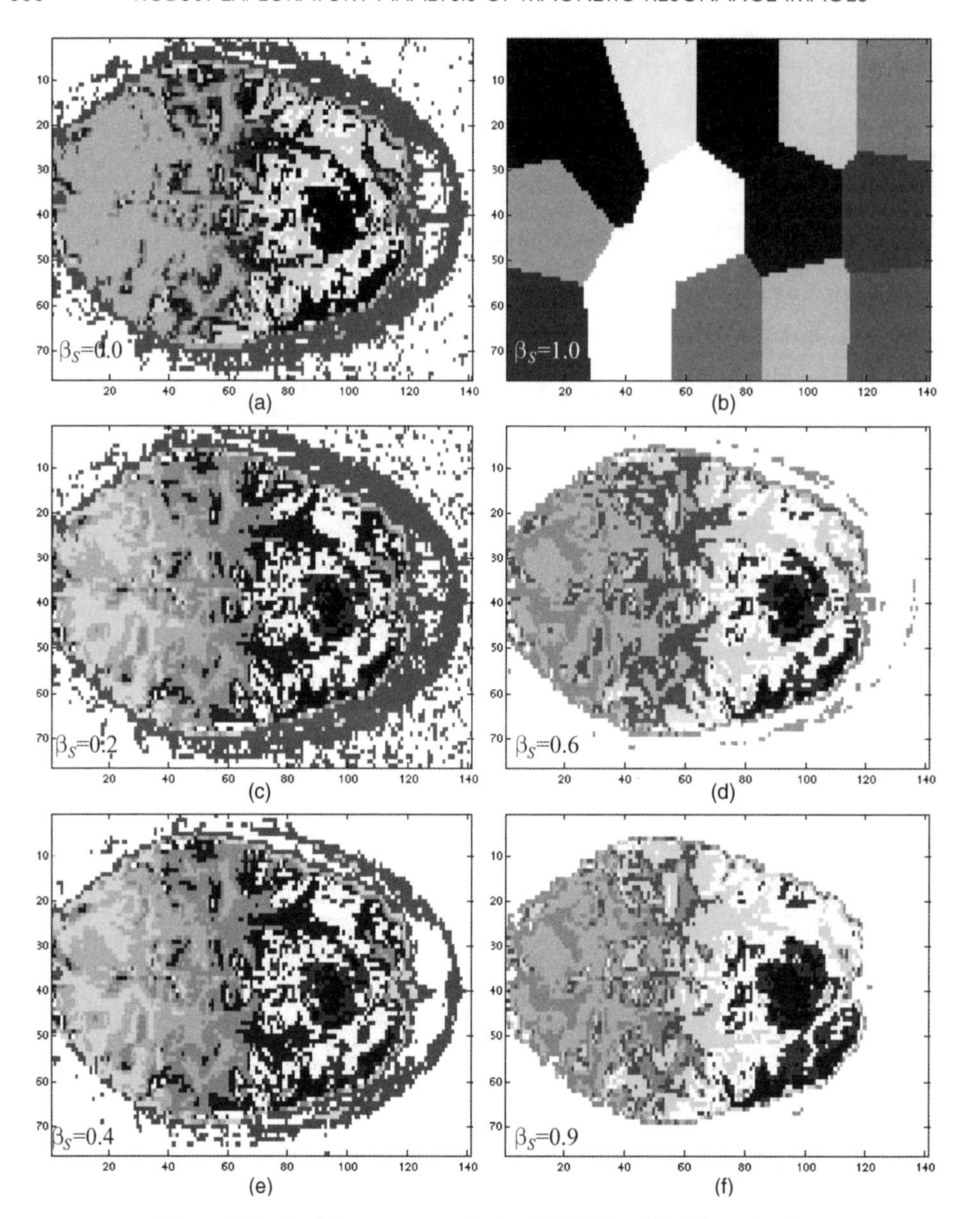

Figure 18.9 Spatial maps generated using FCMP ($C = 15$) with varying β_S.

18.6.2 Visual Stimulus

Previously, auth1 was examined at a general level using basic preprocessing operations. It was seen that spatial continuity of the region of interest could be achieved using spatial topology without any consideration for temporal similarity. FCMP will address this by defining spatial and temporal feature partitions. One concern is that it is not obvious how to relate the spatial and temporal domains. This is solved by executing a series of cluster experiments, each with a different weight between the partitions. Figure 18.9 shows a set of spatial maps for auth1 using FCMP with $C = 15$ clusters and varying values

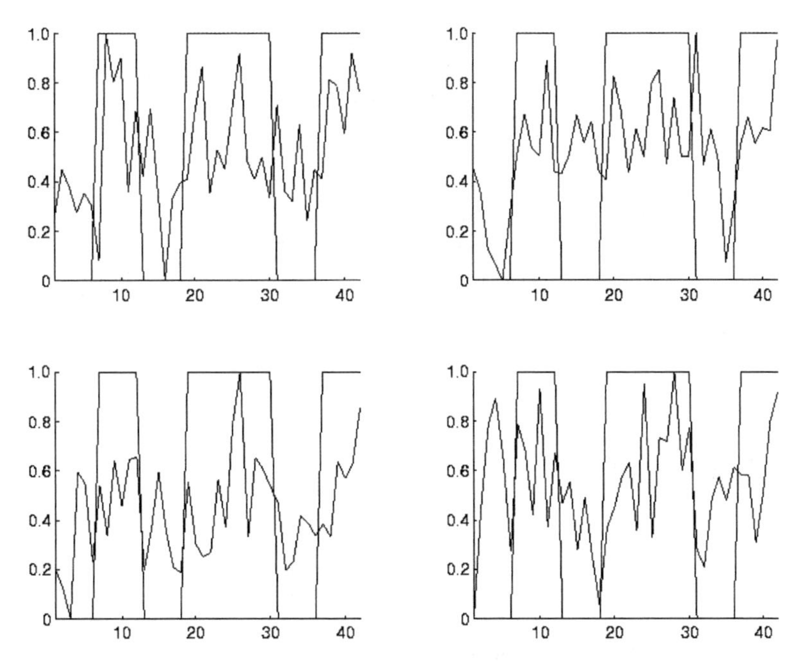

Figure 18.10 False negative time courses that were not selected but near the region of interest.

for β_S. Note that when one feature partition weight is 0 and there are only two weights, the results are exactly that of FCM with the same metric function. When $\beta_S = 0$ one expects to have spatially disconnected regions; when $\beta_S = 1$ one expects to have spatially convex regions. This can be seen in Figure 18.9(a) and (b), respectively. When $0 < \beta_S < 1$ interesting effects can be observed in the resulting

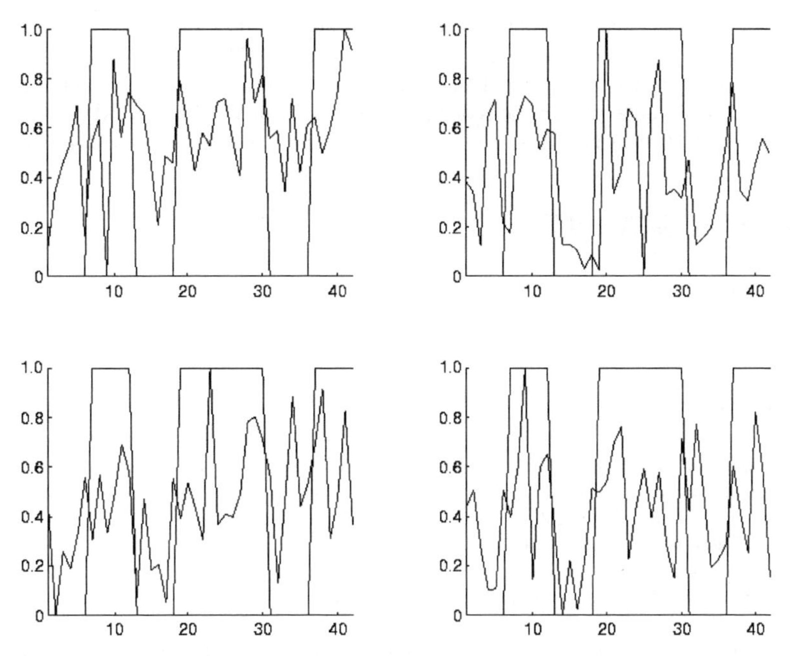

Figure 18.11 False positive time courses that were selected but not near the region of interest.

voxel assignment maps. Temporal similarity and spatial proximity combine to produce spatial regions that exhibit degrees of spatial continuity. The centroids of these regions show corresponding increasing correlation (for regions correlated to the visual cortex) or decreasing correlation (for noise regions). Based on MSE of the centroids to the paradigm, $0.6 < \beta_S < 0.9$ values gave the best results. Examining just the visual cortex region, one can detect the increase in spatial continuity as the spatial feature partition weight increases.

The results of FCMP are now compared to an industry standard system, EvIdent®[26], which uses a version of FCM highly specialized for fMRI analysis. Default parameters were used with Evident®. Figure 18.10 shows four TCs that were spatially proximal to the region of interest (visual cortex) but were not selected as significant by this system. Figure 18.11 shows four TCs that were not spatially proximal to the region of interest but were selected as significant. In both the former case of proximal rejections (false negatives) and the latter case of outlier inclusions (false positives), FCMP correctly clustered the TCs.

18.7 CONCLUSION

The challenges to fMRI data analysis are incorporating spatial information into the temporal fuzzy clustering process (adding spatial context) and discovering novel time courses for further analysis (related to minimizing the mathematical model used in describing the data and exploratory data analysis). With these challenges in mind, a novel fuzzy clustering algorithm was developed, namely FCMP. An FCM extension, FCMP addressed the above issues as follows. FCMP incorporates spatial information by defining multiple feature partitions. Specifically, for fMRI, one is used for temporal features (traditional approach) and one used for spatial features (new contribution). Continuity with advantages and experience accumulated with previous algorithms (FCM variants) can be extended to FCMP. FCMP discovers novel time courses for further analysis by changing the weight given to the different feature partitions. Extreme weight values correspond to considering only a single feature partition. Thus the original FCM analysis was preserved as an option in FCMP analysis. Novelty was defined in terms of results with current industry standard cluster analysis algorithms. The novel time courses discovered by FCMP had significant temporal correlation as well as spatial proximity to regions of interest. FCMP makes robust statements about fMRI data. Examples include: FCMP had a lower false positive rate than an industry standard for the determination of correlated voxel regions (spatial outliers were rejected); FCMP returned more stable spatial regions of interest than the vanilla FCM for fMRI data. This was true with respect to both changes in SNR values for temporal features and distance for spatial features.

ACKNOWLEDGEMENTS

We would like to thank the Natural Sciences and Engineering Research Council of Canada (NSERC) for their financial support of this investigation.

REFERENCES

J.W. Tukey, *Exploratory Data Analysis*, Reading: Addison-Wesley Publishing, 1977.

J.C. Bezdek, 'A Convergence Theorem for the Fuzzy ISODATA Clustering Algorithms', *IEEE Transactions on Pattern Analysis and Machine Intelligence*, **2**, 1–8, 1980.

N.J. Pizzi *et al.* 'Neural network classification of infrared spectra of control and Alzheimer's diseased tissue', *Artificial Intelligence in Medicine*, **7**, 67–79, 1995.

N.J. Pizzi, 'Classification of biomedical spectra using stochastic feature selection', *Neural Network World*, **15**, 257–268, 2005.

G.E.P. Box and G.M. Jenkins, *Time Series Analysis, Forecasting and Control*, Holden-Day: San Francisco, 1970.

P.J. Diggle, *Time Series: A Biostatistical Introduction*, Oxford: Oxford University Press, 1990.

G.C. Reinsel, *Elements of Multivariate Time Series Analysis*, New York: Springer-Verlag, 1997.

F. Melgani and S.B. Serpico, 'A statistical approach to the fusion of spectral and spatio-temporal contextual information for the classification of remote sensing images', *Pattern Recognition Letters*, **23**, 1053–1061, 2002.

I. Couso, S. Montes and P. Gil, 'Stochastic convergence, uniform integrability and convergence in mean on fuzzy measure spaces', *Fuzzy Sets and Systems*, **129**, 95–104, 2002.

P.T. Jackway, 'On the scale-space theorem of Chen and Yan', *IEEE Transactions on Pattern Analysis and Machine Intelligence*, **20**, 351–352, 1998.

A. Lowe, R.W. Jones and M.J. Harrison, 'Temporal pattern matching using fuzzy templates', *Journal of Intelligent Information Systems*, **13**, 27–45, 1999.

S. Roychowdhury and W. Pedrycz, 'Modeling temporal functions with granular regression and fuzzy rules', *Fuzzy Sets and Systems*, **126**, 377–387, 2002.

P. Sincak and J. Vascak (Ed.), *Quo Vadis Computational Intelligence? New Trends and Approaches in Computational Intelligence*, Heidelberg: Springer-Verlag, 2000.

V. Torra, 'Learning weights for the quasi-weighted means', *IEEE Transactions on Fuzzy Systems*, **10**, 653–666, 2002.

M.D. Alexiuk and N.J. Pizzi, 'fuzzy C-means with feature partitions: a spatio-temporal approach to clustering fMRI data', *Pattern Recognition Letters*, **26**, 1039–1046, 2005.

W. Pedrycz, 'Collaborative fuzzy clustering', *Pattern Recognition Letters*, **23**, 1675–1686, 2002.

F. Höppner, F. Klawonn, R. Kruse, and T. Runkler, *Fuzzy Cluster Analysis: Methods for Classification, Data Analysis and Image Recognition*, New York: John Wiley and Sons, Inc., 1999.

M.D. Alexiuk, 'Spatio-temporal fuzzy clustering of functional magnetic resonance imaging data', PhD Thesis, University of Manitoba, 2006.

J. Leski, 'Towards a robust fuzzy clustering', *Fuzzy Sets and Systems*, **137**, 215–233, 2003.

S.J. Farlow (Ed.), *Self-Organizing Methods in Modeling: GMDH Type Algorithms*, New York: Marcel Dekker, 1984.

W. Pedrycz and J. Waletzky, 'Fuzzy clustering with partial supervision', *IEEE Transactions on Systems, Man and Cybernetics*, **27**, 787–795, 1997.

M.A. Brown and R.C. Semelka, *MRI: Basic Principles and Applications*, New York: Wiley-Liss, 1995.

V. Kuperman, *Magnetic Resonance Imaging: Physical Principles and Applications*, New York: Academic Press, 2000.

K.J. Friston *et al.*, 'Statistical parametric maps in functional imaging: a general linear approach', *Human Brain Mapping*, **2**, 189–210, 1995.

M. Jarmasz and R.L. Somorjai, 'EROICA: exploring regions of interest with cluster analysis in large functional magnetic resonance imaging data sets', *Concepts in Magnetic Resonance*, **16**, 50–62, 2003.

N.J. Pizzi, R. Vivanco and R.L. Somorjai, 'EvIdent: a java-based fMRI data analysis application', *Proceedings of SPIE*, **3808**, 761–770, 1999.

19

Concept Induction via Fuzzy C-means Clustering in a High-dimensional Semantic Space

Dawei Song[1], Guihong Cao[2], Peter Bruza[3], and Raymond Lau[4]

[1] *Knowledge Media Institute and Centre for Research in Computing, The Open University, Milton Keynes, UK*
[2] *Department d'Informatique et Recherche Operationnelle, Universite de Montreal, Canada*
[3] *School of Information Technology, Queensland University of Technology, Brisbane, Australia*
[4] *Department of Information Systems, City University of Hong Kong, Kowloon, Hong Kong SAR*

19.1 INTRODUCTION

A human encountering a new concept often derives its meaning via an accumulation of the contexts in which the concept appears. Based on this characteristic, various lexical semantic space models have been investigated. The meaning of a word is captured by examining its co-occurrence patterns with other words in the language use (e.g., a corpus of text). There have been two major classes of semantic space models: document spaces and word spaces. The former represents words as vector spaces of text fragments (e.g., documents, paragraphs, etc.) in which they occur. A notable example is the latent semantic analysis (LSA) (Landauer and Dumais, 1997). The latter represents words as vector spaces of other words, which co-occur with the target words within a certain distance (e.g., a window size). The strength of the association can be inversely proportional to the distance between the context and target words. The hyperspace analog to language (HAL) model employs this scheme (Lund and Burgess, 1996). The dimensionality of semantic spaces is often very high, for example, Lund and Burgess (1996) constructed a

Advances in Fuzzy Clustering and its Applications Edited by J. Valente de Oliveira and W. Pedrycz

70 000 × 70 000 HAL vector space from a 300 million word textual corpus gathered from Usenet. The concepts occurring in the similar contexts tend to be similar to each other in meaning. For example, "nurse" and "doctor" are semantically similar to each other, as they often experience the same contexts, i.e., hospital, patients, etc. The similarity can be measured by the angle (cosine) or Euclidean distance between two word vectors in the semantic space.

Semantic space models can be considered as computational approximations of the conceptual spaces advocated by Gärdenfors (2000), which are built upon geometric structures representing concepts and their properties. At the conceptual level, information is represented geometrically in terms of a dimensional space. In this chapter, we propose to use HAL vectors to prime the geometric representation of concepts. HAL vectors are also interesting because semantic associations computed using these vectors correlate with semantic associations drawn from human subjects (Burgess, Livesay, and Lund, 1998). It has been shown that HAL vectors can be used to simulate semantic, grammatical, and abstract categorizations (Burgess, Livesay, and Lund, 1998). Another advantage of the HAL approach is that it is automatic and computationally tractable.

In a conceptual space, a domain is defined as a set of integral dimensions in the sense that a value in one dimension(s) determines or affects the value in another dimension(s). For example, pitch and volume are integral dimensions representing a domain of "sound." Gärdenfors and Williams (2001) state "the ability to bundle up integral dimensions as a domain is an important part of the conceptual spaces framework." The thrust of Gärdenfors' proposal is that concepts are dimensional objects comprising domains. A domain, in turn, is a vector space with as basis a set of integral dimensions. Properties are represented as regions within a given domain.

By their very nature, conceptual spaces do not offer a hierarchy of concepts as is often the case in ontologies, taxonomies, and suchlike. However, similar objects can be grouped due to their semantic similarity. By way of illustration, a set of feathered objects with wings leads to the grouping "bird." This facet of conceptual space is referred to as *concept induction* in this chapter. One way of gaining operational command of concept induction is by means of clustering of objects in a semantic space. Clustering techniques divide a collection of data into groups or a hierarchy of groups based on similarity of objects (Chuang and Chien, 2005). A well-known clustering algorithm is the K-means method (Steinbach, Karypis, and Kumar, 2000; Cimiano, Hotho, and Staab, 2005), which takes a desirable number of clusters, K, as input parameter, and outputs a partitioning of K clusters on the set of objects. The objective is to minimize the overall intra-cluster dissimilarity, which is measured by the summation of distances between each object and the centroid of the cluster it is assigned to. A cluster centroid represents the mean value of the objects in the cluster. A number of different distance functions, e.g., Euclidean distance, can be used as the dissimilarity measure.

Conventional clustering algorithms normally produce crisp clusters, i.e., one object can only be assigned to one cluster. However, in real applications, there is often no sharp boundary between clusters. For example, depending on the context it occurs, President "Reagan" could belong to a number of different clusters, e.g., one cluster about the US government administration and another about the Iran-contra scandal. The latter reflects the fact that he was involved in the illegal arms sales to Iran during the Iran–Iraq war. Therefore, a membership function can be naturally applied to clustering, in order to model the degree to which an object belongs to a given cluster. Among various existing algorithms for fuzzy cluster analysis (Höppner, Klawonn, Kruse, and Runkler, 1999), a widely used one is the fuzzy C-means (Hathaway, Bezdek, and Hu, 2000; Krishnapuram, Joshi, Nasraoui, and Yi, 2001; Höppner and Klawonn, 2003; Kolen and Hutcheson, 2002, etc.), a fuzzification of the traditional K-means clustering.

The practical implication of the use of fuzzy clustering for conceptual induction is rooted in its ability to exploit the context sensitive semantics of a concept as represented in semantic space. There is a connection here with the field of text mining. Generally speaking, text mining aims at extracting new and previously unknown patterns from unstructured free text (Hearst, 2003; Perrin and Petry, 2003; Srinivasan, 2004). Conceptual space theory and its implementation by means of semantic space models introduced in this chapter provides a cognitively validated dimensional representation of information based on the premise that associations between concepts can be mined (Song and Bruza, 2003).

The goal of this chapter is to introduce the construction of a high-dimensional semantic space via the HAL model (Section 19.2) and address how a fuzzy C-means clustering algorithm, presented in Section 19.3, can be applied to conceptual induction within a HAL space. Its effectiveness is illustrated by a case study included in Section 19.4. Finally, in Section 19.5 we conclude the chapter and highlight some future directions.

19.2 CONSTRUCTING A HIGH-DIMENSIONAL SEMANTIC SPACE VIA HYPERSPACE ANALOG TO LANGUAGE

In this section, we give a brief introduction to the hyperspace analog to language (HAL) model. Given an n-word vocabulary, the HAL space is a word-by-word matrix constructed by moving a window of length l over the corpus by one word increment ignoring punctuation, sentence, and paragraph boundaries. All words within the window are considered as co-occurring with each other with strengths inversely proportional to the distance between them. Given two words, whose distance within the window is d, the weight of association between them is computed by $(l - d + 1)$. After traversing the whole corpus, an accumulated co-occurrence matrix for all the words in a target vocabulary is produced. HAL is direction sensitive: the co-occurrence information for words preceding every word and co-occurrence information for words following it are recorded separately by its row and column vectors. By way of illustration, the HAL space for the example text "the effects of spreading pollution on the population of Atlantic salmon" is depicted below (Table 19.1) using a five word moving window ($l = 5$). Note that, for ease of illustration, in this example we do not remove the stop words such as "the," "of," "on," etc. The stop words are dropped in the experiments reported later. As an illustration, the term "effects" appears ahead of "spreading" in the window and their distance is two-word. The value of cell (spreading, effect) can then be computed as: $5 - 2 + 1 = 4$.

This table shows how the row vectors encode preceding word order and the column vectors encode posterior word order. For the purposes of this chapter, it is unnecessary to preserve order information, so the HAL vector of a word is represented by the addition of its row and column vectors.

The quality of HAL vectors is influenced by the window size; the longer the window, the higher the chance of representing spurious associations between terms. A window size of eight or ten has been used in various studies (Burgess, Livesay, and Lund, 1998; Bruza and Song, 2002; Song and Bruza, 2001; Bai et al., 2005). Accordingly, a window size of eight will also be used in the experiments reported in this chapter.

More formally, a concept[1] c is a vector representation: $c = < w_{cp_1}, w_{cp_2}, \ldots, w_{cp_n} >$ where $p_1, p_2, \ldots, p_n$ are called dimensions of c, n is the dimensionality of the HAL space and w_{cp_i} denotes the weight of p_i in the vector representation of c. In addition, it is useful to identify the so-called *quality properties* of a HAL vector. Intuitively, the quality properties of a concept or term c are those terms which often appear in the same context as c. Quality properties are identified as those dimensions in the HAL

Table 19.1 Example of a HAL space.

	the	effects	of	spreading	pollution	on	population	Atlantic	salmon
the		1	2	3	4	5			
effects	5								
of	8	5		1	2	3	5		
spreading	3	4	5						
pollution	2	3	4	5					
on	1	2	3	4	5				
population	5		1	2	3	4			
Atlantic	3		5		1	2	4		
salmon	2		4			1	3	5	

[1]The term "concept" is used somewhat loosely to emphasize that a HAL space is a primitive realization of a conceptual space.

vector for c which are above a certain threshold (e.g., above the average weight within that vector). A dimension is termed a property if its weight is greater than zero. A property p_i of a concept c is termed a *quality property* iff $w_{cp_i} > \partial$, where ∂ is a nonzero threshold value. From a large corpus, the vector derived may contain much noise. In order to reduce the noise, in many cases only certain quality properties are kept. Let $QP_\partial(c)$ denote the set of quality properties of concept c. $QP_\mu(c)$ will be used to denote the set of quality properties above mean value, and $QP(c)$ is short for $QP_\partial(c)$.

HAL vectors can be normalized to unit length as follows:

$$w_{c_ip_j} = \frac{w_{c_ip_j}}{\sqrt{\sum_k w_{c_ip_k}{}^2}}.$$

For example, the following is the normalized HAL vector for "*spreading*" in the above example (Table 19.1):

spreading $= <$ the: 0.52, effects: 0.35, of: 0.52, pollution: 0.43, on: 0.35, population: 0.17 $>$

In language, word compounds often refer to a single underlying concept. As HAL represents words, it is necessary to address the question of how to represent a concept underpinned by more than a single word. A simple method is to add the vectors of the respective terms in a compound. In this chapter, however, we employ a more sophisticated concept combination heuristic (Bruza and Song, 2002). It can be envisaged as a weighted addition of underlying vectors paralleling the intuition that, in a given concept combination, some terms are more dominant than others. For example, the combination "GATT[2] Talks" is more "GATT-*ish*" than "talk-*ish*." Dominance is determined by the specificity of the term.

In order to deploy the concept combination in an experimental setting, the dominance of a term is determined by its inverse document frequency (*idf*) value. The following equation shows a basic way of computing the *idf* of a term t:

$$idf(t) = \log(N/n)$$

where N is the total number of documents in a collection and n is the number of documents which contain the term t.

More specifically, the terms within a compound can be ranked according to its *idf*. Assume such a ranking of terms: $t_1, \ldots, t_m. (m > 1)$. Terms t_1 and t_2 can be combined using the concept combination heuristic resulting in the combined concept, denoted as $t_1 \oplus t_2$, whereby t_1 dominates t_2 (as it is higher in the ranking). For this combined concept, its degree of dominance is the average of the respective *idf* scores of t_1 and t_2. The process recurs down the ranking resulting in the composed "concept" $((..(t_1 \oplus t_2) \oplus t_3) \oplus \ldots) \oplus t_m)$. If there is only a single term $(m = 1)$, its corresponding normalized HAL vector is used as the combination vector.

We will not give a more detailed description of the concept combination heuristic, which can be found in (Bruza and Song, 2002). Its intuition is summarized as follows:

- Quality properties shared by both concepts are emphasized.
- The weights of the properties in the dominant concept are re-scaled higher.
- The resulting vector from the combination heuristic is normalized to smooth out variations due to the differing number of contexts the respective concepts appear in.

By way of illustration we have the following vector for the concept combination "GATT talks":

gatt $\oplus$ *talks* $= <$ agreement: 0.282, agricultural: 0.106, body: 0.117, china: 0.121, council: 0.109, farm: 0.261, gatt: 0.279, member: 0.108, negotiations: 0.108, round: 0.312, rules: 0.134, talks: 0.360, tariffs: 0.114, trade: 0.432, world: 0.114>

In summary, by constructing a HAL space from text corpus, concepts are represented as weighted vectors in the high-dimensional space, whereby each word in the vocabulary of the corpus gives rise to an axis in the corresponding semantic space. The rest of this chapter will demonstrate how the fuzzy C-means clustering can be applied to conceptual induction and how different contexts are reflected.

[2]General Agreement on Tariffs and Trade is a forum for global trade talks.

19.3 FUZZY C-MEANS CLUSTERING

As the focus of this chapter is not the development of a new clustering algorithm, the fuzzy C-means algorithm we use in our experiment is adapted from some existing studies in the literature (Hathaway, Bezdek, and Hu, 2000; Krishnapuram, Joshi, Nasraoui, and Yi, 2001).

Let $X = \{x_1, x_2, \ldots x_n\}$ be a set of n objects in an *S-dimensional* space. Let $d(x_j, x_i)$ be the distance or dissimilarity between objects x_i and x_j. Let $V = \{v_1, v_2, \ldots, v_K\}$, each v_c be the *prototype* or *mean* of the cth cluster. Let $d(v_c, x_i)$ be the distance or dissimilarity between the object x_i and the mean of the cluster that it belongs to.

The fuzzy clustering partitions these objects into K overlapped clusters based on a computed minimizer of the fuzzy within-group least squares functional:

$$J_m(U, V) = \sum_{c=1}^{K} \sum_{i=1}^{N} U^m(v_c, x_i) d(v_c, x_i) \tag{19.1}$$

where the minimization is performed over all $v_c \in V$, and $U(v_c, x_i)$ is the membership function for the object x_i belonging to the cluster v_c.

To optimize (19.1), we alternate between optimization of $\bar{J}_m(U|V^*)$ over U with V^* fixed and $\bar{J}_m(V|U^*)$ over V with U^* fixed, producing a sequence $\{U^{(p)}, V^{(p)}\}$. Specifically, the $p+1$st value of $V = \{v_1, v_2, \ldots, v_K\}$ is computed using the pth value of U in the right-hand side of:

$$v_c^{(p+1)} = \frac{\sum_{i=1}^{N} x_i{}^* [U^{(p)}(v_c^{(p)}, x_i)]^m}{\sum_{i=1}^{N} [U^{(p)}(v_c^{(p)}, x_i)]^m}. \tag{19.2}$$

Then the updated $p+1$st value of V is used to calculate the $p+1$st value of U via:

$$U^{(p+1)}(v_k^{(p+1)}, x_i) = \frac{d(x_i, v_k^{(p+1)})^{-1/(m-1)}}{\sum_{c=1}^{K} d(x_i, v_c^{(p+1)})^{-1/(m-1)}} \tag{19.3}$$

where $m \in (1, +\infty)$ is the fuzzifier. The greater m is, the fuzzier the clustering is. Krishnapuram, Joshi, Nasraoui, and Yi (2001) recommend a value between 1 and 1.5 for m. In addition, the following constraint holds:

$$\forall i \; i = 1, 2, \ldots, N \quad \sum_{c=1}^{K} U(v_c, x_i) = 1. \tag{19.4}$$

For the sake of efficiency in large data-sets, an alternative method is to use the top $L (L < N)$ objects in the cluster, and the objects are sorted based on descending membership value:

$$v_c^{(p+1)} = \frac{\sum_{i=1}^{L} x_i * [U^{(p)}(v_c^{(p)}, x_i)]^m}{\sum_{i=1}^{N} [U^{(p)}(v_c^{(p)}, x_i)]^m}. \tag{19.5}$$

If the dissimilarity is inner product induced, i.e., square Euclidean measure defined later in Section 19.3.1, it can be proved mathematically that computing V and U iteratively according to Equations (19.2) and (19.3) satisfies the necessary conditions for optima of $J_m(U|V)$ (Bezdek, 1981).

The traditional K-means clustering algorithm, namely, the hard C-means clustering, is a special case of fuzzy C-means clustering by simply replacing Equation (19.3) with:

$$q = \arg\min_c d(v_c, x_i) \qquad U^{(p+1)}(v_c, x_i) = \begin{cases} 1 & \textit{if} \quad c = q \\ 0 & \textit{if} \quad c \neq q \end{cases}.$$

The fuzzy C-means clustering algorithm is detailed as follows.
Fuzzy C-means Algorithm:

Fix the number of clusters K and Max_iter; Set iter = 0;
Pick initial means $V = \{v_1, v_2, \ldots, v_K\}$ from X;

Repeat
Compute memberships $U(v_c, x_i)$ for $c = 1, 2, \ldots, K$ and
$i = 1, 2, \ldots, N$ by using Equation (19.3) (A)
Store the current means, $V^{\text{old}} = V$;
Re-compute the new means v_c for $c = 1, 2, \ldots, K$ by using Equation (19.2)
Iter = Iter + 1; (B)

Until
Iter = Max_iter or
The absolute value of increment of the objective function $|\Delta J(U, V)| < \varepsilon$, where ε is some prescribed tolerance.

19.3.1 Dissimilarity Measures

Several measures can be employed to compute the dissimilarity between two objects (x_j, x_i), as well as between an object and the mean (v_c, x_i). The most frequently used approach is the Lp norm (i.e., Minkowski) distance, which is defined as follows (Hathaway, Bezdek, and Hu, 2000):

$$d(v_c, x_i) = \left(\sum_{j=1}^{S} |x_{i,j} - v_{c,j}|^p \right)^{1/p}$$

where $p \in [1, +\infty)$. This is a generalized dissimilarity measure. By way of illustration, Euclidean distance corresponds to the case when $p = 2$:

$$d(v_c, x_i) = ||x_i - v_c|| = \sqrt{\sum_{j=1}^{S} (x_{i,j} - v_{c,j})^2};$$

If $p = 1$, Manhattan distance results:

$$d(v_c, x_i) = \sum_{j=1}^{S} |x_{i,j} - v_{c,j}|.$$

Moreover, if $p = \infty$:

$$d(v_c, x_i) = \mathop{Max}_{j=1}^{S} |x_{i,j} - v_{c,j}|.$$

Hathaway, Bezdek, and Hu (2000) have shown that $p = 1$ offers the greatest robustness for outlier handling. In addition, other widely used dissimilar measures are:

- Cosine-based dissimilarity:

$$d(v_c, x_i) = e^{-Sim(v_c, x_i)},$$

where $Sim(v_c, x_i)$ is defined as:

$$Sim(v_c, x_i) = \frac{\sum_{j=1}^{S} x_{i,j}{}^* v_{c,j}}{\sqrt{\sum_{j=1}^{S} x_{i,j}^2 \sum_{j=1}^{S} v_{c,j}^2}}.$$

19.3.2 Initialization

Initialization is vital to the performance of the fuzzy C-means algorithm. Though we stated in the beginning of Section 19.3 that the algorithm satisfies the necessary conditions for optima of the objective function $J_m(U|V)$, the fuzzy C-means algorithm is not guaranteed to find the global minimum. Different initialization procedures will produce slightly different clustering results. Nevertheless, appropriate initialization will make the algorithm converge fast. If the K-means are initialized randomly, it is desirable to run the algorithm several times to increase the reliability of the final results. We have experimented with two different ways of initializing the K-means. The first way is to pick all the means candidates randomly. This method is referred to as *Initialization 1*. The second way is to pick the first candidate as the mean over all the items in the space X, and then each successive one will be the most dissimilar (remote) item to all the items that have already been picked. This makes the initial centroid*s* evenly distributed. We refer to this procedure as *Initialization 2*.
Initialization 2 for fuzzy C-means clustering

Fix the number of means $K > 1$;
Compute the first mean v_1:

$$v_1 = \frac{\sum_{i=1}^{N} x_i}{N}$$

Set $V = \{v_1\}, \text{iter} = 1$;

Repeat
iter = iter + 1;

$$v_{\text{iter}} = \max_{\substack{1 \le j \le N \\ x_j \notin V}}\left(\min_{1 \le k \le |V|} d(x_j, v_k)\right) \text{ then } V = V \cup \{v_{\text{iter}}\}$$

Until
iter = K;

For a given data-set, the initial produced by Initialization 2 is fixed. In our experiments, Initialization 2 outperforms Initialization 1 consistently.

19.4 WORD CLUSTERING ON A HAL SPACE – A CASE STUDY

This experiment aims to illustrate the effectiveness of the fuzzy C-means approach for clustering concepts (words) represented as HAL vectors.

19.4.1 HAL Space Construction

We applied the HAL method to the Reuters-21578 collection, which consists of new articles in the late 1980s. The vocabulary is constructed by removing a list of stop words and also dropping some infrequent words which appear less than five times in the collection. The size of final vocabulary is 15 415 words. The window size is set to be eight. A window which is too small leads to loss of potentially relevant correlations between words, whereas a window which is too large may compute irrelevant correlations. We think a window size of eight is reasonable since precision is our major concern. Previous studies in HAL (Lund and Burgess, 1996; Song and Bruza, 2003) have also employed a window size of eight in their experiments.

HAL vectors are normalized to unit length. As an example, Table 19.2 is part of the cosine-normalized HAL vector for "*Iran*" computed from applying the HAL method to the Reuters-21578 collection. This example demonstrates how a word is represented as a weighted vector whose dimensions comprise other words. The weights represent the strengths of association between "Iran" and other words seen in the context of the sliding window: the higher the weight of a word, the more it has lexically co-occurred with "Iran" in the same context(s). The dimensions reflect aspects that were relevant to the respective concepts during the mid to late 1980s. For example, Iran was involved in a war with Iraq, and President Reagan was involved in an arms scandal involving Iran.

19.4.2 Data

The following 20 words were selected from the vocabulary to prime the clustering process: *airbus, boeing, plane, Chernobyl, nuclear, disaster, computer, nec, japan, ibm, contra, industry, iran, iraq, scandal, war, president, reagan, white, house*.

Table 19.3 summarizes a manual clustering of the above words. These words involve approximately the following contexts in the Reuters collection:

(1) aircraft manufacturers;
(2) Chernobyl nuclear leaking disaster in the Soviet Union;
(3) computer companies;
(4) the roles of the White House (i.e., Reagan government) in the middle 1980s (dealing with Iran–Iraq war and trade war against Japan);
(5) the Iran-contra scandal (President Reagan was involved in the illegal arms sales to Iran during the Iran–Iraq war).

Table 19.2 The *Iran* vector.

Iran	
Dimension	Value
arms	0.64
iraq	0.28
scandal	0.22
gulf	0.18
war	0.18
sales	0.18
attack	0.17
oil	0.16
offensive	0.12
missiles	0.10
reagan	0.09
...	...

Table 19.3 Handcrafted result.

Cluster 1	Cluster 2	Cluster 3	Cluster 4	Cluster 5
Airbus	Chernobyl	Computer	White	**Iran**
Boeing	Disaster	Nec	House	Scandal
Plane	Nuclear	Ibm	President	**Contra**
Industry		**Industry**	**Reagan**	**Reagan**
			Iraq	**War**
			War	**Reagan**
			Iran	**Iraq**
			Japan	
			Industry	

Note there is some overlap between clusters. For example, Cluster 4 shares "industry" with Clusters 1 and 3; it also shares "reagan" and "iran" with Cluster 5, etc. .

19.4.3 Fuzzy Clustering of HAL Vectors

In order to find the best performing parameter settings for the fuzzy C-means clustering, we have developed a test bed on which a series of prior studies have been conducted. Cosine combined with fuzzifier 2.0 and Initialization 2 was finally chosen after some initial pilot studies. When the membership value of a word belonging to a cluster is greater than a prior probability (0.2 for this experiment), it is output as a member in the cluster. Table 19.4 lists the result of fuzzy C-means clustering (the number following each word is the membership value of the word belonging to the corresponding cluster).

We also conducted experiments with the K-means algorithm on the same data and the best performing result (via cosine-based dissimilarity function) is depicted in Table 19.5.

19.4.4 Discussions

Table 19.4 shows that the fuzzy clustering results basically reflect the underlying contexts described in Table 19.3, particularly the overlap between Reagan government, Iran–Iraq war, and Iran-contra scandal. However, the K-means clustering result presented in Table 19.5 is less ideal: Cluster 1 contains the words related to industry, either plane-manufacturing or IT; "nuclear" is separated from the "Chernobyl disaster"; "computer" forms a singular cluster; Cluster 4 contains terms related to politics. In short,

Table 19.4 Clustering result of fuzzy C-means algorithm.

Cluster 1	Cluster 2	Cluster 3	Cluster 4	Cluster 5
Airbus: 0.914	Chernobyl: 0.966	Computer: 0.921	White: 0.861	Iraq: 0.869
Boeing: 0.851	**Disaster: 0.302**	Nec: 0.906	House: 0.793	Scandal: 0.814
Plane: 0.852	Nuclear: 0.895	Ibm: 0.897	President: 0.653	**Contra: 0.776**
			Reagan: 0.708	**Iran: 0.725**
			Japan: 0.558	**War: 0.584**
			Industry: 0.494	**Reagan: 0.213**
			Disaster: 0.488	
			War: 0.331	
			Iran: 0.221	
			Contra: 0.203	

Table 19.5 Clustering result from *K*-means algorithm.

Cluster 1	Cluster 2	Cluster 3	Cluster 4	Cluster 5
Airbus	Chernobyl	Computer	Contra	Nuclear
Boeing	Disaster		House	
IBM			Iran	
Industry			Iraq	
Japan			President	
Nec			Reagan	
Plane			Scandal	
			War	
			White	

the results from the case study suggest that fuzzy *K*-means clustering of word "meanings" in a HAL space is promising.

19.5 CONCLUSIONS AND FUTURE WORK

In this chapter, we have introduced a cognitively motivated model, namely, hyperspace analog to language (HAL), to construct a high-dimensional semantic space. The HAL space can be used to realize aspects of Gärdenfors' conceptual space theory dealing with the geometrical representation of information. Within the conceptual space, concepts can be categorized into regions reflecting different contexts. Fuzzy clustering has been investigated in detail as a means for concept induction. We presented a case study on word clustering in the HAL space, which is constructed from the Reuters-21578 corpus. This study, though preliminary, is encouraging and shows that the fuzzy C-means algorithm can produce interesting results.

The work presented in this chapter can potentially be extended to other areas, such as query expansion of information retrieval, web page clustering, etc. Furthermore, we will conduct formal evaluation of the algorithm based on larger collections in the future.

ACKNOWLEDGMENT

This chapter is an extended version of our previous work (Cao, Song, and Bruza, 2004).

REFERENCES

Bai, J. *et al.* (2005) 'Query expansion using term relationships in language models for information retrieval'. In *Proc. of the 14th Int. Conf. on Information and Knowledge Management (CIKM' 2005)*, pp. 688–695.

Bezdek, J.C. (1981) *Pattern Recognition with Fuzzy Objective Function Algorithms*. New York: Plenum.

Bobrowski, J. and Bezdek, J.C. (1991) 'C-means clustering with the L_1 and L_∞ norms'. *IEEE Trans. Syst. Man, Cybern.* **21**, 545–554.

Bruza, P.D. and Song, D. (2002) 'Inferring query models by computing information flow'. In *Proc. of the 12th Int. Conf. on Information and knowledge Management (CIKM2002)*, pp. 260–269.

Burgess, C., Livesay, L. and Lund, K. (1998) 'Explorations in context space: words, sentences, discourse'. In *Quantitative Approaches to Semantic Knowledge Representation* (ed. Foltz, P.W.), *Discourse Processes*, **25**(2&3), 179–210.

Cao, G., Song, D. and Bruza, P.D. (2004) 'Fuzzy K-means clustering on a high dimensional semantic space'. In *Proceedings of the 6th Asia Pacific Web Conference (APWeb'04)*, LNCS 3007, pp. 907–911.

Chuang, S.L. and Chien, L.F. (2005) 'Taxonomy generation for text segments: a practical web-based approach'. *ACM Transactions on Information Systems (TOIS)*, **23**(4), 363–396.

Cimiano, P., Hotho, A. and Staab, S. (2005) 'Learning concept hierarchies from text corpora using formal concept analysis'. *Journal of Artificial Intelligence Research*, **24**, 305–339.

Gärdenfors, P. (2000) *Conceptual Spaces: The Geometry of Thought*. MIT Press.

Gärdenfors, P. and Williams, M. (2001) 'Reasoning about categories in conceptual spaces'. In *Proceedings of 14th International Joint Conference of Artificial Intelligence (IJCAI'2001)*, pp. 385–392.

Hathaway, R.J., Bezdek, J.C. and Hu, Y. (2000) 'Generalized fuzzy C-means clustering strategies using Lp norm distances'. *IEEE Transactions on Fuzzy Systems*, **8**, 576–582.

Hearst, M. (2003) What is text mining? http://www.sims.berkeley.edu/~hearst/text-mining.htmlHöppner, F. and Klawonn, F. (2003) 'A contribution to convergence theory of fuzzy c-means and derivatives'. *IEEE Transactions on Fuzzy Systems*, **11**(5), 682–694.

Höppner, F., Klawonn, F., Kruse, R. and Runkler, T. (1999) *Fuzzy Cluster Analysis*. John Wiley & Sons, Ltd, Chichester, UK.

Kolen, J. and Hutcheson, T. (2002) 'Reducing the time complexity of the fuzzy c-means algorithm'. *IEEE Transactions on Fuzzy Systems*, **10**(2), 263–267.

Krishnapuram, R., Joshi, A., Nasraoui, O. and Yi, Y. (2001) 'Low-complexity fuzzy relational clustering algorithm for web searching'. *IEEE Transactions on Fuzzy Systems*, **9**(4), 595–607.

Landauer, T. and Dumais, S. (1997) 'A solution to Plato's problem: the latent semantic analysis theory of acquisition, induction, and representation of knowledge'. *Psychological Review*, **104**(2), 211–240.

Lund, K. and Burgess, C. (1996) 'Producing high-dimensional semantic spaces from lexical co-occurrence'. *Behavior Research Methods, Instruments and Computers*, **28**(2), 203–208.

Perrin, P. and Petry, F. (2003) 'Extraction and representation of contextual information for knowledge discovery in texts'. *Information Sciences*, **151**, 125–152.

Song, D. and Bruza, P.D. (2001) 'Discovering information flow using a high dimensional conceptual space'. In *Proceedings of the 24th Annual International Conference on Research and Development in Information Retrieval (SIGIR'01)*, 327–333.

Song, D. and Bruza, P.D. (2003) 'Towards context sensitive informational inference'. *Journal of the American Society for Information Science and Technology*, **52**(4), 321–334.

Srinivasan, P. (2004) 'Text mining: generating hypotheses from MEDLINE'. *Journal of the American Society for Information Science and Technology*, **55**(5), 396–413.

Steinbach, M., Karypis, G. and Kumar, V. (2000) 'A comparison of document clustering technique'. In *KDD'2000 Workshop on Text Mining*. Available online: www.cs.cmu.edu/~dunja/KDDpapers/Steinbach_IR.pdf

20
Novel Developments in Fuzzy Clustering for the Classification of Cancerous Cells using FTIR Spectroscopy

Xiao-Ying Wang[1], Jonathan M. Garibaldi[1], Benjamin Bird[2] and Mike W. George[2]

[1] *School of Computer Science and Information Technology, University of Nottingham, UK*
[2] *School of Chemistry, University of Nottingham, UK*

20.1 INTRODUCTION

As a major human health concern, cancer has become a focus for worldwide research. In UK, more than one in three people will be diagnosed with cancer during their lifetime and one in four will die from the disease. The provision of more accurate diagnostic techniques might allow various cancers to be identified at an earlier stage and, hence, allow for earlier application of treatment. Histological evaluation of human tissue is the conventional way for clinicians to diagnose disease. It is relatively unchanged and remains the "gold standard" since its introduction over 140 years ago. This process requires a clinician to remove and examine tissue from suspicious lesions within the patient's body and subsequently, through chemical preparation, to allow thin sections of the tissue to be cut. The addition of contrast inducing dyes allows a trained observer to identify morphological staining patterns within tissue and cells that are characteristic of the onset of disease. Nevertheless, some significant problems remain in traditional histology due to the subjective processes involved. These include missed lesions, perforation of samples, and unsatisfactory levels of inter- and intra-observer discrepancy. This lack of a reliable tool for disease diagnosis has lead to a considerable amount of interest investigating the application of a spectroscopic approach [1,2].

In recent years, Fourier transform infrared (FTIR) spectroscopy has been increasingly applied to the study of biomedical conditions, as it can permit the chemical characterization of microscopic areas in a

Advances in Fuzzy Clustering and its Applications Edited by J. Valente de Oliveira and W. Pedrycz

tissue sample [3–6]. There are two types of FTIR detection: FTIR mapping and FTIR imaging. In mapping, the infrared (IR) spectrum of the samples is collected a point at a time and many separate collections must be made to examine different areas of the tissue sample. However, in the latest imaging, the IR spectrum of the samples is acquired over a wide area in a single collection. Moreover, the imaging technique allows the collection of images in a faster time with higher resolution. In comparison with conventional histology, FTIR spectroscopy has several advantages [7]:

(1) It has the potential for fully automatic measurement and analysis.
(2) It is very sensitive; very small samples are adequate.
(3) It is potentially much quicker and cheaper for large-scale screening procedures.
(4) It has the potential to detect changes in cellular composition prior to such changes being detectable by other means.

In order to analyze the FTIR spectroscopic data from tissue samples, various techniques have been used previously: point spectroscopy, greyscale functional group mapping and digital staining [8], and multivariate analysis methods, for instance principal component analysis [9]. Apart from these, multivariate clustering techniques have often been used to separate sets of unlabeled infrared spectral data into different clusters based on their characteristics (this is an unsupervised process). By examining the underlying structure of a set of spectra data, clustering (also called unsupervised classification) can be performed such that spectra within the same cluster are as similar as possible, and spectra in different clusters are as dissimilar as possible. In this way, different types of cells may be separated within biological tissue. There are many clustering techniques that have been applied in FTIR spectroscopic analysis. These include hierarchical clustering [10–13], *k*-means (KM) clustering [10,14], and fuzzy C-means (FCM) clustering [9,10,13,14]. Studies have shown that, of the various clustering techniques, fuzzy clustering such as FCM can show clear advantages over crisp and probabilistic clustering methods [15]. In this chapter, we review our new techniques in fuzzy clustering, including a simulated annealing based technique [7,16] to classify cancerous cells using FTIR spectroscopy. The aim of this ongoing work is to investigate whether IR spectroscopy can be used as a diagnostic probe to identify early stages of cancer.

The rest of this chapter will have the following structure. In Section 20.2, three clustering algorithms (hierarchical, KM, and FCM) often used in FTIR spectroscopic analysis are described. In Section 20.3, we describe the concept of cluster validity indices as a measure to evaluate the quality of clustering, and introduce the Xie–Beni cluster validity index. In the standard FCM clustering algorithm, the number of clusters has to be specified in advance. Obviously, this is not suitable if the technique is to be transferred into clinical practice. The clustering algorithm developed in Section 20.4 combines the simulated annealing and fuzzy C-means algorithms to achieve the automatic detection of the number of clusters (different types of tissue). Due to the complexity of the tissue sample, clustering results can occasionally obtain an excessive number of clusters. To counteract this problem, we propose an automatic cluster merging method in Section 20.5. Finally, we draw conclusions about the suitability of FTIR analysis for clustering in biological systems.

20.2 CLUSTERING TECHNIQUES

Three unsupervised clustering algorithms that are frequently used in FTIR spectroscopy analysis are hierarchical clustering, *k*-means, and fuzzy C-means. In this section, these three clustering techniques are described in detail. Corresponding experiments were conducted in order to compare their performance.

20.2.1 Hierarchical Clustering

Hierarchical clustering is a way to group the data in a nested series of partitions [17]. It is a "hard" clustering method where each datum can only belong to one cluster. Once a datum is set to a certain

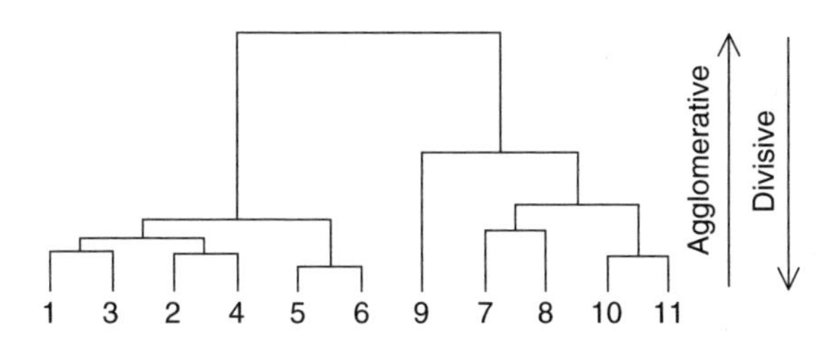

Figure 20.1 Hierarchical clustering dendrogram.

cluster, it is unchangeable in the future. The output of hierarchical clustering is a cluster tree, known as a *dendrogram*. It represents the similarity level between all the patterns. Figure 20.1 shows an example of a dendrogram in which the numbers at the bottom symbolize different data in the data-set.

Based on the algorithm structure and operation, hierarchical clustering can be further categorized into agglomerative and divisive algorithms [17,18]. Agglomerative methods initially consider each datum as an individual cluster and then repeatedly merge the two closest clusters (which are measured based on the corresponding linkage method) to form a new cluster, until all the clusters are merged into one. Therefore, in an agglomerative approach, the dendrogram is generated in a bottom-up procedure. Divisive methods start by considering that all data exist in one cluster and, based on the similarity within the data, the cluster is split into two groups, such that data in the same group have the highest similarity and data in the different groups have the most dissimilarity. This process continues until each cluster only contains single datum. Hence, the divisive approach follows a top-down procedure. These two approaches are illustrated in Figure 20.1.

As part of an agglomerative algorithm there are many different linkage methods that provide a measure of the similarity of clusters based on the data within a cluster [19]. The main linkage methods include single link, complete link and minimum-variant algorithms (Ward's algorithm) [17,18]. Other linkages are derivatives of these three main methods. In the single link algorithm, the distance between two clusters is measured by the two *closest* data within the different clusters. By contrast, in the complete link algorithm, the distance between two clusters is measured by the two *furthest* data within the different clusters. The minimum-variant algorithm is distinct from the other two methods because it uses a variance analysis approach to measure the distance between two clusters. In general, this method attempts to minimize the sum of squares of any two hypothetical clusters that can be generated at each step. This is based on the Euclidean distance between all centroids [20]. Ward's algorithm was adopted in most of the previous applications of FTIR analysis with hierarchical clustering. In some studies (such as [10]), it produced better clustering results.

20.2.2 K-means

K-means (KM) clustering is a nonhierarchical clustering algorithm. It is also a "hard" clustering method because the membership value of each datum to its cluster center is either zero or one, corresponding to whether it belongs to that cluster or not. The procedure for the KM clustering algorithm can be described as follows.

First let $X = \{x_1, x_2, \ldots x_n\}$ represent a vector of real numbers, where n is the number of data. $V = \{v_1, v_2, \ldots v_c\}$ is the corresponding set of centers, where c is the number of clusters. The aim of the *K*-means algorithm is to minimize the squared-error objective function $J(V)$:

$$J(V) = \sum_{i=1}^{c} \sum_{j=1}^{c_i} ||x_{ij} - v_j||^2 \tag{20.1}$$

where $||x_{ij} - v_j||$ is the Euclidean distance between x_{ij} and v_j. c_i is the number of data in cluster i . The ith center v_i can be calculated as:

$$v_i = \frac{1}{c_i}\sum_{j=1}^{c_i} x_{ij}, \qquad i = 1 \ldots c. \tag{20.2}$$

The algorithm then proceeds as follows:

(1) Randomly select c cluster centers.
(2) Calculate the distance between all of the data and each center.
(3) Assign each datum to a cluster based on the minimum distance.
(4) Recalculate the center positions using Equation (20.2).
(5) Recalculate the distance between each datum and each center.
(6) If no data were reassigned then stop, otherwise repeat from step (3).

Further details of KM clustering can be found, for example, in [21].

20.2.3 Fuzzy C-means

FCM clustering is a fuzzy version of KM (hence, it is sometimes also called fuzzy k-means). It was proposed by Bezdek in 1981 [22]. One of the differences between FCM and KM is that FCM contains an additional parameter – the "fuzzifier" $m(1 < m < \infty)$. However, as with the KM algorithm, FCM needs the number of clusters to be specified as an input parameter to the algorithm. In addition, both may suffer premature convergence to local optima [17].

The objective of the FCM algorithm is also to minimize the squared error objective function $J(U, V)$:

$$J(U, V) = \sum_{i=1}^{n}\sum_{j=1}^{c} (\mu_{ij})^m ||x_i - v_j||^2 \tag{20.3}$$

where most of the variables are as above, $U = (\mu_{ij})_{n \times c}$ is a fuzzy partition matrix and μ_{ij} is the membership degree of datum x_i to the cluster center v_j. Parameter m is called the "fuzziness index" or "fuzzifier"; it is used to control the fuzziness of each datum membership, $m \in (1, \infty)$. There is no theoretical basis for the optimal selection of m, and a value of $m = 2.0$ is often chosen [22]. We carried out a brief investigation to examine the effect of varying m in this domain. It was found that as m increased around a value of 2.0, the clustering centers moved slightly, but the cluster assignments were not changed. However, as $m \to \infty$, both the fuzzy objective function J and the Xie–Beni validity index V_{XB} (see Section 20.3) continuously decreased ($\to 0$), and the cluster assignments became unstable and further from the results of clinical analysis. Consequently, we fixed the value of m as 2.0 for all further experiments reported here.

The FCM algorithm is an iterative process to minimize Equation (20.3) while updating the cluster centers v_j and the memberships μ_{ij} by:

$$v_j = \frac{\sum_{i=1}^{N} (\mu_{ij})^m x_i}{\sum_{i=1}^{N} (\mu_{ij})^m}, \forall j = 1, \ldots, c \tag{20.4}$$

$$\mu_{ij} = \frac{1}{\sum_{k=1}^{c} \left(\frac{d_{ij}}{d_{ik}}\right)^{\frac{2}{m-1}}} \tag{20.5}$$

where $d_{ij} = ||x_i - v_j||$, $i = 1 \ldots n$ and $j = 1 \ldots c$.

A suitable termination criterion can be that the difference between the updated and the previous U is less than a predefined minimum threshold or that the objective function is below a certain tolerance value.

Furthermore, the maximum number of iteration cycles can also be a termination criterion. In our experiments we used a minimum threshold of 10^{-7} and a maximum number of iterations of 100.

20.2.4 Application to FTIR Spectroscopy Clustering

Initially, we applied these three clustering techniques to FTIR spectral data sets obtained from previous clinical work provided by Chalmers and colleagues [23] in order to compare their performance. Seven sets of FTIR data, containing tumor (abnormal), stroma (connective tissue), early keratinization, and necrotic specimens, were taken from three oral cancer patients. Parallel sections were cut and stained conventionally to help identify particular regions of interest. After acquisition of the FTIR spectra, the tissue sections were also stained and examined through cytology. Some of the material in the following section has previously appeared in [13,14].

Figure 20.2 (a) shows a 4× magnification visual image from one of Hematoxylin and Eosin stained oral tissue sections. There are two types of cells (stroma and tumor) in this section, clearly identifiable by their light and dark colored stains respectively. Figure 2 (b) shows a 32× magnified visual image from a portion of a parallel, unstained section; the superimposed dashed white lines separate the visually different morphologies. Five single point spectra were recorded from each of the three distinct regions. The locations of these are marked by "+" on Figure 20.2 (b) and numbered as 1–5 for the upper tumor region, 6–10 for the central stroma layer, and 11–15 for the lower tumor region. The 15 FTIR transmission spectra from these positions are recorded as Data-set 1, and the corresponding FTIR spectra are shown in Figure 20.3.

A Nicolet 730 FTIR spectrometer (Nicolet Instrument, Inc., Madison, USA) was used to collect the spectral data. In previous work multivariate analyzes, hierarchical clustering, and principal component analysis (PCA) had been applied in order to facilitate discrimination between different spectral characteristics. First some *basic preprocessing* such as water vapour removal, baseline correction, and normalization had been applied. Then further preprocessing treatments were also performed empirically on the FTIR spectral data, for instance mean-centering, variance scaling, and use of first-derivative spectra. Both the multivariate data analysis and preprocessing of the spectra were undertaken using Infometrix Pirouette® (Infometrix, Inc., Woodinville, WA, USA) software. The spectral range in this study was limited to a 900–1800 cm^{-1} interval. The results from multivariate analysis and cytology showed that accurate clustering could only be achieved by manually applying preprocessing techniques that varied according to the particular sample characteristics and clustering algorithms.

In our experiments, we applied the three previously mentioned clustering techniques to these seven spectral data-sets obtained through conventional cytology [23], but without any extra preprocessing (i.e., only *basic preprocessing*). For hierarchical clustering, the three different types of linkage methods mentioned previously, namely "single," "complete" and "Ward" were utilized. Due to the KM and

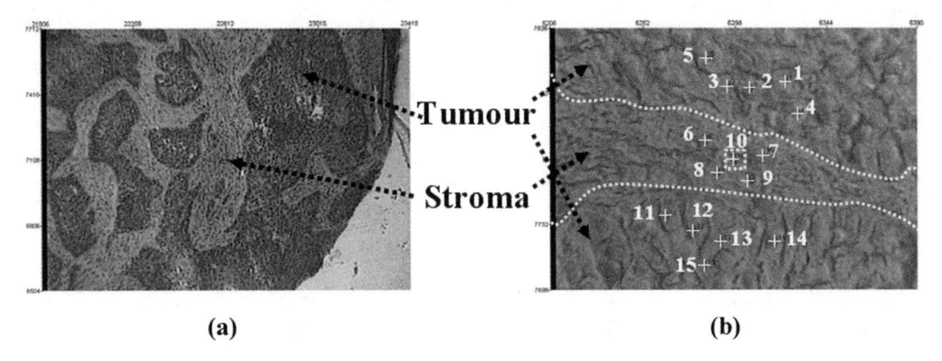

Figure 20.2 Tissue sample from Data-set 1 (a) 4× stained picture; (b) 32× unstained picture.

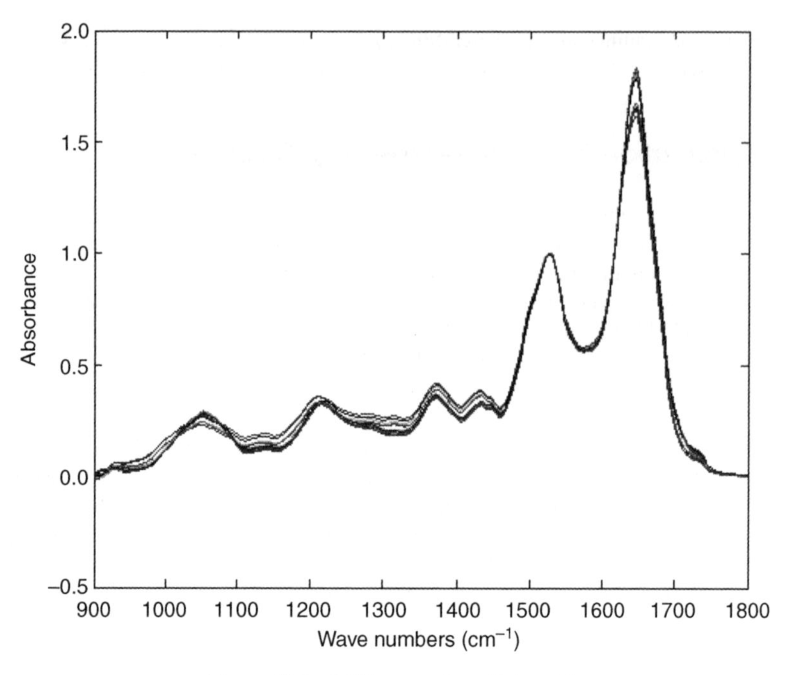

Figure 20.3 FITR spectra from Data-set 1.

FCM algorithms being sensitive to the initial (random) state, we ran each method 10 times. Table 20.1 shows the numbers of different types of tissue identified clinically (column "Clinical study") and as obtained by the three clustering techniques. As mentioned previously, clustering is an unsupervised process and the results of the clustering simply group the data into two or more unlabeled categories. In the results presented below, the clusters were mapped to the actual known classifications in such a way as to minimise the number of disagreements from the clinical labeling in each case.

Table 20.1 Number of the different tissue types identified clinically and as obtained by the clustering techniques.

			Hierarchical clustering								
Data-set names	Tissue types	Clinical study	Single	Complete	Ward	KM			FCM		
Data-set 1	Tumor	10	10	10	10		10			10	
	Stroma	5	5	5	5		5			5	
Data-set 2	Tumor	10	17	9	9		9			9	
	Stroma	8	1	9	9		9			9	
Data-set 3	Tumor	8	4	8	7	3	6	4		4	
	Stroma	3	7	3	4	8	5	7		7	
Data-set 4	Tumor	12	19	12	12	11	19	13	19		11
	Stroma	7	5	7	7	8	5	6	5		8
	Early keratinization	12	7	12	12	12	7	12	7		12
Data-set 5	Tumor	18	29	18	18		14	17		14	
	Stroma	12	1	12	12		16	13		16	
Data-set 6	Tumor	10	10	10	10		10			10	
	Stroma	5	5	5	5		5			5	
Data-set 7	Tumor	21	28	17	15		17			18	
	Stroma	14	13	18	20		18			16	
	Necrotic	7	1	7	7		7			8	

Table 20.2 Comparison results based on the number of disagreements between clinical study and clustering results.

Data-set names	Tissue types	Hierarchical clustering								
		Single	Complete	Ward		KM			FCM	
Data-set 1	Tumor	0	0	0		0			0	
	Stroma	0	0	0		0			0	
Data-set 2	Tumor	7	0	0		0			0	
	Stroma	0	1	1		1			1	
Data-set 3	Tumor	0	0	0	0	0	0		0	
	Stroma	4	5	3	5	2	4		4	
Data-set 4	Tumor	7	3	3	3	7	3	3		7
	Stroma	5	3	3	4	5	2	4		5
	Early keratinization	0	0	0	0	0	0	0		0
Data-set 5	Tumor	12	0	0		0	0	0		
	Stroma	1	0	0		4	1		4	
Data-set 6	Tumor	0	0	0		0			0	
	Stroma	0	0	0		0			0	
Data-set 7	Tumor	7	0	0		0			0	
	Stroma	0	4	6		4			2	
	Necrotic	1	0	0		0			1	

In Table 20.1 it can be seen that, in most of the data-sets, the number of data belonging to the various categories do not exactly match the results from clinical analysis. This is because some of the data that should be classified in the tumor cluster has been misclassified into the stroma cluster and vice versa. For example, in Data-set 2, using the hierarchical clustering single linkage method, the number of data considered as tumor is 17, while one is considered as stroma. We regard the extra data from these clustering techniques as the number of disagreements of classification in comparison with clinical analysis. The results of such comparison are shown in Table 20.2.

After running each clustering technique 10 times, it can be seen that the KM and FCM algorithms obtained more than one clustering result in some data-sets. This is because different initialization may lead to different partitions for both of these algorithms. It can be seen from Tables 20.1 and 20.2 that KM exhibits more variation (in three out of seven data-sets) than FCM (in one out of seven data-sets). The corresponding frequency of the differing clustering partitions obtained (out of 10 runs) is shown in Table 20.3.

In order to further investigate the performance of the different clustering methods, the average number of disagreements for all data-sets was calculated, as shown in Table 20.4. It can be seen that the hierarchical clustering single linkage method has the worst performance, while the complete linkage and Ward methods perform best overall. However, hierarchical clustering techniques are computationally expensive (proportional to n^2, where n is the number of spectral data) and so are not suitable for very large data-sets. KM and FCM have fairly good performance and, for both, the computational effort is approximately linear with n. Hence, compared with hierarchical clustering, these techniques will be far less time-consuming on large data-sets [10]. Moreover, although KM has a slightly better performance than FCM (slightly fewer disagreements, on average), it can be seen from Table 20.3 that KM exhibits far

Table 20.3 Clustering variations for KM and FCM within three data-sets.

Data-set names		KM		FCM	
Data-set 3	2/10	3/10	5/10	-	
Data-set 4	3/10	3/10	4/10	9/10	1/10
Data-set 5	5/10		5/10	-	

Table 20.4 Average number of disagreements obtained in the three classification methods.

	Hierarchical clustering				
	Single	Complete	Ward	KM	FCM
Average number of disagreements	44	16	16	18.8	19.5

more variation in its results than FCM. Hence, the overall conclusion was that FCM is the most suitable clustering method in this context.

20.3 CLUSTER VALIDITY

Clustering validity is a concept to evaluate how good clustering results are. There are many cluster validity indices that have been proposed in the literature for evaluating fuzzy and other clustering techniques. Indices which only use the membership values such as the partition coefficient and partition entropy [24] have the advantage of being easy to compute, but are only useful for a small number of well-separated clusters. Furthermore, they also lack direct connection to the geometrical properties of the data. In order to overcome these problems, Xie and Beni defined a validity index that measures both compactness and separation of clusters [25]. In our study, we selected the Xie–Beni (V_{XB}) cluster validity index to measure the clustering quality as it has been frequently used in recent research [26] and has also been shown to be able to detect the correct number of clusters in several experiments [27]. Some of the material in the following section has previously appeared in [7,16].

20.3.1 Xie-Beni Validity Index

The Xie–Beni validity index, V_{XB}, can be considered as a combination of two parts. The first part is to estimate the compactness of data in the same cluster and the second part is to evaluate the separation of data in different clusters. Let π represent the compactness and s be the separation of the clusters. The Xie–Beni validity index can be expressed as:

$$V_{\mathrm{XB}} = \frac{\pi}{s} \tag{20.6}$$

where

$$\pi = \frac{\sum_{j=1}^{c} \sum_{i=1}^{n} \mu_{ij}^{2} ||x_i - v_j||^2}{n} \tag{20.7}$$

and $s = (d_{\min})^2$. $d_{\min}$ is the minimum distance between cluster centers, given by $d_{\min} = \min_{ij} ||v_i - v_j||$. From these expressions, it can be seen that smaller values of π indicate that the clusters are more compact, while larger values of s indicate that the clusters are well separated. As a result, a smaller value of V_{XB} means that the clusters have greater separation from each other and the data are more compact within each cluster.

20.3.2 Application in FTIR Spectral Clustering Analysis

As mentioned in Section 20.2.3, for the FCM algorithm the number of clusters has to be specified a priori, which is obviously not suitable for real-world applications. However, with the use of cluster validity indices, it is possible to discover the "optimal" number of clusters within a given data-set [28]. This can be achieved by evaluating all of the possible clusters with the validity index; the optimal number of clusters can be determined by selecting the minimum value (or maximum, depending on the validity index used) of the index. This procedure can be described by an *FCM based selection algorithm (FBSA)* [15,28,29].

This algorithm is based on the standard FCM clustering method whereby $c_{\min}$ and $c_{\max}$ represent the minimal and maximal number of clusters that the data-set may contain. The best data structure (C) is returned, based on the optimal cluster validity index value. It can be described in the following steps:

(1) Set $c_{\min}$ and $c_{\max}$.
(2) For $c = c_{\min}$ to $c_{\max}$
  (2.1) Initialize the cluster centers.
  (2.2) Apply the standard FCM algorithm and obtain the new center and new fuzzy partition matrix.
  (2.3) After the FCM reaches its stop criteria, calculate the cluster validity (e.g., V_{XB}).
(3) Return the best data structure (C), which corresponds to the optimal cluster validity value (e.g., minimal V_{XB}).

20.4 SIMULATED ANNEALING FUZZY CLUSTERING ALGORITHM

There have been many clustering methods that have been developed in an attempt to determine automatically the optimal number of clusters. Recently, Bandyopadhyay proposed a *variable string length fuzzy clustering using simulated annealing* (VFC-SA) algorithm [30]. The proposed model was based on a simulated annealing algorithm whereby the cluster validity index measure was used as the energy function. This has the advantage that, by using simulated annealing, the algorithm can escape local optima and, therefore, may be able to find globally optimal solutions. The Xie–Beni index (V_{XB}) was used as the cluster validity index to evaluate the quality of the solutions. Hence this VFC-SA algorithm can generally avoid the limitations that exist in the standard FCM algorithm. However, when we implemented this proposed algorithm, it was found that sub-optimal solutions could be obtained in certain circumstances. In order to overcome this limitation, we extended the original VFC-SA algorithm to produce the simulated annealing fuzzy clustering (SAFC) algorithm. In this section, we will describe the original VFC-SA and the extended SAFC algorithm in detail. Most of the material in this section has appeared previously in [7,16].

20.4.1 VFC-SA Algorithm

In this algorithm, all of the cluster centers were encoded using a variable length string to which simulated annealing was applied. At a given temperature, the new state (string encoding) was accepted with a probability: $1/\{1 + \exp[-(E_{\mathrm{n}} - E_{\mathrm{c}})/T]\}$, where E_{n} and E_{c} represent the new energy and current energy respectively. T is the current temperature.

The V_{XB} index was used to evaluate the solution quality. The initial state of the VFC-SA was generated by randomly choosing c points from the data-sets where c is an integer within the range $[c_{\min}, c_{\max}]$. The values $c_{\min} = 2$ and $c_{\max} = \sqrt{n}$ (where n is the number of data points) were used, following the suggestion proposed by Bezkek in [24]. The initial temperature T was set to a high temperature$T_{\max}$. A neighbor of the solution was produced by making a random alteration to the string describing the cluster centers (as described below) and then the energy of the new solution was calculated. The new solution was kept if it satisfied the simulated annealing acceptance requirement. This process was repeated for a certain number of iterations, k, at the given temperature. A cooling rate, r, where $0 < r < 1$, decreased the current temperature by $T = rT$. This was repeated until T reached the termination criteria temperature $T_{\min}$, at which point the current solution was returned. The whole VFC-SA algorithm process is summarized in the following steps:

(1) Set parameters $T_{\max}, T_{\min}, c, k, r$.
(2) Initialize the string by randomly choosing c data points from the data-set to be cluster centers.
(3) Compute the corresponding membership values using Equation (20.5).
(4) Calculate the initial energy E_c using V_{XB} index from Equation (20.6).
(5) Set the current temperature $T = T_{\max}$.
(6) While $T \geq T_{\min}$
  (6.1) for $i = 1$ to k
    (6.1.1) Alter a current center in the string.
    (6.1.2) Compute the corresponding membership values using Equation (20.5).
    (6.1.3) Compute the corresponding centers with the Equation (20.4).
    (6.1.4) Calculate the new energy E_n from the new string.
    (6.1.5) If $E_n < E_c$ or $E_n > E_c$ with accept probability $>$ a random number between [0, 1], accept the new string and set it as current string.
    (6.1.6) Else, reject it.
  (6.2) End for.
  (6.3) $T = rT$.
(7) End while.
(8) Return the current string as the final solution.

The process of altering a current cluster center (step (6.1.1)) comprised three functions. They are: perturbing an existing center (*perturb center*), splitting an existing center (*split center*) and deleting an existing center (*delete center*). At each iteration, one of the three functions was randomly chosen. When splitting or deleting a center, the sizes of clusters were used to select which cluster to affect. The size, C_j, of a cluster, j, can be expressed by:

$$|C_j| = \sum_{i=1}^{n} \mu_{ij}, \forall j = 1, \ldots c \tag{20.8}$$

where c is the number of clusters.

The three functions are described below.

Perturb Center. A random center in the string is selected. This center position is then modified through addition of the change rate $cr[d] = r \cdot pr \cdot v[d]$, where v is the current chosen center and $d = 1, \ldots, N$, where N is the number of dimensions. r is a random number between $[-1, 1]$ and pr is the perturbation rate that was set through initial experimentation as 0.007 as this gave the best trade off between the quality of the solutions produced and the time taken to achieve them. *Perturb center* can then be expressed as $v_{\text{new}}[d] = v_{\text{current}}[d] + cr[d]$, where $v_{\text{current}}[d]$ and $v_{\text{new}}[d]$ represent the current and new centers respectively.

Split Center. The center of the biggest cluster is chosen by using Equation (20.8). This center is then replaced by two new centers that are created by the following procedure. The point with the highest membership value less than 0.5 to the selected center is identified as the reference point w. Then the distance between this reference point and the current chosen center is calculated using: $\text{dist}[d] = |v_{\text{current}}[d] - w[d]|$. Finally, the two new centers are then obtained by $v_{\text{new}}[d] = v_{\text{current}}[d] \pm \text{dist}[d]$.

Delete Center. The smallest cluster is identified and its center deleted from the string encoding.

20.4.2 SAFC Algorithm

When we implemented the original VFC-SA algorithm on a wider set of test cases than originally used by Bandyopadhyay [30], it was found to suffer from several difficulties. In order to overcome these

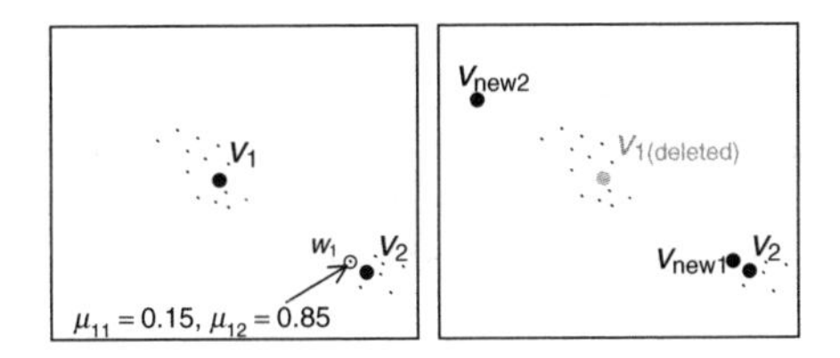

Figure 20.4 An illustration of *split center* from the original algorithm with distinct clusters (where μ_{11} and μ_{12} represent the membership degree of w_1 to the centers v_1 and v_2 respectively).

difficulties, four extensions to the algorithm were developed. In this section, the focus is placed on the extensions to VFC-SA in order to describe the proposed SAFC algorithm.

The first extension is in the initialization of the string. Instead of the original initialization in which random data points were chosen as initial cluster centers, the FCM clustering algorithm was applied using the random integer $c \in [c_{\min}, c_{\max}]$ as the number of clusters. The cluster centers obtained from the FCM clustering are then utilized as the initial cluster centers for SAFC.

The second extension is in *perturb center*. The method of choosing a center in the VFC-SA algorithm is to randomly select a center from the current string. However, this means that even a "good" center can be altered. In contrast, if the weakest (smallest) center is chosen, the situation in which an already good (large) center is destabilized is avoided. Ultimately, this can lead to a quicker and more productive search as improvements to the poorer regions of a solution can be concentrated upon.

The third extension is in *split center*. If the boundary between the biggest cluster and the other clusters is not obvious (not very marked), then a suitable approach is to choose a reference point with a membership degree that is less than, but close to, 0.5. That is to say there are some data points whose membership degree to the chosen center is close to 0.5. There is another situation that can also occur in the process of splitting center: the biggest cluster is separate and distinct from the other clusters. For example, let there be two clusters in a set of data points that are separated, with a clear boundary between them. v_1 and v_2 are the corresponding cluster centers at a specific time in the search as shown in Figure 20.4 (shown in two dimensions). The biggest cluster is chosen, say v_1. Then a data point whose membership degree is closest to but less than 0.5 can only be chosen from the data points that belong to v_2 (where the data points have membership degrees to v_1 less than 0.5). So, for example, the data point w_1 (which is closest to v_1) is chosen as the reference data point. The new centers will then move to v_{new1} and v_{new2}. Obviously these centers are far from the ideal solution. Although the new centers are likely to be changed by the *perturb center* function afterward, it will inevitably take a longer time to "repair" the solutions. In the modified approach, two new centers are created within the biggest cluster. The same data-set as in Figure 20.4 is used to illustrate this process. A data point is chosen, w_1, for which the membership value is closest to the mean of all the membership values above 0.5. Then two new centers v_{new1} and v_{new2} are created according the distance between v_1 and w_1. This is shown in Figure 20.5. Obviously the new centers are better than the ones in Figure 20.4 and therefore better solutions are likely to be found in the same time (number of iterations).

The fourth extension is in the final step of the algorithm (return the current solution as the final solution). In the SAFC algorithm, the best center positions (with the best V_{XB} index value) that have been

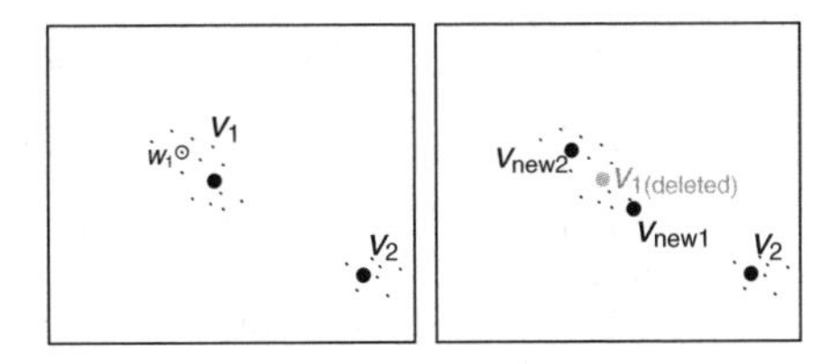

Figure 20.5 The new *split center* applied to the same data-set as Figure 20.4 (where w_1 is now the data point that is closest to the mean value of the membership degree above 0.5).

encountered are stored throughout the search. At the end of the search, rather than returning the current solution, the best solution seen throughout the whole duration of the search is returned.

Aside from these four extensions, we also ensure that the number of clusters always remains within the range $[c_{min}, c_{max}]$. Therefore when splitting a center, if the number of clusters has reached c_{max}, then the operation is disallowed. Similarly, deleting a center is not allowed if the number of clusters in the current solution is c_{min}.

20.4.3 Application to FTIR Spectroscopy Clustering

In order to compare the relative performance of the FCM, VFC-SA, and SAFC algorithms, the following experiments were conducted. The same seven oral cancer data-sets were used in this study. The number of different types of cell in each tissue section from clinical analysis was considered as the number of clusters to be referenced. This number was also used as the number of clusters for the FCM algorithm. The V_{XB} index was utilized throughout to evaluate the quality of the classification for the three algorithms. The parameters for VFC-SA and SAFC are: $T_{min} = 10^{-5}$, $k = 40$ and $r = 0.9$. T_{max} was set as 3 in all cases. That is because the maximum temperature has a direct impact on how much worse a solution can be accepted at the beginning. If the T_{max} value is set too high, this may result in the earlier stages of the search being less productive because simulated annealing will accept almost all of the solutions and, therefore, will behave like a random search. It was empirically determined that when the initial temperature was 3, the percentage of worse solutions that were accepted was around 60 %. In 1996, Rayward-Smith and colleagues discussed starting temperatures for simulated annealing search procedures and concluded that a starting temperature that results in 60 % of worse solutions being accepted yields a good balance between the usefulness of the initial search and overall search time (i.e., high enough to allow some worse solutions, but low enough to avoid conducting a random walk through the search space and wasting search time) [31]. Therefore, the initial temperature was chosen based on this observation.

Solutions for the seven FTIR data-sets were generated by using the FCM, VFC-SA, and SAFC algorithms. Ten runs of each method were performed on each data-set. As mentioned at the beginning of this section, the number of clusters was predetermined for FCM through clinical analysis. The outputs of FCM (centers and membership degrees) were then used to compute the corresponding V_{XB} index value. VFC-SA and SAFC automatically found the number of clusters by choosing the solution with the smallest V_{XB} index value. Table 20.5 shows the average V_{XB} index values obtained after 10 runs of each algorithm (best average is shown in bold).

In Table 20.5, it can be seen that in all of these seven data sets, the average V_{XB} values of the solutions found by SAFC are smaller than both VFC-SA and FCM. This means that the clusters obtained by SAFC have, on average, better V_{XB} index values than the other two approaches. Put another way, it may also indicate that SAFC is able to escape suboptimal solutions better than the other two methods.

Table 20.5 Average of the V_{XB} index values obtained when using the FCM, VFC-SA and SAFC algorithms.

	Average V_{XB} index value		
Data-set	FCM	VFC-SA	SAFC
1	0.048036	0.047837	0.047729
2	0.078896	0.078880	0.078076
3	0.291699	0.282852	0.077935
4	0.416011	0.046125	0.046108
5	0.295937	0.251705	0.212153
6	0.071460	0.070533	0.070512
7	0.140328	0.149508	0.135858

Table 20.6 Comparison of the number of clusters obtained by clinical analysis, VFC-SA and the SAFC methods.

	Number of clusters obtained		
Data-set	Clinical	VFC-SA	SAFC
1	2	2(10)	2(10)
2	2	2(10)	2(10)
3	2	2(10)	3(10)
4	3	2(10)	2(10)
5	2	2(5), 3(5)	3(10)
6	2	2(10)	2(10)
7	3	3(9), 4(1)	3(10)

In the Data-sets 1, 2, 4 and 6, the average of V_{XB} index values in SAFC is only slightly smaller than that obtained using VFC-SA. Nevertheless, when a Mann-Whitney test [32] was conducted on the results of these two algorithms, the V_{XB} index for SAFC was found to be statistically significantly lower (with $p < 0.01$) than that for VFC-SA for all data-sets.

The number of clusters obtained by VFC-SA and SAFC for each data-set is presented in Table 20.6. The numbers in parentheses indicate the number of runs for which that particular cluster number was returned. For example, on Data-set 5, the VFC-SA algorithm found two clusters in five runs and three clusters in the other five runs. The number of clusters identified by clinical analysis is also shown for comparative purposes.

In Table 20.6, it can be observed that in Data-sets 3, 4, 5, and 7, either one or both of VFC-SA and SAFC obtain solutions with a differing number of clusters than provided by clinical analysis. In fact, with Data-sets 5 and 7, VFC-SA produced a variable number of clusters within the 10 runs. Returning to the V_{XB} index values of Table 20.5, it was found that all the average V_{XB} index values obtained by SAFC are better.

It can be observed that the average V_{XB} index obtained by SAFC is much smaller than that of FCM for Data-sets 3 and 4. These two data-sets are also the data-sets in which SAFC obtained a different number of clusters to clinical analysis. In Data-set 3, the average V_{XB} index value for SAFC is also much smaller than for VFC-SA. This is because the number of clusters obtained from these two algorithms is different (see Table 20.6). Obviously a different number of clusters leads to a different cluster structure, and so there can be a big difference in the validity index. In Data-sets 5 and 7, the differences of V_{XB} index values are noticeable, though not as big as Data-sets 3 and 4.

In order to examine the results further, the data have been plotted using the first and second principal components in two dimensions. These have been extracted using the principal component analysis (PCA) technique [33,34]. The data have been plotted in this way because, although the FTIR spectra are limited to within $900\,\text{cm}^{-1}$–$1800\,\text{cm}^{-1}$, there are still 901 absorbance values corresponding to each wave number for each datum. The first and second principal components are the components that have the most variance in the original data. Therefore, although the data have multidimensional, the principal components can be plotted to give an approximate visualization of the solutions that have been achieved. Figures 20.6 (a)–(d) show the results for Data-sets 3, 4, 5, and 7 respectively using SAFC (the data in each cluster are depicted using different markers and each cluster center is presented by a star). The first and second principal components in Data-sets 3, 4, 5, and 7 contain 89.76, 93.57, 79.28, and 82.64 % of the variances in the original data, respectively.

There are three possible explanations for the differences between the clustering results and clinical analysis. First, the clinical analysis *may not* be correct – this could potentially be caused by the different types of cells in the tissue sample not being noticed by the clinical observers or the cells within each sample could have been mixed with others. Secondly, it could be that although a smaller V_{XB} index value was obtained, indicating a "better" solution in technical terms, the V_{XB} index is not accurately capturing

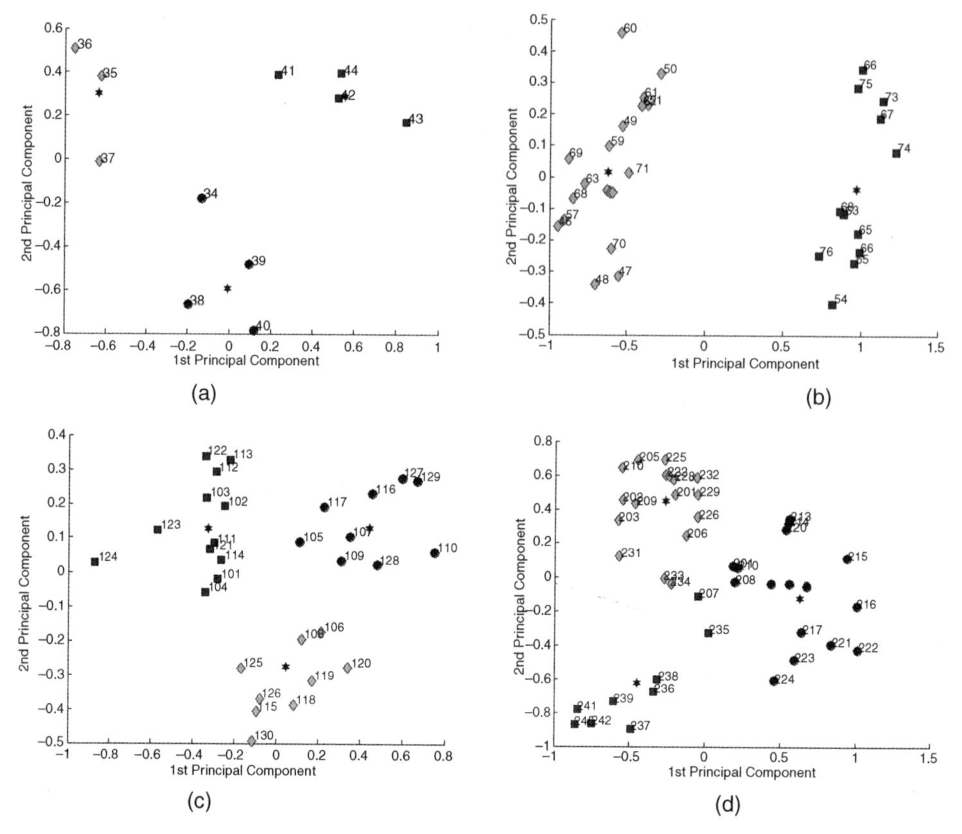

Figure 20.6 SAFC cluster results for Data-sets (a) 3; (b) 4; (c) 5 and (d) 7.

the real validity of the clusters. Put another way, although the SAFC finds the better solution in terms of the V_{XB} index, this is not actually the best set of clusters in practice. A third possibility is that the FTIR spectroscopic data have not extracted the required information necessary in order to permit a correct determination of cluster numbers, i.e., there is a methodological problem with the technique itself. None of these explanations of the difference between SAFC and VFC-SA algorithms detracts from the fact that SAFC produces better solutions in that it consistently finds better (statistically lower) values of the objective function (V_{XB} index).

20.5 AUTOMATIC CLUSTER MERGING METHOD

In Section 20.4, it can be seen that the SAFC algorithm performed well on the seven data-sets. However, these are relatively small and, as the size of the data-set increases, SAFC will become time-consuming. For this reason, when large data-sets are analyzed, an FCM based selection algorithm (FBSA – see Section 20.3.2) can be used to find the optimal number of clusters. In our studies, it has been found that both SAFC and FBSA occasionally identified an excessive number of clusters. This was partly due to the FCM algorithm and partly due to the cluster validity index, where all distances between data points and cluster centers are calculated using the Euclidean distance. This means that when the shapes of the clusters were significantly different from spherical, the clustering and validity measures were not effective. In addition, the complexity and range of different cell types (e.g., healthy, precancerous, and mature cancer) may also result in an excessive number of clusters being identified. Nevertheless, at this stage of the study, we only focus on grouping the cells with the same clinical diagnosis into one cluster, so that the main types of the tissue can be explored through further clinical analysis. In order to achieve this, it is required to combine the most similar clusters together, such as within the suspected precancerous and mature cancer cell types. Although they represent different stages of cancer, they may exhibit certain properties that make them similar to one another. This information may be contained in the existing IR spectra. Most of material in this section has previously appeared in [29].

A new method is proposed that will enable similar separated clusters, produced via the FBSA, to be merged together into one cluster. In order to achieve this, we again used the V_{XB} index as the measure of cluster quality. A set of IR spectra from one tissue section were clustered and the best data structure obtained with the corresponding minimal value of cluster validity (V_{XB}). As mentioned above, the FCM algorithm can occasionally identify an excessive number of clusters in comparison with clinical analysis. Based on this problem, a method that can find and merge two similar clusters has been developed. In the rest of this section, the following questions will be answered: "given a set of c clusters, which two clusters are the most similar" and "How should they be merged?" (specifically, how to calculate the new cluster center).

In this study, two large lymph node cancer data-sets were used. There are 821 absorbance values corresponding to each wave number for each datum in the given IR spectral data-sets. In order to reduce the number of variables in the clustering analysis, we initially identified the first 10 principal components (PCs) that were extracted using PCA. Within these first 10 PCs, the majority of the variance (approximately 99 %) is represented from the original data. The first two PCs incorporate around 80 % of the variance in the original data. Thus by plotting the original data in these two PC dimensions, the approximate data distribution can be visualized.

20.5.1 Finding the Two Most Similar Clusters

Previously, many algorithms have been proposed for merging clusters, for instance [20, 35]. The different approaches can generally be divided into two groups. The first group are those that select the clusters which are "closest" to each other [20], and the second are those that choose the "worst" two clusters (judging by some cluster validity function) [35]. However, neither of these is suitable for solving the problem described in this research, illustrated in Figure 20.7.

A set of IR spectral data was plotted in first and second PCs after applying the FCM-based model selection algorithm. Four clusters (C1, C2, C3 and C4) were formed as shown in Figure 20.7. In clinical analysis, C1 and C3 are all cancer cells although they were taken from different areas of the tissue section that may contain different stages of cancer. C2 and C4 are normal nodal and reticulum cells respectively. Obviously then, the two similar clusters that need to be merged together are C1 and C3. Referring to the two types of techniques that merge clusters, the closest two clusters in the data-set are C1 and C2 (see the distances between each cluster center). Normally, a good cluster is defined by the data points within the cluster being tightly condensed around the center (compactness). In the sample data-set, clusters C1 and

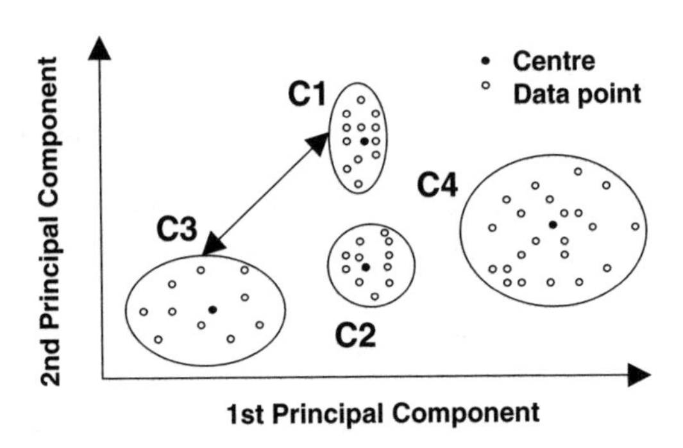

Figure 20.7 A set of clusters obtained from an IR data-set, plotted against the first and second principal components.

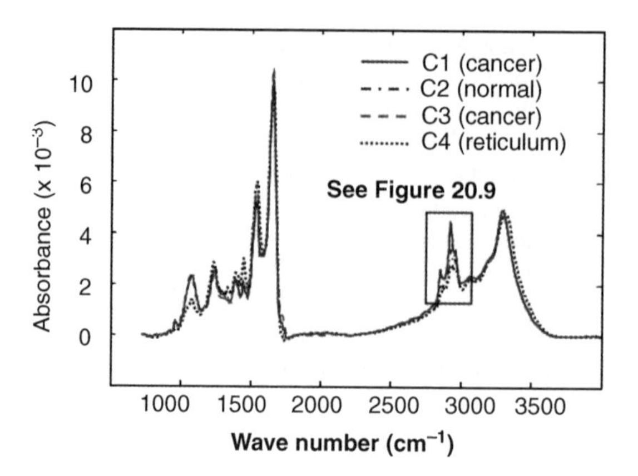

Figure 20.8 Example of mean IR spectra obtained from different clusters.

C2 are more compact than the other two, so the worst two clusters are C3 and C4. So, neither technique chooses the desired two clusters, C1 and C3.

In order to tackle this problem, we examined the original IR spectra rather than searching for a relationship using the data structure in the PCA plot. Plotting the mean spectra from the separate clusters allowed the major differences between them to be more clearly visualized. The similarity between clusters was more obvious at the wavelength where the biggest difference between any two mean spectra was located. Based on this observation, the proposed method to find two similar clusters was as follows:

(1) Obtain the clustering results from the FCM-based model selection algorithm.
(2) Calculate the mean spectra $\overline{A_i}$ for each cluster,

$$\overline{A_i} = \frac{1}{N_i}\sum_{j=1}^{N_i} A_{ij} \qquad (i = 1 \ldots c) \tag{20.9}$$

where N_i is the number of data points in the cluster i, A_{ij} is the absorbance of the spectrum for each data point j in cluster i and c is the number of clusters. The size of $\overline{A_i}$ is p, the number of wave numbers in each spectrum (each mean spectrum is a vector of p elements).
(3) Compute the vector of pair-wise squared differences D_{ij} between all mean spectra,

$$D_{ij} = (\overline{A_i} - \overline{A_j})^2 \qquad (i = 1 \ldots c, j = 1 \ldots c). \tag{20.10}$$

(4) Find the largest single element, $d_{\max}$, within the set of vectors D.
(5) Determine the wave number corresponding to the maximal element $d_{\max}$.
(6) At the wave number identified in step (5), find the two mean spectra with minimal difference. The clusters corresponding to these spectra are merged.

The mean spectra for the four clusters are displayed in Figure 20.8. We calculated the set of differences, D, between each pair of mean spectra using Equation (20.10). The largest difference $d_{\max}$ exists between C1 and C4 as shown in Figure 20.9. The wave number that corresponds to $d_{\max}$ is 2924 cm^{-1}. Examining the four absorbance values at this wave number we can clearly see that the two closest spectra are those belonging to clusters C1 and C3. Hence these are the two clusters selected to be merged.

20.5.2 Automatic Cluster Merging

After finding the two most similar clusters, the next stage is to merge them together. First, all the data points within each of the clusters are assigned to their center with a membership (μ) of one. Note that the

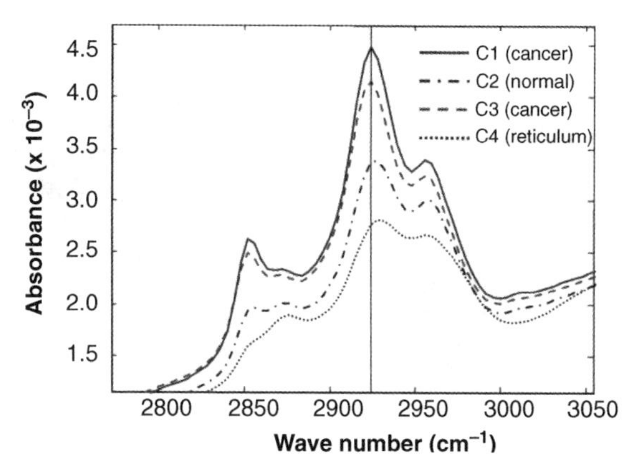

Figure 20.9 Enlarged region of Figure 20.8.

two chosen clusters outlined above (C1 and C3) are now considered as one cluster. Data points within a particular cluster are assigned a membership value of zero to other cluster centers. The following was used to calculate the centers [22]:

$$v_j = \frac{\sum_{i=1}^{n} \mu_{ij} x_i}{\sum_{i=1}^{n} \mu_{ij}}, \forall j = 1, \ldots, c. \tag{20.11}$$

where c is the number of clusters and n is the number of data points.

20.5.3 Application in FTIR Spectroscopic Clustering Analysis

In these experiments, we initially analyzed two lymph node tissue sections named LNII5 and LNII7. The *FBSA* algorithm was first used to generate initial clusters with good data structure. These clusters were then further combined using the new cluster merging method. Due to the fact that different initialization states may lead to different clustering results, we ran the *FBSA* algorithm 10 times on both data-sets. The results were fairly stable, and are shown in Figures 20.10 and 20.11.

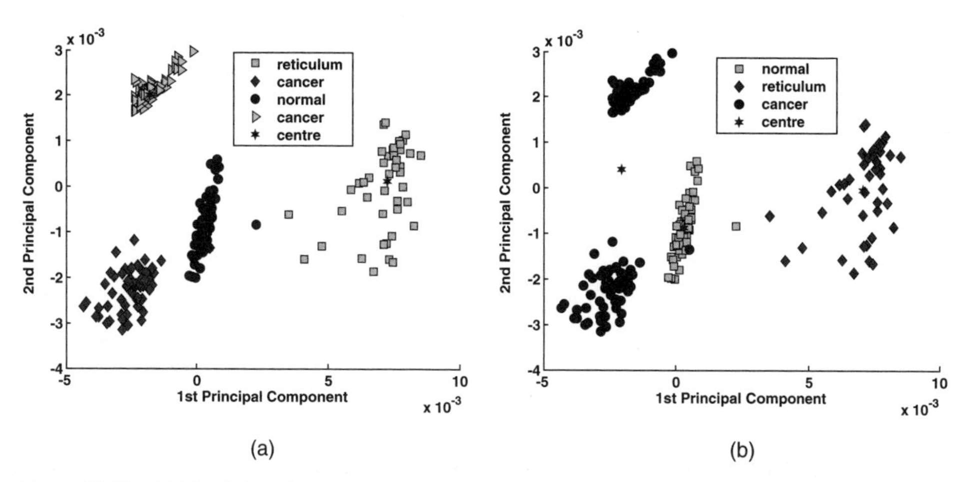

Figure 20.10 (a) LNII5 clustering results obtained from the *FBSA* algorithm (b) LNII5 merged clusters results.

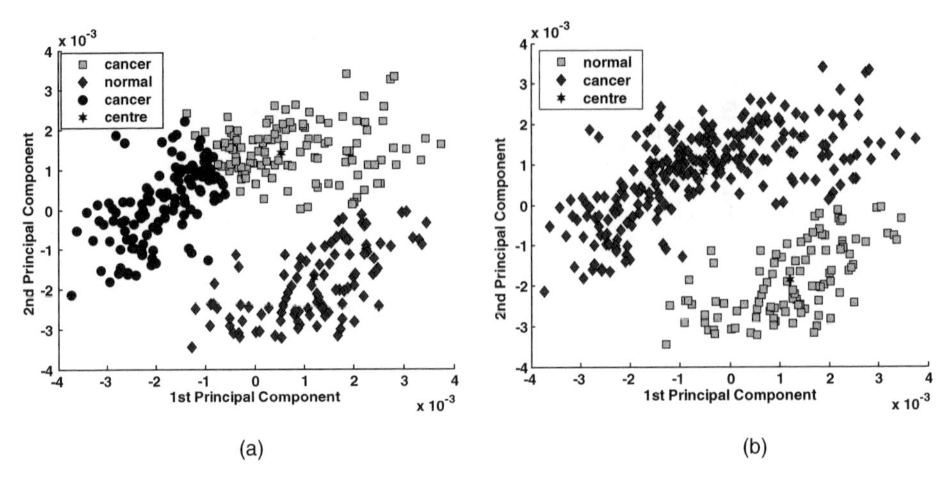

Figure 20.11 (a) LNII7 clustering results obtain from *FBSA* algorithm; (b) LNII7 merged clusters results.

Figures 20.10(a) and 20.11(a) show the initial clustering results obtained from the *FBSA* algorithm for LNII5 and LNII7, respectively, while Figures 20.10(b) and 20.11(b) show the clusters obtained after merging. The results clearly show that the separate cancer clusters have now been correctly merged using the new approach. In order to verify the cluster merging algorithm, the method was further applied to the oral cancer data-sets described earlier, in which the SAFC clustering algorithm had obtained three clusters, rather than the two found by histological analysis. The corresponding results are shown in Figures 20.12 and 20.13.

Altogether then, the proposed method was applied to four separate IR spectral data-sets (for which previous approaches could not obtain the correct number of clusters). For each data-set, the proposed method identified clusters that best matched clinical analysis. It should be noted that after merging clusters, there were still some misclassified data points (approximately three to four for LNII7 and

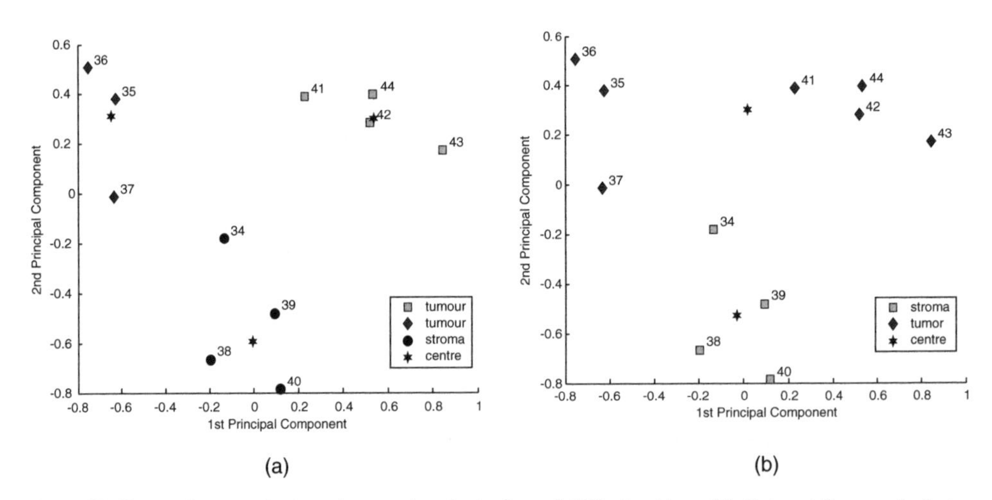

Figure 20.12 (a) Data-set 3 clustering results obtain from *SAFC* algorithm; (b) Data-set 3 merged clusters results.

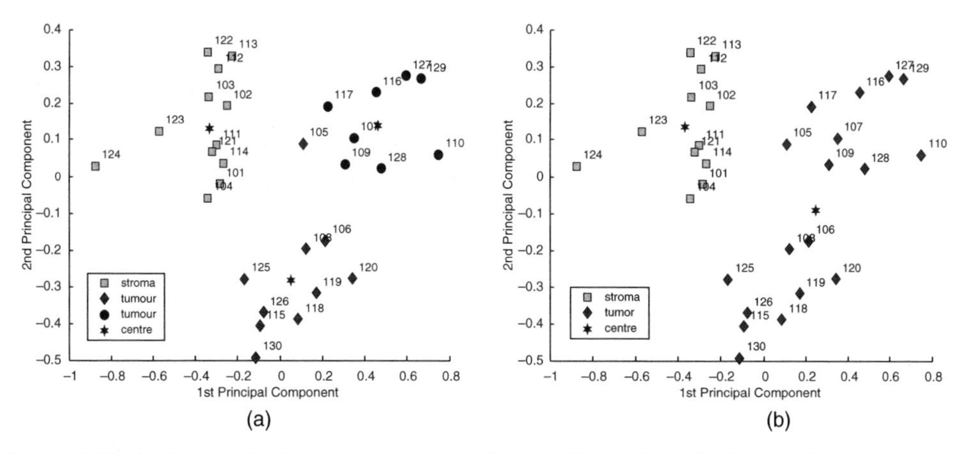

Figure 20.13 (a) Data-set 5 clustering results obtain from *SAFC* algorithm; (b) Data-set 5 merged clusters results.

one to two for the remaining data-sets). However, overall the clustering accuracy was significantly improved.

20.6 CONCLUSION

In this chapter, we have summarized our recent work on the analysis of non-preprocessed FTIR spectra data utilizing fuzzy clustering techniques. Three frequently used methods in FTIR clustering analysis, namely hierarchical clustering, *k*-means, and fuzzy C-means clustering were described in Section 20.2. Each of the clustering techniques was applied to seven data-sets containing FTIR spectra taken from oral cancer cells and the clustering results were compared. It can be seen from the results that the hierarchical complete linkage and Ward methods obtained the lowest number of disagreements in comparison with clinical analysis. However, in practice, these are computationally expensive when large data-sets are analyzed. *K*-means generated good results, but the variation in clustering results obtained from multiple runs implies that it may not be generally suitable in this context. In comparison with *k*-means and hierarchical clustering, fuzzy C-means provided reasonable performance, did not yield much variation in results and is not time-consuming in analysing large data-sets. It also obtained fairly stable results throughout our experimentation.

In Section 20.3, one of the most frequently used cluster validity indices, the Xie–Beni index, was introduced. As was discussed, for fuzzy C-means it is important to prespecify the number of clusters in advance. However, with the help of such a cluster validity index, it is possible to find the "optimal" number of clusters in the given data-sets without it being defined before the algorithm is executed. We have termed this method the "fuzzy C-means based selection algorithm." To achieve a similar purpose, the proposed simulated annealing fuzzy clustering combined both simulated annealing and fuzzy C-means techniques in order to detect automatically the optimal number of clusters from the data-sets. Due, perhaps, to the complexity of biochemical systems, both techniques can occasionally obtain an excessive number of clusters compared with clinical diagnosis. In order to attempt to solve this problem, a newly developed cluster merging method was introduced. Experiments showed that this method can find the two clusters with the most similar biochemical characteristics and then merge them together. This merging technique resulted in much improved agreement with clinical analysis. In the future, we are trying to collect a wider source of sample data for which the number of classifications is known, from a number of clinical domains, such as cervical cancer smear test screening. Establishing the techniques necessary to develop clinically useful diagnosis tools based on the automated interpretation of FTIR spectra across a range of medical domains is the ultimate goal of this research.

ACKNOWLEDGEMENTS

The authors would like to thank Professor H. Barr, Dr N. Stone and Dr J. Smith from the Gloucestershire Royal Hospital for their aid in sample collection and diagnosis. We are particularly grateful to Professor M. A. Chester and Mr J. M. Chalmers for many helpful discussions. The authors would also like to thank the anonymous referees for their comments which have been helpful in improving this chapter.

REFERENCES

[1] Johnson, K.S. *et al.* (2004) 'Elastic scattering spectroscopy for intraoperative determination of sentinel lymph node status in the breast'. *Journal of Biomedical Optics*, **9** (6), 1122–1128.

[2] Godavarty, A. *et al.* (2004) 'Diagnostic imaging of breast cancer using fluorescence-enhanced optical tomography: phantom studies'. *Journal of Biomedical Optics*, **9** (3), 486–496.

[3] Fernandez, D.C. *et al.* (2005) 'Infrared spectroscopic imaging for histopathologic recognition'. *Nature Biotechnology*, **23** 469–474.

[4] Lasch, P. *et al.* (2002) 'Characterization of colorectal adenocarcinoma sections by spatially resolved FT-IR microspectroscopy'. *Applied Spectroscopy*, **56** (1), 1–9.

[5] Chiriboga, L. *et al.* (1998) 'Infrared spectroscopy of human tissue: IV. Detection of dysplastic and neoplastic changes of human cervical tissue via infrared microscopy', *Cellular and Molecular Biology*, **44** *(1), 219–230.*

[6] Choo, L.P. *et al.* (1996) 'In situ characterization of beta-amyloid in Alzheimer's diseased tissue by synchrotron Fourier transform infrared microspectroscopy'. *Biophysical Journal*, **71** (4), 1672–1679.

[7] Wang, X.Y. and Garibaldi, J. (2005) 'Simulated annealing fuzzy clustering in cancer diagnosis'. *Informatica*, **29** (1), 61–70.

[8] McIntosh, L. *et al.* (1999) 'Analysis and interpretation of infrared microscopic maps: visualization and classification of skin components by digital staining and multivariate analysis'. *Biospectroscopy*, **5**, 265–275.

[9] Richter, T. *et al.* (2002) 'Identification of tumor tissue by FTIR spectroscopy in combination with positron emission tomography'. *Vibrational Spectroscopy*, **28** 103–110.

[10] Lasch, P. *et al.* (2004) 'Imaging of colorectal adenocarcinoma using FT-IR microspectroscopy and cluster analysis'. *Biochimica et Biophysica Acta (BBA) - Molecular Basis of Disease*, **1688** (2), 176–186.

[11] Romeo, M.J. and Diem, M. (2005) 'Infrared spectral imaging of lymph nodes: strategies for analysis and artifact reduction'. *Vibrational Spectroscopy*, **38** 115–119.

[12] Wood, B.R. *et al.* (2004) 'Fourier transform infrared (FTIR) spectral mapping of the cervical transformation zone', and dysplastic squamous epithelium. *Gynecologic Oncology*, **93** (1), 59–68.

[13] Wang, X.Y., Garibaldi, J. and Ozen, T. (2003) 'Application of the fuzzy C-means clustering method on the ananlysis of non pre-processed ftir data for cancer diagnosis', in *Proceedings of the 8th Australian and New Zealand Conference on Intelligent Information Systems*, 233–238.

[14] Wang, X.Y. and Garibaldi, J.M. (2005) 'A comparison of fuzzy and non-fuzzy clustering techniques in cancer diagnosis', in *Proceedings of the Second International Conference in Computational Intelligence in Medicine and Healthcare The Biopattern Conference*, 250–256.

[15] Sun, H., Wang, S. and Jiang, Q. (2004) 'FCM-based model selection algorithms for determining the number of clusters'. *Pattern Recognition*, **37**, 2027–2037.

[16] Wang, X.Y., Whitwell, G. and Garibaldi, J. (2004) 'The application of a simulated annealing fuzzy clustering algorithm for cancer diagnosis', in *Proceedings of the IEEE 4th International Conference on Intelligent System Design and Application*, 467–472.

[17] Jain, A.K., Murty, M.N. and Flynn, P.J. (1999) 'Data clustering: a review', *ACM Computing Surveys*, **31** (3), 264–323.

[18] Jiang, D., Tang, C. and Zhang, A. (2004) 'Cluster analysis for gene express data: a survey', *IEEE Transactions on Knowledge and Data Engineering*, **16** (11), 1370–1386.

[19] Garrett-Mayer, E. and Parmigiani, G. (2004) 'Clustering and classification methods for gene expression data analysis', *Johns Hopkins University, Dept. of Biostatistics Working Papers, Johns Hopkins University*, The Berkeley Electronic Press(bepress),

[20] Ward, J.H. (1963) 'Hierarchical grouping to optimize an objective function'. *Journal of the American Statistical Association*, **58** (301), 236–244.

[21] MacQueen, J.B. (1967) 'Some methods of classification and analysis of multivariate observations', in *Proceedings of Fifth Berkeley Symposium on Mathematical Statistics and Probability*, 281–297.
[22] Bezdek, J. (1981) *Pattern Recognition With Fuzzy Objective Function Algorithms*. New York, Plenum.
[23] Allibone, R. *et al.* (2002) 'FT-IR microscopy of oral and cervical tissue samples', *Internal Report*, Derby City General Hospital.
[24] Bezdek, J. (1998) *Pattern Recognition in Handbook of Fuzzy Computation.* Boston, NY, IOP Publishing Ltd.
[25] Xie, X.L. and Beni, G. (1991) 'A validity measure for fuzzy clustering'. *IEEE Trans. Pattern Analysis and Machine Intelligence*, **13** (8), 841–847.
[26] Kim, D.W., Lee, K.H. and Lee, D. (2004) 'On cluster validity index for estimation of the optimal number of fuzzy clusters'. *Pattern Recognition*, **37**, 2009–2025.
[27] Pal, N.R. and Bezdek, J. (1995) 'On cluster validity for the fuzzy C-means model'. *IEEE Trans. Fuzzy System*, **3**, 370–379.
[28] Rezaee, M.R., Lelieveldt, B.P.F. and ReiBer, J.H.C. (1998) 'A new cluster validity index for the fuzzy C-means'. *Pattern Recognition Letters*, **19**, 237–246.
[29] Wang, X.Y. *et al.* (2005) 'Fuzzy clustering in biochemical analysis of cancer cells', in *Proceedings of Fourth Conference of the European Society for Fuzzy Logic and Technology (EUSFLAT 2005)*, 1118–1123.
[30] Bandyopadhyay, S. (2003) Simulated annealing for fuzzy clustering: variable representation, evolution of the number of clusters and remote sensing applications, unpublished, private communication.
[31] Rayward-Smith, V.J. *et al.* (1996) *Modern Heuristic Search Methods* John Wiley & Sons, Ltd, Chichester, UK.
[32] Conover, W.J. (1999) Practical Nonparametric Statistics, John Wiley & Sons, Inc., New York, USA.
[33] Causton, D.R. (1987) *A Biologist's Advanced Mathematics.* London, Allen & Unwin.
[34] Jolliffe, I.T. (1986) *Principal Component Analysis*, New York, Springer-Verlag.
[35] Xie, Y., Raghavan, V.V. and Zhao, X. (2002) 3M Algorithm: Finding an Optimal Fuzzy Cluster Scheme for Proximity Data, **1**, 627–632.

Index

Index compiled by Dave Tyler